**Modern Heterogeneous
Oxidation Catalysis**

*Edited by
Noritaka Mizuno*

Related Titles

U.S. Ozkan (Ed.)

Design of Heterogeneous Catalysts

New Approaches Based on Synthesis, Characterization and Modeling

2009

ISBN: 978-3-527-32079-0

K. Ding, Y. Uozumi (Eds.)

Handbook of Asymmetric Heterogeneous Catalysis

2008

ISBN: 978-3-527-31913-8

G. Ertl, H. Knözinger, F. Schüth, J. Weitkamp (Eds.)

Handbook of Heterogeneous Catalysis

8 Volumes

2008

ISBN: 978-3-527-31241-2

I. Chorkendorff, J.W. Niemantsverdriet

Concepts of Modern Catalysis and Kinetics

2007

ISBN: 978-3-527-31672-4

R.A. Sheldon, I. Arends, U. Hanefeld

Green Chemistry and Catalysis

2007

ISBN: 978-3-527-30715-9

B. Cornils, W.A. Herrmann, M. Muhler, C.-H. Wong (Eds.)

Catalysis from A to Z

A Concise Encyclopedia

2007

ISBN: 978-3-527-31438-6

R.A. van Santen, M. Neurock

Molecular Heterogeneous Catalysis

A Conceptual and Computational Approach

2006

ISBN: 978-3-527-29662-0

Modern Heterogeneous Oxidation Catalysis

Design, Reactions and Characterization

Edited by
Noritaka Mizuno

WILEY-VCH Verlag GmbH & Co. KGaA

The Editor

Prof. Dr. Noritaka Mizuno
The University of Tokyo
Department of Applied Chemistry
7-3-1 Hongo, Bunkyo-ku
Tokyo 113-8656
Japan

All books published by Wiley-VCH are carefully produced. Nevertheless, authors, editors, and publisher do not warrant the information contained in these books, including this book, to be free of errors. Readers are advised to keep in mind that statements, data, illustrations, procedural details or other items may inadvertently be inaccurate.

Library of Congress Card No.: applied for

British Library Cataloguing-in-Publication Data
A catalogue record for this book is available from the British Library.

Bibliographic information published by the Deutsche Nationalbibliothek
The Deutsche Nationalbibliothek lists this publication in the Deutsche Nationalbibliografie; detailed bibliographic data are available on the Internet at http://dnb.d-nb.de

© 2009 WILEY-VCH Verlag GmbH & Co. KGaA, Weinheim

All rights reserved (including those of translation into other languages). No part of this book may be reproduced in any form – by photoprinting, microfilm, or any other means – nor transmitted or translated into a machine language without written permission from the publishers. Registered names, trademarks, etc. used in this book, even when not specifically marked as such, are not to be considered unprotected by law.

Printed in the Federal Republic of Germany
Printed on acid-free paper

Cover design   wmx design, Heidelberg
Typesetting   Thomson Digital, Noida, India
Printing   Strauss GmbH, Mörlenbach
Bookbinding   Litges & Dopf Buchbinderei GmbH, Heppenheim

ISBN: 978-3-527-31859-9

Contents

Preface *XI*

List of Contributors *XIII*

1	**Concepts in Selective Oxidation of Small Alkane Molecules** *1*	
	Robert Schlögl	
1.1	Introduction *1*	
1.2	The Research Field *4*	
1.3	Substrate Activation *7*	
1.4	Active Oxygen Species *15*	
1.5	Catalyst Material Science *22*	
1.6	Conclusion *34*	
	References *35*	
2	**Active Ensemble Structures for Selective Oxidation Catalyses at Surfaces** *43*	
	Mizuki Tada and Yasuhiro Iwasawa	
2.1	Introduction *43*	
2.2	Chiral Self-Dimerization of Vanadium Schiff-Base Complexes on SiO_2 and Their Catalytic Performances for Asymmetric Oxidative Coupling of 2-Naphthol *44*	
2.2.1	Asymmetric Heterogeneous Catalysis Using Supported Metal Complexes *44*	
2.2.2	Chiral V-Dimer Structure on a SiO_2 Surface *45*	
2.2.3	Asymmetric Catalysis for Oxidative Coupling of 2-Naphthol to BINOL *49*	
2.3	Low-Temperature Preferential Oxidation of CO in Excess H_2 on Cu-Clusters Dispersed on CeO_2 *51*	
2.3.1	Preferential Oxidation (PROX) of CO in Excess H_2 on Novel Metal Catalysts *51*	
2.3.2	Characterization and Performance of a Novel Cu Cluster/CeO_2 Catalyst *52*	

Modern Heterogeneous Oxidation Catalysis: Design, Reactions and Characterization
Edited by Noritaka Mizuno
Copyright © 2009 WILEY-VCH Verlag GmbH & Co. KGaA, Weinheim
ISBN: 978-3-527-31859-9

2.4	Direct Phenol Synthesis from Benzene and Molecular Oxygen on a Novel N-Interstitial Re_{10}-Cluster/HZSM-5 Catalyst 57
2.4.1	Phenol Production from Benzene with N_2O, $H_2 + O_2$, and O_2 57
2.4.1.1	Benzene to Phenol with N_2O 58
2.4.1.2	Benzene to Phenol with $H_2 + O_2$ 60
2.4.1.3	Benzene to Phenol with O_2 62
2.4.2	Novel Re/HZSM-5 Catalyst for Direct Benzene-to-Phenol Synthesis with O_2 64
2.4.3	Active Re Clusters Entrapped in ZSM-5 Pores 66
2.4.4	Structural Dynamics of the Active Re_{10} Cluster 68
2.5	Conclusion 71
	References 71
3	**Unique Catalytic Performance of Supported Gold Nanoparticles in Oxidation** 77
	Yunbo Yu, Jiahui Huang, Tamao Ishida, and Masatake Haruta
3.1	Introduction 77
3.2	Low-Temperature CO Oxidation 79
3.2.1	Low-Temperature CO Oxidation in Air 79
3.2.1.1	Junction Perimeter Between Au Particles and the Support 79
3.2.1.2	Selection of Suitable Supports 81
3.2.1.3	Sensitivity to the Size of the Gold Particles 82
3.2.2	Low-Temperature CO Oxidation in H_2 84
3.2.3	Mechanism for CO Oxidation Over Supported Gold Nanoparticles 87
3.2.3.1	Mechanisms Involving Junction Perimeter Between Gold and the Metal-Oxide Supports 87
3.2.3.2	Mechanisms Involving Specific Size or Thickness of Gold Clusters or Thin Layers 90
3.2.3.3	Mechanisms Involving Cationic Gold 92
3.3	Complete Oxidation of Volatile Organic Compounds 92
3.4	Gas-Phase Selective Oxidation of Organic Compounds 97
3.4.1	Gas-Phase Selective Oxidation of Aliphatic Alkanes 97
3.4.2	Gas-Phase Selective Oxidation of Alcohols 98
3.4.3	Gas-Phase Propylene Epoxidation 100
3.4.3.1	Introduction 100
3.4.3.2	Gas-Phase Propylene Epoxidation with Hydrogen–Oxygen Mixtures on Au/TiO_2 101
3.4.3.3	Gas-Phase Propylene Epoxidation with Hydrogen–Oxygen Mixtures on $Au/Ti-SiO_2$ 103
3.5	Liquid-Phase Selective Oxidation of Organic Compounds 106
3.5.1	Oxidation of Mono-Alcohols 107
3.5.2	Oxidation of Diols 112
3.5.3	Oxidation of Glycerol 113
3.5.4	Aerobic Oxidation of Glucose 115
3.5.5	Oxidation of Alkanes and Alkenes 116

3.6	Conclusions *116*	
	References *118*	
4	**Metal-Substituted Zeolites as Heterogeneous Oxidation Catalysts** *125*	
	Takashi Tatsumi	
4.1	Introduction – Two Ways to Introduce Hetero-Metals into Zeolites *125*	
4.2	Titanium-Containing Zeolites *126*	
4.2.1	TS-1 *126*	
4.2.2	Ti-Beta *136*	
4.2.3	Ti-MWW *137*	
4.2.4	Other Titanium-Containing Zeolites *145*	
4.2.5	Solvent Effects and Reaction Intermediate *145*	
4.3	Other Metal-Containing Zeolites *150*	
4.4	Conclusion *151*	
	References *151*	
5	**Design of Well-Defined Active Sites on Crystalline Materials for Liquid-Phase Oxidations** *157*	
	Kiyotomi Kaneda and Takato Mitsudome	
5.1	Introduction *157*	
5.2	Oxidation of Alcohols *157*	
5.2.1	Ru Catalyst *158*	
5.2.2	Pd Catalyst *163*	
5.2.3	Au Catalyst *164*	
5.2.4	Au-Pd Catalyst *166*	
5.3	Epoxidation of Olefins *166*	
5.3.1	Epoxidation with Hydrogen Peroxide *167*	
5.3.1.1	Titanium-Based Catalysts *167*	
5.3.1.2	Tungsten-Based Catalysts *167*	
5.3.1.3	Base Catalyst *169*	
5.3.2	Epoxidation with Molecular Oxygen *172*	
5.4	Cis-Dihydroxylation *173*	
5.5	Baeyer–Villiger Oxidation *175*	
5.6	C–H Activation Using Molecular Oxygen *177*	
5.7	Conclusions *178*	
	References *178*	
6	**Liquid-Phase Oxidations with Hydrogen Peroxide and Molecular Oxygen Catalyzed by Polyoxometalate-Based Compounds** *185*	
	Noritaka Mizuno, Keigo Kamata, Sayaka Uchida, and Kazuya Yamaguchi	
6.1	Introduction *185*	
6.2	Molecular Design of Polyoxometalates for H_2O_2- and O_2-Based Oxidations *186*	
6.2.1	Isopoly- and Heteropolyoxometalates *188*	

6.2.2	Peroxometalates 189
6.2.3	Lacunary Polyoxometalates 190
6.2.4	Transition-Metal-Substituted Polyoxometalates 192
6.3	Heterogenization of Polyoxometalates 193
6.3.1	Solidification of Polyoxometalates with Appropriate Cations 200
6.3.1.1	Metal and Alkylammonium Cations 200
6.3.1.2	Polycations 201
6.3.1.3	Cationic Organometallic Complexes 203
6.3.2	Immobilization of Polyoxometalate-Based Compounds 205
6.3.2.1	Wet Impregnation 205
6.3.2.2	Solvent-Anchoring and Covalent Linkage 207
6.3.2.3	Anion Exchange 208
6.4	Conclusion 210
	References 211

7	**Nitrous Oxide as an Oxygen Donor in Oxidation Chemistry and Catalysis** 217
	Gennady I. Panov, Konstantin A. Dubkov, and Alexander S. Kharitonov
7.1	Introduction 217
7.2	Molecular Structure and Physical Properties of Nitrous Oxide 218
7.3	Catalytic Oxidation by Nitrous Oxide in the Gas Phase 220
7.3.1	Oxidation of Lower Alkanes Over Oxide Catalysts 220
7.3.2	Oxidation Over Zeolites 222
7.3.2.1	Oxidation by Dioxygen 222
7.3.2.2	Oxidation of Benzene to Phenol by N_2O 223
7.3.2.3	Nature of Zeolite Activity, α-Sites 224
7.3.2.4	N_2O specificity, α-Oxygen and its Stoichiometric Reactions 227
7.3.2.5	Hydroxylation of Alkanes and Benzene Derivatives 229
7.3.2.6	Other Types of Oxidation Reactions 230
7.4	Catalytic Oxidation by N_2O in the Liquid Phase 230
7.5	Non-Catalytic Oxidations by N_2O 231
7.5.1	Liquid-Phase Oxidation of Alkenes 231
7.5.1.1	Linear Alkenes 232
7.5.1.2	Cyclic Alkenes 234
7.5.1.3	Cyclodienes 237
7.5.1.4	Bicyclic Alkenes 238
7.5.1.5	Heterocyclic Alkenes 238
7.5.2	Carboxidation of Polymers 240
7.5.2.1	Carboxidation of Polyethylene 240
7.5.2.2	Carboxidation of Polybutadiene Rubber 241
7.6	Economic Aspects of N_2O as Oxidant 244
7.6.1	Recovery of N_2O From Off-Gases 244
7.6.2	Deliberate Preparation of N_2O 245
7.7	Conclusion 246
	References 247

8	**Direct Synthesis of Hydrogen Peroxide: Recent Advances** *253*	
	Gabriele Centi, Siglinda Perathoner, and Salvatore Abate	
8.1	Introduction *253*	
8.1.1	Industrial Production *253*	
8.1.2	Outlook for H_2O_2 Production *254*	
8.1.3	Uses of Hydrogen Peroxide *255*	
8.2	Direct Synthesis of H_2O_2 from an Industrial Perspective *257*	
8.2.1	Status of Development and Perspectives of Industrial Production *257*	
8.2.2	Recent Patents on the Direct Synthesis of H_2O_2 *262*	
8.3	Fundamental Studies *270*	
8.3.1	Intrinsically Safe Operations and Microreactors *271*	
8.3.2	Nature of the Catalyst and Reaction Network *275*	
8.3.3	Role of the Solvent and of Promoters *281*	
8.4	Conclusion *282*	
	References *283*	
9	**Recent Achievements and Challenges for a Greener Chemical Industry** *289*	
	Fabrizio Cavani and Nicola Ballarini	
9.1	Introduction: Old and New Challenges for Oxidation Catalysis in Industry *289*	
9.2	Recent Successful Examples of Alkanes Oxidation *290*	
9.2.1	Oxidation of Ethane to Acetic Acid *290*	
9.2.2	Ammoxidation of Propane to Acrylonitrile *294*	
9.3	New Oxidation Technologies: Oxidative Desulfurization (ODS) of Gas Oil *301*	
9.4	Process Intensification in Catalytic Oxidation *304*	
9.5	An Alternative Approach: Anaerobic Oxidation with Metal Oxides in a Cycle Process (from an Oxidation Catalyst to a Reusable Stoichiometric Oxidant) *306*	
9.5.1	Anaerobic Oxidation of Propene to Acrolein in a CFBR Reactor *309*	
9.5.2	Anaerobic Synthesis of 2-Methyl-1,4-Naphthoquinone (Menadione) *310*	
9.5.3	Anaerobic Oxidative Dehydrogenation of Propane to Propene *311*	
9.5.4	Production of Hydrogen from Methane with Oxide Materials and Inherent Segregation of Carbon Dioxide *313*	
9.6	Current and Developing Processes for the Transformation of Bioplatform Molecules into Chemicals by Catalytic Oxidation *316*	
9.6.1	Glycerol: A Versatile Building Block *320*	
9.7	Conclusion *321*	
	References *323*	

Index *333*

Preface

The range of chemical products is enormous and these products contribute greatly to the quality of our lives. The manufacturing processes of chemicals also lead to vast amounts of wastes, and the reduction or elimination of these wastes is now our central issue. To minimize wastes in chemical manufacturing, the catalytic method is a reliable solution, replacing synthetic processes of low atom efficiency using hazardous stoichiometric reagents. Especially for fine chemicals production, antiquated methodologies with stoichiometric oxidations such as permanganate or dichromate reagents are still widely used. But, times are changing! In the manufacture of large-scale petrochemicals as well as laboratory-scale syntheses, the environmentally-unfriendly processes should be replaced with cleaner and greener oxidants such as O_2, H_2O_2 and N_2O in combination with heterogeneous catalysts. In the last few decades, many efficient heterogeneous oxidation catalysts and methods have been developed and the editor feels that it is necessary to survey the recent developments in heterogeneous oxidation catalysis. This book will discuss mainly the case studies of recent developments in heterogeneous oxidation catalysis and will be directed towards chemists engaged in catalyst preparation and design, catalysis and catalytic organic synthesis, both in academia as well as industry.

In this book, both gas- and liquid-phase oxidations are included although the latter are more numerous. Chapter 1 deals with the concepts in selective oxidation of small alkane molecules. Chapters 2–6 then review the strategies of catalyts design with metal oxides, metal nanoparticles, zeolites, hydroxyapatites, hydrotalcites, montmorillonites and polyoxometalates. The chemistry and application of N_2O as an oxidant are discussed in Chapter 7. Chapters 8 and 9 deal with recent developments in the direct synthesis of H_2O_2 and greener industrial processes, respectively. Each chapter contains extensive references covering the very important and principal literature through to the beginning of 2000.

Finally, the editor would like to express sincere thanks to colleagues and friends who have contributed such fine chapters. He also thanks Dr Kazuya Yamaguchi (The University of Tokyo) and Dr Stefanie Volk (Wiley-VCH) for their help in preparing this book.

March 2009 *Noritaka Mizuno*

Modern Heterogeneous Oxidation Catalysis: Design, Reactions and Characterization
Edited by Noritaka Mizuno
Copyright © 2009 WILEY-VCH Verlag GmbH & Co. KGaA, Weinheim
ISBN: 978-3-527-31859-9

List of Contributors

Salvatore Abate
University of Messina
Dipartimento di Chimica Industriale
ed Ingegneria dei Materiali
Salita Sperone, 39
98166 Messina
Italy

Nicola Ballarini
Università di Bologna
Dipartimento di Chimica Industriale
e dei Materiali
Viale Risorgimento 4
40136 Bologna
Italy

Fabrizio Cavani
Università di Bologna
Dipartimento di Chimica Industriale
e dei Materiali
Viale Risorgimento 4
40136 Bologna
Italy

Konstantin A. Dubkov
Boreskov Institute of Catalysis
Novosibirsk 630090
Russia

Gabriele Centi
University of Messina
Dipartimento di Chimica Industriale
ed Ingegneria dei Materiali
Salita Sperone, 39
98166 Messina
Italy

Masatake Haruta
Tokyo Metropolitan University
Graduate School of Urban
Environmental Sciences
Department of Applied Chemistry
1.1 Minami-osawa
Hachioji
Tokyo 192-0397
Japan
and
Japan Science and Technology Agency
CREST
Kawaguchi
Saitama 332-0012
Japan

List of Contributors

Jiahui Huang
Tokyo Metropolitan University
Graduate School of Urban
Environmental Sciences
1.1 Minami-osawa
Hachioji
Tokyo 192-0397
Japan
and
Japan Science and Technology Agency
CREST
Kawaguchi
Saitama 332-0012
Japan

Tomao Ishida
Tokyo Metropolitan University
Graduate School of Urban
Environmental Sciences
1.1 Minami-osawa
Hachioji
Tokyo 192-0397
Japan
and
Japan Science and Technology Agency
CREST
Kawaguchi
Saitama 332-0012
Japan

Yasuhiro Iwasawa
The University of Tokyo
Department of Chemistry
Graduate School of Science
7-3-1 Hongo
Bunkyo-ku
Tokyo 113-0033
Japan

Keigo Kamata
The University of Tokyo
Department of Applied Chemistry
School of Engineering
7-3-1 Hongo
Bunkyo-ku
Tokyo 113-8656
Japan

Kiyotomi Kaneda
Osaka University
Department of Materials Engineering
Science
Graduate School of Engineering Science
1-3 Machikaneyama
Toyonaka
Osaka 560-8531
Japan

Alexander S. Kharitonov
Boreskov Institute of Catalysis
Novosibirsk 630090
Russia

Takato Mitsudome
Osaka University
Department of Materials Engineering
Science
Graduate School of Engineering Science
1-3 Machikaneyama
Toyonaka
Osaka 560-8531
Japan

Noritaka Mizuno
The University of Tokyo
Department of Applied Chemistry
School of Engineering
7-3-1 Hongo
Bunkyo-ku
Tokyo 113-8656
Japan

List of Contributors

Gennady I. Panov
Boreskov Institute of Catalysis
Novosibirsk 630090
Russia

Siglinda Perathoner
University of Messina
Dipartimento di Chimica Industriale
ed Ingegneria dei Materiali
Salita Sperone, 39
98166 Messina
Italy

Robert Schlögl
Fritz-Haber-Institut der
Max-Planck-Gesellschaft
Department of Inorganic Chemistry
Faradayweg 4–6
14195 Berlin
Germany

Mizuki Tada
Institute of Molecular Science
38 Nishigo-Naka
Myodaiji
Okazaki 444-8585
Japan

Takashi Tatsumi
Tokyo Institute of Technology
Division of Catalytic Chemistry
Chemical Resources Laboratory
4259-R1-9 Nagatsuta-cho
Midori-ku
Yokohama 226-8503
Japan

Sayaka Uchida
The University of Tokyo
Department of Applied Chemistry
School of Engineering
7-3-1 Hongo
Bunkyo-ku
Tokyo 113-8656
Japan

Kazuya Yamaguchi
The University of Tokyo
Department of Applied Chemistry
School of Engineering
7-3-1 Hongo
Bunkyo-ku
Tokyo 113-8656
Japan

Yunbo Yu
Tokyo Metropolitan University
Graduate School of Urban
Environmental Sciences
1.1 Minami-osawa
Hachioji
Tokyo 192-0397
Japan
and
Research Center for Eco-Environmental
Sciences
Chinese Academy of Sciences
18 Shuangqing Road, Haidian District
Beijing 100085
China

1
Concepts in Selective Oxidation of Small Alkane Molecules
Robert Schlögl

1.1
Introduction

The subject of heterogeneously catalyzed selective oxidation has been reviewed many times. Under the keyword combination "selective catalytic oxidation" the ISI database reports about 5400 papers. Over 100 reviews on the topic have been published. In the present discussion, the subjects of methane activation and model studies of unselective CO oxidation, which represent large fields, are excluded. Homogeneously or biologically catalyzed selective oxidation, a combined field that is about 10-fold larger in scientific coverage, is also excluded from this chapter.

Instead, the present chapter deals mainly with the activation of C_2, C_3 and C_4 hydrocarbons focussing on oxidative dehydrogenation and oxo-functionalization as target reactions. This seemingly limited field of research encompasses a central entry port to commodity molecules used in chemical industry. The issues of selectivity and energy conversion are of enormous practical relevance as the potential is great for making the chemical industry more sustainable is in this small area of catalytic chemistry. However, the still limited success in performing these reactions effectively sheds light on the level of our scientific understanding of these reactions. The science is based upon a set of phenomenological concepts referred as "principles" in the literature, enabling the discovery and optimization of the present catalytic materials.

By far the most influential principle is that of "lattice oxygen" [1–3]. It states at its core that atomic oxygen that can selectively oxidize a hydrocarbon has to come from a lattice position of the catalyst. The reduction of the metal centers is thought to arise from oxygen anion transport from deeper layers of the catalyst to its surface. Gas-phase oxygen, being detrimental as reagent with organics, re-oxidizes the catalytic material in a spatio-temporal separation between the hydrocarbon redox chemistry and the catalyst redox chemistry. This separation is widely referred to as the "Mars–van Krevelen type (MvK)" reaction "mechanism" [4, 5]. The original derivation by the authors Mars and van Krevelen [6] did, however, not contain any interpretation of their finding in SO_2 oxidation over vanadium oxides. In their

Modern Heterogeneous Oxidation Catalysis: Design, Reactions and Characterization
Edited by Noritaka Mizuno
Copyright © 2009 WILEY-VCH Verlag GmbH & Co. KGaA, Weinheim
ISBN: 978-3-527-31859-9

kinetic work a second term had to be added to the conventional Langmuir equation to explain the experimental finding. Only much later was it realized that this term should describe the "slow re-oxidation of the catalyst," for which there is little experimental evidence [7]. The postulated general validity of the principle is questioned by the operation of monolayer oxide catalysts [8], which have little ability to deliver lattice oxygen, and by numerous findings in homogeneous catalytic oxidation where molecular oxides such as RuO_4 afford excellent [9] selective oxidation.

Another strongly prevailing principle is that of "phase cooperation" [10–13]. This states that high-performance oxidation catalysts must be of the multi-phase type, as the different functions required in performing the selective oxidation of a hydrocarbon are only adequately optimized when independent functionally optimized phases cooperate in the catalytic cycle. This intuitively appealing concept renders it almost impossible applying the tools of functional analysis to selective oxidation as the chemical and structural complexity arising from a multi-element-multi-phase oxide (MMO) overwhelms all analytical possibilities of today's experimentation and theory. Recent developments in catalyst synthesis have shown on the model level [14–16], as well as with high-performance systems [17–21], that substantial catalytic action can be obtained from proven single-phase systems.

At present, the field is still without an unchallenged scientific base despite the enormous effort invested into the issue and the many papers written. This situation is unsatisfactory with respect to the enormous relevance of selectivity in large-scale industrial processes. In the opinion of the author this is not due to a lack of excellent ideas – almost all possible ideas about the function of these catalysts have been proposed. It is a rational experimental approach beyond the principles described in the literature that is needed to unify the concepts into a scientific foundation for better catalyst developments. A parallel search for such catalysts has been advocated and performed [22] with great effort but apparently no success in solving the challenges of the field. This statement is not negative against high-throughput experimentation as such but shows that this technique also requires a scientifically sound basis for its useful application [23].

The above critical remarks do not diminish the enormous success in the development of oxidation reactions that seem close to impossible within the framework of homogeneous chemistry. The most successful process is the one-step reaction of butane to maleic anhydride (MA):

$$C_4H_{10} + 7/2\, O_2 \rightarrow C_4H_2O_3 + 4H_2O$$

The process solves the problem of activating the poorly reactive butane and then abstracting eight protons without cleaving a carbon–carbon bond. In addition, three oxygen atoms are attached without forming any carbon oxides. This is only possible as the product MA is kinetically stable against further oxidation at the remaining protons, as the locations of further attack are strongly bound to the cyclic carbon skeleton. The reaction occurs in many intermediate steps that are all stabilized against oxidative attack due to chemisorption at an active site that has to become progressively less active during the whole transformation. We assume that the active site stores electrons from the substrate in its structure and becomes progressively reduced, which in turn reduces its ability to activate C–H bonds. With release of the product we expect concomitant release of water from stored OH groups and re-oxidation by oxygen from the feed. As there are no split oxygen molecules at hand (excluding special oxidants) the active site has to accommodate one extra oxygen atom for every two-center bond being made. The fact that per turnover 14 electrons have to be stored at the active site and that the highly suitable vanadium cation changes its oxidation states between 5 and 3 would call for an active site consisting of seven vanadium centers and a suitable combination of terminal and bridging oxygen atoms. A minimum size of an active site of vanadium species would be two centers as they can store the four electrons required to activate one molecule of oxygen. Such a cluster is hardly a "lattice." It also can not be a section of an oxide lattice, as then the reacting site would not be isolated electronically, as through oxygen ion diffusion and electronic conductivity the oxidation state of the active site would be kept as constant as possible so as to minimize the free energy of the system. The active site is thus seen as a cluster supported on a matrix of a foreign material (supported catalyst) or of a compound from which the cluster originates by segregation (self-supported). The cluster exhibits a structure capable of adsorbing activated oxygen and of holding protons as intermediate hydroxyl groups. The adsorbed oxygen, being part of the coordination geometry of the cluster in its regenerated state, has been called "surface lattice oxygen" [8, 24, 25], a term for which the author sees no need as this oxygen is an adsorbate situated at a high energy position of the structure.

The general principle of site isolation [26] is the consequence of many observations in selective catalysis that are not confined to selective oxidation. High performing catalysts exhibit isolated sites, minimizing the influx of electrons and activated reactants (oxygen, hydrogen) during conversion of a substrate molecule. This principle of site isolation [1, 2, 26–32] is the most powerful rule for finding selective oxidation catalysts. It is, however, mostly applied in a crystallographic manner, meaning that structures are sought that exhibit strong variations in atom density in their motifs by combining locally dense building blocks with linking polyhedra, leaving substantial empty space in the unit cell "channels." A particular, instructive example of site isolation is the idealized crystal structure of salts of heteropolyacids (HPA), a family of compounds [18, 33–35] capable of performing many selective oxidation reactions.

It is apparent that logical clashes occur when in the literature the principle of site isolation is applied in conjunction with phase cooperation and lattice oxygen dynamics. Both concepts require for their operation a close contact of the active site with its environment and suggest a continuing exchange of reactants and electrons during the conversion of a substrate molecule. Elaborate sets of assumptions are made to remedy these clashes. The purpose of this review is to suggest a reconciliation of these conflicting concepts as all of them are based on undeniable observations. Table 1.1 gives a (incomplete) list of reviews that highlight the complexity of the arguments that arose from the applications of empirical principles. The list is split into two groups: articles covering reactions and those dealing with the structure and function of catalysts. There is no ordering within the groups.

Table 1.1 Some relevant reviews for the field of selective oxidation.

Year	Topic	Reference
Reactions and processes		
2006	Kinetics of C_2,C_3 ODH reactions	[36]
2001	Propane oxidation	[37]
1997	Propane ammoxidation	[38]
1999	Ethane activation	[39]
1995	ODH reactions	[40]
1997	ODH of alkanes	[41]
2000	Epoxidation	[42]
1979	Selective oxidation reactions	[43]
2002	C_4 activation	[44]
1985	C_4 oxidation	[45]
2000	General review, 30 years of selective oxidation	[46]
1999	Advances in selective oxidation	[31]
Mechanism, catalysts and methods		
2002	Molecular structures of catalysts	[47]
2006	Molecular structures of monolayer oxides	[48]
2001	Heteropolyacid systems	[33]
1996	Key aspects for catalyst design	[49]
2003	*In situ* studies	[50]
2001	*In situ* studies	[51]
2003	Active oxygen species	[52]
1992	Di-oxygen as selective oxidant	[2]
1991	V monolayer catalysts	[53]
2000	Family of oxide monolayer catalysts	[54]
1991	Promoter effects in oxidation	[55]
1992	Defects in oxidation catalysts	[56]
1992	Phase cooperation in oxidation	[57]
2000	Solid state chemistry	[58]
2004	Nanostructures	[59]
2003	Lattice oxygen	[60]
1996	IR studies in selective oxidation	[61]
1991	Theory, a generalized concept	[62]
2007	Theory of propane activation	[63]

1.2
The Research Field

The field is defined here around the activation of butane, propane and ethane plus the oxidation of propene. The reason for this boundary is the similarity of the chemistry and the great need to understand the mechanism of selectivity of activated oxygen in these multi-step reactions. The processes cannot be conducted at high temperatures such as with methane activation as the target products are not stable under conditions where alkane activation is fast. The selective oxidation of ethylbenzene to styrene and that of o-xylene are also not covered as the initial substrates are already activated. The reader is referred to the literature on xylene oxidation [31, 64] and ethylbenzene [65–69] dehydrogenation.

The set of reactions defines the materials largely contained within this set of scientific studies. It deals mostly with vanadium oxides and vanadium phosphates followed by complex MMO phases and HPA. Figure 1.1 shows some relevant trends from the ISI database statistics.

Figure 1.1(a) reveals that the field has been recognized (cited) for about 20 years, although the roots go back about 20 years longer [46]. By looking at the derivative of the growth of citations in Figure 1.1(b) one can state that interest is growing above the trends for more complete coverage of the literature and for the general rapid growth of scientific communication. Following the initial accumulation in the early 1990s a first evolution in activity can be observed. After a period of scientific disappointment steady growth since 2000 is seen, triggered by the advent of model catalysis, nanoscience and the families of complex oxides on the material side. Further motivation may be the pressing need to utilize alkane feedstock on the process side. Some extra trends can be seen upon sorting the total citations into subjects. There is no mega-trend in the field as no individual subject covers more than 7% of all papers. A broad distribution of topics can be found in the literature dealing mainly with phenomenological reports and to a surprising extent with the derivation and discussion of the "principles" mentioned above. Few papers discuss the mode of operation of selective oxidation based upon *in situ* observations or upon theory covering in a fragmented view only part of a catalytic cycle. The largest subset in citations is captured by vanadium oxide systems, showing also a more-than-average increase in citations and in paper output. A "trend" in citations can be stated for gold catalysts (excluding the CO oxidation literature) driven initially by epoxidation and since 2004 by oxidation of bio-feedstock. Titania as the second most frequently reported-on subject (not shown in Figure 1.1) profits both from the gold rush and the increase in vanadia research where it is used as substrate and co-catalyst. Figure 1.1(c) compares the popularity of the main reaction substrates propane and butane by their normalized citations evolving in time. There is a parallel trend for both substrates, with butane having some periods of excess popularity. These periods coincide with the increase in communication activity seen in Figure 1.1(b), allowing to conclude that butane oxidation seems to be a more attractive topic than propane oxidation as judged from citation behavior. It is stated that the number of citations for propane is more than twice as frequent as the numbers for butane in every year analyzed. The reviewer notes that the distribution of topics in the paper literature does not agree well with technical practice nor industrial needs in the future. Although neither conference contributions nor patent literature are added to the picture, the open literature deals with the subject of academic relevance in far greater intensity (citations) than with hard technical issues. This is in contrast to other areas of technology (semiconductors, materials, automation) where science and technology reflect each other much more closely in open scientific communication.

Figure 1.2 reports the regional distribution of active groups as obtained from the locations of the authors of papers. There is broad worldwide interest in the topic. Three nations contribute over 12% each to the communication production; a group of

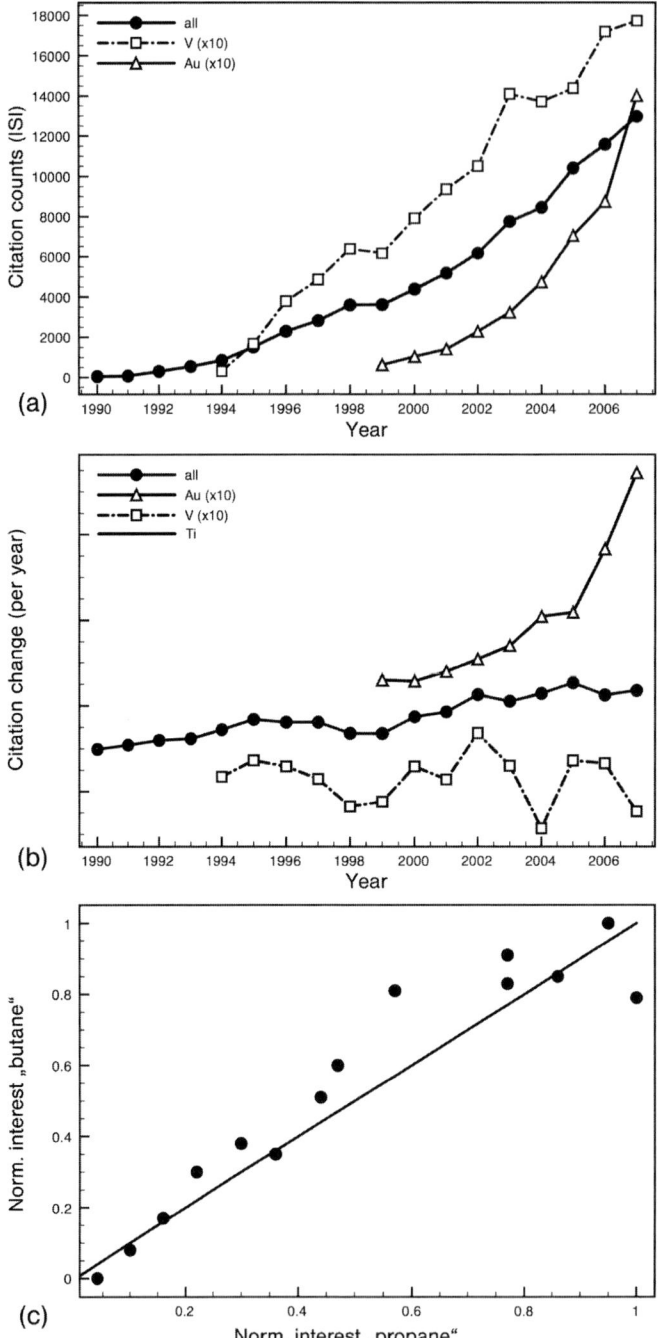

Figure 1.1 Analysis of the scientific literature for selective oxidation. (a) Number of citations over time; (b) derivatives of (a); (c) correlation plot for the substrates butane and propane on a normalized scale.

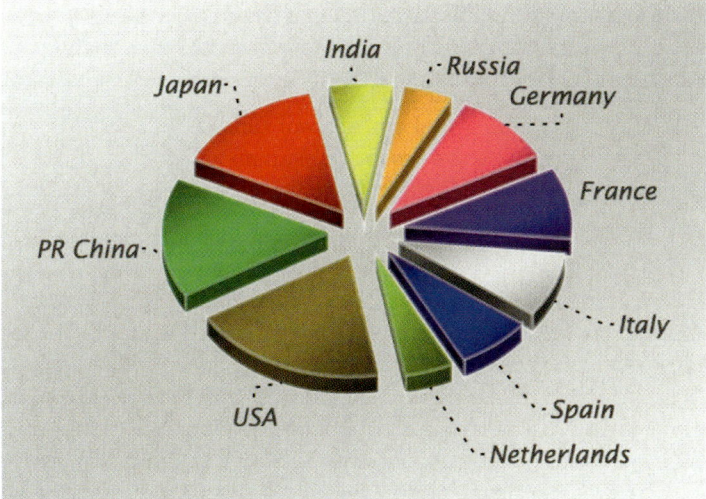

Figure 1.2 Regional distribution of 5300 papers on selective oxidation created from the author affiliations.

seven more nations contribute more than 3% each to the field. It is perhaps no surprise with respect to the technical relevance and resource allocation in the world that Asia is the region of development followed by the USA. If Europe were to cooperate in a common research effort it would outnumber all other nations, an observation that may be surprising given that neither resources nor consumption of the products of selective oxidation coincide with this trend. What would be consistent, however, is the strong support that fundamental science finds in Europe, a direction from which substantial breakthroughs are to be expected in the future.

1.3
Substrate Activation

A fundamental issue in selective oxidation is the activation of C–H bonds that is always required for ODH (oxidative dehydrogenation) and oxo-functionalization and is detrimental for epoxidation. A particular case is silver [70] as catalyst, which can achieve highly selective epoxidation of ethene as well as highly selective dehydrogenation of methanol to formaldehyde although it is notably in both cases "only" the same metallic catalyst. We will return to this case in the next section, which deals with the multiplicity of active oxygen species.

The issue of C–H activation has been addressed many times in the literature. It is common ground to state that the initial C–H activation should be the rate-determining step in the overall process of selective oxidation. This statement can be verified under conditions of proper kinetic studies when limited conversions and small concentrations of reducing products leave the catalyst in its original and active state and when no substantial re-adsorption or site-blocking of products occur.

A prototype study for this issue was performed for the conversion of ethane to acetic acid [71] and the same group highlighted in an earlier comparative study of C_3 oxidation [54] that, although initial propane activation is a difficult step, subsequent reactions associated with either excessive residence times of intermediates or with branching of reaction sequences into total oxidation may interfere with the overall selectivity to partial oxidation products.

On the basis of the general reasoning that selectivity in partial oxidation is limited by the fact that a catalyst capable of activating the first C—H bond of an alkane will also activate C—H bonds of the more reactive products and intermediates, a semi-empirical rule was established [72] linking partial bond strength to kinetic selectivity compiled from the open literature. This approach [45] defines a difference in bond energy between the bond to be activated and the strength of the successively activated C—H or C—C bond. An empirical rule often used in the literature states that promising selectivity can only be reached if this energy difference is less than 40 kJ mol^{-1}. A "universal" curve of many reported observations against the energy difference shows a surprising correlation. The correlation is so surprising as, principally, there is no immediate relation between thermodynamic quantities such as the bond energy of molecules at 0 K and kinetic observations of selectivity of partial oxidations at high temperatures, ambient pressure and not always well-specified conditions. One paper that explicitly discussed the case of propane activation stated that other kinetic factors [72] than the main reaction to the target molecule, such as branching and extended residence times of intermediates and products, can strongly affect the observed performance. The reviewer supports the view that the universal relation of bond energies and selectivity, although seemingly built upon a physical concept (bond strength variations), is merely an empirical correlation rather than a physics-based structure–function relation. However, this view does not deny the usefulness of the relation in discussing kinetic results with respect to likely boundary selectivity that can be expected for a given product.

One of the first studies that shed light on the details of the activation of propane over a single site vanadium catalyst [63] was performed without any a-priori assumptions or speculative geometries [73] of active sites. This was partly possible as the type of monolayer-grafted vanadia catalyst [8, 74] used is one of the best studied systems [54, 75] and it is clearly established – despite some problematic spectroscopic interpretations [48, 76] – that no extended ordering such as oxide crystallites is present in the catalyst's selective form. Although the work did not model a complete catalytic cycle it clearly shows that under the assumed reaction conditions, including realistic reaction temperatures, an energetically well-conceivable pathway of activation exists that involves an early surface radical structure that converts into propene via either a second activation or an addition–elimination of a hydroxyl species. This pathway was not enforced upon the system but found from a large selection of geometrically possible options and under full consideration of the electronic open-shell character of catalyst and substrate during reaction. The reaction path is redrawn from the data of ref. [63] in Figure 1.3. The paper discusses the transferability of the results for cases of defective bulk oxide systems of vanadium and molybdenum oxides. The only conceivable uncertainty in the selection of reaction pathways made concerns the absence of co-adsorbates and of solvation by water being present as a required or adventitious component in all oxidation reaction atmospheres.

From Figure 1.3 the first step is the abstraction of one proton, clearly by the action of a vanadyl (possibly also molybdenyl) group. The resulting radical intermediate has four possible reaction pathways that are energetically at least conceivable. The most facile route is abstraction of the second hydrogen through a fresh second group (pathway A). This has the advantage of not having to reduce the vanadium site to oxidation state + III but requires the close proximity of the second site or violates the concept of site isolation. Careful kinetic control of an ODH process or a desired further oxidation would favor such a reaction scenario. Under prevailing site isolation conditions, reaction pathways B–D will also lead to propene with, however, higher energy barriers. Pathways B and C require the participation of a V-O–substrate bond in the reaction and pathway C suggests the production of isopropanol as intermediate, which would be an excellent precursor for deeper oxidation. Pathway D is the most direct route. The resulting intermediate (5) is the same as obtained from pathway B assuming stabilization by water solvation.

The resulting hydrated reduced active site is intuitively the most plausible product. In the literature it is assumed [63] that this species is rapidly re-oxidized. This view is supported by numerous experimental observations and speculative statements [77–82] on various systems but it has never been shown directly that the monomeric species is rapidly re-oxidized. In Figure 1.3 the steps connected by colored arrows deal with re-oxidation; a disproportionation reaction liberates one active site from the dual site pathway A. In any case the reduced species is re-oxidized by a di-oxygen molecule that can only react to a give peroxo-species that has been verified to exist [83] on a relevant V(III) vanadia model system. Using *in situ* photoluminescence spectroscopy it was concluded that a highly electrophilic oxygen species, which here would be assigned to a decomposing peroxo group, can regenerate [84, 85] pre-reduced highly disperse vanadium. Regeneration of the active site would require thermal recombinative action of two adjacent sites or sacrificial oxidation of the product, which would limit the selectivity of the overall process to 8/9 of all propene produced if only reaction with the product is assumed to regenerate the active site. Interestingly, a mono-oxo oxidant (N_2O) that can re-oxidize a spent active site without participation of other sites or of additional redox partners gives for highly disperse systems an increased selectivity that was ascribed to an "overall lower degree of oxidation" [80] of the active surface. The structures (4 and 5 in Figure 1.3) allow explaining a direct electronic effect [54] of the support R in the process as the polarity of the support-oxygen bridge to the active V atom will affect the energy barrier for these reaction routes. The alkoxy-intermediates (2 and 3 in Figure 1.3) that react through high barriers are the intuitive starting structures for consecutive deeper oxidation to oxygenates or eventually CO_x. Their participation is frequently discussed [3, 72, 86–88] in the literature and agreement exists that this intermediate is most relevant for obtaining selective oxidation products.

There is no reason to assume that the ODH of alkanes, be it as stand-alone reaction or as initial process for deeper oxidation, will not proceed as shown schematically in Figure 1.3. The problem of re-oxidation of the active site, which has not yet been

1 Concepts in Selective Oxidation of Small Alkane Molecules

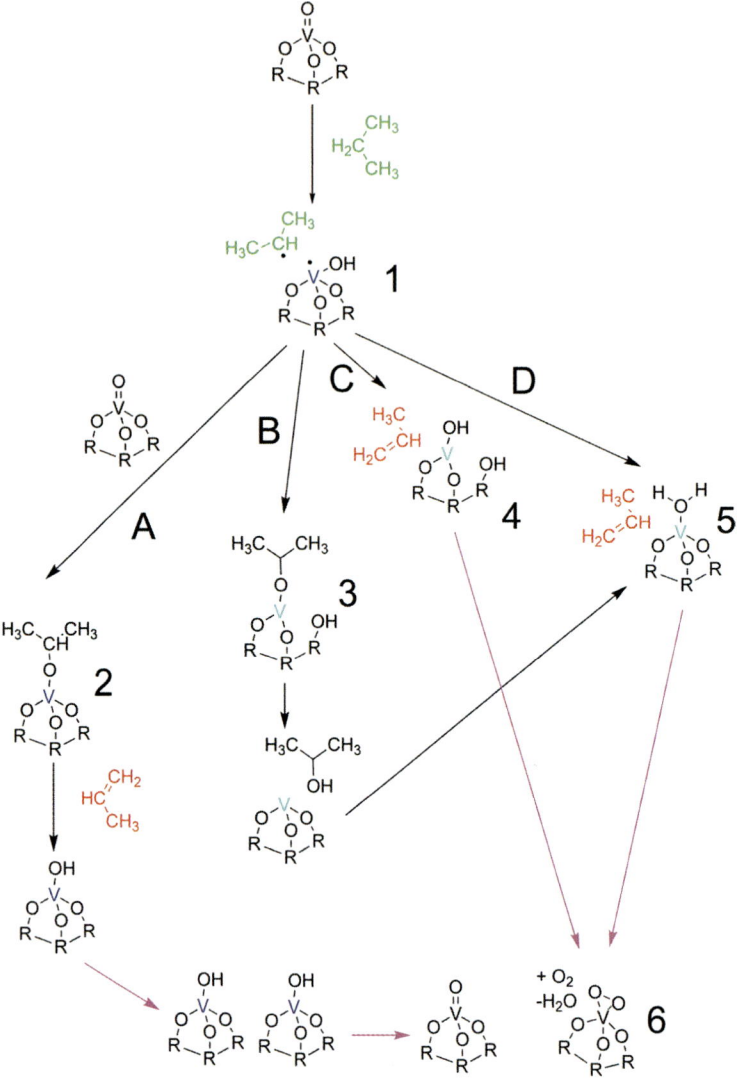

Figure 1.3 Activation of propane on a single-site vanadium oxide system. The figure is redrawn from the results of ref. [63].

treated rigorously, would not occur, however, if one assumed dimers as active sites. On bulk oxides no information about the size of isolated sites under reaction conditions is available. On the monolayer systems [76, 81, 89] some dimeric or oligomeric sites co-exist with monomeric sites. Until a rigorous theoretical study of the mode of operation of such a site has been carried out, the intuitive suggestions displayed in Figure 1.4 may serve to illustrate both the comp-lexity of the reaction pathways and also the possibilities of achieving selective operation without chemi-

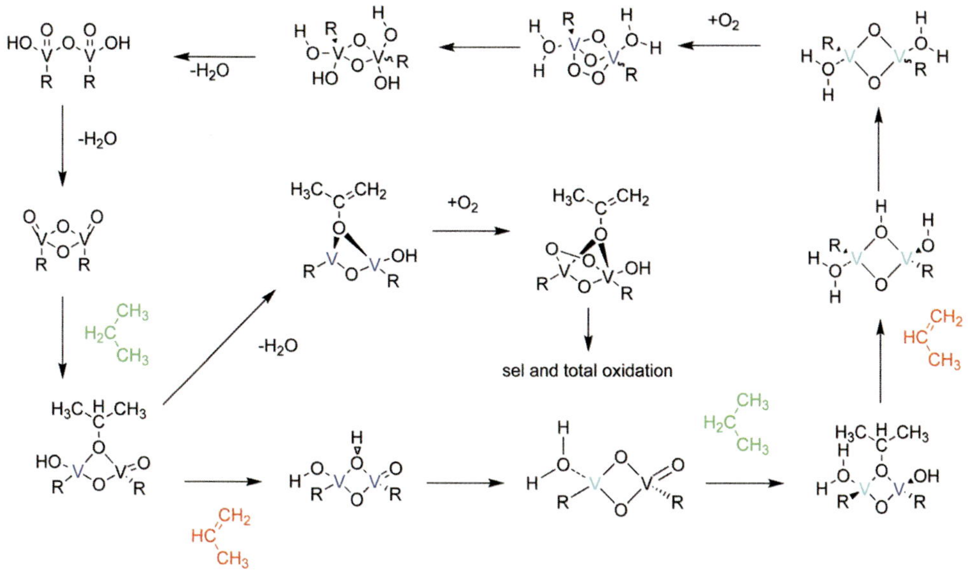

Figure 1.4 Hypothetical sequence of events of propane oxidation on a dimeric active site of vanadium oxide. The steps are extrapolated from the results shown in Figure 1.3.

cally unexpected structures. We know [90, 91] that dimeric sites will be present in an initially dehydrated state at the reaction temperature. They can perform, without mechanistically required losses in selectivity, oxidative dehydrogenation as shown in the outer cycle of Figure 1.4. The dimeric nature allows the active site to accumulate the four electrons that are required for activation of one di-oxygen molecule. Perfect site isolation is conserved in this cycle. It is of course speculative that all intermediates take the form sketched in Figure 1.4. In the presence of a suitable chemical potential of oxygen, the regeneration of the partly reduced site will begin while alkoxides are still attached to the site. This path (center of Figure 1.4) will lead to the creation of a different type of reactive oxygen, being less strongly reduced (more electrophilic) than the oxygen constituents of the active site. Such species are well-suited precursors to react with alkoxides to give oxygenates (aldehydes and carboxylic acids).

The fact that the active site is already partly reduced will diminish its ability (nucleophilicity) to activate all C−H bonds. In this way the oxygen activation on a site that is partly reduced will create a situation in which oxygen transfer can occur selectively without simultaneous activation of many reactive sites at the alkoxide. Obviously, for such a fortunate situation no external regeneration of the active site by lattice oxygen or by withdrawal of electrons to distant electron sinks (phase cooperation) must occur. The concept of site isolation finds in such an interpretation a natural cause: a catalytic site must be constructed in such a way that its electronic structure is allowed to fluctuate between a highly active initial state and moderate consecutive states as the conversion of the substrate molecule proceeds. The site is catalytic if it is ensured that at the end of the transformation the reactivity of the site is so low that the converted product can desorb, leaving space for electronic and structural reorganization that may be associated with dehydration of the site. This property will be referred to as "adaptive."

On monolayer oxidation catalysts it is quite possible that a distribution of monomeric and oligomeric sites can co-exist. Such catalysts can perform complex selective oxidation reactions [8, 92–94] without having to rely on multi-element systems. In bulk catalysts such a concept seems more problematic as, despite contrary statements [10, 26, 95], it is incorrect that certain elements of a structural motif are electronically isolated from other elements; active bulk redox catalysts are either metals or narrow band-gap semiconductors and exhibit electronic structures with substantial covalent bonding character distributed over all atoms in the unit cell. A boundary case may be mixed oxides where multivalent metal centers are completely surrounded by main group element oxo groups that would confine the electronic interaction to certain regions of the unit cell. In the MMO systems this is not the case, as anisotropic but extended electronic interactions operate throughout the crystal. Even if these extended interactions are small, in the highest oxidation state of the system they will inevitably increase the filling of metal d-bands during partial reduction upon catalytic action. An exception in the class of bulk catalysts are the heteropolyacid systems being used as partially salified crystalline and stable materials with an electronic structure of a molecular solid [18, 33, 96].

For such catalysts it is postulated that segregation driven by free energy minimization at the termination of the bulk will occur, thereby separating the active surface in a matrix and in isolated embedded sites [73, 97] that are by constitution not dissimilar to metal-oxo clusters present on supported systems. In such a case, one would expect two types of sites competing for reactants: those from the surface matrix without site isolation and those from the isolated active sites. Consequently, one would expect parallel pathways of reaction for selective and for unselective oxidation. The non-isolated sites remain electronically activated through exchange with the bulk and act unselectively. The isolated sites can fluctuate in their electronic structure, so allowing for partial oxidation following the above arguments of the need for adaptive sites to cope with the increasing reactivity of the substrate with progressing activation. Observations about crystal face selectivity [27, 88] need to be taken with care as, unlike in metal single crystals, the fracture of an oxide is not a method [98] for producing a uniquely bulk-terminated surface. The material basis for such a working catalyst is further discussed below.

The oxidation of propane to acrylic acid is a prominent example of where the splitting of the reaction over more than one site is discussed. A literature compilation [72] of data suggests two sites and a competition between partial and total oxidation via direct and intermediate deep oxidation. Data from the same system compiled in Figure 1.5 support this view. In Figure 1.5(a) one observes, under kinetically relevant conditions of below 10% conversion, opposing trends for total and partial oxidation where not only the sign of the trend but also the shape of the curve indicate that different sites and different limiting factors operate in this reaction. The side product acetic acid is also not a direct precursor to total combustion. From a set of such measurements a tentative reaction network based upon kinetic parameters was constructed (Figure 1.5b). The MMO catalyst is a suitable system as it exhibits a high barrier towards direct combustion of the feed. The selective oxidation path exhibits a kinetically difficult step after propane activation leading to detectable amounts of propene, which is in agreement with the above discussion of Figure 1.3. The difficult step is the oxidation of the activated alkane for which "acrolein" stands only as proxy, as the structure of the real intermediate is unknown. The desired product occurs with the lowest apparent kinetic barriers in the whole network. This observation helps in understanding the finite yield of an activated molecule. The low numbers indicate, according to expectation, that these barriers are "apparent" activation energies and will not allow any conclusion about the rate-controlling elementary step. The overall low productivity of the system may be traced back to a low initial production of activated alkane that is shared by three parallel reaction channels. Re-adsorption and conversion of initially desorbed propene may make a substantial, but in these experiments undetectable, contribution to selective oxidation. Such a scenario is in line with perceptions that more than one active site must cooperate to first create an olefin and then to oxidize it. Also in agreement with Figure 1.3 it is conceivable that, via the intermediate iso-propanol that cannot be detected in the gas phase, further oxidation to the detectable acetic acid occurs, contributing to the CO_x formation.

The present discussion shows that more than one type of activated oxygen operates in the network of alkane activation. Besides the ubiquitous "lattice oxygen" there are alternative concepts of reactive oxygen species that will be described in the next section. A critical material property linking alkane activation and alkane oxidation is that of acidity. Selective oxidation always requires the management of protons, thus calling for carefully designed acid–base properties. This is also advocated in the literature [8, 92, 99–101] but only limited hard information can be found on this difficult issue – hampered severely by a practical definition of acid–base properties on a solid when no redox-active probe molecules [102] can be applied. The concept of initial activation by strongly acidic sites [103, 104] (super-acids) can operate in a homogeneous fluid phase. The often discussed C–H activation of alkanes over solid acids such as zeolites is a complex issue as, often, additional metal species are present [105, 106] or the reactivity is short-lived [107] with radical species involved.

A typical study in which the lattice oxygen issue and the surface acid–base have been compared deals with two model reactions probing C–C bond breaking of methyl ethyl ketone vs. dehydration of 2-butanol [108] over TiO_2-supported monolayer catalysts. The study, employing mainly temperature-programmed techniques of probe molecule adsorption, found that the lattice-oxygen modifier antimony has simultaneously a strong moderating effect on the surface acidity, thereby highlighting that the two effects are hard to separate. The reviewer notes that the use of the term "lattice oxygen" here differs somewhat from the initial meaning of oxygen atoms diffusing through the bulk as on a monolayer catalyst at around 573 K it would be hard to activate "true" lattice oxygen. The mobile oxygen species detected should be ascribed to adsorbates located at defects of the ideal termination.

In the literature [109] on homogeneous C–H bond activation substantial evidence exists for selective radical activation using nitrogen-containing non-metallic

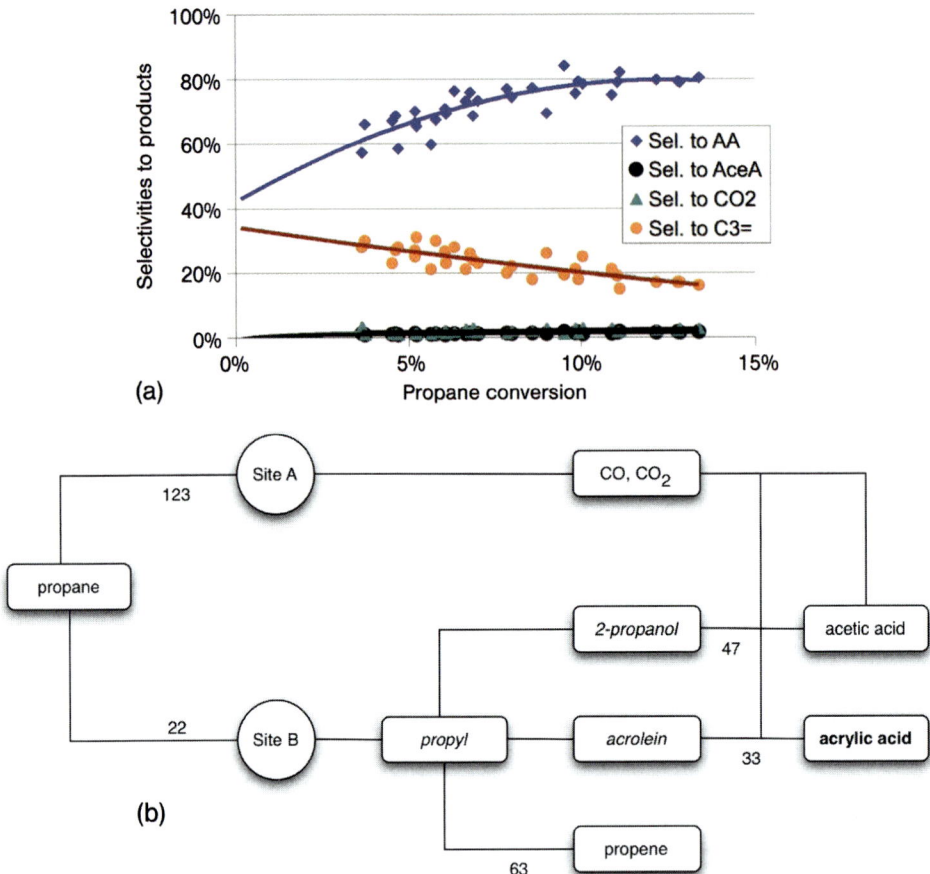

Figure 1.5 Performance (a) and experimentally derived reaction network (b) of a MoVTe catalyst (phase mixture 70% M1, 30% M2 phase, prepared by the slurry method, chemical composition as in the literature) under kinetic conditions ensuring no more than 10% conversion. A parallel microreactor "Nanoflow" was used and the catalyst was diluted 1:1 with SiC. The numbers in the reaction network are apparent activation energies in kJ mol^{-1}. Data kindly provided by S.B. Abd Hamid [32].

catalysts. Such pathways, observed under mild conditions in the presence of moderating solvents, work well with activated alkanes and with strong nucleophiles such as bromine as initial activators. In gas-phase heterogeneous catalysis the participation of such pathways seems less likely, in particular when product selectivity is taken into consideration. There are, however, reports [110, 111] about heterogeneously created radicals performing catalysis in the homogeneous gas phase. This issue, extensively studied with methane activation [112], still needs further careful studies.

1.4
Active Oxygen Species

The extensive and controversial discussion about the nature of active oxygen species may benefit from a separation of two issues: (1) the number and chemical nature of active oxygen species and (2) the location of active oxygen species. The present chapter refrains from reporting again the different views on both issues as this has been done several times in the reviews mentioned in Section 1.1.

Active oxygen species occur through reductive activation of di-oxygen (generally) at catalyst surfaces having to present four free-electrons per oxygen molecule. From this, oxides in their highest oxidation states clearly cannot activate oxygen and, for such materials, defects encompassing lower-valent metal centers need to be present. Successive activation via hyperoxo, peroxo and mono-oxo species renders the activated oxygen consecutively more nucleophilic and less electrophilic with respect to its reactivity towards co-adsorbed organic substrates. For semiconducting oxides the metal sites of oxygen activation become progressively more oxidized. Consequently, the bonding interaction between oxygen and metal site becomes progressively stronger, which is another way of explaining the change in reactivity from electrophilic to nucleophilic. It is important to realize that the gradual adaptation of reactivity with extent of activation occurs only at isolated sites. This interconversion is key to understanding the multiple reactivity of "simple" [87, 113] activated oxygen, which in turn is the key to understanding selectivity in oxidation catalysis. This mode of operation requires pre-formed defects in high-valent oxides and operates at low temperatures. The peroxo intermediate has been well-characterized by surface science [83] both in structure and its reactivity, decomposing in vacuum substantially below ambient temperature. This does not mean that such a species would not be reactive at conventional oxidation temperatures and pressures; its resulting low stationary coverage asks for rapid consumption as an electrophilic species or for a low kinetic probability of the reaction channel. The critical need for defects explains also the observation that catalyst treatments that intentionally create lattice defects [114] seem to be beneficial for catalyst preparation. Studies of the VPO system [115–117] suggest this with a non-uniform picture about the origin and stability of the activation. Several reports on the action of MMO systems [99, 118–121] also clearly find that lattice defects are beneficial for catalytic action [32] and may even be a prerequisite [56] for activity.

The undisputed relevance of lattice defects for catalytic function is, however, often associated with the possibility that reducing gas-phase molecules create active oxygen in an oxide catalyst through defect formation. Active oxygen may stem from the lattice and diffuse to the surface where it undergoes, by recombinative desorption, gradually the same interconversions from initially very nucleophilic (O^{2-}) oxo-anion to the more electrophilic charged di-oxygen species. This process has been observed to exist on oxidation catalysts numerous times with multiple techniques, as reported in the reviews mentioned in Table 1.1. The often quoted [3, 13, 46, 122] unambiguous relevance of lattice oxygen as reactant for selective oxidation in general is, however, to be considered with some reservations. Many studies were carried out under gas phase conditions with a low chemical potential of oxygen; thermodynamics thus predict a tendency of oxides to release oxygen according to the potential gradient from bulk to surface. Experiments in which pulses of organic feed only or in which reduced pressures (high vacuum) are applied will thus always show the involvement of lattice oxygen [123] and also support isotopic scrambling experiments [4, 33, 71, 124–127] between gas pulses and bulk oxygen. The process of sequential reduction of a catalyst by organic feed followed by a re-oxidation in a separate reactor (riser concept [123]) is a macroscopic proof of this action of lattice oxygen.

The quantification of diffusion processes was a major application of the analytical technique TAP (temporal analysis of product) [80, 87]. An example [128] of the use of TAP experiments at various pressure levels and by studying qualitatively the response function of a complex MMO catalyst can be found in a study of the mechanism of propane oxidation to acrylic acid. The paper relies on temporal evolution curves and draws far-reaching conclusions from the observation of hysteresis loops that can have many different origins other than lattice oxygen participation. The temperature at which such oxygen diffusion sets in will depend on the type of oxide and on the defect status. Densely packed oxides exhibit oxygen diffusion [129, 130] typically above 773 K, whereas typical MMO systems characterized by complex structural motifs with substantial void fractions in their unit cell, and hence anisotropic pathways for oxygen diffusion associated with a non-dalton composition in oxygen [4, 125], do show the onset of oxygen diffusion at about 650 K, a typical temperature for selective oxidation reactions.

If a catalytic cycle should be maintained, oxygen diffusion out to the surface must be complemented by an inward diffusions of surface-activated oxygen resulting from accumulation of reduced metal centers required to activate gas-phase oxygen. Not all studies mentioned here ensured in their experiments that the conditions of "lattice oxygen" catalysis were such as to fulfill the conditions of cyclic reversibility [34, 51, 82, 131, 132] as opposed to stoichiometric and irreversible reduction [133] caused by a structural phase transition. As long as complex MMO oxides are being used and the extent of reduction is kept to levels where no bulk transformation can be detected this condition can be verified [20, 99, 118, 121, 134, 135]. The kinetics of re-oxidation of partly reduced oxide catalysts was found to be rapid [77, 78, 80, 82] and always faster than its reduction.

Nanostructuring [32, 59, 94, 136] and arrays of defects [137] (sheer defects) create a partition into nanodomains of larger crystals, resulting from the typical high-temperature calcination reactions required [19, 121] to form defined phases of MMO. The resulting grain boundaries and sheer planes create pathways [99] for enhanced oxygen transport [56] and thus will support [118] lattice oxygen participation. A suitable structure for lattice oxygen participation is salts of heteropolyacids, where the loose packing of the individual molecules given by the intrinsic size of the Keggin units provides stabilization [138] for defective molecules (lacunary species) and for transport of oxygen to the outer surface. This structural property may be one reason for the enormous and versatile [33] catalytic activity of HPA catalysts in selective oxidation. In contrast, if no long-range structure is present then amorphous [115, 118, 134, 139, 140] oxides may become too strong as donors, giving rise to excess active oxygen. This will lead to over-oxidation with the resulting energy release triggering re-crystallization of the amorphous oxide and hence causing deactivation; hence there is a clear optimum of defects in a catalyst that seems to be much closer to nano-sized well-ordered entities rather than to highly disordered or amorphous solids.

The enormous body of experimental observations [16, 25, 54, 80, 87, 113, 118, 141–143] on oxygen ion diffusion in conjunction with selective oxidation does not, however, allow us to conclude directly that this oxygen is relevant for catalytic selective oxidation under conditions of an oxygen chemical potential that is equal to or larger than the decomposition potential of the catalyst; such conditions always apply when gas-phase oxygen is co-fed with the organic substrate, and usually in stoichiometric excess over the desired extent of oxidation. The almost general extrapolation in mechanistic discussions of the role of lattice oxygen determined under conditions of under-potential oxygen abundance [123] to situations with oxygen over-potential [141] with respect to catalyst decomposition seems, despite its frequent occurrence, not to be justified. The next section presents a hypothesis that explains the participation of oxygen mobility [141, 144–146] in the generation of active sites rather than in the conversion of the substrate that can equally well explain the experimental observations without having to assume unjustified extrapolations. The reviewer states that lattice oxygen can and does perform selective oxidation and that all preparative steps enhancing oxygen diffusivity can be useful for enhancing catalytic action up to the point where excess lattice oxygen becomes detrimental. All this does not mean, however, that the role of the diffusing oxygen [147] is the same in the two fundamentally different regime of reacting atmosphere with oxygen deficit and oxygen surplus.

Irrespective of how oxygen becomes "activated" there is broad agreement [1, 31, 33, 40, 46, 54, 142, 148–153] that several species of oxygen must co-exist to explain selective oxidation and substrate activation. There is also broad agreement that species may be categorized into nucleophilic and electrophilic with respect to substrates. Theory has ascribed this differentiation to the strength and polarity of the metal-to-oxygen bond. Semi-empirical [143] and quantum chemical approaches [16, 88, 154–156] at different levels have been used to assure the variability of the metal–oxygen interaction as a function of local chemical composition and structural coordination. We can be certain that a substantial differentiation of the metal–oxygen bonding exists on a structurally diverse [1, 5, 131, 142] oxide surface. It is reassuring that today we can state a substantial unification between the concept of heterogeneous sites at oxides and its counterpart on metals [98, 152, 157, 158] as being the relevant locations for catalytic reactivity. Tacitly, the assumption that every site at a surface capable of adsorbing a species can also induce its reaction has been replaced by the idea that many sites adsorb species but few of them react [159–161] the substrates to give products.

Figure 1.6 depicts the fundamental reactivity of nucleophilic and electrophilic oxygen. On oxides we can assign nucleophilic oxygen to strongly bound oxygen with a metal–oxygen bond polarized towards the oxygen atom. Electrophilic oxygen is usually more weakly bound (it has to be if it is to be transferred to a substrate) with

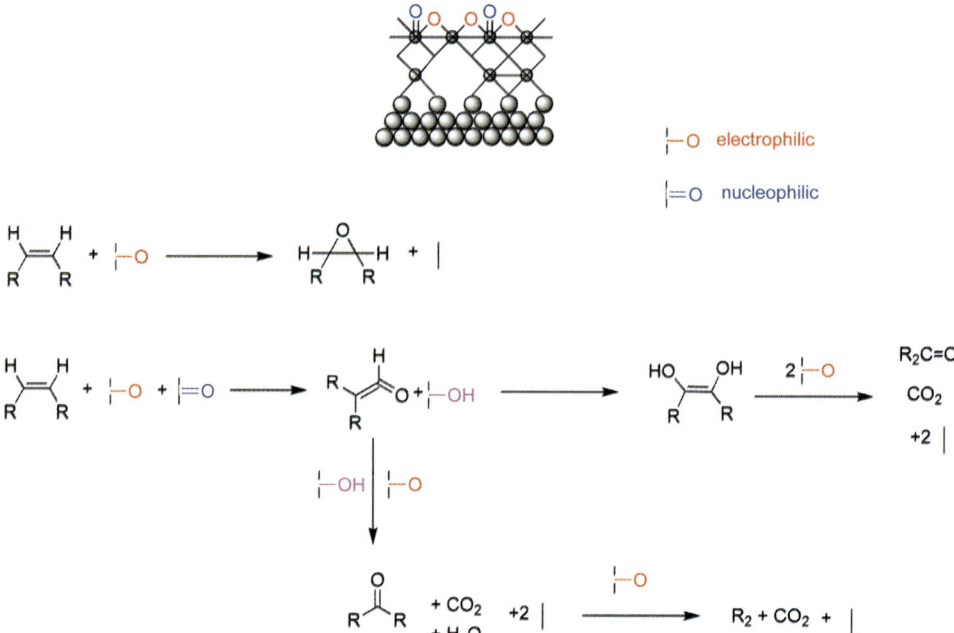

Figure 1.6 Schematic representation of location and reactivity of nucleophilic and electrophilic oxygen. Vertical lines denote active surface sites.

a bond polarization leaving a positive partial charge on the oxygen. The top scheme in Figure 1.6 gives one possible location on a defective oxide surface. The scheme also illustrates that the designation of "lattice oxygen" is not correct as we consider only reactive oxygen that must be site-isolated from the bulk lattice. Semantically a "surface lattice" describes the situation correctly but there is no need to discriminate this location from a chemisorption of oxygen on a defect site (where the bridging oxygen atoms are exposed to the surface) with different charge states of the metal sites. The oxygen species must specifically not be localized as drawn in Figure 1.6 for a situation with low chemical potential of oxygen. In compounds with multiple cation types (MMO) it was suggested that the electronegativity of M in M=O groups would create the bond polarization [77, 162, 163]. Adsorbates with formal O^- electron configurations can be particularly electrophilic [1, 2, 27, 142]. It is not excluded that all three situations co-exist on a complex oxide, thus allowing not for only two idealized oxygen species but for a distribution of reactivity of the prototype "nucleophilic" and "electrophilic". This serves as the basis for assuming that multiple reaction pathways act in parallel on the same substrate.

The electrophilic reaction pathway leads, with an olefin, to the epoxide and a free site as symbolized in Figure 1.6. A purely nucleophilic site would catalyze oxidative dehydrogenation as discussed above. The frequent case of co-existence of nucleophilicity and electrophilicity leads to selective or total oxidation as seen in the lower part of Figure 1.6. It is also seen that, in qualitative agreement with the assumed detrimental nature of gas-phase oxygen leading to weakly bound O^-, it is the kinetic availability of electrophilic oxygen that determines the selectivity to partial or total oxidation. The central quantity in designing selective oxidation catalysts is thus the control of the kinetic availability (local chemical potential) of weakly bound oxygen during a catalytic cycle.

This has been frequently recognized and the challenge arises of determining the chemical reactivity and surface abundance of activated oxygen. The usual solution of this problem is the application of an isotopic exchange reaction and kinetic analysis, as TPR/TPO cycling does not discriminate between labile reactive and labile non-reactive (sub-surface) oxygen species in a polycrystalline sample. Several studies on monolayer oxide catalysts [144, 164, 165] and on bulk phases [126, 166, 167] have revealed [147] that, depending on the redox potential over a sample, molecular (electrophilic) or atomic (nucleophilic) isotope exchange can occur. In the literature it was not made sufficiently clear that it is not the type of catalysts, as assumed, but the quantity of the oxygen chemical potential (temperature, redox state, defect state, partial pressure in gas phase) that controls the type of isotopic exchange. As the chemical potential is hard to determine with catalytic processes, one observes material discriminations [147] that are only "snapshots" of a common reactivity pattern detected under not-comparable potential conditions (usually only the gas-phase pressure is considered).

In this situation it is desirable to have other means of identification examining the nature of active oxygen. The standard surface-science tool [168] for this would be temperature-programmed desorption (TPD) or temperature-programmed reaction [169, 170] with probe molecules (CO for electrophilic oxygen, methanol for nucleophilic oxygen). A recent case study with RuO_2 as model catalyst showed that X-ray photoelectron spectroscopy (XPS) at high resolution [171, 172] can directly resolve the two types of functional oxygen even when located on an oxide bulk. These studies rest in their interpretation on a large body of data [52, 70, 166, 173, 174] that discriminate on metallic silver the reactivity of activated oxygen for either ethylene epoxidation (electrophilic) or methanol dehydrogenation (nucleophilic), both reactions of technical relevance.

Photoemission and TPD are excellent tools for analyzing activated oxygen as they can give information on the reactivity and on the quantification of the atomic oxygen species. On oxide surfaces it is, however, difficult to discriminate the active species from the lattice oxygen. It is to be expected that only a fraction of the surface holds the reactive oxygen, translating into a sensitivity problem as only a small percentage of the whole signal can be expected to carry the relevant information. By profile analysis alone it is almost impossible on a high-performance catalyst to find the reactive oxygen. Figure 1.7 shows how by applying *in situ* techniques one can prepare the catalyst surface such as to allow for the identification of reactive oxygen. Figure 1.7(a) shows the high-resolution synchrotron difference XPS spectra of a VPO catalyst [175–177] in two states of productivity. The *in situ* oxidation of butane was performed at 473 and at 673 K, marking the onset and maximum, respectively, of stable catalytic production of maleic anhydride. The difference spectrum between

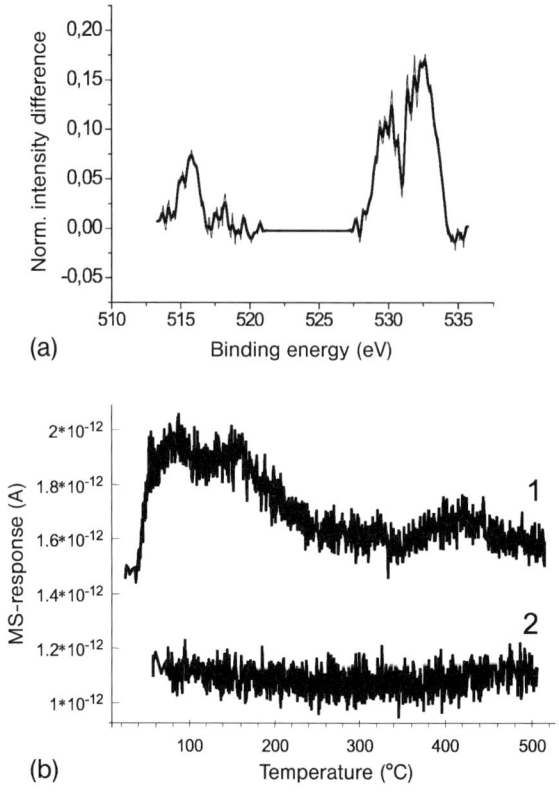

Figure 1.7 Active oxygen on a working VPO [175] catalyst. (a) XPS *in situ* difference spectra between phases of low and high performance. (b) Oxygen TPD spectrum after preparing the surface of polycrystalline VPO by *in situ* butane oxidation.

the two states reveals the species that are more prominent at the surface when high productivity is observed. In the region of vanadium 2p 3/2 we see a signature for vanadium +III with an abundance of less than 3% of that of the main vanadium content. In the oxygen 1s region we observe two features consistent with electrophilic oxygen (530 eV) and with water or phosphoric acid (533 eV). As we also see, constantly, a small excess of phosphate over its bulk stoichiometric value we would state that reduced vanadium, electrophilic oxygen and segregated phosphate are the signs of a working VPO catalyst. This is consistent with the observations from *in situ* TPD shown in Figure 1.7(b). Only after equilibrating and performing the butane oxidation *in situ* was it possible to study the presence of the relevant reactive oxygen; in cases where unreacted samples or pre-evacuated samples were used, much larger signals were found that relate to bulk-diffusion of oxygen as they will not disappear upon repetitive TDS heating cycles. Reaction-prepared VPO shows two signals at low and high temperature (trace 1 in Figure 1.7b), both of which are only present as surface species as repeated heating (trace 2) removes the signal. The assignment to electrophilic (labile) and nucleophilic (stable) oxygen is tentative but the fact that desorption of the nucleophilic species coincides with the reaction temperature and the knowledge that both species can be found by spectroscopy [155, 175] lend some support to this interpretation.

The recent enormous progress in preparing and studying surface science qualities of relevant systems [178–180] such as vanadium oxides plus their defects has given further clear evidence that electrophilic weakly bound oxygen [83] as well as more stable defect-related oxygen [48, 181–183] do exist and exhibit a reactivity much in parallel to high-performance catalysts of the monolayer type [184, 185].

One of the most significant results from the advent of these surface science studies on oxides relevant for the present catalytic applications is the fact that oxides can be multiply terminated and that they are not terminated [154, 180, 186–190] in cuts through the bulk structure. This is not unexpected in general [98, 156, 179] but it is of great value to know this in attempts to understand the mechanisms that activate oxides for catalysis. These rigorous studies must be differentiated from more empirical studies carried out on termination issue with qualitative methods and without predictive power but with the still invaluable advantage that they can be applied [97, 191–193] to complex MMO catalyst systems. Such studies can be used to probe the surface reactivity, to address the issue of segregation of, for example, vanadium out of an MMO system and to compare different qualities of the nominally same material with speculative assumptions about the influence of defects.

As a function of the chemical potential of oxygen (partial pressure, reductive adsorbates, water, sticking coefficient, surface defects) "simple" oxides such as the transition metal sesquioxides (M_2O_3) can adopt different [194] and eventually co-existing terminations. At high potential, prevailing under conditions such as the air calcination used in synthesis under practical pressures of oxygen, such oxides are always terminated by a metal–oxygen double bond ("-yl group") irrespective of their bulk formal oxidation state. This seems to hold also in cases when such compounds do not exist as three-dimensional bulk structures. As the oxygen chemical potential decreases, defects in the -yl termination occur, allowing rearrangement of the local coordination and binding of hydroxyl species. At even lower potential a metal-terminated surface is most stable, allowing the binding of molecular oxygen [83] and other electrophilic species. These surfaces are quite different from the expected geometry of a bulk-cut and can produce complex and also metastable surface structures.

For the present discussion it is obvious that extended surfaces of oxides in contact with practical pressures of oxygen will be inert for catalytic reactions as the surface exposes essentially oxygen electronic states and gives no access to metal d-states carrying electrons for chemisorption and activation of adsorbates. Structural defects such as steps and kinks ("roughness"), chemical effects such as pre-reduced sites or "designed" weak spots in a complex structure such as main group element atoms [26, 195, 196] in a transition metal ion matrix are pre-requisites to initiate catalytic [197] reaction. Only then can the organic substrate redox chemistry discussed above generate the stationary concentration of reduced (defective) sites [1, 114, 131, 162, 178] required for steady state operation. The kinetic barrier at which such a steady state defect formation will occur depends substantially on the state of long-range ordering: a nanocluster of oxide with almost no translational ordering of its "surface" will adjust much more quickly in its termination according to the gas-phase potential than a semi-infinite (micron) sized well-ordered crystal face [16, 182, 190, 198] often seen in VPO or MMO catalysts.

1.5
Catalyst Material Science

Selective oxidation materials fall into two broad categories: supported systems and bulk systems. The latter are of more practical relevance although one intermediary system, namely vanadia on titania [92, 199–201], is of substantial technical relevance. This system is intermediary as titania may not be considered an inert support but rather as a co-catalysts [202] capable of, for example, delivering lattice oxygen to the surface. The bulk systems [100, 121, 135, 203] all consist of structurally complex oxides such as vanadyl phosphates, molybdates with main group components (BiMo), molybdo-vanadates, molybdo-ferrates and heteropolyacids based on Mo and W (sometimes with a broad variation of chemical composition). The reviews mentioned in Table 1.1 deal with many of these material classes.

The fact that most supported catalysts, in particular with refractory oxides, are not as efficient in selective oxidation as bulk systems is circumstantial evidence for the necessity of having a bulk delivering lattice oxygen to the active surface. Supported systems have only their "surface lattice" oxygen; however, as discussed [54, 71, 75], the electronic structure of the support will influence substantially the redox chemistry and thus the surface lattice oxygen availability of the active component. Studies comparing catalytic functions of such systems to be carried out with great care, as it is not clear that different supports always hold the same structure of the active oxide although this was strongly [204] suggested. The chemistry of metal oxide supports considered as large ligands is sufficiently different that it is not likely that even when the same synthesis method is applied one will obtain the same [76] active structure. The option of designing the support as oxygen delivery system [205, 206], which is possible with, for example, titania, ceria, zirconia and their mixtures, is counter-affected by the tendency of these supports to react with the active surface phase and form multernary compounds (e.g., spinels and perovskites) that are often difficult to identify due to their low average abundance.

On classical systems of vanadia [53] on silica and alumina the principle of site isolation [8, 47, 74] is easily fulfilled with adequate loadings of active component. In MMO systems this is much more difficult, as the electronic structure of these oxides is not that of a salt with separated anion and cation electronic interactions but is highly covalent with electronic states connecting all participating atoms [16, 190, 198, 207], in particular when the catalyst is slightly reduced [208, 209] such as when in contact with an organic feed (also when containing oxygen gas).

Bulk materials used as oxidation catalysts not only allow for oxygen transport but also accommodate a wide and homogeneous modification of their electronic structure, ranging from wide-band semiconducting highly oxidized states with electronic formally d-0 states at the metal centers to a broad range of narrow-band semiconducting materials with d-n states at the metal sites. The result is a tunable electronic structure with the provision of chemically reactive electrons at or near the Fermi edge. Their concentration will depend on the extent of reduction that can be formulated as the effective oxygen chemical potential given by the reductive power of the organic substrate present under catalysis conditions. This tunability is designated as "adaptiveness". No statement is made as to whether this tunable electronic structure that we can detect is accompanied by geometric structural transformations such as defect structures [34, 82] and their order varieties. As indicated above it is thus suggested that the site isolating properties of a supported system and the adaptiveness of a bulk MMO system can be combined in the concept of a self-supported catalyst.

Figure 1.8 presents an example of such adaptiveness for a simple reference system, namely V_2O_5. A single crystal was studied by *in situ* XPS at various pressures of oxygen [175] at 673 K, a typical operation temperature of a MMO catalyst. Figure 1.8(a) reveals the quantitative analysis of core level spectra. With synchrotron XPS we can perform [210, 211] non-destructive depth profiling and study the surface and sub-surface regimes separately. The depth calibration is only semi-quantitative as the electron scattering will change with changing valence electronic structure of the sample. It can be seen that the surface stays stable in its chemical composition whereas the sub-surface regime loses some oxygen in accordance with the lattice oxygen concept. This implies that the oxide surface is structurally stable [133, 183, 212] and yet transmits oxygen from the bulk into the gas phase without significant kinetic hindrance. The high structural stability is expected from theoretical work [16, 131, 188] showing that creation of surface defects in an ordered oxide is a energetically very costly process that will not happen by thermal action except for "compensation" by a chemical reaction (such as water or CO_2 formation). An estimation of the extent of oxygen transport along the gradient perpendicular to the surface can be judged from the time evolution of the oxygen loss. The consequence of the lost oxygen must be a reduction of the oxidation state. It can be seen that the bulk approximates the expected value of 5.0 almost too accurately whereas the surface is defective, as derived from the slightly reduced value. The fact that the outer surface is enriched in oxygen with respect to the bulk value and that we do not lose this oxygen indicates a restructuring with vanadyl species as the reason for the stability. In agreement with single-crystal experiments [180] the oxidation state of the vanadium reduces to lower values without losing these vanadyl species, in disagreement with chemical intuition expecting a loss of V=O (+5) through thermolysis. Within expectation, the extent of reduction of the vanadyl-stabilized surface is faster than that of the bulk which is buffered in its local electronic structure through the site-interacting chemical bonding. The result of the process studied will be a chemical heterogeneity in two dimensions, a patchy surface of vanadyl groups on a sub-surface of strongly reduced vanadium with a gradient back to a, on average, weakly reduced bulk.

In Figure 1.8(b) the spectroscopic consequences of the heating in various chemical potentials of oxygen can be seen directly. The reduction causes a donor state band to

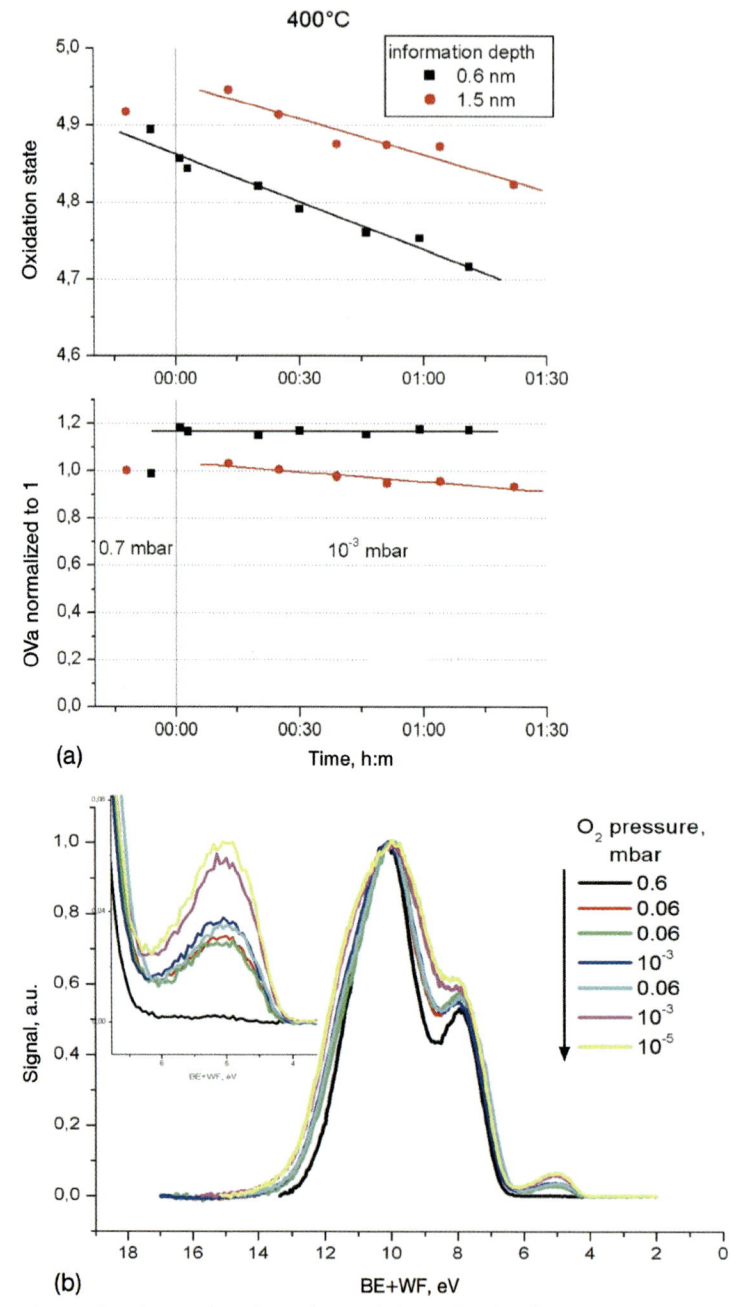

Figure 1.8 *In situ* XPS data of a single crystal of V_2O_5 (basal surface orientation) heated under different oxygen partial pressures.
(a) Quantification of XPS data obtained at two levels of depth information during constant heating of the crystal at 673 K.
(b) Valence band XPS data as function of oxygen partial pressure at 673 K. For experimental details see Figure 1.7 and text.

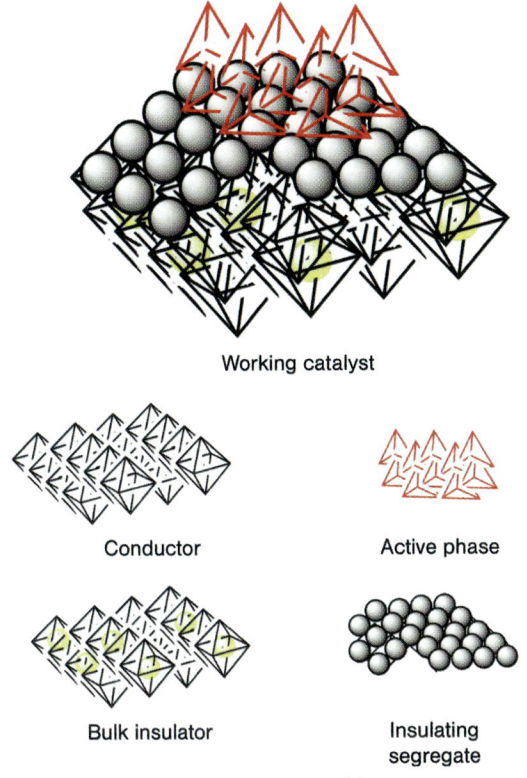

Figure 1.9 Schematic representation of the microstructure of a working oxide catalyst for selective oxidation.

fill the gap of states above the valence band of the (transparent) parent crystal. Only at high oxygen potentials is this structure stable. In agreement with the adaptiveness, the filling of the valence gap with (predominantly [182, 198]) V 3d states increases with a gradual reduction of oxygen potential (achieved in this model experiment by lowering the pressure). The gradual reduction also causes a change in the geometric structure, forming ordered or disordered sub-oxides [87, 133, 139, 183, 184], much in agreement with observations on VPO catalysts [56, 213–215] discernible by spectroscopy at the surface in the change of the shape of the main valence band transitions.

We propose as target material property for a working selective oxidation catalyst a particular microstructure as presented in Figure 1.9. The general type is always "supported," whereby the support for bulk catalysts is the whole of the prepared material indicated in the figure as "bulk insulator." As this phase is multi-component, at the surface we expect that segregation will occur driven by the tendency to minimize the free energy. A conducting oxide will separate from a mobile segregate. In VPO catalysts this could be a pyrophosphate with a mixture of vanadyl phosphate (as most stable phase) as bulk insulator and a VxOy thin layer as conducting oxide plus some free phosphoric acid (to account for the compositional and electron balance) as mobile segregate. In other classes of MMO systems bismuth, antimony and tellurium carry the function of the mobile segregate. In Figure 1.9 it is assumed that the active component (a cluster of MxOy) is separated from the conductive layer by the segregate, to which the function of site isolation agent is assigned. In this picture the synthesized non-activated pre-catalyst is support and reservoir for the active phase. A microstructure of working catalysts as depicted in Figure 1.9 would explain the enormous sensitivity of the catalyst material to its activation and equilibration [95, 137, 141, 148, 150, 176, 216] conditions and the role of structural promoters [55, 213] (limiting the surface segregation against complete conversion of the pre-catalyst into several phases under harsh reaction conditions). Furthermore, the success of a supported catalyst in mimicking bulk catalyst performances would be logical [8, 74, 84, 85, 94, 154, 217–219] as both types are derived from the same fundamental building principle but represent different material realizations.

Evidently, if a pre-catalyst can undergo a material transformation as indicated in Figure 1.9, there will be a high tendency to stabilize the system through crystal growth into phase-separated stable components such as binary oxides. We have to assume that this also will be happening in proportion to the performance, as catalysis will provide the energy source driving the transformation. This simple concept can explain the experimental observation of the apparently beneficial character of several phases cooperating in a reaction: the detection of a phase mixture with the extent of mixing being proportional to the activity is an indication of the progression of the transformation of the initial single phase pre-catalyst via its active state into a deactivated multi-phase form. The concepts of phase cooperation [57, 220, 221] and of remote control [13] would find as correct experimental observations their explanation as post-mortem analysis artifact. The literature has to the knowledge of the author no report of the beneficial effect of phase cooperation witnessed by *in situ* functional analysis, which would contradict the hypothesis put forward of phase cooperation being a proxy for structural dynamics.

The fact that catalysts have a long lifetime in contrast to the estimated stability of oxo-clusters held on segregated oxides or on supports calls for a living catalyst, that is, for a concept in which this adaptive site will be regenerated during catalytic operation. This idea was formulated during *in situ* structural studies [138, 139] of HPA salt catalysts; the beneficial role of water addition [33, 162, 222, 223], despite its site-blocking function, was partly ascribed to its action [33, 96, 224] as co-catalyst for the structural regeneration at comparatively low temperatures of operation. An element of structural dynamics was also seen in the activation sequence of VPP catalysts [148, 150] and again water addition [225, 226] is an important factor here. One product of VPO activation is phases of $VOPO_4$ that exhibit a very rapid phase dynamics that are strongly, and unexpectedly, affected by the chemical potential of hydrocarbon [132] species. Other authors have advocated [162, 214, 227, 228], in attempts to describe the molecular nature of active sites at VPO systems, some elements of transformations leading to structures that differ from cuts through the bulk crystal structure.

The author suggests considering working catalysts for selective oxidation reactions as dynamic materials. By this term the ability of the pre-catalyst to transform gradually into a surface microstructure producing site-isolated active sites is controlled by the chemical potential of the reagents. The activation is thus different from a normal (surface) phase transformation [229] as it follows in its course and kinetics not only the potential and energy gradients given by the solid material. It is rather the combination of adverse influences of oxygen and water on one side and hydrocarbons on the other side that control the structural result [132] and the kinetics of the transformation process. In this way a solid state conversion is achieved, orienting its progression on the effectiveness of the catalyst in responding to the surrounding chemical potential: bad catalysts will either transform rapidly into the deactivated forms or will not start in their transformation. Good catalysts will quickly start the transformation but then progress only gradually to extend the lifetime of the active microstructure as intermediate.

This definition of catalyst dynamics, being more precise than suggestions of a "living" catalyst quoted above [148], needs to be discriminated against other perceptions of catalyst dynamics. First, in other oxidation reactions catalyst dynamics defined in the above way have been observed by forcing temporal conversion patterns with profiles of chemical potential; the action of lattice oxygen was made observable in oxides and metals [152, 206]. The structural dynamics were studied in Mo-V-W systems [126, 134] by changing the abundance of stabilizing cations and observing the effect on catalytic performance. In a review on *in situ* structural studies of high-performance catalysts [161] the relevance of structural dynamics in the above sense, with the proof of reversibility, was put forward for a metal-on-oxide support system. The still rare observations of material dynamics [82, 230–234] in chemical oscillating reactions are inspired by the work of kinetic oscillations seen in surface reactions under controlled conditions [235, 236] that are special cases of reactor oscillations [237–240] seen in studies of oxidation processes.

Catalyst transformations seen in complex selective oxidation catalysts are irreversible within the conditions of catalyst operation: they may be reversed with the addition of special agents such as high water pressures and/or oxygen but such effects have not yet been studied systematically. The operation of a riser reactor [123, 241] also requires reversible transformation of the catalyst, being discharged and re-charged with active oxygen in much the same way as a battery. The transformation affects the whole bulk of the material and occurs on forced timescales of minutes. This kind of oxygen pumping is well known [206, 242, 243] from combustion catalysts but will in most catalyst systems not be sufficiently controllable to achieve selectivity as the combined solid-state and reaction product control of the process is not operative in these systems trimmed for storage capacity.

The concept of structural dynamics is clearly demonstrated in the VPO catalyst system. The high-resolution TEM in Figure 1.10 shows an activated VPO catalyst (no promoters, made by the alcohol route) that exhibits a typical [215, 225, 244] termination with little structural order supported on a perfectly ordered crystal consistent with the pyrophosphate structure (and with other structures of the VPO family [245, 246]). The lack of long-range order is not seen due to the operation

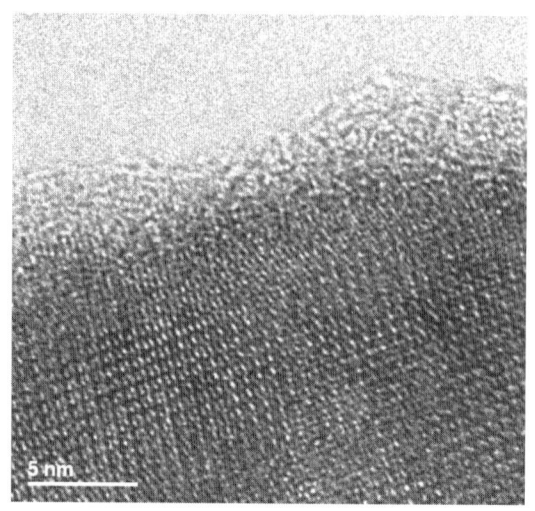

Figure 1.10 TEM image of a VPO catalyst (no promoters, prepared by the alcohol method after 100 h equilibration in feed); as 200 keV electrons at a low dose of below 10 A cm^{-2} were used the beam damage can account for only approximately half of the overlayer seen above the (100) lattice image of VPO.

of an amorphous [244] active phase that is not the consequence of TEM artifacts, which tend to superimpose on this termination feature by thickening the disordered boundary layer. The poor surface ordering is the consequence of the metastability of the microstructure depicted in Figure 1.9 after cooling and transfer in uncontrolled atmospheres: the disorder is seen as the consequence of the decomposition of the active phase. Similar terminations [19, 32, 116, 133] were also seen on other MMO catalysts.

The author regrets that only limited discussion [115, 118, 137, 140, 201, 244, 247, 248] about termination issues in high-performance catalysts can be found in the literature, which may be due to difficulties in observation. There is no common conclusion about the effect of amorphous structuring as no clear distinction is made between surface disorder being not an artifact of observation ("good") and bulk amorphicity as lack of crystallinity of the whole material ("bad"). The model catalysts community is much more advanced here and we can learn that highly ordered surfaces exist with chemically stable terminations of the -yl type; when the model systems are made reactive, massive reductions (by sputtering or electron irradiation) need to be applied [178, 235] that destroy the long-range structural ordering of the model surface. These observations are fully consistent with experiments describing the catalytic dehydrogenation of ethylbenzene to styrene over model MMO iron oxide surfaces [68] where with the onset of reaction a complete loss was documented of the long-range structural ordering.

The application of *in situ* surface analysis has given some experimental hints [155, 175–177, 244, 249] on the operation of material dynamics in working VPO catalysts. Both XPS and XAS have been applied as detection methods for discriminating the chemical state of the reacting surface from the bulk. The whole catalyst material [132] also undergoes structural transformations [101, 102, 139, 214, 225, 250] as seen, for example, in Figure 1.9 with a less reactive sample than VPO. The TEM observation shows that the active surface without any beam damage is about 1 nm thick and thus surface analytical tools need to be used that give information about this volume as otherwise substantial differences between the active surface and the underlying bulk might not be detected. Using surface-sensitive XPS [210, 251] the evolution of the phosphate coverage was studied for the same pre-activated VPO sample [175] as employed for Figure 1.10. With heating under reaction conditions at a total pressure of about 1 mbar the phosphate became mobile and was lost with a time evolution shown in Figure 1.11(a). The presence of reducing or

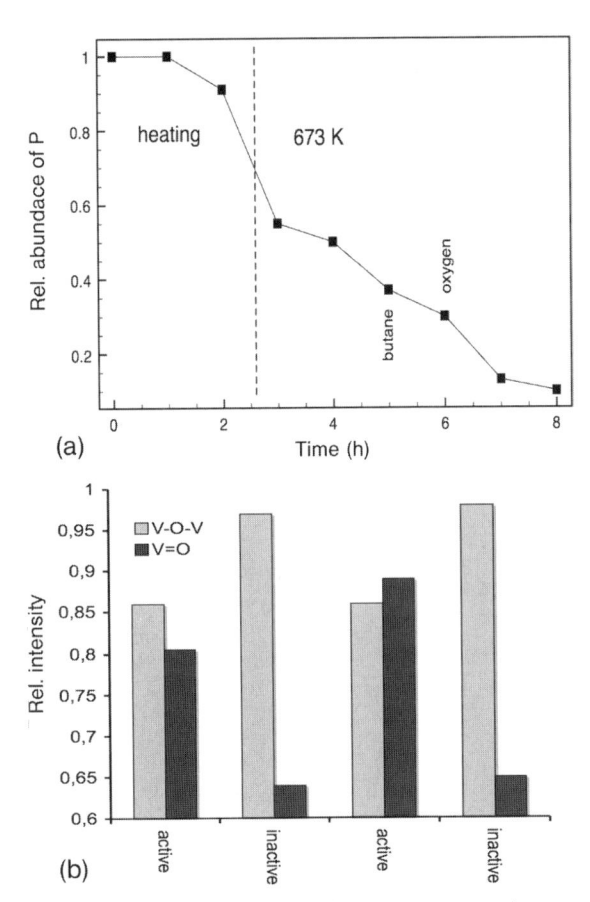

Figure 1.11 (a) Loss of phosphate from a VPO catalyst (see Figure 1.7) during heating in feed and at instances of pure oxygen or pure butane. The total pressure in the XPS chamber during in situ XPS was 0.22 mbar; production of MA at steady state was ensured. (b) Evolution of integrated intensities of characteristic features for V—O—V and V=O bonds from V L_{III} edge XAS spectrum as function of the operation and inactivity of the surface: for details see ref. [155].

oxidizing atmospheres had, under the mild conditions of the experiment, a small but not decisive effect on the evolution. We conclude that phosphate is a mobile species under reaction conditions. The substantial loss of phosphate is partly attributed to the low total pressure of the experiment and considered in its extent as an experimental artifact. The graph demonstrates, however, that within short timescales enormous re-distributions of phosphate are possible. This was held responsible [252] for the generation of either active or selective surface active layers on bulk VPO catalysts. This work uses the general approach of forcing the evolution of the structural dynamics by changes in the chemical potential of the gas phase. With the limited pressure of the *in situ* XAS experiments [155, 177] it is more practical to change the potential by the surface temperature rather than by the pressure. Figure 1.11(b) displays the result of such a forcing experiment whereby every episode was kept at steady state for 2 h. Characteristic signatures of the two bonding entities V-O-V and V=O were detected and followed in position and relative intensities with the forcing of the catalytic function. Clearly, there is a correlation of the intensity evolution with catalytic function. The convergence of relative intensities can be seen as an indication that the local electron density at V=O is reduced during operation whereas the electron density at the bridge site is enhanced. Both observations are in good qualitative agreement with the mechanistic expectations discussed with Figures 1.3 and 1.4. In the non-operating state the electron densities, as indicated by the XAS intensity changes, fit more to a fully oxidized V=O site and a more electron-rich V-O-V entity whose M–O bond distance (about 175 pm), as determined from the absolute energy position relative to that of V_2O_5, is much longer than expected [253] in the VPO structure.

Based on our experimental information and the consistent opinion of the literature that VPO is a bulk pre-catalyst that undergoes transformations at least in the surface-near volume into an active phase that is not identical with the projections of the bulk VPO structure to its terminations, we attempt to construct a diagram highlighting the elements and its interrelations that control this vital process. The complexity of the situation acknowledged in the literature is apparent from the complexity of Figure 1.12. It is thought that all known facts can be accommodated by the reaction network of Figure 1.12, which represents a "wiring diagram" of butane oxidation. It is different from all previous attempts to analyze the situation as it deliberately links the surface chemistry with its gas-phase controls to the solid state chemistry with its controls via real structure (nano-size, defects, compositional fluctuations). The linkers are the products H_2O and MA, which exert quite different and adverse effects. For simplicity, the energy flow and its enabling functions are omitted from the diagram. The obvious fact that an explanation of the catalytic operation without its controlling influences on the nature of the active sites here is impossible manifests the "gap" phenomenon [254] encountered in many metal-catalyzed reactions when surface science data (deliberately omitting the material chemistry) are used to predict catalytic performances.

The central shaded part of Figure 1.12 describes the catalytic operation of active sites. These are considered as isolated sites of a cluster of vanadium oxo species (Figures 1.3 and 1.4). They are adaptive in the sense that they need first to be activated

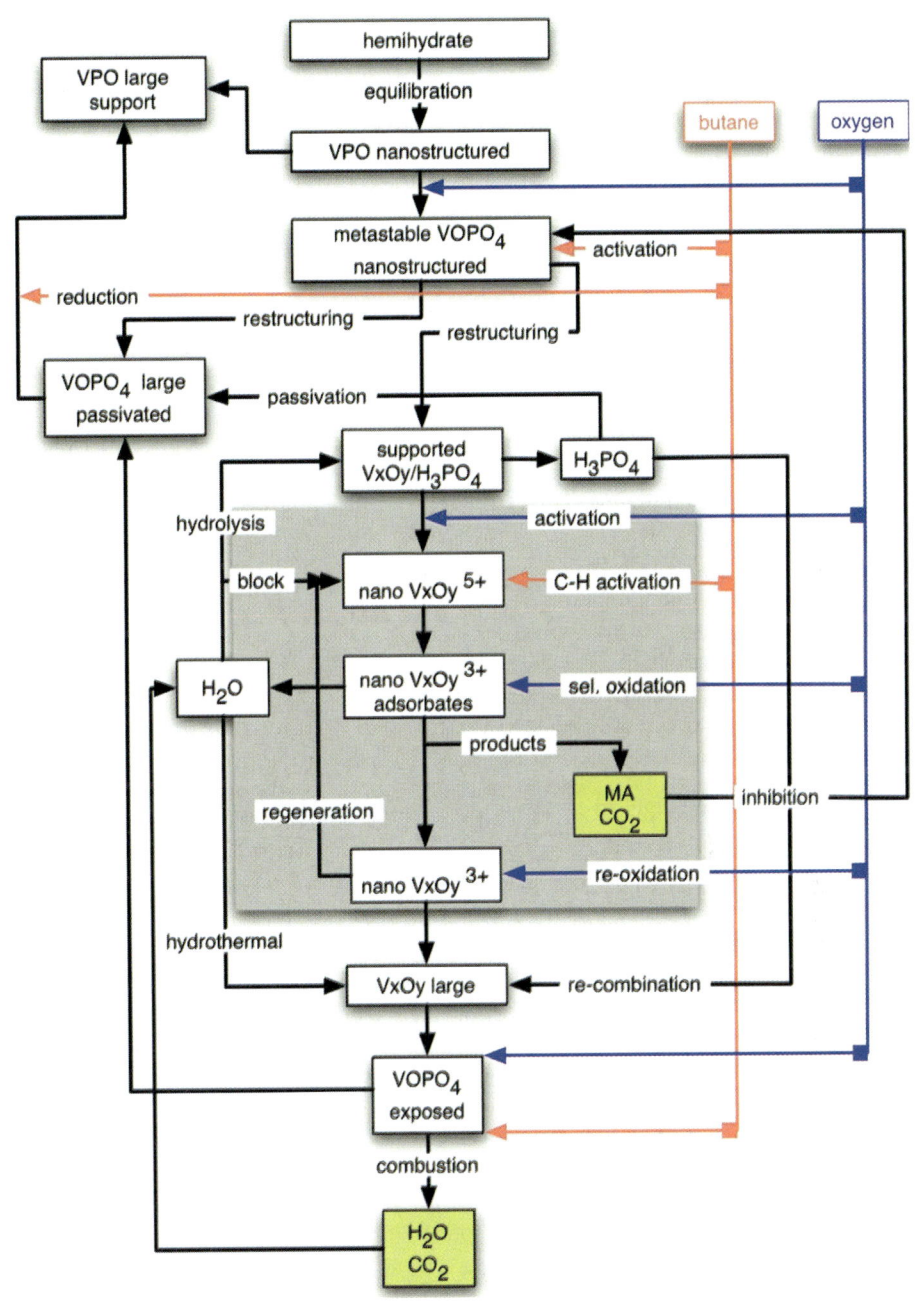

Figure 1.12 Wiring diagram of the mode of operation of the VPO system under steady state operation. The central shaded part represents the catalytic operation, the outer parts highlight the complexity of the structural dynamics creating a steady state abundance of isolated active sites.

(oxidized) to react with the alkane, producing strongly adsorbed intermediates that are successively oxidized on the same site which being, however, partly reduced and thus becomes less active in C–H activation. This adaptiveness, for which the oxidations states in Figure 1.12 are only proxies, resolves the problem of having to deal with consecutive reactions on increasingly reactive substrates and still conserving selectivity. The intensive discussion about the "right oxidation state" [255, 256] can be resolved by the fact that on the supporting tetravalent pyrophosphate small amounts of $5+$ and $3+$ species coexist; their relative kinetic stability is different from that of the support, giving, upon careful analysis, rise to the often reported slight deviation from the tetravalent state. It is conceivable from the literature reports that attempts to clean the surface for an intended accurate analysis [192] may have destroyed all active sites and thus giving rise to the conclusion that the best catalyst is purely tetravalent; the preparation has brought about the pure precursor of the system. For the central issue of selectivity in oxidation the quality of site isolation is essential and no compromise should be tolerated in coupling the electronic structure of the active site to reservoirs of electrons. This isolation reduces the overall velocity of the reaction. Total oxidation sites (or materials) can have a coupling of their sites to the electron reservoir of the bulk and thus be fast catalysts.

Oxidative regeneration of the site is in competition with its deactivation by the combined action of hydrolysis and of mobile phosphate. Both are enhanced with growing performance – an argument for operating selective oxidation catalysts with comparatively low loads. The deactivation leads to growth of the active material into three-dimensional particles in much the same way as the pre-catalyst is grown from phosphoric acid, vanadium oxide and some reducing agent. It is anticipated that first a clean surface of $VOPO_4$ will result that may act as combustion catalyst (no site isolation) and thus enhance the energy supply to the system and augment the chemical potential of hydrolyzing water. Both effects in conjunction with the reducing potential of the gas phase tend to decompose the vanadium phosphates and create free mobile phosphate, passivating the non-selective surface.

The whole process needs some initiation during the equilibration phase of the practical VPO system. The bulk precursor "hemihydrates" are converted into VPO, which is metastable in oxygen and temporarily converts into $VOPO_4$, as frequently [192, 214, 215, 227, 245, 246, 257] found in the literature. Conflicting reports [45, 87, 258] about the role of this transformation can be resolved as follows. The initial equilibration leads to a nanostructured intergrown mixture of reactive crystallites of VPO and $VOPO_4$. This material undergoes segregating structural transformations under the action of feed: a large part re-crystallizes into stable $VOPO_4$ that is a combustion catalyst that transforms slowly into a passivated form due to phosphate coverage as described above. Some of it is reduced to large crystals of VPO, which act as support for the main active phase.

The transformation of the initial defective $VOPO_4$ is thus not a side effect but the central step enabling the active phase. The defect structure controlled by addition of promoters like Co, Ga, Fe and others will affect the partitioning between large crystalline material and still nanostructured VPO that is the reactive precursor to hydrolysis leading to VxOy and mobile phosphate (presumably as phosphoric acid, as seen in Figure 1.11). This mixture is the material of the 1–2-nm thick surface layer seen in Figure 1.10 and frequently described [175, 176, 192, 214, 215, 227] in the literature. It provides the site-isolated VxOy species that are well separated by phosphate from each other and from the underlying bulk. This material can be considered as the intermediate phase necessary to achieve the overall stabilization of the metastable VPO solid under the destabilizing chemical potential of the feed into the higher oxidized $VOPO_4$ – a non-topotactic solid-state redox transformation.

It is now critical that the whole process is dynamic, that is, responds in its kinetics to the reaction. Otherwise a continuous transformation would, with a linear rate law, transform all VPO into $VOPO_4$. The gas phase has a dual controlling influence on the process: The educt butane accelerates [132] at least part of the re-crystallization through the intermediate omega phase but the product MA inhibits this step and thus stabilizes the precursor; as the transformation of the $VOPO_4$ is a pre-requisite for forming the active phase that in turn leads to the chemical potential of the stabilizing MA. We see how a negative feedback loop stabilizes the catalyst for a slow transformation through a long-standing intermediate concentration of the reactive mixture VxOy/mobile phosphate.

This loop is, however, affected by the availability of the reactant oxygen, which in surplus destroys the precursor VPO. Further, oxygen is positively needed to activate and re-oxidize the VxOy sites but leads also to more water formation that in turn hydrothermally deactivates the active mass. Likewise, water is needed to separate, via hydrolysis, the vanadium phosphate into VxOy and mobile phosphate. The multiplicity of the feedback loops is at first puzzling but explains the apparent stable steady state that can be reached with a catalyst undergoing so many chemical and microstructural transformations; the multiplicity of controls prevents one single factor becoming dominant and thus potentially destabilizing the whole process.

The idea that the equilibrating transformations should be completed before the catalyst becomes active is not right; if one drives the reaction to completion, which may, depending on the chemical potential chosen, either be well-crystalline VPO or stable $VOPO_4$, then one has taken out of the system all structural dynamics and thus prevented the system from generating new active sites. Literature statements about the stable phases thus address the deactivated forms of the catalyst that may have some activity in the final terminating layer left and so, during short episodes of *in situ* testing, still show some (misleading) productivity. A stable long-term operation of the system requires the ongoing transformation initiated in the equilibration and inhibited by the product MA. In this context the enigmatic influence of the reduction agent used to generate the precursor "hemihydrate" acquires a rational dimension, as it on one hand affects with its chemical reduction potential the kinetics of the formation of the precursor and thus its defect structure while on the other hand the molecular structure of the reducing agent will stabilize the initial solid against structural transformation. A conceivable mechanistic concept is a complexing action that prevents the relative motion of polyhedra – being the molecular nature [116, 259, 260] of this whole transformation circus.

1.6
Conclusion

The substantial and fragmented literature on selective oxidation and the *ad hoc* character of the established principles of operation of the relevant catalysts leave the general impression of a still poorly understood research field. This does not devalue the remarkable success in finding and optimizing selective oxidation catalysts but it explains the apparent difficulty in expanding the number of reactions at productivities needed for sustainable commodity production. The present chapter makes an effort to draw many facts into a consistent picture. Not all aspects of nanostructuring, ionic transport, surface acidity and local electronic structure – the physical ingredients in explaining the function of an active site – are rigorously discussed here; in most cases, complete description of the phenomena has not yet been achieved. The complex and still mostly undiscovered role of carbon deposits [251] in modifying the performance of oxidation catalysts is a subject requiring more experimental evidence before it can be integrated into a general picture. Nanostructured carbon [261] originating from side reactions on oxides was found to exhibit a quite substantial performance in oxidation catalysis.

Central is the postulation of isolated sites that consist of sufficient metal ions to accommodate the redox equivalents required for one transformation and of sufficient oxygen ions to store the protons to be converted finally into water molecules. These sites need to be isolated from the supply of electrons to realize a shortage of activated oxygen. Concomitant with this kinetic argument for selectivity is the requirement for adaptiveness; the active site must be most active for the first C–H activation and must then become moderated in this function as the activation of substrate gradually proceeds. This can be realized by a gradual reduction of the oxidation state of the active site that must not be compensated for during the transformation of the substrate. In complex transformations such as butane oxidation to MA, some redox equivalents might be used during transformation to activate gas-phase oxygen into electrophilic reagents but this action must be strictly confined to stoichiometric conditions so as not to destroy the selectivity of the reaction.

The isolated and adaptive sites are not prepared during manufacture of the catalyst but occur in a dynamic process through co-operation of the pre-catalyst reactivity with the chemical potential of the reactant gas phase. In this way a gradually interconverting pre-catalyst generates the more reactive active phase that eventually transforms the whole pre-catalyst into a deactivated stable state consisting of a mixture of well-crystallized phases. In oxide systems this transformation requires the mobility of oxygen species and thus explains the well-documented fact that lattice oxygen mobility is related to catalytic action; the phenomenon is the proxy for the structural transformation creating the active sites and is thus positively correlated with performance. This interpretation leaves unaffected the fact that active oxygen located at lattice defects (of unknown ordered termination) formally seems to complement the coordinative undersaturation of the defect site and may for this reason be considered as "surface lattice oxygen" although it is a conventional adsorbate on a high energy site.

The combination of rigorous surface science experiments on sufficiently complex structural models with the broader application of *in situ* functional analysis is essential to further substantiate this notion of oxidation catalysts as being governed by the same rules as "simpler" metal systems. The generation and deactivation of high energy "Taylor sites" in a sea of adsorbing "Langmuir sites" is the general model proposed for selective oxidation catalysts. It can be detected by the method of forcing gradients in chemical potential and analyzing the spectral response of the system.

References

1 Haber, J. (1997) in: 3rd World Congress on Oxidation Catalysis, Studies in Surface Science and Catalysis, Vol. 110, 1–17, Elsevier.
2 Haber, J. and Mlodnicka, T. (1992) *Journal of Molecular Catalysis*, **74**, 131–141.
3 Haber, J. and Serwicka, E.M. (1987) *Reaction Kinetics and Catalysis Letters*, **35**, 369–379.
4 Heracleous, E. and Lemonidou, A.A. (2006) *Journal of Catalysis*, **237**, 175–189.
5 Todorova, T.K., Ganduglia-Pirovano, M.V. and Sauer, J. (2007) *Journal of Physical Chemistry C*, **111**, 5141–5153.
6 Mars, P. and van Krevelen, D.W. (1954) *Chemical Engineering Science*, **8**, 41–59.
7 Schlögl, R., Knop-Gericke, A., Hävecker, M. et al. (2001) *Topics in Catalysis*, **15**, 219–228.
8 Wachs, I.E. and Weckhuysen, B.M. (1997) *Applied Catalysis A – General*, **157**, 67–90.
9 Plietker, B. (2005) *Synthesis – Stuttgart*, 2453–2472.
10 DeSanto, P., Buttrey, D.J., Grasselli, R.K. et al. (2003) *Topics in Catalysis*, **23**, 23–38.
11 DeSanto, P., Buttrey, D.J., Grasselli, R.K. et al. (2004) *Zeitschrift für Kristallographie*, **219**, 152–165.
12 Grasselli, R.K., Buttrey, D.J., DeSanto, P. et al. (2004) *Catalysis Today*, **91-2**, 251–258.
13 Delmon, B., Ruiz, P., Carrazan, S.R.G. et al. (1996) in Catalysts in Petroleum Refining and Petrochemical Industries 1995, Studies in Surface Science and Catalysis, Vol. 100, 1–25, Elsevier, Amsterdam.
14 Bohme, D.K. and Schwarz, H. (2005) *Angewandte Chemie – International Edition*, **44**, 2336–2354.
15 Knobl, S., Zenkovets, G.A., Kryukova, G.N. et al. (2003) *Physical Chemistry Chemical Physics*, **5**, 5343–5348.
16 Tokarz-Sobieraj, R., Hermann, K., Witko, M. et al. (2001) *Surface Science*, **489**, 107–125.
17 Ballarini, N., Battisti, A., Cavani, F. et al. (2006) *Applied Catalysis A – General*, **307**, 148–155.
18 Berndt, S., Herein, D., Zemlin, F. et al. (1998) *Berichte der Bunsen-Gesellschaft – Physical Chemistry, Chemical Physics*, **102**, 763–774.
19 Knobl, S., Zenkovets, G.A., Kryukova, G.N. et al. (2003) *Journal of Catalysis*, **215**, 177–187.
20 Ueda, W. and Oshihara, K. (2000) *Applied Catalysis A – General*, **200**, 135–143.
21 Vitry, D., Dubois, J.L. and Ueda, W. (2004) *Journal of Molecular Catalysis A – Chemical*, **220**, 67–76.
22 Guram, A., Hagemeyer, A., Lugmair, C.G. et al. (2004) *Advanced Synthesis & Catalysis*, **346**, 215–230.
23 Schlögl, R. (1998) *Angewandte Chemie – International Edition*, **37**, 2333–2336.
24 Wachs, I.E., Deo, G., Jehng, J.M. et al. (1996) in Heterogeneous Hydrocarbon Oxidation, ACS Symposium Series, ACS Vol. 638, 292–299.
25 Zhao, C.L. and Wachs, I.E. (2006) *Catalysis Today*, **118**, 332–343.

26 Brazdil, J.F., Teller, R.G., Grasselli, R.K. and Kostiner, E. (1985) *ACS Symposium Series*, **279**, 57–74.
27 Haber, J., Tokarz, R. and Witko, M. (1996) in Heterogeneous Hydrocarbon Oxidation, *ACS Symposium Series*, ACS, Vol. 638, 249–258.
28 Holmberg, J., Grasselli, R.K. and Andersson, A. (2003) *Topics in Catalysis*, **23**, 55–63.
29 Holmberg, J., Hansen, S., Grasselli, R.K. and Andersson, A. (2006) *Topics in Catalysis*, **38**, 17–29.
30 Holmberg, J., Grasselli, R.K. and Andersson, A. (2004) *Applied Catalysis A – General*, **270**, 121–134.
31 Grasselli, R.K. (1999) *Catalysis Today*, **49**, 141–153.
32 Wagner, J.B., Timpe, O., Hamid, F.A. et al. (2006) *Topics in Catalysis*, **38**, 51–58.
33 Okuhara, T., Mizuno, N. and Misono, M. (2001) *Applied Catalysis A – General*, **222**, 63–77.
34 Wienold, J., Timpe, O. and Ressler, T. (2003) *Chemistry – A European Journal*, **9**, 6007–6017.
35 Villabrille, P., Romanelli, G., Gassa, L. et al. (2007) *Applied Catalysis A – General*, **324**, 69–76.
36 Grabowski, R. (2006) *Catalysis Reviews – Science and Engineering*, **48**, 199–268.
37 Lin, M.M. (2001) *Applied Catalysis A – General*, **207**, 1–16.
38 Centi, G., Perathoner, S. and Trifiro, F. (1997) *Applied Catalysis A – General*, **157**, 143–172.
39 Banares, M.A. (1999) *Catalysis Today*, **51**, 319–348.
40 Mamedov, E.A. and Corberan, V.C. (1995) *Applied Catalysis A – General*, **127**, 1–40.
41 Kung, H.H. and Kung, M.C. (1997) *Applied Catalysis A – General*, **157**, 105–116.
42 Dusi, M., Mallat, T. and Baiker, A. (2000) *Catalysis Reviews – Science and Engineering*, **42**, 213–278.
43 Dadyburjor, D.B., Jewur, S.S. and Ruckenstein, E. (1979) *Catalysis Reviews – Science and Engineering*, **19**, 293–350.
44 Madeira, L.M. and Portela, M.F. (2002) *Catalysis Reviews – Science and Engineering*, **44**, 247–286.
45 Hodnett, B.K. (1985) *Catalysis Reviews – Science and Engineering*, **27**, 373–424.
46 Grzybowska-Swierkosz, B. (2000) *Topics in Catalysis*, **11**, 23–42.
47 Banares, M.A. and Wachs, I.E. (2002) *Journal of Raman Spectroscopy*, **33**, 359–380.
48 Guimond, S., Abu Haija, M., Kaya, S. et al. (2006) *Topics in Catalysis*, **38**, 117–125.
49 Albonetti, S., Cavani, F. and Trifiro, F. (1996) *Catalysis Reviews – Science and Engineering*, **38**, 413–438.
50 Brückner, A. (2003) *Catalysis Reviews – Science and Engineering*, **45**, 97–150.
51 Schlögl, R., Knop-Gericke, A., Hävecker, M. et al. (2001) *Topics in Catalysis*, **15**, 219–228.
52 Bukhtiyarov, V.I., Hävecker, M., Kaichev, V.V. et al. (2003) *Physical Review B*, Vol. 67, 235422, 1–12.
53 Bond, G.C. and Tahir, S.F. (1991) *Applied Catalysis*, **71**, 1–31.
54 Chen, K.D., Bell, A.T. and Iglesia, E. (2000) *Journal of Physical Chemistry B*, **104**, 1292–1299.
55 Hutchings, G.J. (1991) *Applied Catalysis*, **72**, 1–32.
56 Gaiboyes, P.L. (1992) *Catalysis Reviews – Science and Engineering*, **34**, 1–54.
57 Weng, L.T. and Delmon, B. (1992) *Applied Catalysis A – General*, **81**, 141–213.
58 Gellings, P.J. and Bouwmeester, H.J.M. (2000) *Catalysis Today*, **58**, 1–53.
59 Schlögl, R. and Abd Hamid, S.B. (2004) *Angewandte Chemie – International Edition*, **43**, 1628–1637.
60 Moro-oka, Y., Ueda, W. and Lee, K.H. (2003) *Journal of Molecular Catalysis A – Chemical*, **199**, 139–148.
61 Busca, G. (1996) *Catalysis Today*, **27**, 457–496.
62 Witko, M. (1991) *Journal of Molecular Catalysis*, **70**, 277–333.
63 Rozanska, X., Fortrie, R. and Sauer, J. (2007) *Journal of Physical Chemistry C*, **111**, 6041–6050.

64 Ishii, Y., Sakaguchi, S. and Iwahama, T. (2001) *Advanced Synthesis & Catalysis*, **343**, 393–427.
65 Muhler, M., Schlögl, R. and Ertl, G. (1992) *Journal of Catalysis*, **138**, 413–444.
66 Hirano, T. (1986) *Applied Catalysis*, **26**, 65–79.
67 Murakami, Y., Ivayama, K., Uchida, H. *et al.* (1981) *Journal of Catalysis*, **71**, 257–269.
68 Shekhah, O., Ranke, W. and Schlögl, R. (2004) *Journal of Catalysis*, **225**, 56–68.
69 Chang, J.S., Hong, D.Y., Vislovskiy, V.P. and Park, S.E. (2007) *Catalysis Surveys from Asia*, **11**, 59–69.
70 Nagy, A.J., Mestl, G. and Schlögl, R. (1999) *Journal of Catalysis*, **188**, 58–68.
71 Li, X.B. and Iglesia, E. (2007) *Chemistry – A European Journal*, **13**, 9324–9330.
72 Costine, A. and Hodnett, B.K. (2005) *Applied Catalysis A – General*, **290**, 9–16.
73 Brazdova, V., Ganduglia-Pirovano, M.V. and Sauer, J. (2005) *Journal of Physical Chemistry B*, **109**, 23532–23542.
74 Deo, G., Wachs, I.E. and Haber, J. (1994) *Critical Reviews in Surface Chemistry*, **4**, 141–187.
75 Khodakov, A., Olthof, B., Bell, A.T. and Iglesia, E. (1999) *Journal of Catalysis*, **181**, 205–216.
76 Magg, N., Immaraporn, B., Giorgi, J.B. *et al.* (2004) *Journal of Catalysis*, **226**, 88–100.
77 Andersen, P.J., Kung, H.H., Sinev, M. *et al.* (1993) *Studies in Surface Science and Catalysis*, **75**, 205–217.
78 Melsheimer, J., Mahmoud, S.S., Mestl, G. and Schlögl, R. (1999) *Catalysis Letters*, **60**, 103–111.
79 Kondratenko, E.V., Ovsitser, O., Radnik, J. *et al.* (2007) *Applied Catalysis A – General*, **319**, 98–110.
80 Kondratenko, E., Cherian, M. and Baerns, M. (2006) *Catalysis Today*, **112**, 60–63.
81 Venkov, T.V., Hess, C. and Jentoft, F.C. (2007) *Langmuir*, **23**, 1768–1777.
82 Ressler, T., Timpe, O., Neisius, T. *et al.* (2000) *Journal of Catalysis*, **191**, 75–85.
83 Abu Haija, M., Guimond, S., Romanyshyn, Y. *et al.* (2006) *Surface Science*, **600**, 1497–1503.
84 Launay, H., Loridant, S., Nguyen, D.L. *et al.* (2007) *Catalysis Today*, **128**, 176–182.
85 Nguyen, L.D., Loridant, S., Launay, H. *et al.* (2006) *Journal of Catalysis*, **237**, 38–48.
86 Oyama, S.T., Desikan, A.N. and Zhang, W. (1993) *ACS Symposium Series*, **523**, 16–30.
87 Kartheuser, B., Hodnett, B.K., Zanthoff, H. and Baerns, M. (1993) *Catalysis Letters*, **21**, 209–214.
88 Thompson, D.J., Ciobica, I.M., Hodnett, B.K. *et al.* (2004) *Catalysis Today*, **91–92**, 177–180.
89 Hess, C., Wild, U. and Schlögl, R. (2006) *Microporous and Mesoporous Materials*, **95**, 339–349.
90 Hess, C., Tzolova-Muller, G. and Herbert, R. (2007) *Journal of Physical Chemistry C*, **111**, 9471–9479.
91 Hess, C. (2007) *Journal of Catalysis*, **248**, 120–123.
92 Derewinski, M., Haber, J., Kozlowski, R. *et al.* (1991) *Bulletin of the Polish Academy of Sciences – Chemistry*, **39**, 403–409.
93 Samson, K. and Grzybowska-Swierkosz, B. (2007) *Polish Journal of Chemistry*, **81**, 1345–1354.
94 Hess, C., Looi, M.H., Abd Hamid, S.B. and Schlögl, R. (2006) *Chemical Communications*, 451–453.
95 Agaskar, P.A., DeCaul, L. and Grasselli, R.K. (1994) *Catalysis Letters*, **23**, 339–351.
96 Klokishner, S., Melsheimer, J., Ahmad, R. *et al.* (2002) *Spectrochimica Acta Part A – Molecular and Biomolecular, Spectroscopy*, **58**, 1–15.
97 Thompson, D.J., Fanning, M.O. and Hodnett, B. (2003) *Journal of Molecular Catalysis A – Chemical*, **198**, 125–137.
98 Ertl, G. and Freund, H.J. (1999) *Pysics Today*, **52**, 32–38.
99 Ovsitser, O., Uchida, Y., Mestl, G. *et al.* (2002) *Journal of Molecular Catalysis A – Chemical*, **185**, 291–303.
100 Glaum, R., Benser, E. and Hibst, H. (2007) *Chemie Ingenieur Technik*, **79**, 843–850.
101 Lopez, J.M. (2006) *Topics in Catalysis*, **41**, 3–15.

102 Zhang, Y., Martin, A., Wolf, G.U. et al. (1996) *Chemistry of Materials*, **8**, 1135–1140.

103 Ahmad, R., Melsheimer, J., Jentoft, F.C. and Schlögl, R. (2003) *Journal of Catalysis*, **218**, 365–374.

104 Jentoft, F.C. and Gates, B.C. (1997) *Topics in Catalysis*, **4**, 1–13.

105 Bellussi, G. and Rigutto, M.S. (1994) *Advanced Zeolite Science and Applications*, **85**, 177–213.

106 Guisnet, M., Gnep, N.S. and Alario, F. (1992) *Applied Catalysis A – General*, **89**, 1–30.

107 Guisnet, M. and Magnoux, P. (1997) *Catalysis Today*, **36**, 477–483.

108 Suprun, W.Y., Machold, T. and Papp, H. (2008) *Zeitschrift für Physikalische Chemie – International Journal of Research in Physical Chemistry & Chemical Physics*, **222**, 129–151.

109 Fokin, A.A. and Schreiner, P.R. (2003) *Advanced Synthesis & Catalysis*, **315**, 1035–1052.

110 Daniel, C., Monnier, J.R. and Keulks, G.W. (1973) *Journal of Catalysis*, **31**, 360–368.

111 Daniel, C. and Keulks, G.W. (1972) *Journal of Catalysis*, **24**, 529–535.

112 Lunsford, J.H. (1995) *Angewandte Chemie – International Edition*, **34**, 970–980.

113 Sexton, A.W., Kartheuser, B., Batiot, C. et al. (1998) *Catalysis Today*, **40**, 245–250.

114 Mestl, G., Herzog, B., Schlögl, R. and Knozinger, H. (1995) *Langmuir*, **11**, 3027–3034.

115 Ayub, I., Su, D., Willinger, M. et al. (2003) *Physical Chemistry Chemical Physics*, **5**, 970–978.

116 Su, D.S., Roddatis, V., Willinger, M. et al. (2001) *Catalysis Letters*, **74**, 169–175.

117 Taufiq-Yap, Y.H., Goh, C.K., Hutchings, G.J. et al. (2006) *Journal of Molecular Catalysis A – Chemical*, **260**, 24–31.

118 Werner, H., Timpe, O., Herein, D. et al. (1997) *Catalysis Letters*, **44**, 153–163.

119 Mestl, G., Linsmeier, C., Gottschall, R. et al. (2000) *Journal of Molecular Catalysis A – Chemical*, **162**, 455–484.

120 Uchida, Y., Mestl, G., Ovsitser, O. et al. (2002) *Journal of Molecular Catalysis A – Chemical*, **187**, 247–257.

121 Dieterle, A., Mestl, G., Jager, J. et al. (2001) *Journal of Molecular Catalysis A – Chemical*, **174**, 169–185.

122 Blasco, T. and Nieto, J.M.L. (1997) *Applied Catalysis A – General*, **157**, 117–142.

123 Delmon, B. (1997) Catalyst Deactivation 1997, Studies in Surface Science and Catalysis, Vol. 111, 39–51, Elsevier, Amsterdam.

124 Haber, J. and Turek, W. (2000) *Journal of Catalysis*, **190**, 320–326.

125 Klissurski, D. and Rives, V. (1994) *Applied Catalysis A – General*, **109**, 1–44.

126 Kampe, P., Giebeler, L., Samuelis, D. et al. (2007) *Physical Chemistry Chemical Physics*, **9**, 3577–3589.

127 Buyevskaya, O.V., Kubik, M. and Baerns, M. (1996) in Heterogeneous Hydrocarbon Oxidation, ACS Symposium Series, ACS, Vol. 638, 155–169.

128 Fushimi, R., Shekhtman, S.O., Gaffney, A. et al. (2005) *Industrial & Engineering Chemistry Research*, **44**, 6310–6319.

129 Maier, J. (1993) *Angewandte Chemie*, **105**, 333–354.

130 Maier, J. and Månch, W. (1996) *Journal of the Chemical Society – Faraday Transactions*, **92**, 2143–2149.

131 Ganduglia-Pirovano, M.V. and Sauer, J. (2005) *Journal of Physical Chemistry B*, **109**, 374–380.

132 Conte, M., Budroni, G., Bartley, J.K. et al. (2006) *Science*, **313**, 1270–1273.

133 Su, D.S. and Schlögl, R. (2002) *Catalysis Letters*, **83**, 115–119.

134 Giebeler, L., Kampe, P., Wirth, A. et al. (2006) *Journal of Molecular Catalysis A – Chemical*, **259**, 309–318.

135 Li, X.K., Zhao, J., Ji, W.J. et al. (2006) *Journal of Catalysis*, **237**, 58–66.

136 Abd Hamid, S.B., Othman, D., Abdullah, N. et al. (2003) *Topics in Catalysis*, **24**, 87–95.

137 Duvauchelle, N., Kesteman, E., Oudet, F. and Bordes, E. (1998) *Journal of Solid State Chemistry*, **137**, 311–324.

138 Ilkenhans, T., Herzog, B., Braun, T. and Schlögl, R. (1995) *Journal of Catalysis*, **153**, 275–292.
139 Mestl, G., Ilkenhans, T., Spielbauer, D. et al. (2001) *Applied Catalysis A – General*, **210**, 13–34.
140 Ruth, K., Burch, R. and Kieffer, R. (1998) *Journal of Catalysis*, **175**, 27–39.
141 Taufiq-Yap, Y.H., Hasbi, A.R.M., Hussein, M.Z. et al. (2006) *Catalysis Letters*, **106**, 177–181.
142 Haber, J. (1996) in Heterogeneous Hydrocarbon Oxidation, *ACS Symposium Series*, ACS, Vol. 638, 20–34.
143 Moriceau, P., Lebouteiller, A., Bordes, E. and Courtine, P. (1999) *Physical Chemistry Chemical Physics*, **1**, 5735–5744.
144 Rice, G.L. and Scott, S.L. (1997) *Journal of Molecular Catalysis A – Chemical*, **125**, 73–79.
145 Tsuji, H. and Hattori, H. (2004) *ChemPhysChem*, **5**, 733–736.
146 Kondratenko, E.V., Cherian, M., Baerns, M. et al. (2005) *Journal of Catalysis*, **234**, 131–142.
147 Doornkamp, C., Clement, M., Gao, X. et al. (1999) *Journal of Catalysis*, **185**, 415–422.
148 Centi, G. (1993) *Catalysis Today*, **16**, 5–26.
149 Oyama, S.T., Desikan, A.N. and Hightower, J.W. (1993) *ACS Symposium Series*, **523**, 1–14.
150 Busca, G., Cavani, F., Centi, G. and Trifiro, F. (1986) *Journal of Catalysis*, **99**, 400–414.
151 Kail, B.W., Perez, L.M., Zaric, S.D. et al. (2006) *Chemistry – A European Journal*, **12**, 7501–7509.
152 Carley, A.F., Davies, P.R. and Roberts, M.W. (2002) *Catalysis Letters*, **80**, 25–34.
153 Shetti, V.N., Rani, M.J., Srinivas, D. and Ratnasamy, P. (2006) *Journal of Physical Chemistry B*, **110**, 677–679.
154 Todorova, T.K., Ganduglia-Pirovano, M.V. and Sauer, J. (2005) *Journal of Physical Chemistry B*, **109**, 23523–23531.
155 Hävecker, M., Knop-Gericke, A., Mayer, R.W. et al. (2002) *Journal of Electron Spectroscopy and Related Phenomena*, **125**, 79–87.
156 Reuter, K. and Scheffler, M. (2006) *Physical Review B*, **73**, 045433.
157 Bao, X., Barth, J.V., Lehmpfuhl, G. et al. (1993) *Surface Science*, **284**, 14–22.
158 Bowker, M. and Bennett, R.A. (2001) *Topics in Catalysis*, **14**, 85–94.
159 Norskov, J.K. and Stoltze, P. (1987) *Surface Science*, **189/190**, 91–105.
160 Hellman, A., Baerends, E.J., Biczysko, M. et al. (2006) *Journal of Physical Chemistry B*, **110**, 17719–17735.
161 Topsoe, K., Ovesen, C.V., Clausen, B.S. et al. (1997) in Dynamics of Surfaces and Reaction Kinetics in Heterogeneous Catalysis, Studies in Surface Science and Catalysis, Vol. 109, 121–139, Elsevier, Amsterdam.
162 Vedrine, J.C., Millet, J.M.M. and Volta, J.C. (1996) *Catalysis Today*, **32**, 115–123.
163 Volta, J.C. (1996) *Catalysis Today*, **32**, 29–36.
164 Chen, K.D., Khodakov, A., Yang, J. et al. (1999) *Journal of Catalysis*, **186**, 325–333.
165 Fierro, S., Nagel, T., Baltruschat, H. and Comninellis, C. (2007) *Electrochemistry Communications*, **9**, 1969–1974.
166 Stegelmann, C., Schiodt, N.C., Campbell, C.T. and Stoltze, P. (2004) *Journal of Catalysis*, **221**, 630–649.
167 Au-Yeung, J., Chen, K.D., Bell, A.T. and Iglesia, E. (1999) *Journal of Catalysis*, **188**, 132–139.
168 Bondzie, V.A., Parker, S.C. and Campbell, C.T. (1999) *Journal of Vacuum Science & Technology A – Vacuum, Surfaces and Films*, **17**, 1717–1720.
169 Bottcher, A. and Niehus, H. (1999) *Physica Status Solidi A – Applied Research*, **173**, 101–107.
170 Bottcher, A. and Niehus, H. (1999) *Physical Review B*, **60**, 14396–14404.
171 Blume, R., Hävecker, M., Zafeiratos, S. et al. (2007) *Catalysis Today*, **124**, 71–79.
172 Blume, R., Hävecker, M., Zafeiratos, S. et al. (2006) *Journal of Catalysis*, **239**, 354–361.

173 Bao, X., Muhler, M., Pettinger, B. et al. (1993) *Catalysis Letters*, **22**, 215–225.
174 Bao, X., Muhler, M., SchedelNiedrig, T. and Schlögl, R. (1996) *Physical Review B*, **54**, 2249–2262.
175 Kleimenov, E., Bluhm, H., Hävecker, M. et al. (2005) *Surface Science*, **575**, 181–188.
176 Hävecker, M., Mayer, R.W., Knop-Gericke, A. et al. (2003) *Journal of Physical Chemistry B*, **107**, 4587–4596.
177 Hävecker, M., Knop-Gericke, A., Bluhm, H. et al. (2004) *Applied Surface Science*, **230**, 272–282.
178 Bandara, A., Abu-Haija, M., Hobel, F. et al. (2007) *Topics in Catalysis*, **46**, 223–230.
179 Freund, H.J. (2005) *Catalysis Today*, **100**, 3–9.
180 Kolczewski, C., Hermann, K., Gulmond, S. et al. (2007) *Surface Science*, **601**, 5394–5402.
181 Witko, M., Grybos, R. and Tokarz-Sobieraj, R. (2006) *Topics in Catalysis*, **38**, 105–115.
182 Witko, M., Hermann, K. and Tokarz, R. (1999) *Catalysis Today*, **50**, 553–565.
183 Surnev, S., Ramsey, M.G. and Netzer, F.P. (2003) *Progress in Surface Science*, **73**, 117–165.
184 Goodrow, A. and Bell, A.T. (2007) *Journal of Physical Chemistry C*, **111**, 14753–14761.
185 Bronkema, J.L. and Bell, A.T. (2008) *Journal of Physical Chemistry C*, **112**, 6404–6412.
186 Kresse, G., Surnev, S., Schoiswohl, J. and Netzer, F.P. (2004) *Surface Science*, **555**, 118–134.
187 Dupuis, A.C., Abu Haija, M., Richter, B. et al. (2003) *Surface Science*, **539**, 99–112.
188 Kroger, E.A., Sayago, D.I., Allegretti, F. et al. (2007) *Surface Science*, **601**, 3350–3360.
189 Sayede, A.D., Mathieu, C., Khelifa, B. and Aourag, H. (2003) *Materials Chemistry and Physics*, **81**, 183–190.
190 Czekaj, I., Hermann, K. and Witko, M. (2003) *Surface Science*, **545**, 85–98.
191 Bettahar, M.M., Costentin, G., Savary, L. and Lavalley, J.C. (1996) *Applied Catalysis A – General*, **145**, 1–48.
192 Guliants, V.V., Holmes, S.A., Benziger, J.B. et al. (2001) *Journal of Molecular Catalysis A – Chemical*, **172**, 265–276.
193 Wachs, I.E., Jehng, J.M. and Ueda, W. (2005) *Journal of Physical Chemistry B*, **109**, 2275–2284.
194 Wang, X.G., Weiss, W., Shaikhutdinov, S.K. et al. (1998) *Physical Review Letters*, **81**, 1038–1041.
195 Gaigneaux, E.M., Dieterle, M., Ruiz, P. et al. (1998) *Journal of Physical Chemistry B*, **102**, 10542–10555.
196 Grasselli, R.K., Centi, G. and Trifiro, F. (1990) *Applied Catalysis*, **57**, 149–166.
197 Haber, J. and Lalik, E. (1997) *Catalysis Today*, **33**, 119–137.
198 Kroger, E.A., Sayago, D.I., Allegretti, F. et al. (2008) *Surface Science*, **602**, 1267–1279.
199 GrzybowskaSwierkosz, B. (1997) *Applied Catalysis A – General*, **157**, 263–310.
200 Viparelli, P., Ciambelli, P., Volta, J.C. and Herrmann, J.M. (1999) *Applied Catalysis A – General*, **182**, 165–173.
201 Tessier, L., Bordes, E. and Gubelmann-Bonneau, M. (1995) *Catalysis Today*, **24**, 335–340.
202 Krstajic, N.V., Vracar, L.M., Radmilovic, V.R. et al. (2007) *Surface Science*, **601**, 1949–1966.
203 Benser, E., Glaum, R., Dross, T. and Hibst, H. (2007) *Chemistry of Materials*, **19**, 4341–4348.
204 Keller, D.E., Airaksinen, S.M.K., Krause, A.O. et al. (2007) *Journal of the American Chemical Society*, **129**, 3189–3197.
205 Pozdnyakova, O., Teschner, D., Wootsch, A. et al. (2006) *Journal of Catalysis*, **237**, 17–28.
206 Boaro, M., de Leitenburg, C., Dolcetti, G. and Trovarelli, A. (2000) *Journal of Catalysis*, **193**, 338–347.
207 Witko, M., Hermann, K. and Tokarz, R. (1994) *Journal of Electron Spectroscopy and Related Phenomena*, **69**, 89–98.
208 Henrich, V.E. (1979) *Progress in Surface Science*, **9**, 143–164.
209 Henrich, V.E., Cox, P.A., Henrich, V.H. and Cox, P.A. (1994) *MoO 3 and MoO 3-x*, University Press, Cambridge.

210 Salmeron, M. and Schlögl, R. (2008) *Surface Science Reports*, **63**, 169–199.
211 Bluhm, H., Hävecker, M., Knop-Gericke, A. et al. (2004) *Journal of Physical Chemistry B*, **108**, 14340–14347.
212 Devriendt, K., Poelman, H. and Fiermans, L. (1999) *Surface Science*, **435**, 734–739.
213 Sajip, S., Bartley, J.K., Burrows, A. et al. (2001) *New Journal of Chemistry*, **25**, 125–130.
214 Hutchings, G.J., Desmartinchomel, A., Olier, R. and Volta, J.C. (1994) *Nature*, **368**, 41–45.
215 Kiely, C.J., Sajip, S., Ellison, I.J. et al. (1995) *Catalysis Letters*, **33**, 357–368.
216 Sartoni, L., Delimitis, A., Bartley, J.K. et al. (2006) *Journal of Materials Chemistry*, **16**, 4348–4360.
217 Reddy, K.M., Balaraju, M., Prasad, P.S.S. et al. (2007) *Catalysis Letters*, **119**, 304–310.
218 Lin, Y.C., Chang, C.H., Chen, C.C. et al. (2008) *Catalysis Communications*, **9**, 675–679.
219 Liu, J., Zhao, Z., Xu, C.M. et al. (2006) *Catalysis Today*, **118**, 315–322.
220 Grasselli, R.K. (2001) *Topics in Catalysis*, **15**, 93–101.
221 Grasselli, R.K. (2002) *Topics in Catalysis*, **21**, 79–88.
222 Melsheimer, J., Krohnert, J., Ahmad, R. et al. (2002) *Physical Chemistry Chemical Physics*, **4**, 2398–2408.
223 Uchida, S. and Mizuno, N. (2003) *Chemistry – A European Journal*, **9**, 5850–5857.
224 Lee, J.K., Russo, V., Melsheimer, J. et al. (2000) *Physical Chemistry Chemical Physics*, **2**, 2977–2983.
225 Sajip, S., Bartley, J.K., Burrows, A. et al. (2001) *Physical Chemistry Chemical Physics*, **3**, 2143–2147.
226 Landi, G., Lisi, L. and Volta, J.C. (2004) *Catalysis Today*, **91** (2), 275–279.
227 SananesSchulz, M.T., Tuel, A., Hutchings, G.J. and Volta, J.C. (1997) *Journal of Catalysis*, **166**, 388–392.
228 de Oliveira, P.G.P., Eon, J.G., Chavant, M. et al. (2000) *Catalysis Today*, **57**, 177–186.
229 Huang, X.M., Tang, X.F., Xu, Y.D. and Shen, W.J. (2006) *Chinese Journal of Catalysis*, **27**, 324–328.
230 Ionescu, N.I., Jaeger, N.I., Plath, P.J. and Liauw, M.A. (2000) *Journal of Thermal Analysis and Calorimetry*, **61**, 995–1003.
231 Bîttger, I., Pettinger, B., Schedel-Niedrig, T. et al. (2001) *Self Sustained Oscillations Over Copper in the Catalytic Oxidation of Methanol*, Elsevier, Amsterdam.
232 Werner, H., Herein, D., Schulz, G. et al. (1997) *Catalysis Letters*, **49**, 109–119.
233 Slin'ko, M.M. and Jaeger, N.I. (1994) *Physicochemical Basis for the Appearance of Self-Sustained Oscillations in Heterogeneous Catalytic Systems*, Elsevier.
234 Vanneer, F.J.R., Vanderlinden, B. and Bliek, A. (1997) *Catalysis Today*, **38**, 115–128.
235 Libuda, J. and Freund, H.J. (2005) *Surface Science Reports*, **57**, 157–298.
236 Bar, M., Bangia, A.K., Kevrekidis, I.G. et al. (1996) *Journal of Physical Chemistry*, **100**, 19106–19117.
237 Zerkle, D.K., Allendorf, M.D., Wolf, M. and Deutschmann, O. (2000) *Journal of Catalysis*, **196**, 18–39.
238 Dudukovic, M.P., Larachi, F. and Mills, P.L. (2002) *Catalysis Reviews – Science and Engineering*, **44**, 123–246.
239 Kellow, J.C. and Wolf, E.E. (1990) *Chemical Engineering Science*, **45**, 2597–2602.
240 Sundmacher, K., Schultz, T., Zhou, S. et al. (2001) *Chemical Engineering Science*, **56**, 333–341.
241 Theologos, K.N. and Markatos, N.C. (1993) *Aiche Journal*, **39**, 1007–1017.
242 Stobbe, E.R., de Boer, B.A. and Geus, J.W. (1999) *Catalysis Today*, **47**, 161–167.
243 Madier, Y., Descorme, C., Le Govic, A.M. and Duprez, D. (1999) *Journal of Physical Chemistry B*, **103**, 10999–11006.
244 Hutchings, G.J., Lopez-Sanchez, J.A., Bartley, J.K. et al. (2002) *Journal of Catalysis*, **208**, 197–210.
245 Abon, M. and Volta, J.C. (1997) *Applied Catalysis A – General*, **157**, 173–193.

246 Abon, M., Herrmann, J.M. and Volta, J.C. (2001) *Catalysis Today*, **71**, 121–128.
247 Bordes, E. (1993) *Catalysis Today*, **16**, 27–38.
248 Liu, Y.M., Feng, W.L., Li, T.C. *et al.* (2006) *Journal of Catalysis*, **239**, 125–136.
249 Bluhm, H., Hävecker, M., Kleimenov, E. *et al.* (2003) *Topics in Catalysis*, **23**, 99–107.
250 Guerrero-Perez, M.O. and Banares, M.A. (2007) *Journal of Physical Chemistry C*, **111**, 1315–1322.
251 Vass, E.M., Hävecker, M., Zafeiratos, S. *et al.* (2008) *Journal of Physics – Condensed Matter*, **20**, Art. Nr. 184016.
252 Ballarini, N., Cavani, F., Cortelli, C. *et al.* (2006) *Catalysis Today*, **117**, 174–179.
253 Nguyen, P.T., Hoffman, R.D. and Sleight, A.W. (1995) *Materials Research Bulletin*, **30**, 1055–1063.
254 Somorjai, G.A. and Park, J.Y. (2007) *Catalysis Letters*, **115**, 87–98.
255 Coulston, G.W., Bare, S.R., Kung, H. *et al.* (1997) *Science*, **275**, 191–193.
256 Ebner, J.R., Thompson, M.R., Grasselli, R.K. and Sleight, A.W. (1991) *Key Structure-Activity Relationships in the Vanadium Phosphorus Oxide Catalyst System*, Elsevier, Amsterdam.
257 Mota, S., Abon, M., Volta, J.C. and Dalmon, J.A. (2000) *Journal of Catalysis*, **193**, 308–318.
258 O'Mahony, L., Sutton, D. and Hodnett, B.K. (2004) *Catalysis Today*, **91–92**, 185–189.
259 Hebert, C., Willinger, M., Su, D.S. *et al.* (2002) *European Physical Journal B*, **28**, 407–414.
260 Willinger, M.G., Su, D.S. and Schlögl, R. (2005) *Physical Review B*, **71**, 155118, 1–8.
261 Su, D.S., Maksimova, N., Delgado, J.J. *et al.* (2005) *Catalysis Today*, **102**, 110–114.

2
Active Ensemble Structures for Selective Oxidation Catalyses at Surfaces
Mizuki Tada and Yasuhiro Iwasawa

2.1
Introduction

Molecular-level design of catalytic ensemble structures on surfaces in a controllable manner, based on new chemical concepts and strategies regarding composition or structure, provides a promising opportunity for the development of novel and efficient catalysts active for selective oxidation. Novel strategies and concepts for the creation of active ensemble structures on flat and porous surfaces may emerge from self-assembly and *in situ* transformation of precursors immobilized on the surfaces, with the aid of *in situ* characterization by sophisticated physical techniques [1–6].

The unique catalytic properties of supported metal complexes and clusters, which differ from metals and metal oxides, have been applied to various chemical processes. Such designed surfaces provide new catalytic systems with the integrated advantages of both heterogeneous and homogeneous catalysts for target reactions in green, sustainable processes [7]. Furthermore, tremendous catalysis can appear with self-organized assembly and regulation on surfaces, which may be hard to produce in homogeneous systems and conventional heterogeneous systems. Ordered micropores may also provide reaction spaces that different from solutions and flat surfaces due to the concave surface, confinement effect and 3D acidity, resulting in the creation of unique oxidation catalysis. *In situ* characterization is, inevitably, important for the development of this new class of efficient ensemble catalysts [8–10].

Ligands coordinated to metal center significantly modify and regulate not only the reactivity of the metal electronically but also the geometrical reaction space around the metal [1, 11]. As a result, marked selectivity can be achieved on metal-complex catalysts that may be difficult to obtain on metal particles and metal single-crystal surfaces. In the case of immobilized metal complexes on flat and concave oxide surfaces, making direct metal–surface bonding, the oxide surfaces can also modify and regulate the catalytic properties of the metal sites in addition to the electronic and geometric effects of the ligands.

While it has been recognized that the design of active ensembles and control of their reactivity based on *in situ* characterization are critical in achieving high catalytic performance with good selectivity, these issues have not been addressed adequately and are still a serious challenge in catalysis and materials research. In this chapter we summarize three recent examples of our work: the chemical design of catalytically active ensembles by chiral self-dimerization of metal complexes immobilized on SiO_2 surfaces, self-assembly during hydrothermal synthesis, and *in situ* ensemble synthesis in zeolite pores. The obtained ensemble structures were utilized, respectively, as catalysts for asymmetric oxidative coupling of 2-naphthol, preferential oxidation (PROX) of CO in excess H_2, and direct phenol synthesis from benzene with molecular O_2. This chapter also documents how such active ensemble structures with tremendous catalytic performances are characterized by physical techniques, including XAFS, ESR, solid-state NMR, XPS, FT-IR, DFT, and so on.

2.2
Chiral Self-Dimerization of Vanadium Schiff-Base Complexes on SiO_2 and Their Catalytic Performances for Asymmetric Oxidative Coupling of 2-Naphthol

2.2.1
Asymmetric Heterogeneous Catalysis Using Supported Metal Complexes

In the last three decades, we have designed and successfully prepared various supported metal complexes on oxide surfaces that exhibit unique catalytic activities and selectivities that are different from those of their homogeneous analogues [3, 4, 9, 12–15]. With the aid of several sophisticated spectroscopic techniques, the structures and roles of catalytically active species on surfaces have been characterized and identified [3, 4, 9, 12–25]. Chemical interactions between metal complexes and oxide surfaces can provide new reactivity of metal species by the construction of a spatially controlled reaction environment and the formation of unsaturated active metal species, leading to high catalytic activity, selectivity and durability [21–25].

The detailed design of chiral reaction spaces for asymmetric catalysis on surfaces is still a serious challenge [1, 26, 27]. The chemistry of homogeneous metal complexes in solution can not straightforwardly be transferred to chemistry on solid surfaces, and both catalytic activity and enantioselectivity in homogeneous systems often decrease upon simple immobilization of the homogeneous meal complexes on surfaces.

Thomas *et al.* have proposed the immobilization of chiral metal complexes on a mesoporous silica utilizing a special restriction imposed by the concave surface at which active metal complexes are located [27, 28]. In contrast, immobilization on a flat silica surface and subsequent surface functionalization using achiral organic molecules provides a heterogeneous enantioselective catalyst for Diels–Alder reaction [29, 30]. Immobilization of chiral metal complexes on ligand-bonded surfaces has also been achieved for several asymmetric reactions, affording enantioselectivities similar to those of the homogeneous analogues [31–33].

Chiral self-dimerization is a technique for producing spontaneously a chiral dimer structure from monomer precursors by their chemical attachment on a SiO$_2$ surface [34, 35]. We found that two V-monomer Schiff-base complexes spontaneously dimerize to produce a novel asymmetric reaction space between two coordinatively-unsaturated V sites [34, 35]. Surface silanol groups are suggested to have crucial roles in not only attaching the vanadyl complexes but also in their structural reconstruction on the surface: the chiral self-dimerization. Thus, the surface creates a new chirality of the vanadyl complexes, which is not possible in homogeneous systems because of the quite low energy difference between chiral isomers in solutions [36–38]. The chiral reaction space created between two chiral vanadium centers is highly enantioselective for the asymmetric oxidative coupling of 2-naphthol to BINOL (1,1′-binaphthol) with 96% conversion, 100% selectivity to BINOL, and 90% ee, while the V-monomer precursor itself is inactive for the oxidative coupling [34, 35].

BINOL and its derivatives have been utilized as versatile chiral sources for asymmetric catalysis, and efficient catalysts for their syntheses are, ultimately, required in many chemical fields [39–42]. The oxidative coupling of 2-naphthols is a direct synthesis of BINOL derivatives [43, 44], and some transition metals such as copper [45, 46], iron [46, 47] and manganese [48] are known as active metals for the reaction. However, few studies on homogeneous metal complexes have been reported for the asymmetric coupling of 2-naphthols [49–56]. The chiral self-dimerized V dimers on SiO$_2$ is the first heterogeneous catalyst for the asymmetric oxidative coupling of 2-naphthol.

2.2.2
Chiral V-Dimer Structure on a SiO$_2$ Surface

Several V-monomer precursors (**1**) with Schiff-base ligands obtained from α-amino acids (L-valine, L-isoleucine, L-leucine, L-*tert*-leucine and L-phenylalanine) (Figure 2.1) were attached on oxide surfaces by impregnation of each V complex (**1**) in dehydrated ethanol. The V-monomer precursors (**1**) selectively reacted with surface silanols, leading to the structural reconstruction of the tridentate Schiff-ligand coordination, as characterized by FT-IR, ESR, XAFS, XPS, XRF, UV/Vis and DFT calculations at each stage during the attachment of the V complex precursor (**1**) and the oxidative coupling of 2-naphthol. The Ph–O moiety of the Schiff-base ligand selectively transformed into the Ph–OH configuration via a surface reaction with surface Si–OH, resulting in a coordinatively unsaturated vanadium conformation on the SiO$_2$ surface (Figure 2.1).

Vanadium K-edge XANES measured in transmission mode at 15 K showed that the attached V complex (**2**) maintained its square pyramidal conformation with a V=O bond (Figure 2.2a). Curve-fitting analysis of V K-edge EXAFS Fourier transforms (Figure 2.2b) provided local structure information on the supported V complex (**2**) with an unsaturated conformation, which differs from that of the V-monomer precursor (**1**). The EXAFS curve-fitting was performed in the *R*-space with two shells: short V=O and long V–O bonds. A V=O bond was observed at 0.157 ± 0.001 nm,

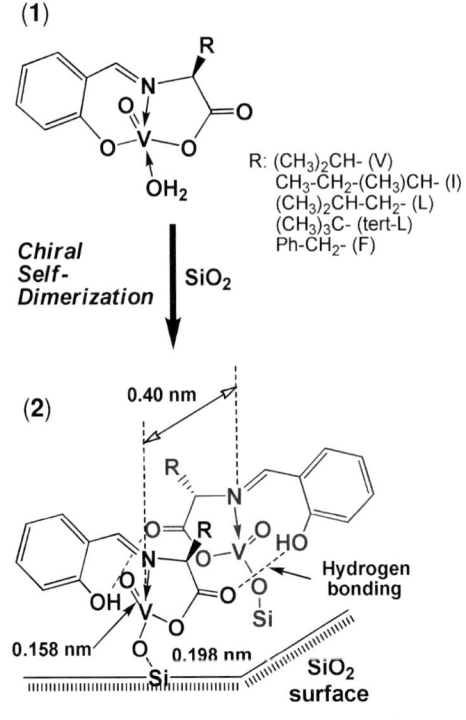

Figure 2.1 Chiral self-dimerization of V Schiff-base monomers (1) on SiO_2.

which is similar to that of (1), while the coordination number (CN) of V−O bonds at 0.199 ± 0.002 nm decreased from 3.8 ± 0.4 to 2.8 ± 0.5 after attachment of the V precursor on the SiO_2 surface, suggesting the formation of unsaturated V sites on the surface. The CN of V−O by the support also decreased for 0.8, 1.6 and 3.4 V wt% samples, independent of V loadings. Longer-distance bonding was not observed with all the catalysts, which indicates that there was no direct V−V bonding in the supported V catalysts (2).

EXAFS and FT-IR spectra of the supported V complexes demonstrate the unique surface-attachment reaction with surface Si−OH groups. All coordination sites of the Schiff-base ligand to the V^{4+} center possess infrared-active functional groups, Ph−O, Ph ring, C=N and COO (Figure 2.1): three frequencies (1598, 1373 and 1362 cm^{-1}) are assigned as one $v_{asym(COO)}$ and two $v_{sym(COO)}$; the 1629 cm^{-1} peak is assigned to $v_{(C=N)}$; the four peaks at 1547, 1470, 1447 and 1436 cm^{-1} are ascribed to $v_{(Ph)}$; a strong peak of 1290 cm^{-1} is attributed to $v_{(Ph-O)}$. A small difference between $v_{asym(COO)}$ and $v_{sym(COO)}$ (~230 cm^{-1}) indicates the delocalization of electron density on the C=O bond to O−CO, which may be caused by the formation of a hydrogen bond on C=O with a Ph−OH group of an adjacent V complex, as discussed below.

The $v_{(C=N)}$ was very similar to that of the V precursor (1), indicating no significant change in the original coordination of its moiety upon support. In contrast, the four $v_{(Ph)}$ peaks differ from those of the precursor (1) in both their positions and relative

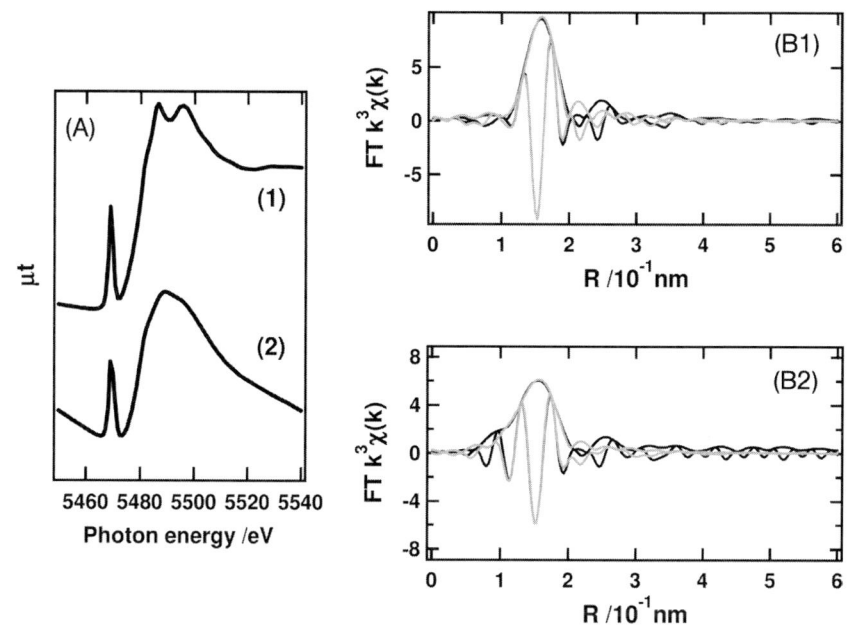

Figure 2.2 V K-edge (a) XANES and (b) EXAFS Fourier transforms for a L-leucine V precursor (1) and its SiO$_2$-supported V complex (V 3.4 wt%) (2). Black and gray lines represent observed and fitted spectra, respectively.

intensities, and the $v_{(Ph-O)}$ at 1290 cm^{-1} for the precursor dramatically shifted to 1391 cm^{-1}. Such a large shift was observed with an ionized molecule Ph–O$^-$. Evidently, therefore, the large shift of $v_{(Ph-O)}$ and the changes in the $v_{(Ph)}$ intensity ratio were caused by the structural reconstruction of the Ph–O coordination to produce a new Ph–OH moiety that was promoted by proton transfer from Si–OH. These results demonstrate that the PhO$^-$ coordination is severed from the V center by the reaction of the V precursor complex (1) with Si–OH to form a Ph–OH group (Figure 2.1) [19, 20].

Figure 2.3 shows ESR spectra measured at 6 K for the SiO$_2$-supported V complex (2) in the presence and absence of O$_2$. The V-monomer complex (1) showed hyperfine signals attributed to d^1 configuration of a V^{4+}=O complex, while a broad peak was also observed on the hyperfine signals for the supported V complex (2) (Figure 2.3b). The broad signal is attributed to dipole–dipole coupling between adjacent d^1 species. The ESR and FT-IR data demonstrate that another V complex is located near a V complex to form a dimeric assembly in such a way that there is no direct V–V bonding is seen in the EXAFS data.

The oxidative coupling reaction of 2-naphthol proceeded in the presence of O$_2$, and we found that the ESR signal intensity greatly increased after O$_2$ adsorption (Figure 2.3b) and that the half-band signal was also detected (Figure 2.3a). These changes in ESR were reversible (Figure 2.3c). Judging from the d$_1$ electron on V^{4+}, these results indicate the formation of triplet state (strong interaction) by the

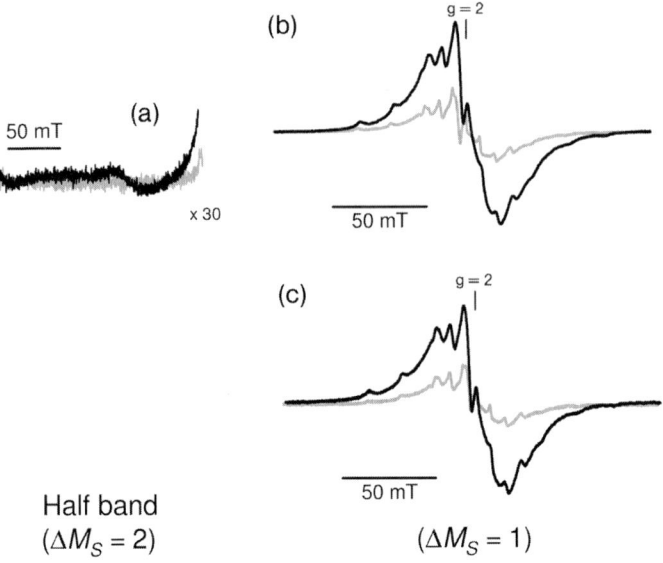

Figure 2.3 ESR spectra at 6 K for the SiO_2-supported V complex (V: 3.4 wt%) (**2**) in the absence (gray line) and presence (black line) of O_2. (a) Half-band for the first O_2 adsorption; (b) main signal for the first O_2 adsorption; and (c) main signal for the second O_2 adsorption after the evacuation of adsorbed O_2 in the first run.

insertion of O_2 between two V–V sites on the SiO_2. The V–V distance in the V dimer produced by the self-assembly on the surface was estimated to be 0.40 ± 0.05 nm from the relative intensity of the forbidden half-field transition ($|\Delta Ms| = 2$) to the allowed transition ($|\Delta Ms| = 1$) (Equation (2.1)):

$$\frac{\text{Intensity}(|\Delta Ms| = 2)}{\text{Intensity}(|\Delta Ms| = 1)} = C/r^6 \tag{2.1}$$

where r is the V–V distance and C is constant [21].

After evacuation of the O_2-adsorbed sample, the intensity returned completely to the original value and the change in ESR signal occurred reversibly, indicating the reversible adsorption of O_2 molecules on the V dimer. Thus, the supported V complex (**2**) possesses the capability for O_2 activation that is indispensable for the oxidative coupling reaction. The broad signal and the behavior for oxygen molecules were observed similarly with all the catalysts independent of the V loadings in the loading range 0.3–3.4 wt%. Thus, notably, the chiral self-dimerization of the V precursors (**1**) occurs commonly with small V loadings up to a full monolayer of the V complexes on the SiO_2 surface.

The Ph–OH moiety formed by the surface reaction of the V monomer precursor (**1**) with surface OH groups undergoes hydrogen bonding with the C=O group of the ligand in the adjacent V complex to assemble the supported V complexes on the surface (**2**, Figure 2.1). A V-monomer precursor (**1**) selectively reacts with a surface Si–OH group, followed by the chiral self-dimerization to form a novel V dimer stabilized by hydrogen bonding between Ph–OH in a V complex and C=O in another V complex on the surface.

DFT calculations for the structure of the chiral self-dimerized V dimer on SiO_2 reveal that two V=O bonds are directed to opposite sides of the principal molecular plane of the V complex, making chiral V sites, similar to a structural unit in the crystallographic structure of a V complex single-crystal [57]. The DFT modeling also indicates that a favorable reaction space for the oxidative coupling of two 2-naphthols in the chiral pocket is created between the two unsaturated chiral V centers. The two sets of hydrogen bonding between two vanadyl complexes stabilized the unsaturated vanadium centers. The cross section of the supported V dimers on the surface was evaluated by DFT. The V loading of 3.4 wt% corresponded to full coverage, based on the calculated cross section (0.52 nm^2 per vanadyl complex) on the SiO_2 surface (200 $m^2 g^{-1}$).

2.2.3
Asymmetric Catalysis for Oxidative Coupling of 2-Naphthol to BINOL

The supported V-dimer catalysts (**2**) were active for the coupling of 2-naphthol under aerobic conditions, while the homogeneous V precursor (**1**) is inactive for the reaction (Table 2.1). On the SiO_2-supported V catalysts (**2**), the conversion reaches a maximum 96%, and the reaction rate in toluene is 1.3 times higher than in $CHCl_3$. The supported V catalyst (**2**) is perfectly selective (100% selectivity) and reusable for the BINOL synthesis. In contrast, Al_2O_3 and TiO_2 were not suitable as supports for the V precursor (**1**), leading to low activities and selectivities for BINOL.

Enantioselectivity was not modified by the chiral alkyl groups of Schiff-base ligands: there are no significant differences in the performances between catalysts derived from L-isoleucine (51% ee), L-leucine (54% ee) and L-phenylalanine (56% ee) in $CHCl_3$. The most bulky ligand, L-*tert*-leucine, has a *tert*-butyl group neighboring the V reaction site, which reduced the enantioselectivity. These two alkyl groups overhang outside the V complexes as shown in Figure 2.1 and seemingly they did not affect the asymmetric coupling of 2-naphthol positively.

The vanadium loading on SiO_2 including an L-leucine segment was varied in the V loading range 0.3–3.4 wt% to maximize enantioselectivity for the coupling reaction (Table 2.1). No leaching of supported V complexes was observed for all the supported catalysts in toluene. With an increase in V loading, 0.3, 0.8, 1.6 and 3.4 wt%, the enantioselectivity dramatically increased: 32, 39, 48 and 90% ee, respectively (Table 2.1). The 90% ee for the 3.4 wt% vanadium catalyst, which corresponds to full coverage of the V complexes, is comparable to the best performance for the coupling of 2-naphthol on a homogeneous catalyst reported thus far [23–26]. Furthermore, the supported V catalysts can be reused after filtration and exhibited similar catalytic performances (Table 2.1). At full coverage of the V dimer complexes

Table 2.1 Catalytic performances of homogeneous and heterogeneous V catalysts for asymmetric oxidative coupling of 2-naphthol. The V dimer/2-naphthol ratio was 1 : 36 and 100 mg of supported catalysts were used in 5 mL of toluene.

Catalyst ligand[a]	Temperature (K)	Time (day)	Solvent	Conversion (%)	Selectivity (%)	Ee% (R)
Precursor-L[b]	293	5	CHCl$_3$	0	0	—
Precursor-L[b,c]	293	3	CHCl$_3$	15	73	8
Precursor-L[b,c]	263	9	CHCl$_3$	0	0	—
V-L/SiO$_2$ 0.3 wt%	263	5	CHCl$_3$	9	100	54
V-L/SiO$_2$ 0.3 wt%	293	5	Toluene	96	100	13
V-L/Al$_2$O$_3$ 1.7 wt%	293	5	CHCl$_3$	69	53	−2
V-L/TiO$_2$ 0.8 wt%	293	5	CHCl$_3$	52	0	—
V-V/SiO$_2$ 0.3 wt%	293	2	CHCl$_3$	26	100	12
V-V/SiO$_2$ 0.3 wt%	293	5	Toluene	99	100	5
V-V/SiO$_2$ 0.3 wt%	263	6	Toluene	12	100	14
V-I/SiO$_2$ 0.3 wt%	293	3	CHCl$_3$	37	100	17
V-I/SiO$_2$ 0.3 wt%	263	5	CHCl$_3$	6	100	51
V-I/SiO$_2$ 0.3 wt%	293	2	Toluene	41	100	21
V-L/SiO$_2$ 0.3 wt%[d]	293	2	Toluene	40	100	13
V-L/SiO$_2$ 0.3 wt%[d]	263	6	Toluene	9	100	31
V-tert-L/SiO$_2$ 0.3 wt%	263	5	Toluene	11	100	12
V-F/SiO$_2$ 0.3 wt%	293	5	CHCl$_3$	81	100	10
V-F/SiO$_2$ 0.3 wt%	263	5	CHCl$_3$	9	100	56
V-L/SiO$_2$ 0.3 wt%	263	5	Toluene	11	100	32
V-L/SiO$_2$ 0.3 wt%[e]	263	5	Toluene	10	100	33
V-L/SiO$_2$ 0.8 wt%	263	5	Toluene	33	100	39
V-L/SiO$_2$ 1.6 wt%	263	5	Toluene	42	100	48
V-L/SiO$_2$ 3.4 wt%	263	5	Toluene	93	100	90
V-L/SiO$_2$ 3.4 wt%[e]	263	5	Toluene	91	100	89

[a] L: L-leucine, V: L-valine, I: L-isoleucine, tert-L: L-tert-leucine and F: L-phenylalanine.
[b] Homogeneous reaction.
[c] Chlorotrimethylsilane was added as acid.
[d] Hydroxynaphthaldehyde was coordinated instead of salicylaldehyde.
[e] Reused.

on the SiO$_2$ surface both the configuration and reaction environment of the V dimer are regulated rigidly for the achievement of higher enantioselectivity than obtained with lower V loadings.

The enantioselectivity of the coupling reaction is determined by the chiral conformation on the V center rather than the chirality of Schiff-base ligands. The chirality of the ligands sterically affects the chiral self-dimerization of V complexes on the surface, but the chiral ligands themselves do not determine the enantioselectivity for the 2-naphthol coupling. The increase in V loading on SiO$_2$ resulted in higher regulation of the mobility of assembled V species on the surfaces, resulting in the high enantioselectivity (90% ee) at 93% conversion on the V 3.4 wt% catalyst compared to the case of the V 0.3 wt% catalyst with the same V structure (Table 2.1). The chiral self-dimerization of the V complex precursors at the SiO$_2$ surface produced a unique reaction space for the selective oxidation.

Notably, even if the V monomer precursors (**1**) are racemic at the V center in solution, only a particular arrangement of two V monomers among possible structural arrangements is allowed as a stable structure at the SiO$_2$ surface, creating a new chiral V center (**2**) to provide a chiral reaction space for the enantioselective catalysis. Chiral self-assembly is a common phenomenon that is applied to other catalytic materials on surfaces.

2.3
Low-Temperature Preferential Oxidation of CO in Excess H$_2$ on Cu-Clusters Dispersed on CeO$_2$

2.3.1
Preferential Oxidation (PROX) of CO in Excess H$_2$ on Novel Metal Catalysts

Hydrogen as the most efficient and cleanest energy source for fuel-cell power is produced by partial oxidation followed by the water gas-shift reaction and reforming of hydrocarbons or methanol [58]. A small amount of CO (0.3–1%) in the so-produced H$_2$ must be selectively removed because CO greatly poisons Pt/C and Pt-M/C electrocatalysts in proton-exchange-membrane (PEM) fuel cells [59, 60]. PROX of CO in excess H$_2$ is a key reaction in the practical use of H$_2$ in PEM fuel-cell systems.

The most important requirement for PROX catalysts is high CO oxidation activity while prohibiting undesirable H$_2$ oxidation. Various precious metal catalysts such as Pt [61–73], Au [68, 74–83], Rh [84], Ru [69, 84–86], Pt-Sn [70, 71], Pt-Fe [72], and so on with good PROX activities have been utilized as candidates of PROX catalysts. However, there are no precious metal catalysts with high performances under practical fuel cell operating conditions in the presence of H$_2$O and CO$_2$. Furthermore, few non-precious metal catalysts possess high enough PROX activities at low temperatures. CuO/CeO$_2$ was reported to be an active catalyst at >150 °C [68, 87–89], but non-precious metals still do not show good CO conversions at ≤120 °C under the practical PEM fuel-cell operating conditions.

2.3.2
Characterization and Performance of a Novel Cu Cluster/CeO$_2$ Catalyst

We have prepared Cu nanocluster catalysts supported on CeO$_2$ and MoO$_2$ materials by hydrothermal syntheses using a surfactant, cetyltrimethylammonium bromide (CTAB) [90–92]. The catalysts exhibited unique catalytic performances for PROX reactions (Cu$^+$/CeO$_2$) [90] and methanol dehydrogenation (Cu^0/MoO$_2$) [91, 92]. The novel CeO$_2$-supported Cu$^+$-cluster catalyst was the first non-precious catalyst to be active and selective for the PROX reaction below 120 °C under PEM fuel-cell operation conditions [90], while the MoO$_2$-supported Cu^0-cluster catalyst was a tremendously selective catalyst for H$_2$ production from methanol [91, 92].

The supported Cu/CeO$_2$ catalyst (denoted here as Cu/Ce-CTAB) was hydrothermally prepared using Ce and Cu nitrates as precursors with a surfactant, CTAB [90]. In a typical synthesis method, Ce(NO$_3$)$_3$ 6H$_2$O was dissolved in hot distilled water, to which Cu(NO$_3$)$_2$ 3H$_2$O in H$_2$O was added dropwise. Then, CTAB was dissolved in a mixture of H$_2$O and ethanol, and the obtained solution was added to the Cu + Ce solution. The typical molar composition is Cu/CTAB/H$_2$O = 1.0 : 0.55 : 325. The homogeneous slurry mixture was hydrothermally treated at 175 °C for 24 h in a Teflon-lined autoclave vessel under an autogeneous pressure. The resultant product was washed with distilled H$_2$O and EtOH, and dried at ambient temperature for 10 h and then at 100 °C for 8 h, followed by heating at 500 °C for 6 h under a He flow. The Cu contents of the obtained solid catalysts were determined by XRF.

The metallic Cu clusters on MoO$_2$ (denoted Cu/Mo-CTAB) were completely inactive for CO PROX at 90 °C (Table 2.2). Metallic Cu clusters on ZnO (Cu/Zn-CTAB) and SiO$_2$ (Cu/Si-CTAB) were active for the methanol dehydrogenation but they were inactive for the PROX reaction. Similarly prepared Cu/Zr-CTAB, Cu/Fe-CTAB and Cu/Al-CTAB catalysts were also inactive for the PROX reaction. In contrast, the new Cu/Ce-CTAB catalyst exhibited tremendous activity with the feed CO/O$_2$/H$_2$/He = 1 : 1 : 50 : 48 (mol.%) (Table 2.2), whereas the activities of conventional impregnated Cu/CeO$_2$ and Cu/Ce$_2$O$_3$ catalysts and co-precipitated Cu-Ce catalysts were much lower.

XRD showed the formation of fluorite CeO$_2$, while no XRD peaks attributed to Cu metal or oxides were observed (Figure 2.4), indicating that Cu species were dispersed as small units on the CeO$_2$ surfaces. A XPS Cu 2p$_{3/2}$ peak was observed at 932.4 eV (referred to C 1s of 284.6 eV); the binding energy revealed that the Cu^{2+} precursor was reduced to Cu^0 or Cu$^+$ under the hydrothermal conditions (XPS can not discriminate the oxidation states of Cu due to similar binding energies). The Cu^{2+}-nitrate precursor is similarly reduced on other oxide surfaces such as MoO$_2$, ZnO and SiO$_2$ as reported previously. In contrast to these previous supports, however, the Cu^{2+} precursor along with the Ce-nitrate precursor was not fully reduced to the metallic Cu^0 state under the hydrothermal synthesis conditions. The Cu atoms of Cu/Ce-CTAB were bonded to oxygen atoms, with Cu–O at 0.194 nm besides Cu–Cu bonding at 0.256 nm as proved by Cu K-edge EXAFS. The XRD, XPS and XAFS data evidence the formation of Cu$^+$ clusters rather than Cu^0 metallic clusters on CeO$_2$ in the Cu/Ce-CTAB catalyst.

Table 2.2 CO-PROX catalytic performances of Cu and Ce catalysts in excess H_2 at 90 °C[a].

Catalyst	Cu (wt%)	CO (conversion %)	O_2 (selectivity %)	Specific rate ($\mu mol_{CO}\, g_{Cu}^{-1}\, s^{-1}$)
CeO_2	0	0	—	0
Ce_2O_3	0	0	—	0
Ce-CTAB	0	0	—	0
Cu-CTAB	100	0	—	0
Cu/Ce_2O_3 (impreg.)[b]	10	8.9	100	1.1
Cu/Ce_2O_3 (impreg.)[b,c]	10	12.8	100	1.6
Cu/CeO_2 (impreg.)[b]	10	7.2	100	0.9
Cu/CeO_2 (post-impreg. with CTAB)[b]	7.5	0	—	0
$CuBr/CeO_2$ (impreg.)	7.5	0	—	0
Cu-Ce (co-precip.)[d]	10	5.6	100	0.7
Cu-Ce (co-precip.)[c,d]	10	7.3	100	0.9
Cu/Ce-no CTAB[e]	9.6	10.2	100	1.3
Cu/Ce-no CTAB[c,e]	9.6	15.9	100	2.1
Cu/Ce-CTAB	7.5	91.9	99.8	15.2
Cu/Ce-CTAB[f]	7.5	72.1	100	23.8
Cu/Ce-CTAB[f]	4.5	32.0	100	17.6
Cu/Ce-CTAB[f]	15	53.9	100	8.9
Pt-Cu/Ce-CTAB[f,g]	7.5	16.5	100	5.4
Pd-Cu/Ce-CTAB[f,g]	7.5	14.8	100	4.9
Au-Cu/Ce-CTAB[f,g]	7.5	58.4	100	19.3
In-Cu/Ce-CTAB[f,g]	7.5	4.2	100	1.4
Cu/Ce-Pluorunic[h]	7.5	0	—	0
Cu/Ce-dodecyl sulfate[i]	7.5	0	—	0
Cu/Zr-CTAB[j]	10	0	—	0
Cu/Fe-CTAB[k]	10	0	—	0
Cu/Mo-CTAB[l]	7.8	0	—	0
Cu/Zn-CTAB[m]	3.7	0	—	0
Cu/Al-CTAB[n]	5.3	0	—	0
Cu/Si-CTAB[o]	8.9	0	—	0

[a] Catalyst: 0.4 g, time on stream: 2 h, W/F = 2.24 $g_{cat}\, h\, mol^{-1}$, $CO/O_2/H_2/He = 1:1:50:48$ (mol.%).
[b] Impregnation of Cu nitrate.
[c] Reduced with H_2 (15% in He) at 500 °C.
[d] Co-precipitation.
[e] Hydrothermal synthesis without CTAB.
[f] Catalyst: 0.2 g, W/F = 1.12 $g_{cat}\, h\, mol^{-1}$, 80 °C.
[g] An additional metal was impregnated on Cu/Ce-CTAB.
[h] Pluorunic was used as a neutral surfactant.
[i] Dodecyl sulfate was used as an anionic surfactant.
[j] Prepared with Zr nitrate.
[k] Prepared with Fe nitrate.
[l] Prepared with $(NH_4)_6Mo_7O_{24}$.
[m] Prepared with Zn nitrate.
[n] Prepared with $Al_2(SO_4)_3$.
[o] Prepared with $Si(OC_2H_5)_4$.

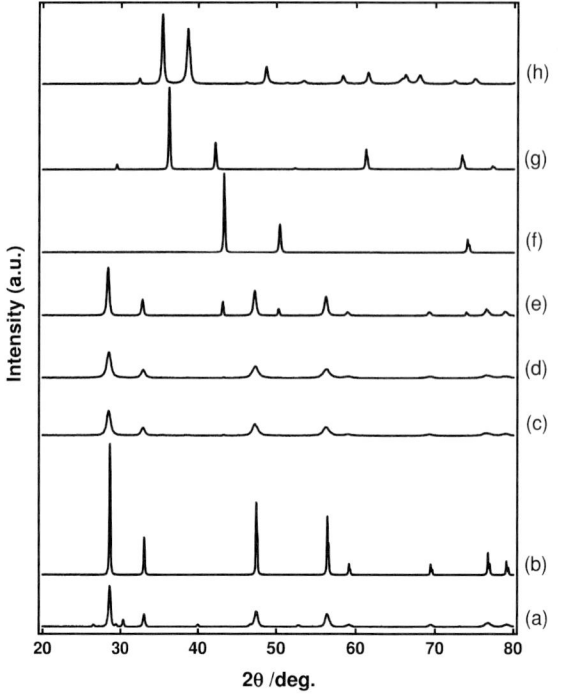

Figure 2.4 XRD patterns of Cu/Ce-CTAB catalysts and references: (a) CeO_2, (b) Ce_2O_3, (c) fresh Cu/Ce-CTAB (Cu 7.5 wt%), (d) Cu/Ce-CTAB after PROX reaction (Cu 7.5 wt%), (e) fresh Cu/Ce-CTAB (Cu 15 wt%), (f) Cu, (g) Cu_2O and (h) CuO.

The Cu/Ce-CTAB catalyst hydrothermally prepared in the presence of the CTAB surfactant showed a single-phase morphology, reflecting the assembly effect of the surfactant as imaged by SEM in Figure 2.5a, which was completely different from the inhomogeneous disordered morphology of a conventional Cu/Ce-oxide catalyst (Figure 2.5b). The SEM image did not change after the PROX reaction, which indicates a high stability of the new catalyst under the PROX conditions.

EXAFS analysis provided structural parameters (bond distance and coordination number) for Cu–O and Cu–Cu. The small coordination numbers of the Cu–Cu (0.9) and Cu–O bonds (2.4) indicate that the hydrothermal synthesis prohibits the growth of Cu species and produced small Cu^+-oxide clusters, which did not significantly change in size after the PROX reaction [90]. CO, 5.75×10^{-4} mol adsorbed on 1 g of Cu/Ce-CTAB (0.49 CO/Cu) was present, but no CO_2 formation was observed. The results indicate that neither the water-gas shift reaction nor CO oxidation with lattice oxygen proceeded on the Cu/Ce-CTAB catalyst. In contrast, with 2.40×10^{-4} mol O_2 adsorbed on 1 g of the fresh Cu/Ce-CTAB catalyst (0.20 O_2/Cu) a stoichiometric amount of CO_2 (0.39 CO_2 per Cu) was produced when this surface was subsequently exposed to CO, which suggests the high oxidation activity of the Cu^+-oxide cluster species on the CeO_2 surface. XRF analysis showed that the small amount

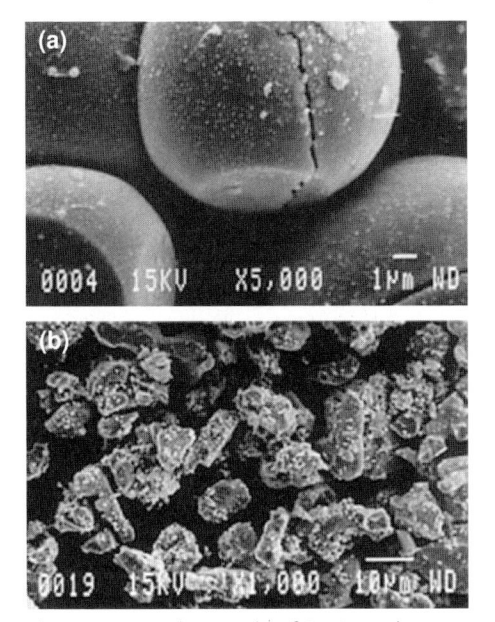

Figure 2.5 SEM photographs of Cu/Ce catalysts.
(a) Fresh Cu/Ce-CTAB (Cu 7.5 wt%); (b) impregnated Cu/Ce$_2$O$_3$ (Cu 10 wt%).

of Br (Br/Cu atomic ratio less than 0.26), derived from the surfactant CTAB, remained on the Cu/Ce-CTAB catalysts hydrothermally prepared. However, no Cu–Br contribution was observed by Cu K-edge EXAFS [90].

The detailed mechanism for the formation of reduced Cu$^+$ species under the hydrothermal synthesis conditions in the presence of CTAB without any additional reducing reagent is not clear at present, but the degree of reduction of the Cu- and oxide-precursors may depend on the oxophilicity of metal oxides: Cu oxide (most reducible) < Mo oxide < Zn oxide < Si oxide < Al oxide ∼ Zr oxide ∼ Ce oxide (hard to reduce). Further, chemical interaction of the Cu$^+$ clusters with the CeO$_2$ surface may also be the key to stabilizing the Cu$^+$ clusters on the support.

Table 2.2 shows the catalytic performances of various Cu and Ce catalysts for CO PROX reactions in excess H$_2$ at 90 °C. Ce oxides, Ce-CTAB and Cu-CTAB were completely inactive for CO oxidation at 90 °C. In contrast, the hydrothermally-prepared Cu/Ce-CTAB catalyst (7.5 wt%Cu) exhibited good catalytic performance for the CO PROX, with 91.9–96.1% CO conversion and 99.4–99.8% O$_2$ selectivity at 90 °C in a feed of CO/O$_2$/H$_2$ = 1 : 1 : 50 (Table 2.2). Table 2.3 summarizes the performance of the Cu/Ce-CTAB catalyst under various reaction conditions, different W/F, reaction temperatures and feed compositions. Notably, high CO conversions and O$_2$ selectivities were also achieved in reactant feeds containing substantial amounts of H$_2$O and CO$_2$. The CO conversions and O$_2$ selectivities at W/F = 2.24 g$_{cat}$ h mol^{-1} and 90 °C were 85.7% and 98.7%, respectively when H$_2$O (10%) existed, and 81.4% and 98.2%, respectively when H$_2$O (10%) and CO$_2$ (20%) co-existed.

Table 2.3 CO-PROX performances (conversion % and O_2 selectivity %) of the 7.5 wt% Cu/Ce-CTAB catalyst at different W/F (g_{cat} h mol^{-1}), reaction temperature and reactant composition.

W/F (g_{cat} h mol^{-1})	Temperature (°C)	CO/O_2 (1:1) conversion (%)	$CO/O_2/H_2$ (1:1:50) conversion (selectivity) (%)	$CO/O_2/H_2/H_2O$ (1:1:50:10) conversion (selectivity) (%)	$CO/O_2/H_2/H_2O/CO_2$ (1:1:50:10:20) conversion (selectivity) (%)
0.56	80	76.8	64.5 (100)	54.1 (100)	
	90	82.4	71.9 (100)	66.7 (100)	
	100	94.0	87.5 (100)	82.3 (100)	
1.12	80	83.4	72.1 (100)	64.8 (100)	55.3 (100)
	90	91.2	82.5 (100)	75.3 (100)	62.9 (100)
	100	99.2	93.2 (100)	88.2 (100)	83.9 (99.4)
	110	100	96.3 (100)	93.8 (99.2)	89.3 (98.9)
	120	100	100 (99.2)	100 (97.1)	97.3 (96.9)
2.24	80	87.9	75.8 (100)	64.9 (100)	
	90	97.8	91.9 (99.8)	85.7 (98.7)	81.4 (98.2)
	100	100	98.5 (99.8)	91.5 (96.5)	
3.36	80	89.2	77.9 (99.5)	67.2 (98.5)	
	90	100	96.1 (99.4)	88.4 (95.1)	
	100	100	100 (98.2)	93.7 (90.2)	

The CO conversion increased significantly with an increase in W/F, while the O_2 selectivity decreased a little from 100%. The catalytic performances of the Cu/Ce-CTAB catalyst (7.5 wt% Cu) were examined more systematically at W/F = 1.12 g_{cat} h mol^{-1}. The conversion increased with increasing temperature from 80 to 120 °C, while maintaining good selectivities (96.9–100%); for example, the conversions (selectivities) at 90 °C were 82.5% (100%), 75.3% (100%) and 62.9% (100%) at the feeds of $CO/O_2/H_2$ (1:1:50), $CO/O_2/H_2/H_2O$ (1:1:50:10) and $CO/O_2/H_2/H_2O/CO_2$ (1:1:50:10:20), respectively, and at 120 °C the performances increased, respectively, to 100% (99.2%), 100% (97.1%) and 97.3% (96.9%). Au/Fe_2O_3 catalysts are highly active for CO oxidation at 80 °C, but O_2 selectivity is as low as 51% [74] (<60%) [68, 76] in the presence of H_2O and CO_2. A conventional CuO/CeO_2 catalyst is less selective, and high temperature (>150 °C) is required for sufficient CO conversion (90%) [68, 87, 88]. Pt/Al_2O_3 [62, 64, 68] exhibits average performances. The Cu/Ce-CTAB catalyst is the first example of a non-precious metal catalyst with remarkable performances in both activity and selectivity for CO PROX with H_2O and CO_2 at low temperatures ($\leq$120 °C) [90].

Cu/Ce-noCTAB hydrothermally prepared in the absence of CTAB was much less active (10.2–15.9% CO conversion) – similar to conventional impregnated and co-precipitated catalysts (Table 2.2). The reduction of these catalysts with hydrogen did not efficiently improve their catalytic activities. The high PROX activity was characteristic of the hydrothermally synthesized material, where reduced Cu^+ species were dispersed on CeO_2. The conventional impregnated Cu/CeO_2 catalyst was post-impregnated with CTAB, but no activity appeared at 90 °C (Table 2.2) Other surfactants, neutral Pluorunic and anionic dodecyl sulfate, did not produce any catalytically active Cu species by the hydrothermal procedure (Table 2.2). Additional metals (Pt, Pd, Au or In) supported on the Cu/Ce-CTAB catalyst inhibited the PROX activity of the Cu/Ce-CTAB catalyst more or less. The hydrothermally-prepared Cu^+-oxide clusters on CeO_2 using as surfactant CTAB gave remarkably good performances for CO PROX reaction under the PEM fuel cell operating conditions. This is the first example of a non-precious metal catalyst with good performances for the CO PROX in the presence of H_2O and CO_2 at $\leq$120 °C. The catalytic performances were maintained for at least 10 h in the presence of H_2O and CO_2. The PROX performances of the Cu^+-oxide clusters may be promoted by traces of bromides, as suggested by LDI (laser-induced desorption ionization) MASS [90].

2.4
Direct Phenol Synthesis from Benzene and Molecular Oxygen on a Novel N-Interstitial Re_{10}-Cluster/HZSM-5 Catalyst

2.4.1
Phenol Production from Benzene with N_2O, $H_2 + O_2$, and O_2

Phenol is a major industrial product, and its production exceeds 7.2 million tons per year in the world. The phenol market is still growing because of an increase in the use of phenolic resins, which are used in plywood, construction, automotive and appliance industries, and of caprolactam and bisphenol A, which are intermediates in the manufacture of nylon and epoxy resins, respectively [93]. About 95% of the worldwide production of phenol is processed in three well-known steps, the so-called cumene process, whose initial source is benzene, which is easily obtained by petroleum reforming. The cumene process is performed by (1) the initial reaction of benzene with propene on acid catalysts like MCM-22, (2) auto-oxidation of the obtained cumene to form explosive cumene hydroperoxide in basic solution and (3) the decomposition of cumene hydroperoxide to phenol and acetone in sulfuric acid:

The three-step cumene process, including the liquid-phase reactions, is energy-consuming, environmentally unfavorable and disadvantageous for practical operation. The process also produces the unnecessary by-product acetone stoichiometrically. Furthermore, the intermediate, cumene hydroperoxide, is explosive and cannot be concentrated in the final step, resulting in low phenol yield (~5%, based on the amount of benzene initially used). Thus, direct phenol synthesis from benzene in a one-step reaction with high benzene conversion and high phenol selectivity is most desirable from the viewpoints of environment-friendly green process and economical efficiency.

2.4.1.1 Benzene to Phenol with N_2O

It was reported independently by three research groups that MFI-type zeolites selectively catalyze the reaction of N_2O with benzene to give phenol: $C_6H_6 + N_2O \rightarrow C_6H_5OH + N_2$ [93–96]. Fe/ZSM-5 shows remarkable performance in benzene hydroxylation to phenol with N_2O as oxidant, which is the first example of a successful gas phase direct phenol synthesis from benzene [97]. No other catalysts show similar high performances to the Fe/ZSM-5 catalyst. At present, iron is the sole element capable of catalyzing the benzene-to-phenol reaction [98]. Direct oxidation of benzene to phenol by N_2O has been commercialized in the so-called AlphOx process in Solutia Inc., USA, where N_2O is obtained as a by-product in adipic acid production with nitric acid [97, 99, 100]; a selectivity >95% to phenol is achieved at >40% conversion at around 400 °C. But the process is cost-effective only if N_2O can be obtained cheaply as a by-product in adipic acid production.

Considerable efforts have been made to understand ZSM-5-based catalysts for the selective oxidation of benzene to phenol by nitrous oxide. However, the nature of the active species remains unclear. The most important proposals for the active species are extraframework Fe species [101], Brønsted acid sites [102] and Lewis and Al sites [103, 104]. The activity is usually interpreted in terms of very small, possibly dinuclear, Fe species stabilized by the negative extraframework charge of the zeolite. However, other groups have found correlations between the amount of extraframework Al species and the catalytic performance [100, 103]. It was also suggested that neither Brønsted nor Lewis acid Al sites take part in the selective oxidation reaction [105, 106]. Hensen et al. suggested that the active centers consist of ferrous ions stabilized by extraframework Al species. The activity increased with increasing Al content in MFI zeolite. Steaming at high temperature promoted removal of the Fe and Al ions from the zeolite framework and improved the formation of active sites. UV/Vis spectra suggest that almost all lattice-Fe^{3+} species have migrated to extraframework positions during steaming. In situ Fe K-edge XANES revealed a distinct difference in reducibility of iron in Fe-silicate and Fe-ZSM-5 with aluminium upon dehydration [107–111]. Whereas a noticeable part of Fe^{3+} was reduced to ferrous ions in the Fe-containing aluminosilicalite, the ferric species in the ferrosilicalite maintained their oxidation state. Dehydration of steam-activated [Fe, Al]MFI resulted in extensive iron reduction as evidenced by the large shift in the edge energy (7124.2 → 7120.0 eV). The decrease was strongest in the temperature range 473–573 K. The ability to selectively oxidize benzene to phenol with N_2O in ferroaluminosilicalite was suggested to be related to these Fe^{2+} centers. A strong correlation was observed between the initial rate of phenol formation and the intensity of the infrared band at 1635 cm^{-1} due to N_2O interacting with extraframework mixed Fe-Al oxide species. A small amount of iron may be involved in the selective oxidation of benzene to phenol [107]. Iron atoms in [Fe, Al]MFI zeolite may migrate to zeolite defect sites (hydroxyl nests), forming highly active Fe^{3+} isolated species in square pyramidal or highly distorted octahedral coordination, as suggested by UV/Vis diffuse reflectance spectroscopy. The process occurred during the catalyst pre-treatment as well as during the initial half an hour of time on stream, leading to an increase in phenol productivity [111].

It was suggested that a considerable portion of the Fe ions in [Fe, Al]MFI is present in a different, more symmetrical coordination. This could be related to the presence of cationic mononuclear [112, 113] or dinuclear [105, 113] iron complexes generated by the auto-reduction of Fe^{3+} during catalyst activation, but it could also be related to neutral extraframework Fe-O-Al species [107]. The [Fe, Al]MFI catalysts with low Fe concentration (0.075–0.6 wt%) contain about 90% of iron in the high-spin Fe^{2+} state. In the presence of N_2O at 623 K, most Fe^{2+} ions (>90%) were oxidized to Fe^{3+} ions, as deduced from Mössbauer spectroscopy. In the presence of benzene, subsequent reduction of Fe^{3+} to Fe^{2+} took place. However, not all the Fe^{3+} ions could selectively oxidize benzene to phenol. This indicates that only a fraction of iron is catalytically active. For [Fe, Al]MFI catalysts with relatively high Fe concentration, most of the extraframework Fe species were inactive in the direct oxidation of benzene to phenol [114]. It has been demonstrated that the active sites of Fe/MFI zeolite (α sites) contain a $(-Fe^{2+}-O)_n$-Al-fragment of a low nuclearity, in the extraframework positions in the zeolite micropores, that is capable of the redox transformation $Fe^{2+} \leftrightarrow Fe^{3+}$ in the presence of N_2O [105, 115–117]. On N_2O dissociation, these sites generate a radical form of surface oxygen O^-, designated as O_α^- (α-oxygen). Mössbauer spectroscopy showed that one O_α atom performs a one-electron oxidation of the active iron; $Fe_\alpha^{2+} + N_2O \rightarrow Fe_\alpha^{3+} - O_\alpha^- + N_2$, which identified α-oxygen as a mono-charged radical species, O_α^-. The radical character of O^- was supported by theoretical calculations [118]. A linear dependence of the reaction rate on the concentration of independently identified active sites generating O^- radicals (αsite) was obtained. This is interpreted as convincing evidence of the O^- involvement in the catalytic oxidation of benzene to phenol with N_2O [119].

No kinetic H/D isotope effect in benzene oxidation by O_α was observed, and the mechanism shown below was proposed [120, 121]:

benzene + OFe/ZSM-5 ⟶ benzene oxide ⟶ phenol (OH)

Another pathway on an isolated surface iron-oxo species rather than the arene-epoxide pathway has also been proposed by DFT calculations as follows:

$$ZSM\text{-}5\text{-}[FeO_2]^+ + C_6H_6 \rightarrow ZSM\text{-}5\text{-}[FeO_2]^+(C_6H_6)$$

$$\rightarrow ZSM\text{-}5\text{-}[FeO]^+(C_6H_5OH)$$

where the most favorable pathway involves the direct oxidation of benzene without the epoxide intermediate. N_2O dissociation is favored over $ZSM\text{-}5\text{-}[FcO]^+$ to form $ZSM\text{-}5\text{-}[FeO_2]^+$. In this mechanism the kinetic isotope effect for the oxidation of 1,3,5-d_3-benzene is near unity, in agreement with experimental observations [122]. The other pathway via hydrogen abstraction from benzene to form $[Fe(OH)(C_6H_5)]$ has also been proposed [123]. The presence of Brønsted acid sites was suggested to contribute to the activation of N_2O. DFT confirmed the possible formation of protonated N_2O, leading to a Wheland-type intermediate, and thus supporting an electrophilic aromatic substitution assisted by the confined environment provided by the active zeolite framework [124].

2.4.1.2 Benzene to Phenol with $H_2 + O_2$

H_2O_2 has been a popular oxidizing agent for decades in the textile, electronics and pulp and paper industries [125]. Applications of H_2O_2 in bulk chemistry have been reviewed in a few papers [126, 127]. In the H_2O_2 oxidation of benzene to phenol a selectivity of 90 + % has been reported with catalysts based on TS-1 [128] and/or Ti-MCM-41 [129] at reasonable conversion levels. However, H_2O_2 is too expensive compared to the frequently applied air and/or oxygen, and so there are no drivers for economically sustainable and economic application.

The direct hydroxylation of benzene and aromatics with a mixture of O_2 and H_2 have been performed by simultaneously mixing benzene, oxygen and hydrogen in the liquid phase using a very complicated system containing a multi-component catalyst, a solvent and some additives. Besides the possibility of an explosive gas reaction, these hydroxylations gave only very low yields, 0.0014–0.69% of phenol and aromatic alcohols. For example, Pd-containing titanium silicalite zeolites catalyzed the hydroxylation of benzene by O_2–H_2 under mild conditions to give phenol. [130]. A SiO_2-supported Pt catalyst exhibited phenol production with a selectivity of 16.1% promoted with 20 ppm of V^{3+} acetylacetonate, $V(acac)_3$. It was estimated that oxygen activated on the vanadium species was transferred to platinum sites and, as a result, the rate of phenol formation was promoted. [131]. The electrochemically assisted benzene oxidation in a H_2–O_2 proton exchange membrane fuel cell has been studied, but the maximum phenol yield was as little as 0.35% at 100 mA cm^{-2} at 80 °C.

The vapor phase hydroxylation of benzene to phenol with a mixture of O_2 and H_2 on SiO_2-supported bicomponent catalysts containing group VIII metals and heteropoly compounds has been investigated [132]. A Pd-Cu composite catalyst also showed gas phase oxidation of benzene to phenol with a high selectivity at a low conversion, accompanied with a lot of water formation from H_2 and O_2 [133]. A Pt-VO$_x$/SiO$_2$ catalyst was also active for the benzene-to-phenol reaction, with 97% selectivity at 1.0% conversion at 413 K, whereas a Pd-VO$_x$/SiO$_2$ catalyst reached a conversion of only 0.2% conversion with 86% selectivity [134]. Benzene can also be hydroxylated over Pt/Pd-containing acid catalysts such as zeolites, Amberlyst and Nafion/SiO$_2$ composites. The best results were obtained over Nafion/silica composites [135].

In contrast to the simultaneous mixing of benzene, O_2 and H_2, Niwa et al. developed the direct hydroxylation of phenol from benzene through a membrane reactor system in which H_2 and O_2 are separately supplied or in which H_2 is fed into a mixed gas stream of benzene and oxygen through a metallic thin layer [136]. The membranes were prepared by coating a porous α-alumina tube (NOK Corporation) with a thin layer of palladium (1 μm thick, 100 mm long) by means of a metalloorganic chemical vapor deposition technique using palladium(II) acetate. The palladium membrane reactor worked well under mild conditions below 250 °C for the direct benzene oxidation to phenol. Phenol was produced in high selectivities of over 90% at low benzene conversion (<3.0%) and over 80% at high conversions (10–15%).

Active oxygen species that are electrophilic tend to prefer the ring to the methyl group, and oxygen species that have a radical character (O$^-$ radical) prefer the methyl group of toluene to the ring. Among the various active oxygen species, for example, atomic oxygen (O*) (^{3}P ground state), HCOO$^{\bullet}$, HO$^{\bullet}$, O^{2-}, O$^-$, OH$^-$, and so on, negatively charged species are unlikely to be the active species. The first three species can act as electrophiles. Hydrogen is dissociated in permeating through the palladium membrane, and the dissociated hydrogen appearing on the surface of the opposite site of the membrane immediately reacts with oxygen to give HCOO$^{\bullet}$ and H_2O_2. Then H_2O_2 decomposes to HO$^{\bullet}$, atomic oxygen and water. The ^{3}P oxygen atom has been known to easily add to carbon–carbon double bonds, including conjugated ones such as benzene [137]. This type of addition is 10^3 times faster than hydrogen abstraction from the methyl group. Thus, it is not unreasonable to consider that ^{3}P O is largely responsible for the phenol synthesis in the membrane process. However, it is difficult to specify the real active species from HCOO$^{\bullet}$, HO$^{\bullet}$ and ^{3}P O at present, although it seems that the active oxygen species is derived from HCOO$^{\bullet}$ and H_2O_2.

Dihydroxy compounds and quinones were detected in trace amounts, whereas hydrogenated compounds such as cyclohexane, cyclohexanol and cyclohexanone were hardly produced [136]. The same group suggested that oxidation and hydrogenation of benzene and/or phenol are competitive but that under standard operating conditions the oxidation would predominate [138]. It was also reported that the hydrogen efficiency was quite low since the production rate of water was 500–1000× that of phenol, which needs to be improved for practical utilization of the membrane system.

The performance of membrane reactors was also evaluated by Vulpescu et al., who reported much worse performance of the Pd membrane, 3.8% conversion and 4.1% selectivity [139], than that reported by Niwa et al. [136]. The difference between the two reports is large. This might be due to inhomogeneities in the Pd membrane, with different properties at different places in the membrane reactor and also to differences in oxygen concentration at different places, and hence differences in the kind and concentration of active oxygen species. The surface state of Pd during the reaction could be divided into two major regions: the oxidized region near the gas entrance of the reactor, which favored complete oxidation, and the reduced region near the gas exit, which favored hydrogenation [140, 141]. These two surface states depend on the concentrations of oxygen and hydrogen both inside and outside the membrane during the reaction. Hydroxylation occurred only in a limited region in the membrane reactor where the oxygen concentration was low [142]. For effective hydroxylation, the reaction atmosphere on the Pd surface is, therefore, crucial. Consequently, the interesting Pd-based catalytic membrane reactor is still far from commercial application [143].

2.4.1.3 Benzene to Phenol with O_2

Utilization of molecular O_2 as a sole oxidant is ideal. Direct phenol synthesis from benzene with O_2 has been studied extensively to date, but it shows low activity and selectivity with CuO/Al_2O_3 [143] and Cu/ZSM-5 [144, 145]. Several conditions of reactions were optimized with Cu/ZSM-5 catalysts, but the maximum yield was 4.9% with about 30% selectivity at 673 K [144, 145].

Copper loading influenced the nature of the Cu/ZSM-5 catalyst. At low Cu loadings, the catalyst was more selective to phenol, while at high Cu loadings CO_2 was the major product. In situ H_2-TPR XAFS studies revealed that at low Cu loadings Cu existed as isolated pentacoordinated ions, with four equatorial oxygens, Cu–O 1.94 Å, and a more distant axial oxygen (Cu–O = 2.34 Å). The results suggest that the isolated Cu sites are the active sites responsible for phenol formation [146]. The amount of isolated Cu^{2+} species with a square-pyramidal configuration on the Cu/HZSM-5 catalyst estimated from ESR observation also correlated with the yield of phenol [145].

In contrast to the previous reports [144–146], a ZSM-5 catalyst prepared by using a solid-state ion-exchange procedure with a 100% modification level of copper (Cu/Al = 0.5) showed a good phenol yield that increased with temperature up to 400 °C, approaching a high value of about 10% (taken at 20–30 min on stream), though no data for selectivity were reported [147]. HZSM-5 was modified by introducing Cu^{2+} ions using the contact-induced ion-exchange procedure. Benzene was introduced by a syringe pump, which provided a constant volume rate (0.0068 mol h^{-1}).

Reaction conditions were as follows: Bz/O_2/He = 10:5:85 (mol.%); W/F = 37 g_{cat} h (g mol_{Bz})$^{-1}$. The conversion of benzene was taken at 20–30 min on stream. Using this catalyst [146], the phenol yield at 20–30 min on stream was approximately twice that reported previously [144, 145]. UV/Vis and Raman data indicated that the production of phenol was maximized in the presence of copper-polymeric (size-limited) species, though isolated copper species such as cations and dimers also catalyzed the benzene-to-phenol transformation.

Copper-modified Ca-phosphate catalysts were also used to convert benzene into phenol in the presence of oxygen and ammonia at 450 °C, where N_2O was formed under the reaction conditions [148]. The addition of ammonia to the O_2–H_2O feed promoted the phenol yield, but the yield was small (<1.3%). The use of N_2O as oxidant was better than the O_2–NH_3 feed.

A Pd(OAc)$_2$/phenanthroline catalytic system was reported to catalyze the benzene-to-phenol conversion in the presence of CO as a sacrificial reagent in an autoclave at 180 °C [149]. Similarly, phenol was produced from benzene with air (10–15 atm) in the presence of CO (10 atm) as a sacrificial reagent by using molybdovanadophosphoric acids as catalysts in a liquid phase, involving acetic acid at 90 °C, while no reaction occurred in the absence of CO [150].

The coexistence of reducing agents such as ascorbic acid has been examined in catalytic systems such as MCM-41-supported Cu catalysts [151], CuO/Al_2O_3 [152] and VO$_x$/CuSBA-15 [153]. The last of these catalyst showed a high yield of phenol of about 27% with selectivity of nearly 100% [153]. It was claimed that benzene hydroxylation occurred through a hydroxyl radical pathway, and the catalytic activation of molecular oxygen to form hydroxyl radicals would be a rate-determining step. Typical reaction conditions for the VO$_x$/CuSBA-15 system were complicated, involving 50 mg of the catalyst (20.2% Cu and 12.8% V loadings), 1 mL (11.3 mmol) of benzene, 36 mL of solvent [acetic acid/water = 2:1 (v/v)], 11.9 mmol of ascorbic acid, a reaction temperature of 80 °C, 0.7 MPa oxygen, and 5 h reaction in an airtight stainless steel reactor with a magnetic stirrer. A VO_2(V^{4+}) state, characterized by XPS, may be effective for the selective oxidation of benzene to phenol [154].

Liquid phase oxidation of benzene with O_2 over an iron-heteropoly acid system has been studied in a stainless steel autoclave reactor [155] and a Pd(OAc)$_2$ + V/ZrO$_2$ system involving acetic acid, LiOAc and sulfolane in air [156]. The Cu-substituted polyoxotungstate compound [(C_4H_9)$_4$N]$_5$[PW$_{11}$CuO$_{39}$(H_2O)] was also used as catalyst for the liquid-phase oxidation of benzene to phenol by O_2 with ascorbic acid as a reducing agent in an acetone/sulfolane/water mixed solvent. It showed 9.2% benzene conversion (TON = 25.8 after 12 h at 323 K) and 91.8% selectivity to phenol [157]. Heteropolyacid (HPA) and Pd(OAc)$_2$ were immobilized in functional organic group modified solid surfaces of hexagonal mesoporous silica (HMS) and polyimine (PIM). Comparison of the activity of the system with the homogeneous catalytic system containing HPA and Pd(OAc)$_2$ revealed the following order:

$$\text{HPA} + \text{Pd(OAc)}_2 > \text{HMS-HPA} + \text{Pd(OAc)}_2 > \text{PIM-HPA} + \text{Pd(OAc)}_2 > \text{HPA} + \text{HMS-Pd(OAc)}_2 > \text{HMS-HPA} + \text{HMS-Pd(OAc)}_2.$$

This solution system involves LiOAc, acetic acid and sulfolane in addition to the catalyst, benzene and O_2. [158].

In conclusion no catalysts with good performances (>5% conversion and >50% selectivity, simultaneously) have been discovered to date. New selective catalysts for direct phenol synthesis from benzene with O_2 are essential for the novel industrial process replacing the cumene process – there are many problems to be resolved.

2.4.2
Novel Re/HZSM-5 Catalyst for Direct Benzene-to-Phenol Synthesis with O_2

Economically and environmentally favorable benzene–O_2 catalytic systems with high selectivity for phenol synthesis have not been discovered to date because molecular oxygen is difficult to activate selectively to oxidize benzene to phenol. Rhenium is known to be oxophilic with various valences [159, 160]. Various oxide compounds of V, Mn, Mo and W are widely used as catalysts for selective oxidation using O_2, while exploitation of Re catalysts for selective oxidation is still undeveloped because of the easy sublimation of Re oxides like Re_2O_7 under oxidation reaction conditions. Several arrangements of ReO_x species with chemical interaction with various oxide supports, different from those on single Re oxides, have been reported to provide their unique catalytic properties [161–164]. For example, ReO_x supported on Fe_2O_3 behaved as a highly selective catalyst for one-step methylal synthesis from three molecules of methanol and molecular oxygen [165, 166]. Octahedral ReO_x clusters in zeolite pores were active for propene oxidation/ammoxidation in the presence of NH_3 [167, 168], indicating that Re species can work as oxidation catalysts under a reductive atmosphere.

A HZSM-5 zeolite-supported novel N-interstitial Re_{10} cluster was found to be active for the direct phenol synthesis from benzene and O_2 in the presence of NH_3 [169, 170]. The acidity and pore structure of HZSM-5 led to the self-limited formation of the novel N-interstitial Re_{10} cluster in the pore, which cannot be produced on other oxide surfaces or in solutions, and direct phenol synthesis using O_2 as a sole oxidant was achieved with impressive results (10% conversion and 94% selectivity) for the first time [169–171].

Zeolite-supported Re catalysts have been synthesized by chemical vapor deposition (CVD) of MTO (CH_3ReO_3) (3) on various zeolites such as HZSM-5, H-Beta, H-USY and H-Mordenite. HZSM-5 samples with different Al contents were prepared by a hydrothermal synthesis method. For comparison, conventional impregnated catalysts were also prepared by an impregnation method using an aqueous solution of NH_4ReO_4. All catalysts were pretreated at 673 K in a flow of He before use as catalyst.

Table 2.4 shows the catalytic performances of the Re/zeolite catalysts under steady-state conditions for the selective oxidation of benzene with O_2. A Re-CVD/HZSM-5 catalyst ($SiO_2/Al_2O_3 = 19$) preferentially produced phenol with 87.7% selectivity in the presence of NH_3 (Table 2.4). No other liquid products were detected and the only by-product was gaseous CO_2. The phenol selectivity was highly dependent on the

Table 2.4 Performance of Re/zeolite catalysts for direct phenol synthesis at 553 K under steady-state reaction conditions[a].

Catalyst	SiO$_2$/Al$_2$O$_3$	Method	Re (wt%)	TOF (10^{-5} s^{-1})[b]	PhOH selectivity (%)[c]
HZSM-5	19	—	—	Trace	0
Re/HZSM-5[d]	19	CVD	0.58	Trace	0
Re/HZSM-5	19	CVD	0.58	65.6	87.7
Re/HZSM-5[e]	19	CVD	0.58	51.8	85.6
Re/HZSM-5[f]	19	CVD	2.2	83.8	82.4
Re/HZSM-5[g]	19	CVD	0.58	74.6	93.9
Re/HZM-5[h]	19	CVD	0.58	86.1	90.6
Re/HZSM-5	19	Imp.	0.6	11.8	27.7
Re/HZSM-5[d]	23.8	CVD	0.58	Trace	0
Re/HZSM-5	23.8	CVD	0.58	36.2	68.0
Re/HZSM-5[d]	39.4	CVD	0.59	Trace	0
Re/HZSM-5	39.4	CVD	0.59	31.0	48.0
Re/H-beta	37.1	CVD	0.53	18.5	12.0
Re/H-USY	29	CVD	0.60	Trace	0
Re/H-mordenite	220	CVD	0.55	26.3	23.4
Re/SiO$_2$-Al$_2$O$_3$	19	Imp.	1.2	Trace	0

[a] Catalyst = 0.20 g; W/F = 6.7 g$_{cat}$ h mol^{-1}; He/O$_2$/NH$_3$/benzene = 46.4 : 12.0 : 35.0 : 6.6 (mol.%). Values of the detailed carbon mass balance and material balance examined in most experimental runs were between 97 and 99%.
[b] Consumed benzene/Re/s.
[c] Phenol selectivity in carbon%.
[d] In the absence of NH$_3$.
[e] W/F = 5.2 g$_{cat}$ h mol^{-1}.
[f] W/F = 10.9 g$_{cat}$ h mol^{-1}; He/O$_2$/NH$_3$/benzene = 46.4 : 12.0 : 35.0 : 6.6 (mol.%).
[g] Pulse reaction on the NH$_3$-pretreated catalyst (0.1 g): 1 pulse of benzene + O$_2$ (He/O$_2$/benzene = 81/12/7 (mol %)).
[h] Pulse reaction on the NH$_3$-pretreated catalyst (1.0 g): 1 pulse of benzene + O$_2$ (He/O$_2$/benzene = 81/12/7 (mol %)).

structures, acid strengths and SiO$_2$/Al$_2$O$_3$ ratios of zeolites (Table 2.4): the rate of phenol formation decreased in the order:

HZSM-5(SiO$_2$/Al$_2$O$_3$ = 19) > HZSM-5(SiO$_2$/Al$_2$O$_3$ = 24)

> HZSM-5(SiO$_2$/Al$_2$O$_3$ = 39) ≫ H-Mordenite > H-beta > H-USY.

Thus, HZSM-5 (SiO$_2$/Al$_2$O$_3$ = 19) among the zeolites employed was the most favorable support for the Re species. The results also suggest that Al-OH in the HZSM-5 framework is a coordination site for active Re species (Table 2.4).

Impregnation and physically mixed Re catalysts were much less active and much less selective for the phenol synthesis (Table 2.4). The CVD catalyst was almost 18 times more active than the conventional impregnation catalyst. In the physically mixed and impregnated catalysts, the Re^{7+} precursors partly aggregated as ReO$_x$ like ReO$_2$ in the presence of the NH$_3$ reductant and such ill-defined Re aggregates decreased both activity and phenol selectivity as shown in Table 2.4.

Table 2.5 Catalytic performance of pulse reactions on the CVD-Re/HZSM-5 ($SiO_2/Al_2O_3 = 19$) catalyst (Re: 0.58 wt%) for direct phenol synthesis from benzene and O_2 at 553 K in the absence of NH_3[a].

Length of NH_3 treatment at 553 K (h)	Reactant	TOF ($10^{-5}\,s^{-1}$)	PhOH selectivity (%)
2	Benzene [b]	0	0
2	Benzene + O_2 [c]	74.6	93.9
2[d]	Benzene + O_2 [c]	86.1	90.6
0	Benzene [b]	0	0
0	Benzene + O_2 [c]	0	0

[a] Catalyst amount: 0.10 g.
[b] 1 pulse of benzene [He/benzene = 93.4 : 6.6 (mol.%)].
[c] 1 pulse of benzene + O_2 [He/O_2/benzene = 81.4 : 12.0 : 6.6 (mol.%)].
[d] Catalyst amount: 1.0 g.

The coexistence of NH_3 is indispensable for selective benzene oxidation. Neither benzene oxidation nor combustion proceeded in the absence of NH_3 (Table 2.5). Fe/ZSM-5 has been reported to be active and selective for phenol synthesis from benzene using N_2O as an oxidant [97], but selective benzene oxidation did not proceed with N_2O instead of O_2. The addition of H_2O to the system gave no positive effects on the catalytic performance, either. In addition, other amine compounds such as pyridine and isopropyl amine did not produce phenol. The phenol formation rate and selectivity increased with increasing NH_3 pressure because the coexisting NH_3 produces active Re clusters, as described below, and reached maximum conversion and selectivity at a partial pressure of NH_3 of around 35–42 kPa.

2.4.3
Active Re Clusters Entrapped in ZSM-5 Pores

After the steady-state reaction on the Re-CVD/HZSM-5 ($SiO_2/Al_2O_3 = 19$) catalyst at 553 K, Re monomers were observed, as suggested by Re L_{III}-edge EXAFS. Figure 2.6 shows the oscillation and Fourier transform of the monomers. No Re–Re bond was found in the Re species, but Re=O [CN (coordination number) = 3.5 ± 0.2 at 0.173 ± 0.001 nm] and Re–O (CN = 1.3 ± 0.6 at 0.211 ± 0.002 nm) bonds were observed. The bonding feature suggested by the CNs indicates that the valence of the Re monomer is Re^{7+}, with the structure illustrated in Figure 2.7. The EXAFS analysis data are almost the same as the structural parameters for (3) (Re^{7+}). A distinct pre-edge peak in Re L_I-edge XANES demonstrates that the Re^{7+} monomer has a tetrahedral symmetry. The Re monomer was completely inactive for the selective oxidation of benzene with O_2.

Reaction with NH_3 for 2 h at 553 K dramatically changed the structure of supported Re species: Re L_{III}-edge EXAFS analysis revealed direct Re–Re bonds at 0.276 ± 0.002 nm, whose coordination number (CN) was 5.2 ± 0.3 (Figure 2.6). Two other types of chemical bonds were observed: Re–N/O at 0.204 ± 0.001 nm (CN = 2.8 ± 0.3) and Re=O bonds at 0.172 ± 0.001 nm (CN = 0.3 ± 0.2). The major contribution of

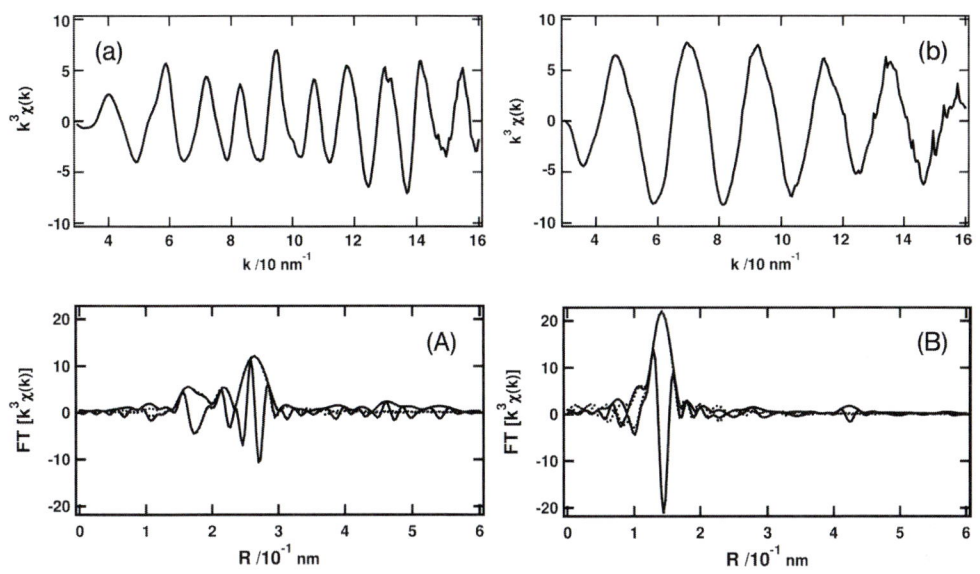

Figure 2.6 Re L_{III}-edge EXAFS oscillations (a and b) and their Fourier transforms (A and B) of the Re-CVD/ZSM-5. (a and A) After the treatment with NH_3 at 553 K for 2 h (**4**, see Figure 2.7); (b and B) after the reaction with benzene and O_2 at 553 K (**5**, see Figure 2.7). Solid and dotted lines in (A and B) represent observed and fitted spectra (both absolute and imaginary parts), respectively.

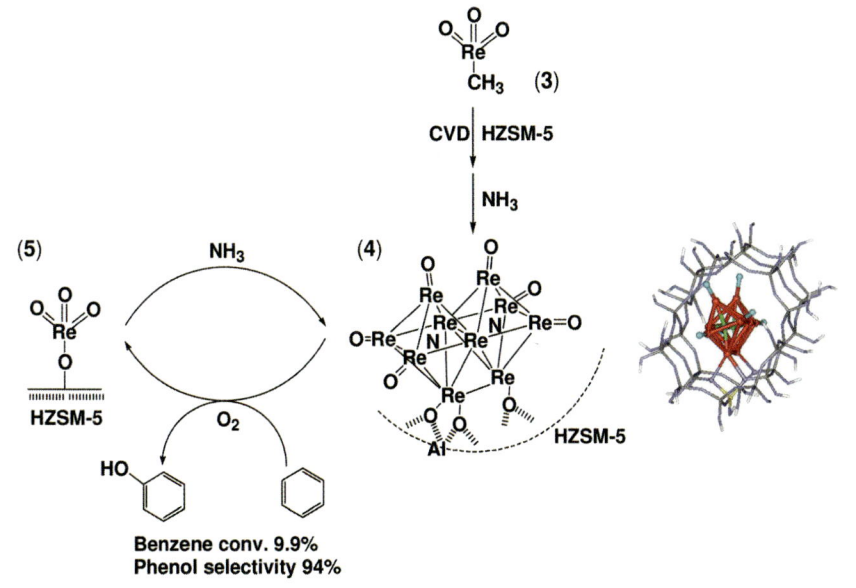

Benzene conv. 9.9%
Phenol selectivity 94%

Figure 2.7 Structural changes in the Re-CVD/HZSM-5 catalyst during direct phenol synthesis from benzene and O_2 and upon treatment with NH_3, and a proposed model structure of the N-interstitial Re$_{10}$ cluster (**4**) embedded in the pore of HZSM-5.

metal–metal bonds indicates the formation of Re clusters in HZSM-5. Assuming that Re-oxide materials tend to have Re_6 octahedral structures, the CN of Re–Re bonds of 5.2 indicates Re_{10} clusters edge-shared with two Re_6 octahedra, whose structure has been simulated by DFT calculations (Figure 2.7). All *ab initio* calculations were performed using a commercially available density functional code (Material Studio Dmol3 ver3.0, Accelrys, USA) [172, 173], where exchange-correlation interaction was treated by the Perdew–Wang 91 functional (PW91) within a generalized gradient approximation [174]. Double-numeric basis sets with polarization functions (DNP), whose quality is comparable to 6-31G*, were adopted. All electrons were explicitly included. Inclusion of the relativistic effect for Re core electrons has no effect on the calculated results [171].

Interestingly, the Re clusters contained the stoichiometric amount of N_2 (N_2 per Re_{10}), which was characterized by temperature-programmed desorption (TPD), with N_2 evolving at 685 K. The interstitial nitrogen atoms are the key element in stabilizing the Re_{10} clusters. The Re cluster framework was not stabilized by the hollow-site nitrogen (Re_3-N), bridged nitrogen (Re-N-Re), or bridge NH species (Re-NH-Re), and no other structures reproduced the Re–Re bond distance of 0.276 ± 0.002 nm determined by EXAFS. Among several tens of Re cluster structures involving the Re_6 octahedral framework examined by DFT calculations, the N-interstitial Re_{10} cluster framework (**4**) was the only stable structure that reproduced the local structure around a Re atom determined by EXAFS and the Re oxidation state suggested by XPS.

Pulse reactions of a mixture of benzene and NH_3 without O_2 on the N-interstitial Re_{10} cluster never proceeded, providing evidence that the lattice oxygen of the Re_{10} cluster is inactive for phenol synthesis. Notably, the N-interstitial Re_{10} cluster (**4**) produced by NH_3 selectively converted benzene into phenol in the presence of O_2 without NH_3. The phenol selectivity was as high as 94% at 10% conversion in the pulse reactions, which is the highest reported phenol selectivity for the direct phenol synthesis processes using O_2 as a sole oxidant. As seen clearly in the pulse reaction results, selective benzene oxidation with molecular oxygen on the active Re_{10} clusters proceeds without NH_3, demonstrating that the Re_{10} cluster is the active species for the direct phenol synthesis with molecular O_2. When molecular O_2 is the active oxygen species for the selective oxidation of benzene to phenol, the reaction is given by:

$$16C_6H_6 + 15O_2 \rightarrow 15C_6H_5OH + 6CO_2 + 3H_2O$$

which indicates a theoretical maximum phenol selectivity of 93.8%. The observed selectivity of 93.9% in the pulse reactions corresponds to this theoretical maximum [171]. NH_3 has a role in producing the active Re_{10} cluster structure but is not relevant to the phenol synthesis. NH_3 promotes the clusterization of the Re species as a reducing reagent, supplying N atoms to stabilize the Re_{10} cluster framework (**4**).

2.4.4
Structural Dynamics of the Active Re_{10} Cluster

A phenol synthesis by one-path benzene conversion of 5.8% proceeded on the Re_{10} cluster/HZSM-5 (**19**) catalyst in the steady-state reaction, but only the inactive

Re monomer was observed after the reaction. This structure–reactivity gap can be explained by the difference in rates between the reduction and oxidation processes of Re species. Oxygen not only oxidizes benzene to phenol but also oxidizes the active Re_{10} clusters (4) to the inactive ReO_4 monomers ((5), Figure 2.7) competitively. Indeed, the amount of phenol molecules produced in the pulse reactions was estimated to be 3% of the amount of Re_{10} clusters (4) [170]. Under steady-state conditions in the presence of NH_3, the population of the active Re_{10} clusters (4) was only 3.7% of the supported Re species, which cannot be captured even by in situ EXAFS under the reaction conditions.

The formation of the active Re_{10} cluster (4) takes 120 min at the same temperature under identical conditions. The rate gap between the cluster formation and disintegration is 1600 times or more, resulting in the low concentration of the active species under the reaction conditions. In situ measurements of energy-dispersive XANES at Re L_{III}-edge in real time revealed that a structural change from the active Re_{10} cluster (4) to the inactive Re monomers (5) accompanied the selective oxidation of benzene with O_2 to phenol. Notably, there were at least three isosbestic points, which indicates that direct change of the Re_{10} cluster to the Re monomer occurs without definite intermediates on this time scale. The real-time DXAFS analysis demonstrated that the Re_{10} cluster (4) disintegration was of first-order reaction kinetics. These results indicate that the first step for the oxidation reaction is the reaction between the Re_{10} cluster (4) and O_2 followed by faster decomposition steps to the Re monomers (5).

The selectivity of phenol may depend on the existence of heterogeneous Re species. If there are some intermediate structures with low phenol selectivity, the total phenol selectivity would decrease. On this supported Re catalyst, the following three phenomena are key to achieving high phenol selectivity: (i) there were no unfavorable intermediates during the decomposition of the active Re_{10} cluster with O_2, (ii) the Re monomer changed to a completely inactive species and (iii) selective interaction between benzene and O_2 to produce phenol occurs on the Re_{10} clusters in the pores [171].

DFT calculations for the benzene–O_2 reaction on the $Re_{10}N_2$ cluster (4) (Figure 2.8) revealed that one oxygen molecule dissociatively adsorbs on the face of a Re_6 octahedron, and the so-activated oxygen can react with adsorbed benzene π-coordinated to the Re_{10} cluster (Figure 2.8). When the direction of these adsorbed molecules overlap on the Re_{10} cluster, insertion of an oxygen into the C–H bond is concerted at the transition state, whose structure is also presented in Figure 2.8. The energy barrier from the adsorbed state to the transition state was 39 kcal mol^{-1}, where the adsorption energy of the benzene π-coordination and end-on adsorption of O_2 was in total 68 kcal mol^{-1}. Hence this concerted mechanism on the novel Re_{10} cluster resulted in a low activation energy for the formation of phenol. The benzene and O_2 activated on the Re_{10} cluster reacted readily to produce phenol and atomic oxygen, releasing 78 kcal mol^{-1}. The atomic oxygen left behind after phenol formation/desorption may cause the oxidative decomposition of the Re_{10} cluster framework by further interaction with other oxygen molecules [171]. After reactions with a mixture of benzene and O_2, the catalytically active Re_{10} cluster (4) is converted into inactive Re monomers (5).

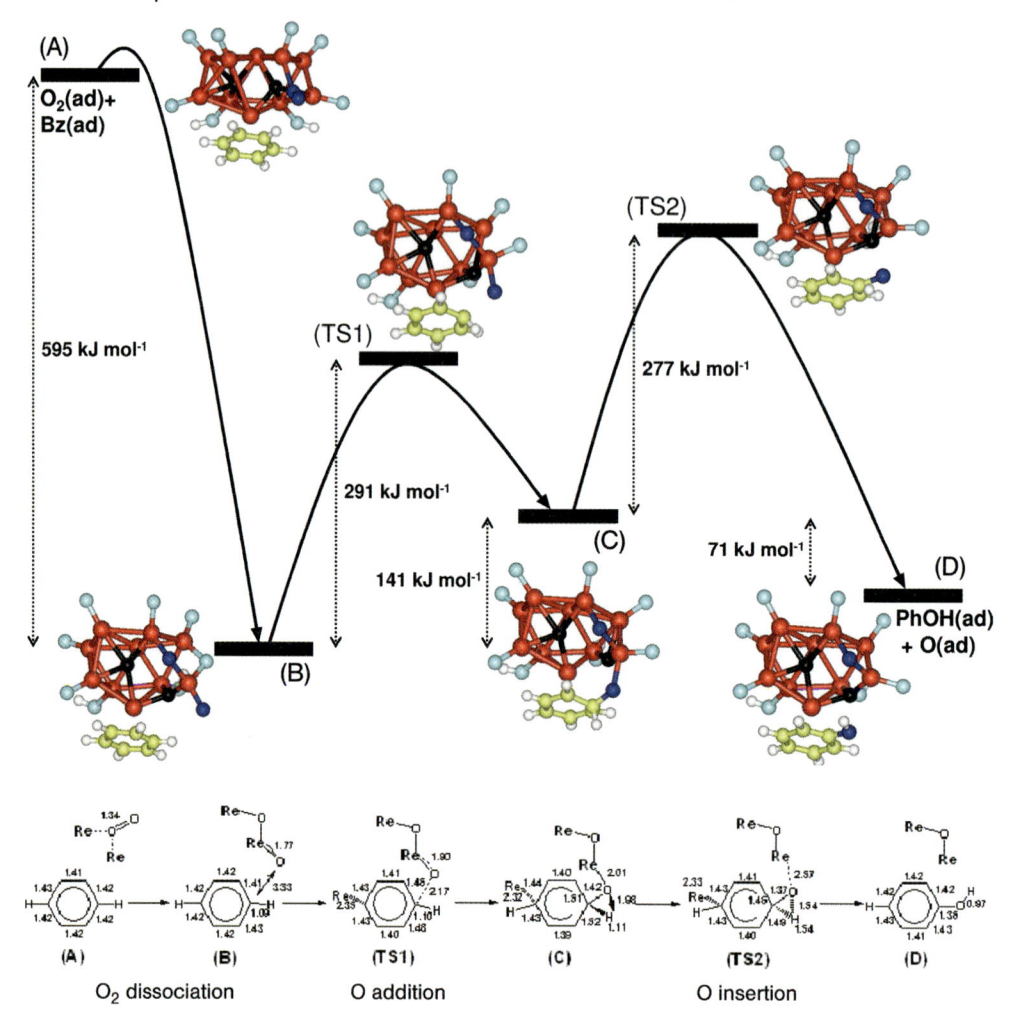

Figure 2.8 Reaction mechanism for phenol synthesis from benzene and O_2 on the Re_{10} cluster/HZSM-5 calculated by DFT. The end-on adsorbed O_2 (a) dissociates to two atomic oxygen on the Re_{10} cluster (b), and a so-produced atomic oxygen adds to a carbon atom of benzene via a transition state (TS1), forming an sp^3 carbon (c). Finally, the oxygen inserts into the C–H bond via a transition state (TS2), producing phenol. Two Re atoms of the Re_{10} cluster cooperatively/concertedly contribute to the reaction processes (TS1) → (C) → (TS2) (1 kcal mol^{-1} = 4.184 kJ mol^{-1}).

Molecular oxygen was the active oxidant for the phenol synthesis and the maximum phenol selectivity (93.9%) was achieved on the N-interstitial Re cluster supported on HZSM-5. Molecular oxygen was activated in the space between the two Re_6 octahedral cores of the Re_{10} cluster and benzene concertedly reacted on the activated O_2 with the very low activation energy of 24 kJ mol^{-1}. The pore size of HZSM-5 is 5.5 Å, which is similar to the size of the Re_6-octahedral cluster framework, resulting in the self-limiting growth of Re species. The HZSM-5 pore structure is supposed to prohibit the further clusterization of the Re catalyst and also undesired subsequent reactions of the produced phenol. The synergy of coordination of the N-interstitial Re_{10} cluster and the micropore of HZSM-5 makes it possible to achieve the highly selective phenol synthesis from benzene and O_2.

The dynamic XAFS study of the catalytically active structures under the reaction conditions should enable us to further develop novel catalysts for direct phenol synthesis from benzene and the efficient activation of molecular oxygen.

2.5
Conclusion

Nowadays increasing attention is being directed towards the development of ultimate catalytic systems with perfect performances (100% yield) from the viewpoints of environment-friendly green process, economical efficiency and minimum consumption of resources. To achieve this, both the catalytically active site and a selective reaction space should be designed precisely so as to regulate the approach of reactant molecules to the active site and their structural change in the space. Natural enzymes realize such wonderful catalyses under mild reaction conditions; however, the preparation of artificial enzyme catalysts as molecular architecture has generally been difficult and remains a long-term challenge. New and distinct materials and chemistry, with stepwise preparation in a controllable manner by using suitable precursors, including organometallic and inorganic complexes, provide an opportunity for the development of efficient catalytic, molecularly-organized surfaces. Key factors in the chemical design of supported catalyst surfaces are composition, structure, oxidation state, distribution, morphology, polarity, and so on, which should be organized at the surface. In the development of novel catalysts, new chemical concepts concerning composition or structure are conceived. Understanding and controlling catalyst surfaces are also key issues for the development of multi-functionalized catalytic sites with molecular-level control at the surfaces. This chapter has presented three examples of selective oxidation reactions, namely, chiral oxidative coupling, PROX reaction and direct phenol synthesis, where the catalytic activities and selectivities are insufficient for practical use in industrial, sustainable green processes. The necessary further developments of those catalysts need the aid of *in situ* characterization and new concepts and methods for catalyst preparation.

References

1 Tada, M. and Iwasawa, Y. (2006) *Chemical Communications*, 2833.
2 Tada, M. and Iwasawa, Y. (2007) *Coordination Chemistry Reviews*, 251, 2702–2716.
3 Iwasawa, Y. (1987) *Advances in Catalysis*, Vol 42, 35, 187.
4 Iwasawa, Y. (1997) *Accounts of Chemical Research*, 30, 103.
5 Iwasawa, Y. (1993) *Catalysis Today*, 18, 21.

6 Gates, B.C. (2001) *Topics in Catalysis*, **14**, 173.
7 Tada, M., Motokura, K. and Iwasawa, Y. (2008) *Topics in Catalysis*, **48**, 32.
8 Iwasawa, Y. (2003) *Journal of Catalysis*, **216**, 165–177.
9 Iwasawa, Y. (1996) *Studies in Surface Science Catalysis*, **101**, 21.
10 Tada, M. and Iwasawa, Y. (2005) *Annual Review of Materials Research*, **35**, 397–426.
11 Avenier, P., Taoufik, M., Lesage, A. et al. (2007) *Science*, **317**, 1056.
12 Iwasawa, Y. (1986) *Tailored Metal Catalysts*, Reidel, Dordrecht.
13 Tada, M. and Iwasawa, Y. (2003) *Journal of Molecular Catalysis A – Chemical*, **199**, 115.
14 Tada, M. and Iwasawa, Y. (2003) *Journal of Molecular Catalysis A – Chemical*, **204–205**, 27.
15 Suzuki, A., Tada, M., Sasaki, T. et al. (2002) *Journal of Molecular Catalysis A – Chemical*, **182–183**, 415.
16 Izumi, Y., Chihara, H., Yamazaki, H. and Iwasawa, Y. (1994) *The Journal of Physical Chemistry*, **98**, 594.
17 Asakura, K., Bando, K.K., Iwasawa, Y. et al. (1990) *Journal of the American Chemical Society*, **112**, 9096.
18 Bando, K.K., Asakura, K., Arakawa, H. et al. (1996) *The Journal of Physical Chemistry*, **100**, 13636.
19 Shido, T., Yamaguchi, A., Asakura, K. and Iwasawa, Y. (2000) *Journal of Molecular Catalysis A – Chemical*, **163**, 67.
20 Asakura, K., Noguchi, Y. and Iwasawa, Y. (1999) *The Journal of Physical Chemistry B*, **103**, 1051.
21 Tada, M., Sasaki, T. and Iwasawa, Y. (2002) *Physical Chemistry Chemical Physics*, **4**, 4561.
22 Tada, M., Sasaki, T., Shido, T. and Iwasawa, Y. (2002) *Physical Chemistry Chemical Physics*, **4**, 5899.
23 Tada, M., Sasaki, T. and Iwasawa, Y. (2002) *Journal of Catalysis*, **211**, 496.
24 Tada, M., Sasaki, T. and Iwasawa, Y. (2004) *The Journal of Physical Chemistry B*, **108**, 2918.
25 Tada, M., Shimamoto, M., Sasaki, T. and Iwasawa, Y. (2004) *Chemical Communications*, 2562.
26 Corma, A. and Garcia, H. (2006) *Advanced Synthesis and Catalysis*, **348**, 1391–1412.
27 Jones, M.D., Raja, R., Thomas, J.M. et al. (2003) *Angewandte Chemie – International Edition*, **42**, 4326.
28 Thomas, J.M., Raja, R. and Lewis, D.W. (2005) *Angewandte Chemie – International Edition*, **44**, 2562.
29 Tada, M., Tanaka, S. and Iwasawa, Y. (2005) *Chemistry Letters*, **34**, 1362.
30 Tanaka, S., Tada, M. and Iwasawa, Y. (2007) *Journal of Catalysis*, **245**, 173.
31 Heitbaum, M., Glorius, F. and Escher, I. (2006) *Angewandte Chemie – International Edition*, **45**, 4732.
32 Fraile, J.M., Garcia, J.I., Mayoral, J.A. and Roldan, M. (2007) *Organic Letters*, **9**, 731.
33 O'Leary, P., Krosveld, N.P., De Jong, K.P. et al. (2004) *Tetrahedron Letters*, **45**, 3177.
34 Tada, M., Taniike, T., Kantam, L.M. and Iwasawa, Y. (2004) *Chemical Communications*, 2542.
35 Tada, M., Kojima, N., Izumi, Y. et al. (2005) *The Journal of Physical Chemistry B*, **109**, 9905.
36 Cundari, T.R., Sisterhen, L.L. and Stylianopoulos, C.L. (1997) *Inorganic Chemistry*, **36**, 4029.
37 Cundari, T.R., Saunders, L. and Stylianopoulos, C.L. (1998) *Journal of Physical Chemistry A*, **102**, 997.
38 Pessoa, J.C., Calhorda, M.J., Cavaco, I. et al. (2002) *Journal of The Chemical Society – Dalton Transactions*, 4407.
39 Bao, J., Wulff, W.D. and Rheingold, A.L. (1993) *Journal of the American Chemical Society*, **115**, 3814.
40 Noyori, R. and Takaya, H. (1990) *Accounts of Chemical Research*, **23**, 345.
41 Blaser, H.-U. (1992) *Chemical Reviews*, **92**, 935.
42 Kaupp, G. (1994) *Angewandte Chemie – International Edition in English*, **33**, 728.
43 Pummerer, R., Rieche, A. and Prell, E. (1926) *Chemische Berichte*, **59**, 2159.

44 Dewar, M.J.S. and Nakaya, T. (1968) *Journal of the American Chemical Society*, **90**, 7134.
45 Feringa, B. and Wynberg, H. (1978) *Bioorganic Chemistry*, **7**, 397.
46 Armengol, E., Corma, A., Garcia, H. and Primo, J. (1999) *European Journal of Organic Chemistry*, **64**, 1915.
47 Kuiling, D., Yang, W., Lijun, Z. and Yangjie, W. (1996) *Tetrahedron*, **52**, 1005.
48 Yamamoto, H., Fukushima, H., Okamoto, Y. et al. (1984) *Journal of the Chemical Society. Chemical Communications*, 1111.
49 Nakajima, M., Miyoshi, I., Kanayama, K. and Hashimoto, S.-I. (1999) *The Journal of Organic Chemistry*, **64**, 2264.
50 Lin, X., Yang, J. and Kozlowski, M.C. (2001) *Organic Letters*, **3**, 1137.
51 Irie, R., Matsutani, K. and Katsuki, T. (2000) *Synlett*, 1433.
52 Chu, C.Y., Hwang, D.R., Wang, S.K. and Uang, B.J. (2001) *Chemical Communications*, 980.
53 Hon, S.W., Li, C.H., Kuo, J.H. et al. (2001) *Organic Letters*, **3**, 869.
54 Luo, Z., Liu, Q., Gong, L. et al. (2002) *Chemical Communications*, 914.
55 Luo, Z., Liu, Q., Gong, L. et al. (2002) *Angewandte Chemie – International Edition*, **41**, 4532.
56 Somei, H., Asano, Y., Yoshida, T. et al. (2004) *Tetrahedron Letters*, **45**, 1841.
57 Rehder, D., Schulzke, C., Dan, H. et al. (2000) *Journal of Inorganic Biochemistry*, **80**, 115.
58 Rostrup-Nielson, J.R. and Rostrup-Nielson, T. (2002) *Cattech*, **6**, 150.
59 Lemons, R.A. (1990) *Journal of Power Sources*, **29**, 251.
60 Igarashi, H., Fujino, T. and Watanabe, M. (1995) *Journal of Electroanalytical Chemistry*, **391**, 119.
61 Kahlich, M.J., Gasteiger, A. and Behm, R.J. (1997) *Journal of Catalysis*, **171**, 93.
62 Korotkikh, O. and Farrauto, R. (2000) *Catalysis Today*, **62**, 249.
63 Kim, D.H. and Lim, M.S. (2002) *Applied Catalysis A*, **224**, 27.
64 Manaslip, A. and Gulari, E. (2002) *Applied Catalysis B*, **37**, 17.
65 Wootsch, A., Descorme, C. and Duprez, D. (2004) *Journal of Catalysis*, **225**, 259.
66 Fukuoka, A. and Ichikawa, M. (2006) *Topics in Catalysis*, **40**, 103.
67 Fukuoka, A., Kimura, J., Oshio, T. et al. (2007) *Journal of the American Chemical Society*, **129**, 10120.
68 Avgouropoulos, G., Ioannides, T., Papadopoulou, C. et al. (2002) *Catalysis Today*, **75**, 157.
69 Schubert, M.M., Kahlich, M.J., Gasteiger, H.A. and Behm, R.J. (1999) *Journal of Power Sources*, **84**, 175.
70 Oh, S.H. and Sinkevitch, R.M. (1993) *Journal of Catalysis*, **142**, 254.
71 Ozkara, S. and Aksoylu, A.E. (2003) *Applied Catalysis A*, **251**, 75.
72 Shubert, M.M., Kahlich, M.J., Feldmeyer, G. et al. (2001) *Physical Chemistry Chemical Physics*, **3**, 1123.
73 Kotobuki, M., Watanabe, A., Uchida, H. et al. (2005) *Journal of Catalysis*, **236**, 262.
74 Landon, P., Fergus on, J., Solsona, B.E. et al. (2005) *Chemical Communications*, 3385.
75 Qiao, B.T. and Deng, Y.Q. (2003) *Chemical Communications*, 2192.
76 Kahlich, M.J., Gasteiger, A. and Berm, R.J. (1992) *Journal of Catalysis*, **182**, 430.
77 Schubert, M.M., Venugopal, A., Kahlich, M.J. et al. (2004) *Journal of Catalysis*, **222**, 32.
78 Qiao, B.T. and Deng, Y.Q. (2003) *Chemical Communications*, 2192.
79 Grigorova, B., Mellor, J., Palazov, A. and Greyling, F.W O Pat 00/59631/2000.
80 Carrettin, S., Conception, P., Corma, A. et al. (2004) *Angewandte Chemie – International Edition*, **43**, 2538.
81 Panzera, G., Modafferi, V., Candamano, S. et al. (2004) *Journal of Power Sources*, **135**, 177.
82 Luengnaruemitchai, A., Osuwan, S. and Gulari, E. (2004) *International Journal of Hydrogen Energy*, **29**, 429.

83 Deng, W., De Jesus, J., Saltsburg, H. and Stephanopoulos, M.F. (2005) *Applied Catalysis A*, **291**, 126.
84 Bethke, G.K. and Kung, H.H. (2000) *Applied Catalysis A*, **194**, 43.
85 Grisel, R.J.H. and Nieuwenhuys, B.E. (2001) *Journal of Catalysis*, **199**, 48.
86 Han, Y.F., Kahlich, M.J., Kinne, M. and Behm, R.J. (2002) *Physical Chemistry Chemical Physics*, **4**, 389.
87 Liu, Y., Fu, L.Q. and Stephanopoulos, M.F. (2004) *Catalysis Today*, **93–95**, 241.
88 Avgouropoulos, G., Ioannides, T., Matralis, H.K. et al. (2001) *Catalysis Letters*, **73**, 33.
89 Luo, M.F., Ma, J.M., Lu, J.Q. et al. (2007) *Journal of Catalysis*, **246**, 52.
90 Tada, M., Bal, R., Mu, X. et al. (2007) *Chemical Communications*, 4689.
91 Bal, R., Tada, M. and Iwasawa, Y. (2005) *Chemical Communications*, 3433.
92 Tada, M., Bal, R., Namba, S. and Iwasawa, Y. (2006) *Applied Catalysis A*, **307**, 78.
93 Centi, G. and Perathoner, S. *Encyclopedia of Catalysis, Selective Oxidation – Section E* (ed. I.T. Horvath), John Wiley & Sons, Inc., New York
94 Gubelmann, H. and Tirel, J. (1988) US Patent 5001280, EP Patent 0341165.
95 Kharitonov, A.S., Alexandrova, T.N., Vostrikova, L.A. et al. (1988) Russian Patent 4445646.
96 Suzuki, E., Nakashiro, K. and Ono, Y. (1988) *Chemistry Letters*, 953.
97 Panov, G.I. (2000) *Cattech*, **4**, 18.
98 Pirutko, L.V., Chernyavsky, V.S., Uriarte, A.K. and Panov, G.I. (2002) *Applied Catalysis A – General*, **227**, 143.
99 Bellussi, G. and Perego, C. (2000) *Cattech*, **4**, 4.
100 Notte, P.P. (2000) *Topics in Catalysis*, **13**, 387.
101 Panov, G.I., Sobolev, V.I. and Kharitonov, A.S. (1990) *Journal of Molecular Catalysis*, **61**, 85.
102 Burch, R. and Howitt, C. (1993) *Applied Catalysis A – General*, **103**, 135.
103 Motz, J.L., Heinrichen, H. and Hölderich, W.F. (1998) *Journal of Molecular Catalysis A – Chemical*, **136**, 175.
104 Kollmer, F., Hausmann, H. and Hölderich, W.F. (2004) *Journal of Catalysis*, **227**, 398.
105 Dubkov, K.A., Ovanesyan, N.S., Shteinman, A.A. et al. (2002) *Journal of Catalysis*, **207**, 341.
106 Kubanek, P., Wiehterlova, B. and Sobalik, Z. (2002) *Journal of Catalysis*, **21**, 109.
107 Hensen, E.J.M., Zhu, Q. and van Santen, R.A. (2005) *Journal of Catalysis*, **233**, 136.
108 Hensen, E.J.M., Zhu, Q., Janssen, R.A.J. et al. (2005) *Journal of Catalysis*, **233**, 123.
109 Zhu, Q., van Teeffelen, R.M., van Santen, R.A. and Hensen, E.J.M. (2002) *Journal of Catalysis*, **221**, 575.
110 Hensen, E., Zhu, Q., Liu, P.-H. et al. (2004) *Journal of Catalysis*, **226**, 466.
111 Centi, G., Perathoner, S., Pino, F. et al. (2005) *Catalysis Today*, **110**, 211.
112 Choi, S.H., Wood, B.R., Ryder, J.A. and Bell, A.T. (2003) *The Journal of Physical Chemistry. B*, **107**, 11843.
113 Jia, J., Pillai, K.S. and Sachtler, W.M.H. (2004) *Journal of Catalysis*, **221**, 119.
114 Taboada, J.B., Hensen, E.J.M., Arends, I.W.C.E. et al. (2005) *Catalysis Today*, **110**, 221.
115 Hensen, E.J.M., Zhu, Q., Hendrix, M.M.R.M. et al. (2004) *Journal of Catalysis*, **221**, 560.
116 Taboada, J.B., Overweg, A.R., Kooyman, P.J. et al. (2005) *Journal of Catalysis*, **231**, 56.
117 Yuranov, I., Bulushev, D.A., Renken, A. and Kiwi-Minsker, L. (2007) *Applied Catalysis A – General*, **319**, 128.
118 Malykhin, S.E., Zilberberg, I.I. and Zhidomirov, G.M. (2005) *Chemical Physics Letters*, **414**, 434.
119 Chernyavsky, V.S., Pirutko, L.V., Uriarte, A.K. et al. (2007) *Journal of Catalysis*, **245**, 466.
120 Dubkov, K.A., Soboov, V.I., Talsi, E.P. et al. (1997) *Journal of Molecular Catalysis A – Chemical*, **123**, 155.

121 Kachurovskaya, N.A., Zhidomirov, G.M. and van Santen, R.A. (2004) *The Journal of Physical Chemistry. B*, **108**, 5944.
122 Ryder, J.A., Chakraborty, A.K. and Bell, A.T. (2003) *Journal of Catalysis*, **220**, 84.
123 Yoshizawa, K., Shiota, Y. and Kamachi, T. (2003) *The Journal of Physical Chemistry B*, **107**, 11404.
124 Esteves, P.M. and Louis, B. (2006) *The Journal of Physical Chemistry B*, **110**, 16793.
125 Othmer, K. (1995) *Hydrogen Peroxide, Encyclopedia of Chemical Technology*, 4th edn, Vol. 13, p. 961.
126 Perego, C., Carati, A., Ingallina, P. et al. (2001) *Applied Catalysis A – General*, **221**, 63.
127 Clerici, M.G. and Ingallina, P. (1998) *Catalysis Today*, **41**, 351.
128 Jing, H., Guo, Z., Ma, H. et al. (2002) *Journal of Catalysis*, **212**, 22.
129 Bhaumik, A., Mukherjee, P. and Kumar, R. (1998) *Journal of Catalysis*, **178**, 101.
130 Tatsumi, T., Yuasa, K. and Tominaga, H. (1992) *Chemical Communications*, 1446.
131 Miyake, T., Hamada, M., Niwa, H. et al. (2002) *Journal of Molecular Catalysis A – Chemical*, **178**, 199.
132 Kuznetsova, N.I., Kuznetsova, L.I., Likholobov, V.A. and Pez, G.P. (2005) *Catalysis Today*, **99**, 193.
133 Kitano, T., Kuroda, Y., Mori, M. et al. (1993) *Journal of the Chemical Society – Perkin Transactions 2*, 981.
134 Ehrich, H., Berndt, H., Pohl, M.-M. et al. (2002) *Applied Catalysis A – General*, **230**, 271.
135 Laufer, W., Niederer, J.P.M. and Hölderich, W.F. (2002) *Advanced Synthesis and Catalysis*, **344**, 1064.
136 Niwa, S., Eswaramoorthy, M., Nair, J. et al. (2002) *Science*, **295**, 105.
137 Sato, S. and Cvetanovit, R.J. (1959) *Journal of the American Chemical Society*, **81**, 3223.
138 Itoh, N., Niwa, S., Mizukami, F. et al. (2003) *Catalysis Communications*, **4**, 243.
139 Vulpescu, G.D., Ruitenbeek, M., van Lieshout, L.L. et al. (2004) *Catalysis Communications*, **5**, 347.
140 Sato, K., Hanaoka, T., Hamakawa, S. et al. (2006) *Catalysis Today*, **118**, 57.
141 Sato, K., Hanaoka, T., Niwa, S. et al. (2005) *Catalysis Today*, **104**, 260.
142 Vulpescu, G.D., Ruitenbeek, M., van Lieshout, L.L. et al. (2004) *Catalysis Communications*, **5**, 347.
143 Miyahara, T., Kanzaki, H., Hamada, R. et al. (2001) *Journal of Molecular Catalysis A – Chemical*, **176**, 141.
144 Hamada, R., Shibata, Y., Nishiyama, S. and Tsuruya, S. (2003) *Physical Chemistry Chemical Physics*, **5**, 956.
145 Shibaya, Y., Hamada, R., Ueda, T. et al. (2005) *Industrial & Engineering Chemistry Research*, **44**, 8765.
146 Castagnola, N.B., Kropf, A.J. and Marshall, C.L. (2005) *Applied Catalysis A – General*, **290**, 110.
147 Kubacka, A., Wang, Z., Sulikowskim, B. and Corberan, V.C. (2007) *Journal of Catalysis*, **250**, 184.
148 Bahidsky, M. and Hronec, M. (2004) *Catalysis Today*, **91–92**, 13.
149 Jintoku, T., Takaki, K., Fujiwara, Y. et al. (1990) *Bulletin of the Chemical Society of Japan*, **63**, 438.
150 Tani, M., Salamoto, T., Mita, S. et al. (2005) *Angewandte Chemie – International Edition*, **44**, 2586.
151 Okumura, J., Nishiyama, S., Tsuruya, S. and Masai, M. (1998) *Journal of Molecular Catalysis A – Chemical*, **135**, 133.
152 Kanzaki, H., Kitamura, T., Hamada, R. et al. (2004) *Journal of Molecular Catalysis A – Chemical*, **208**, 203.
153 Gu, Y.-Y., Zhao, X.-H., Zhang, G.-R. et al. (2007) *Applied Catalysis A – General*, **328**, 150.
154 Gao, X. and Xu, J. (2006) *Catalysis Letters*, **111**, 203.
155 Seo, Y.-J., Mukai, Y., Tagawa, T. and Goto, S. (1997) *Journal of Molecular Catalysis A – Chemical*, **120**, 149.
156 Murata, K., Yanyong, R. and Inaba, M. (2005) *Catalysis Letters*, **102**, 143.
157 Liu, Y., Murata, K. and Inaba, M. (2005) *Catalysis Communications*, **6**, 679.

158 Liu, Y., Murata, K. and Inaba, M. (2006) *Journal of Molecular Catalysis A – Chemical*, **256**, 247.

159 Romao, C.C., Kuhn, F.E. and Herrmann, W.A. (1997) *Chemical Reviews*, **97**, 3197.

160 Oal, J. and Baran, J. (2001) *Journal of Catalysis*, **203**, 466.

161 Mandelli, D., van Vliet, M.C.A., Arnold, U. et al. (2001) *Journal of Molecular Catalysis A – Chemical*, **168**, 165.

162 Salameh, A., Coperet, C., Basset, J.M. et al. (2007) *Advanced Synthesis and Catalysis*, **349**, 238.

163 Lo, H.C., Han, H.N., D'Souza, L.J. et al. (2007) *Journal of the American Chemical Society*, **129**, 1246.

164 Onaka, M. and Oikawa, T. (2002) *Chemistry Letters*, **8**, 850.

165 Yuan, Y., Shido, T. and Iwasawa, Y. (2000) *Chemical Communications*, 1421.

166 Yuan, Y. and Iwasawa, Y. (2002) *The Journal of Physical Chemistry. B*, **106**, 4441.

167 Viswanadham, N., Shido, T. and Iwasawa, Y. (2001) *Applied Catalysis A – General*, **219**, 223.

168 Viswanadham, N., Shido, T., Sasaki, T. and Iwasawa, Y. (2002) *The Journal of Physical Chemistry B*, **106**, 10955.

169 Bal, R., Tada, M., Sasaki, T. and Iwasawa, Y. (2006) *Angewandte Chemie – International Edition*, **45**, 448.

170 Tada, M., Bal, R. and Iwasawa, Y. (2006) *Catalysis Today*, **117**, 171.

171 Tada, M., Bal, R., Sasaki, T. et al. (2007) *J Phys Chem C*, **111**, 10095.

172 Delley, B. (1990) *Journal of Chemical Physics*, **92**, 508.

173 Delly, B. (2000) *Journal of Chemical Physics*, **113**, 7756.

174 Perdew, J.P. and Wang, Y. (1992) *Physical Review B – Condensed Matter*, **45**, 13244.

3
Unique Catalytic Performance of Supported Gold Nanoparticles in Oxidation
Yunbo Yu, Jiahui Huang, Tamao Ishida, and Masatake Haruta

3.1
Introduction

Modern chemical industries are based on petroleum, which is composed of hydrocarbons. Accordingly, selective oxidation of hydrocarbons is of immense importance to produce functional organic compounds containing oxygen, such as epoxides, ketones, aldehydes, alcohols and acids. These organic oxygenates are used for polymers, surfactants, detergents, cosmetics, and so on. Because of this, oxidation is the second largest chemical processes after polymerization and makes up about 30% of total production in the chemical industry [1].

Currently, oxidation processes are mostly based on chlorine or organic peroxides using more than two reaction stages. Chlorinated organic intermediates are neutralized to transform organic oxygenates, by-producing a huge amount of $CaCl_2$ and some amounts of toxic chlorinated organic compounds. Organic peroxides are inevitably expensive and are accompanied by co-products, in quantities that often exceed market demands. Therefore, it is strongly desired to replace these oxidizing agents with molecular oxygen or air. However, activation of molecular oxygen requires about 112 kcal mol^{-1} (the bonding energy of the strongest C—H bond in CH_4 is 107 kcal mol^{-1}) so that, once dissociated, it is hard to control the reactivity of the resultant atomic species. The most common route is combustion (complete oxidation) to produce CO_2 and H_2O, providing us with thermal energy. One way of controlling the reactivity of oxygen species is to use reductive oxygen activation, which means that oxygen activation can be done under milder conditions, by using H_2 or CO, for example, to dissociate O—O bonding [2].

If hydrocarbons can be oxidized efficiently to produce valuable organic oxygenates by using molecular oxygen instead of hazardous chlorine and expensive organic peroxides, it would change the present chemical industry into a green and sustainable one. Catalysis by gold, which was once believed to be almost inert, shows such possibilities [3]. When Au is supported on metal oxides as nanoparticles (NPs) it can catalyze oxygen under milder conditions than Pt group metal catalysts to enable CO

Modern Heterogeneous Oxidation Catalysis: Design, Reactions and Characterization
Edited by Noritaka Mizuno
Copyright © 2009 WILEY-VCH Verlag GmbH & Co. KGaA, Weinheim
ISBN: 978-3-527-31859-9

oxidation at room temperature and to allow efficient selective oxidation of hydrocarbons and alcohols.

The number of papers dealing with catalysis by Au was more or less than 5 a year in the 1980s but reached 700 in 2005 and 600 in 2006. There are three major streams in current research activities on Au catalysts: expansion of applications, especially to liquid-phase organic reactions [4], discussion on the active states of Au [5], and exploration of new forms of Au catalysts. The last stream has emerged recently and is represented by Au submicron tube [6], nanoporous Au [7, 8], polymer stabilized Au colloids [9] and Au on solid polymers [10, 11], which in turn provide valuable information for determining what states of Au are surprisingly active and selective.

Most representative Au catalysts reported so far can be positioned in a ternary system composed of three important factors: metal–oxide junction, water and OH^- (Figure 3.1); a fourth important factor, the size of Au particles, is given in nm [12]. At least two conditions should be fulfilled for Au to exhibit high catalytic activity. Taken into account of these conditions, what is the mechanism for the activation of molecular oxygen over the surfaces of Au NPs and how does the support define the catalytic nature and activity? Clear answers to these questions would provide a valuable guide map for green sustainable chemistry through the control of oxygen

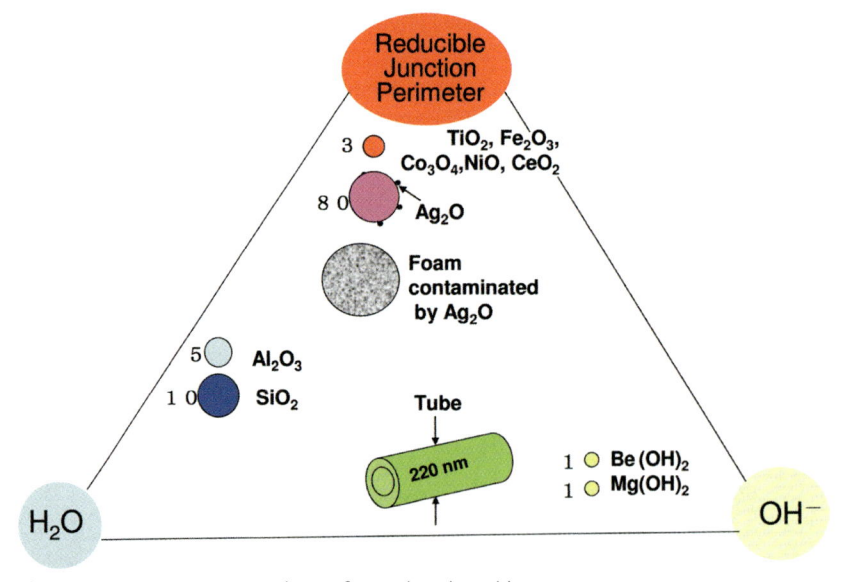

Figure 3.1 Four important conditions for catalysis by gold (metal–oxide junction, water, OH^- and size of the Au particles or tubes). The figure by each circle is the diameter (nm) of the gold particles or tubes. Gold turns out to be catalytically very active, provided that at least two of the four conditions are fulfilled. For example, in CO oxidation at room temperature even unsupported gold is active in the presence of alkaline (OH^-) water (H_2O).

reactivity. Gold seems to be the best prospect for such control since, generally, it is almost inert but exhibits unique catalytic properties when properly prepared.

3.2
Low-Temperature CO Oxidation

The oxidation of CO is the simplest reaction and has been the most intensively studied since Langmuir first presented a theory of adsorption and catalysis for this reaction [13]. Supported Au NPs such as Au/TiO_2, Au/Fe_2O_3 and Au/Co_3O_4 are extraordinarily active in CO oxidation, even at 200 K, and are much more active than the other noble metals catalysts at temperatures below 400 K [14–16]. Gold clusters composed of several atoms can promote the reaction between CO and O_2 to form CO_2 at as low as 40 K [17]. Most recently, Lahr and Ceyer [18] have extended the temperature range at which the activity for CO oxidation is observed to as low as 70 K by using an Au/Ni surface alloy.

Currently, low-temperature CO oxidation over Au catalysts is practically important in connection with air quality control (CO removal from air) and the purification of hydrogen produced by steam reforming of methanol or hydrocarbons for polymer electrolyte fuel cells (CO removal from H_2). Moreover, reaction mechanisms for CO oxidation have been studied most extensively and intensively throughout the history of catalysis research. Many reviews [4, 19–28] and highlight articles [12, 29, 30] have been published on CO oxidation over catalysts. This chapter summarizes of the state of art of low temperature CO oxidation in air and in H_2 over supported Au NPs. The objective is also to overview of mechanisms of CO oxidation catalyzed by Au.

3.2.1
Low-Temperature CO Oxidation in Air

Except for H_2 oxidation and hydrocarbon hydrogenations, most reactions are remarkably structure-sensitive over supported Au catalysts. One typical reaction is CO oxidation, which is remarkably sensitive to the junction perimeter between Au particles and support, the type of support and the size of the Au particles.

3.2.1.1 Junction Perimeter Between Au Particles and the Support
Carefully prepared Au catalysts have a relatively narrow particle size distribution, giving mean diameters in the range 2–10 nm with a standard deviation of about 30%. A major reason why Au particles remain as NPs even after calcination 573 K is the epitaxial contact of Au NPs with the metal oxide supports. Gold particles always expose its most densely packed plane, the (111) plane, in contact with α-Fe_2O_3(110), Co_3O_4(111), anatase TiO_2(112), and rutile TiO_2(110).

Figure 3.2 shows a typical TEM image of Au NPs epitaxially attached to anatase TiO_2 [31]. The surface atomic configuration is better matched for a Au(111) plane sitting on the oxygen layer of anatase TiO_2 than on the Ti layer. Three-dimensional nanostructure analyses by electron holography together with the high-resolution

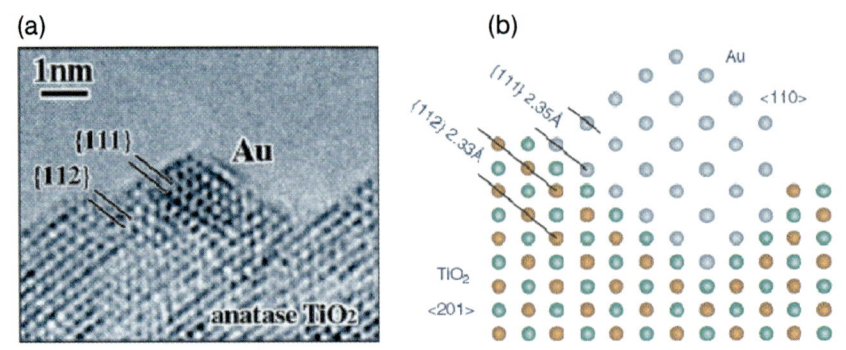

Figure 3.2 (a) TEM micrograph of an Au/TiO$_2$ catalyst prepared by deposition precipitation; (b) schematic model of the interface [31].

TEM shown in Figure 3.3 reveal that smaller hemispherical Au particles less than 2 nm thick have contact angles with the support of below 90° (wet interface), whereas for larger Au particles (5 nm thick) the same angle is above 90° (dry interface) [32]. This difference in the wettability of Au particles may arise from the change in electronic state of the contact interfaces with the particle size. When Au/TiO$_2$ was calcined at temperatures above 573 K, Au particles coagulated to form larger particles, mostly gathered in the valleys at the junctions between the TiO$_2$ particles [33].

Figure 3.4 shows turnover frequencies (TOFs), the reaction rate over one single surface metal atom per second, of CO oxidation at 300 K over Au/TiO$_2$ and Pt/TiO$_2$ catalysts prepared by deposition-precipitation (DP), photocatalytic deposition and impregnation (IMP) methods [34]. The DP method yields hemispherical metal particles with their flat planes strongly attached to the TiO$_2$ support (see Figure 3.2), while photocatalytic deposition and IMP methods yield spherical particles simply loaded on the TiO$_2$ support and, therefore, much larger particles – particularly in the case of Au. Over Pt/TiO$_2$, the reaction of CO with O$_2$ can take place on the Pt surfaces and the metal oxide support is not directly involved in the reaction. This can account for why different preparation methods do not make any appreciable difference in the TOF of Pt catalysts. In contrast, the TOF of Au/TiO$_2$ markedly depends on the contact structure, changing by four orders of magnitude. The TOF of strongly attached hemispherical Au NPs exceeds that of Pt by one order of magnitude.

The sharp contrast between the above two catalysts in CO oxidation suggests that the reactions may take place at the perimeter interfaces around Au NPs. To confirm this hypothesis, Vannice prepared an inversely supported catalyst, namely, TiO$_2$ layers deposited on a Au substrate, and observed appreciable catalytic activity [35]. We prepared the Au/TiO$_2$ catalysts by mechanically mixing a colloidal solution of Au particles 5 nm in diameter with TiO$_2$ powder, followed by calcination in air at different temperatures [36]. Calcination at 873 K promotes the coagulation of Au particles to form larger particles with diameters above 10 nm, but at the same time with stronger contact (observed by TEM), leading to much higher catalytic activity than calcination at 573 K.

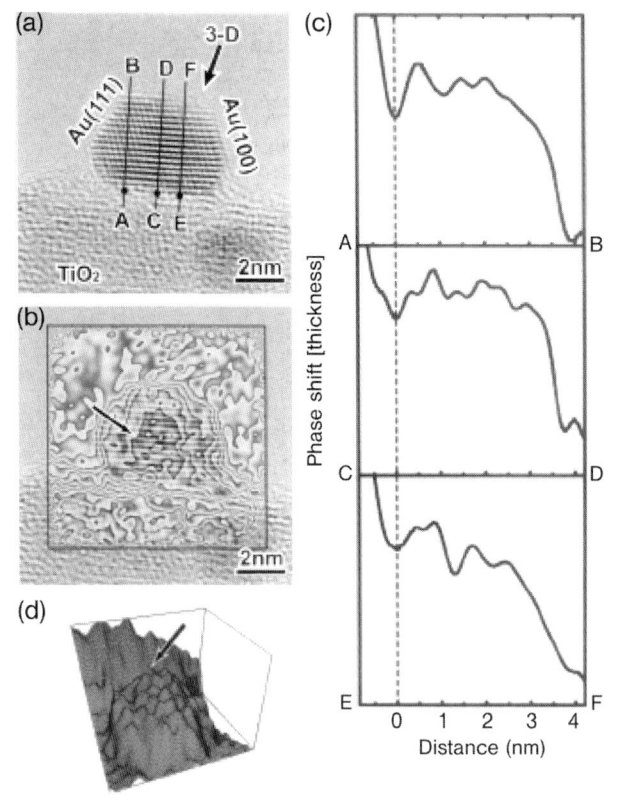

Figure 3.3 (a) HREM image and (b) phase image reconstructed from the hologram in a ~4-nm high Au/TiO$_2$ catalyst. The phase image was amplified ×30 and superimposed on the HREM image. (c) Line profiles of the phase image (b) along lines A–B, C–D and E–F in the HREM image (a). The black dots in the HREM image and the dotted lines in the line profiles indicate the Au/TiO$_2$ interface. (d) 3D wire frame view from the arrow in (a) [32].

3.2.1.2 Selection of Suitable Supports

For CO oxidation, as shown in Figure 3.5, many oxides other than strongly acidic materials such as Al$_2$O$_3$–SiO$_2$, WO$_3$ and activated carbon can be used as a support and induce activity even below 300 K. For Pd and Pt, semiconductor metal oxides lead to enhanced catalytic activities but at temperatures above 300 K. Semiconductive metal oxides such as TiO$_2$, Fe$_2$O$_3$ and NiO provide more stable Au catalysts than do insulating metal oxides such as Al$_2$O$_3$ and SiO$_2$. Among Au supported on Al$_2$O$_3$, SiO$_2$ or TiO$_2$, the TOFs at room temperature in the presence of moisture are nearly equal, indicating that the contributions of the metal oxide supports are more or less similar in intensity [37]. The difference appears in the moisture effect: Al$_2$O$_3$ and SiO$_2$ require a concentration of H$_2$O 10 ppm greater than that for TiO$_2$ for CO oxidation to proceed at room temperature [38]. Alkaline earth metal hydroxides such as Be(OH)$_2$

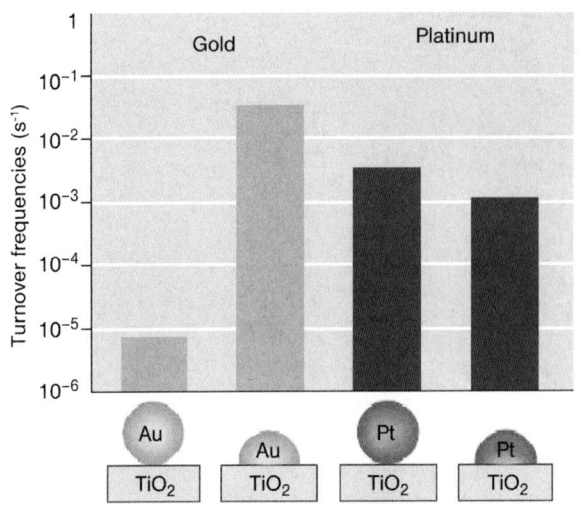

Figure 3.4 Turnover frequencies for CO oxidation at 300 K over spherical and hemispherical particles of Au and Pt supported on TiO_2 [21].

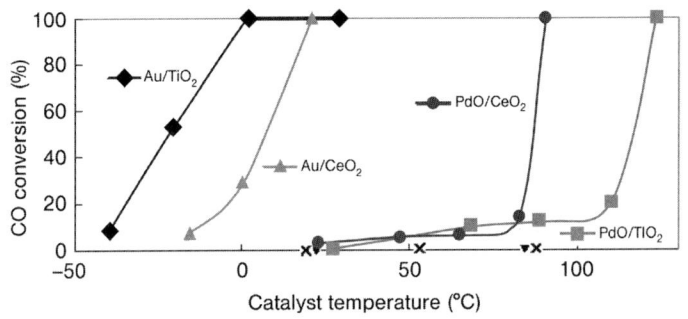

Figure 3.5 Catalytic activity of supported Pd and Au catalysts for CO oxidation. Metal loadings: 1.0 wt%, D-P method, calcined at 573 K, CO 1.0 vol.% in air, SV = 20 000 h^{-1} mL per g-cat.; ($\times$) activated carbon and ($\blacktriangledown$) polymers.

and $Mg(OH)_2$ are excellent choices for demonstrating high activity at a temperature as low as 196 K [39, 40]. In contrast, when an acidic material, such as Al_2O_3–SiO_2, WO_3 or activated carbon, is used as a support Au exhibits poor activity – even above 473 K the conversions are far below 100% [22, 37]. Polymer supports do not allow catalysis by Au in the gas phase even though the particle size is as small as 2 nm [41].

3.2.1.3 Sensitivity to the Size of the Gold Particles

Figure 3.6 plots the TOFs of CO oxidation over Au/TiO_2 at 273 K and Pt/SiO_2 at 437 K as a function of the mean diameter of the Au particles [42]. The TOF increases sharply

3.2 Low-Temperature CO Oxidation

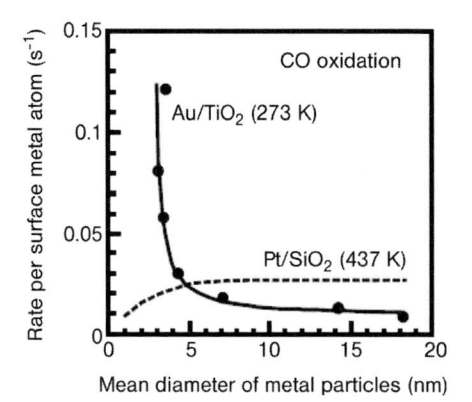

Figure 3.6 Turnover frequency (TOF) for CO oxidation at 273 K over Au/TiO$_2$ as a function of the mean diameter of Au particles [42].

with a decrease in diameter below 5 nm. In contrast, the Pt group of metals shows a decreasing or steady TOF [43]. The rates over Au/TiO$_2$ were about one order of magnitude greater when measured as the temperature was lowered from 353 K than when measured as the temperature was raised from 203 K [21, 44]. This difference is assumed to arise from the accumulation of carbonate species on the surfaces of the support at low temperatures, resulting in the loss of the activating power of the perimeter interfaces for O$_2$. Therefore, the rate over Au/TiO$_2$, which was deactivated during experiments at lower temperatures, is regarded as close to the rate of the CO reaction with O$_2$ over the surfaces of Au particles without the contribution of O$_2$ activation at the perimeter interfaces. The one order of magnitude difference in the rate between fresh (obtained by high temperature measurements) and deactivated Au/TiO$_2$ can be ascribed to the contribution of the TiO$_2$ support. The increase in TOF with a decrease in diameter of Au particles can be explained if the adsorption sites for CO are edge, corner or step sites, and the reaction zone is the periphery around the Au particles, the fractions of which increase in hemispherical Au particles with a decrease in their size [45].

Among supported noble metal catalysts, Au supported on Mg(OH)$_2$ is the most active for CO oxidation at 196 K. However, after 3–4 months it becomes completely inactive. It has been suggested that for the two structures of 13-atom clusters the icosahedron is active whereas the cubo-octahedron is inactive [40]. The active Au/Mg(OH)$_2$ catalyst, which is primarily composed of icosahedral Au clusters of 13 atoms, showed a negative apparent activation energy in the temperature range 196–273 K [46]. This can be explained by the enhanced transformation of the icosahedron into the cubo-octahedron with an increase in reaction temperature. Heiz and coworker prepared model catalysts by depositing size-selected Au anion clusters onto a single crystal of MgO. Although the sizes and structures of the deposited Au clusters were not observed by STM, it was reported that there was appreciable size dependency of CO adsorption and that the highest reactivity to CO was observed for anion clusters consisting of 11 atoms [47]. It was also reported that 8 and 11 are the smallest and the second-smallest number of atoms to exhibit catalytic activity for CO oxidation over the MgO support. A higher activity of Au clusters on defect-rich MgO than on defect-poor MgO was observed. *Ab initio* simulations indicate that partial electron transfer from the surface of the Au clusters and oxygen-vacancy defects in the support play an essential role for the genesis of catalytic activity [48, 49].

3.2.2
Low-Temperature CO Oxidation in H_2

One particular application for which supported Au catalysts may find a niche market is in fuel cells [4, 50] and in particular in polymer electrolyte fuel cells (PEFC), which are used in residential electric power and electric vehicles and operate at about 353–473 K. Polymer electrolyte fuel cells are usually operated by hydrogen produced from methane or methanol by steam reforming followed by water–gas shift reaction. Residual CO (about 1 vol.%) in the reformer output after the shift reaction poisons the Pt anode at a relatively low PEFC operating temperature. To solve this problem, the anode of the fuel cell should be improved to become more CO tolerant (Pt-Ru alloying) and secondly catalytic systems should be developed that can remove even trace amounts of CO from H_2 in the presence of excess CO_2 and water.

Considerable effort has been made to design suitable catalysts for the competitive oxidation of CO in a stream of H_2. The most commonly used noble metal catalysts (alumina supported Pt, Pt/Fe, Ru and Rh) are unsuitable because they are in principle more active for H_2 oxidation than for CO oxidation and require high temperatures (423–473 K) to be effective [51–54]. Pt NPs turn out to be selective to CO oxidation when incorporated into the cages of zeolite [55] or nanotube of mesoporous SiO_2 [56], while Ru/Al_2O_3 catalysts are selective to CO oxidation [57]. It has been found that Pt metal catalysts behave differently for CO oxidation in H_2 and for CO oxidation and H_2 oxidation in air. In contrast, supported Au NPs such as Au/TiO_2 and Au/Al_2O_3 are much more active for CO oxidation than for H_2 oxidation and are active at much lower temperatures [37, 58, 59]. Owing to this intrinsic nature, different kinds of Au catalysts have been investigated to remove a trace amount of CO in the presence of excess H_2 (Table 3.1).

For successful operation a selective CO oxidation catalyst in a reformer–PEFC system must be operated at ca. 353–373 K in a complex feed consisting of CO, O_2, H_2, CO_2, H_2O and N_2, and be capable of reducing CO concentrations from about 1% to below 50 ppm – this is equivalent to a CO conversion of at least 99.5% [4, 54, 60]. In addition, this conversion must be achieved with the addition of equimolar O_2 (twice the stoichiometric amount) and the competitive oxidation of H_2 must be minimized. This is expressed as selectivity, which is defined as the percentage of the oxygen fed consumed in the oxidation of CO; for commercial operation a selectivity of 50% is acceptable, since at this selectivity minimal H_2 is oxidized to water.

Table 3.1 shows the catalytic performance of supported Au catalysts for the preferential oxidation (PROX) of CO in H_2 together with the actual reaction conditions and targeted performances. Au/Al_2O_3 [61–63], Au/Mn_2O_3 [58], Au/Fe_2O_3 [54, 60, 61, 64–66] and Au/CeO_2 [54, 60–62, 67–70] have been reported to

3.2 Low-Temperature CO Oxidation

Table 3.1 Performance of supported Au catalysts for CO oxidation in H_2.

Catalyst supports	Cal (K)	Components of feed gas (%)						T (K)	SV (h^{-1} mL g^{-1})	Conversion (%)	Selectivity (%)	Reference
		CO	O_2	H_2	H_2O	CO_2	Balance					
Research target		0.8–1.0	0.4–1.0	50	2–10	20–24	Inert N_2/He	353–383	—	>99.5	>50	[4, 54, 60]
Al_2O_3	573	2	2	48	—	—	He	383	~2100	~67	—	[63]
Mn_2O_3	573	1.0	1.0	98	—	—	—	323–353	10 000	>95	—	[58]
Fe_2O_3	393	0.9	0.9	50	4.7	22	N_2	353	12 000	92	47	[54, 60]
	393	1	4	50	0	0	He	323	20 000	100	100	[64]
	673	0.1	0.1	66	10	22	—	353	—	a)	29	[65]
	673	1	1.25	50	10	15	—	353	25 000	65	60	[66]
	673	1	1.25	50	10	15	—	100	25 000	99.5	54.5	
CeO_2	673+773	0.9	0.9	50	4.7	22	N_2	353	12 000	99.8	51	[54, 60]
	393	0.9	0.9	50	4.7	22	N_2	353	12 000	68	35	[54, 60]
	393	0.8	0.4	58.4	0	0	—	333	165 000	60	99.7	[67]
	673	1	1.25	50	0	0	He	340	0.03–0.144 g.s.cm^{-3}	96	40	[68]
	673	1	1.25	50	0	15	He	353		68	30	
	673	1	1.25	50	10	15	He	373		65	37	
	773	1	1	40	10	2	He	363	30 000	76	62	[69]
	773	1	1	40	10	2	He	363	30 000	18	56	
	—	1	1.5	48	0	24	He	353	111 000	20	20	
Co/CeO$_2$/TiO$_2$/SnO$_2$	673	0.5	0.8	2	0	0	N_2	353	—	76	85	[70]
	673	0.5	0.8	2	0	0	N_2	373	—	91	91	[73]
CeO$_2$-Co$_3$O$_4$	473	1.0	1.0	50	0	0	N_2	353	30 000	91	51	[74]
MnO$_2$-TiO$_2$ (Mn/Ti = 2:98)	573	1.33	1.33	65.33	0	0	He	353	30 000	98.4	51.3	[75]
(AuPt)A zeolite	773	1	1	40	0	0	He	443	30 000	100	55	[76]
	773	1	1	40	10	10	He	443	30 000	99	54	

a) Activity quoted as 0.003 mol-CO g-Au^{-1} s^{-1}.

be suitable for PROX of CO in excess H_2, and they are usually calcined at high temperature, above 673 K, for higher CO selectivities [58]. Hemispherical metallic Au NPs attached to the support metal oxides at their flat planes are very active for CO oxidation. If cationic Au is active for the water–gas shift reaction as Flytzani-Stephanopoulos reported [71], it may consequently catalyze the reverse reaction:

$$CO_2 + H_2 \rightarrow CO + H_2O$$

Because this reaction causes the formation of CO, depressing overall CO conversion can not be achieved. Accordingly, one design criterion may be to prepare a catalyst consisting of small Au NPs (about 3–7 nm) alone without cationic Au species [60]. As a result, the calcination temperature used to prepare supported Au NPs is of crucial importance and calcination at a high temperature is reasonable. Another more probable hypothesis is based on the knowledge that metallic Au NPs are also responsible for the water–gas shift reaction [72]. Higher temperature calcination enhances the coagulation of small Au clusters and improves the crystallinity of surface layers of the support. These changes may be correlated with the rates of CO oxidation and of reverse water–gas shift reaction. Notably, in the temperature range 353–383 K, the thermodynamic equilibrium in not in favor of the reverse water–gas shift reaction.

Table 3.1 also shows that, in the absence of H_2O and CO_2, multi-component catalysts such as Au-$Co/CeO_2/TiO_2/SnO_2$ [73], Au/CeO_2-Co_3O_4 [74], Au/MnO_2-TiO_2 [75] and AuPt/A zeolite [76] give high CO conversion and selectivity for the PROX of CO in excess H_2. At 353 K, over Au/CeO_2–Co_3O_4 (Ce/Co $= 0.2$) pretreated in oxidative atmosphere, 91% CO conversion and 51% selectivity to CO_2 can last for 260 h [74]. A catalyst of AuPt/A zeolite (at a weight ratio of Au/Pt $= 1:2$) [76] provides 100% CO conversion and 55% CO selectivity at $\sim$443 K in the absence of H_2O and CO_2. Avgouropoulos and coworkers have reported that the presence of CO_2 and H_2O caused a significant decrease in both activity and selectivity of Au/CeO_2 [68]. This decrease in catalytic performance is attributed to the competitive adsorption of CO, CO_2 and H_2O on the catalyst surfaces [53, 70, 77, 78]. The presence of CO_2 will favor the reverse water–gas shift reaction at above 423 K [4, 54, 60]. However, Luengnaruemitchai et al. have recently shown that there is not much difference between the catalytic performance of AuPt/A zeolite in the presence and in the absence of CO_2 and H_2O at 443 K [76]. The catalyst performance needs to be improved in the temperature range compatible with PEFCs (353–373 K).

In a realistic gas stream containing H_2O and CO, Avgouropolous et al. have almost achieved the target performances [66]: their system worked at 373 K by using an Au/Fe_2O_3 catalyst calcined at 673 K, but required excess O_2 and a slightly lower levels of CO_2. Subsequently, Hutchings and coworkers [54, 60] have shown that a Au/Fe_2O_3 catalyst prepared by a two-stage calcination at 673 and 823 K can meet the demanded target (Figure 3.7). Furthermore, they proposed that the most important factor in the design of a successful catalyst for this application is the need to control the activity of the catalyst for the reverse water–gas shift reaction whilst retaining the activity for CO oxidation. Armed with this knowledge we can now anticipate that Au catalysts will be

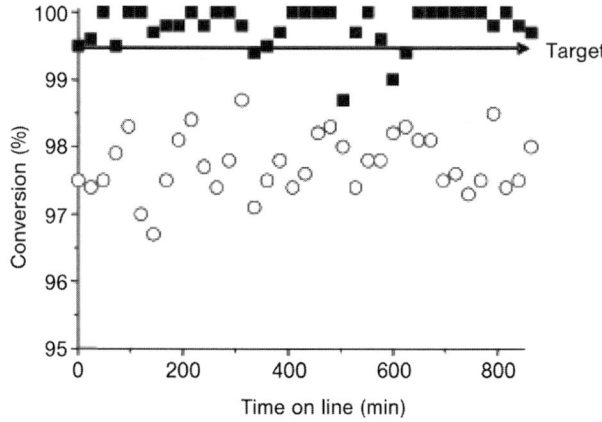

Figure 3.7 Variation of CO (■) and O_2 (○) conversion with time-on-line for a 5% Au/Fe_2O_3 catalyst calcined in air at 673 K + 823 K. Reaction conditions: 80 °C, 0.9% CO, 0.9% O_2, 50% H_2, 22% CO_2, 4.7% H_2O with the balance N_2; 100 mg of catalyst was used with a total flow rate of 20 mL min^{-1}, GHSV = 12 000 h^{-1} [54].

designed with much higher activity and selectivity for the PROX of CO for application to PEFCs [4].

3.2.3
Mechanism for CO Oxidation Over Supported Gold Nanoparticles

Louis has recently discussed this mechanism in detail [79]. Figure 3.8 shows hypotheses for the active states of supported Au catalysts for CO oxidation proposed so far in the order of decreasing size of Au: junction perimeter between Au and the metal-oxide supports, specific size or thickness of Au clusters or thin layers, and cationic Au.

3.2.3.1 Mechanisms Involving Junction Perimeter Between Gold and the Metal-Oxide Supports

Figure 3.9 shows a schematic representation of the most probable mechanism for CO oxidation over supported Au catalysts. This mechanism was first proposed, and later described in more detail, by Haruta [21, 22] based on structural analyses, kinetic and FT-IR studies; it is in line with a Langmuir–Hinshelwood mechanism. A reaction between CO adsorbed on the Au NPs and O_2 activated on the support occurs at the junction perimeter between Au and the metal-oxide supports. In this case, hemispherical Au NPs are much more active than spherical NPs (Figure 3.4), owing to the longer junction perimeter around the NPs.

The rate of CO oxidation over Au/TiO_2, Au/Fe_2O_3 and Au/Co_3O_4 is almost independent of the concentration of CO and is slightly dependent (~0.25 order) on the concentration of O_2 down to a concentration of 0.1 vol.% [42]. For unsupported

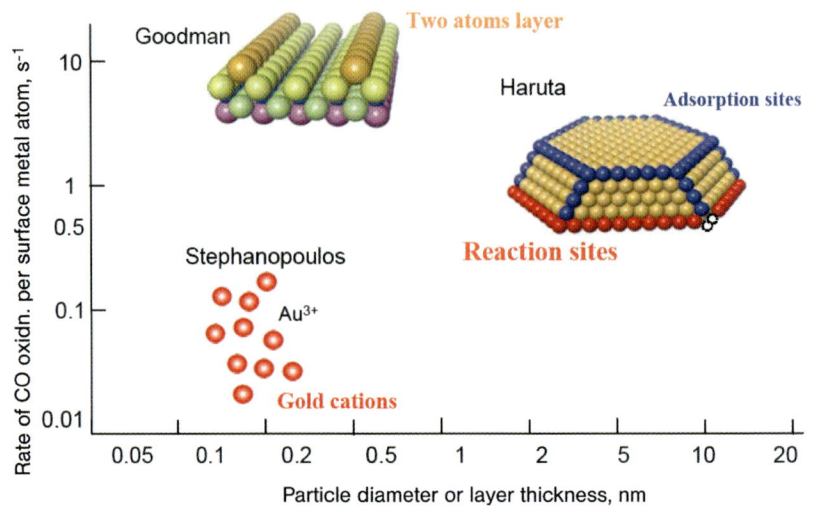

Figure 3.8 Hypotheses for the active states of supported Au catalysts for CO oxidation.

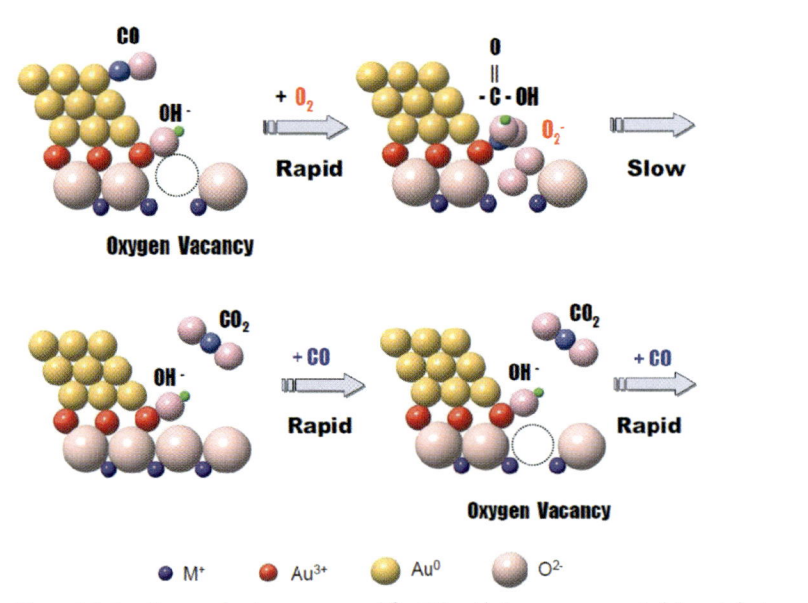

Figure 3.9 Reaction mechanisms proposed for CO oxidation over supported Au catalysts.

Au powder with a mean particle diameter of 17 nm, the rate is almost independent of the concentration of O_2 as well as CO [44]. This independence suggests that CO and O_2 are adsorbed nearly to saturation on the catalyst surfaces and that the reaction of the two adsorbed species is the rate-determining step.

In FT-IR spectra for CO adsorption at 90 K over Au/TiO_2 calcined in air at different temperatures [80], the most active sample (calcined at 573 K, with Au NPs having a mean diameter of 2.4 nm) exhibits the largest peak intensities, at around 2100 cm^{-1}. This can be attributed to the linear adsorption of CO on the metallic Au sites [81]. When the diameter of the Au particles becomes greater than 10 nm (sample calcined at 873 K), the intensity of the peak is markedly reduced, indicating that CO adsorption might only occur on the steps, edges and corners of the Au NPs and not on the smooth surfaces. This agrees well with what has been discussed previously based on the self-consistent density functional theory calculations by Mavrikakis et al. [45].

Gold is indeed active in CO oxidation at much lower temperatures than any Pt group metals. The difference in reactivity may be because Rh, Pd and Pt easily dissociate molecular oxygen at low temperature and bind strongly both atomic oxygen and CO [79]. Conversely, on Au the reactants are loosely bound, especially O_2, which is in favor of CO oxidation at low temperature. As a result, many studied have been performed to determine the structure of adsorbed oxygen species and how oxygen is activated on the surface of Au catalysts. A TAP (temporal analysis of products) study of O_2 adsorption and the reaction of O_2 with CO [82, 83], $^{18}O_2$ isotope experiments [82–84] and ESR measurements [84, 85] indicate that molecularly adsorbed O_2, most likely O_2^- at the perimeter interface, is involved in the oxidation of CO. An FT-IR study on the introduction of $C^{16}O$ at 300 K to Au/TiO_2 preadsorbed with $^{18}O_2$ [80] showed that $C^{16}O_2$ is formed in a quantity comparable to that of $C^{16}O^{18}O$, meaning that the oxygen species (^{16}O) contained in the surface layer of the TiO_2 support is also involved in CO oxidation at room temperature.

Recently, using Raman spectra (Figure 3.10) and TPR techniques Corma et al. have found that η^1-superoxide and peroxide species may possibly be located at one-electron defect sites on the surface of nanocrystalline CeO_{2-x}. The formation

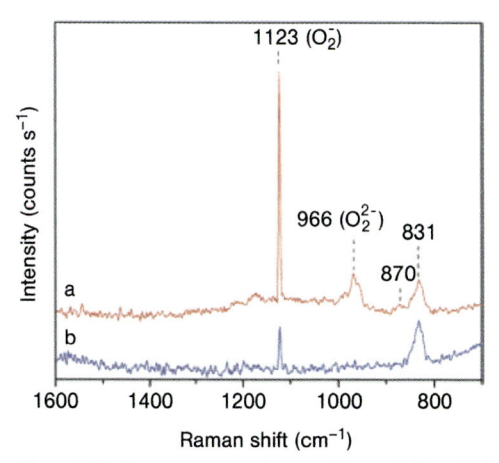

Figure 3.10 Raman spectra characterizing a 1.92 wt% Au/CeO_2 sample (a) at the beginning and (b) at the end of a CO oxidation reaction [86].

of reactive oxygen species on the nanocrystalline support is enhanced by the Au NPs. Moreover, a correlation is observed between the catalytic activity of Au supported on nanocrystalline CeO_{2-x} and the amount of η^1-superoxide and peroxide species. Based on these results, a mechanism involving reaction between CO adsorbed on Au NPs and reactive oxygen species on nanocrystalline support was proposed [86, 87].

By density functional theory (DFT) calculations, Molina and Hammer [23, 24] have studied CO oxidation over model Au/MgO(100) and Au/TiO$_2$(110) catalysts with different types of Au–support interface boundary. They proposed that the hemispherical type of interface boundary is likely to be predominant for medium-sized NPs by providing the optimal degree of low-coordinated Au atoms in the neighborhood of the support. Their result therefore provides a rationale for why the reactivity per site may reach a maximum at a critical particle size, as has been observed experimentally for similar systems. For all Au systems with different types of Au–support interface boundary, they assumed that CO oxidation reaction proceeds via CO adsorption and trapping of O_2, leading to the formation of a metastable CO·O_2 reaction intermediate, which dissociates into CO_2 and adsorbed atomic oxygen. The atomic oxygen may react directly with gas-phase CO. No separate O_2 molecular or dissociative adsorption is found to be favorable. Important differences were found in the reactivity of the various Au-MgO interface boundaries. This is explained in terms of two properties: the Au-Au coordination determining the local reactivity of the Au atoms and the presence of the MgO support that, besides providing excess electrons to the Au clusters, forms ionic bonds to the peroxo part of the CO·O_2 reaction intermediate.

Bond and Thompson [19] also proposed that CO reaction takes place at the junction perimeter of metallic gold (Au^0), oxidized gold (Au^{III}) and the support. The CO adsorbed on Au^0 is attacked by an hydroxyl group either on a support cation or on a peripheral Au^{III} ion, forming a carboxylate group attached to the latter. This is in turn attacked by a superoxide ion (O_2^-), which must be responsible for oxidizing two carboxylate ions: the hydroxyl group returns whence it came and is ready to re-engage in the catalytic cycle. The hypothesis of the presence of Au^{III} species arises from the fact that, for other noble metals, cationic species are located at the metal–support interface as a "chemical glue" to bind the particle to the support.

3.2.3.2 Mechanisms Involving Specific Size or Thickness of Gold Clusters or Thin Layers

Goodman and coworkers reported that, in model catalysts prepared by depositing Au onto TiO$_2$ monolayers grown on Mo single-crystal substrates or onto TiO$_2$(110) single crystals, two atomic gold layers presented especially high catalytic activity in terms of TOF and that a specific thickness of nanoparticles or thin layers was critical for CO oxidation on Au catalysts [26, 27, 88, 89]. Their hypothesis differs from the junction perimeter hypothesis in that it does not take into account the direct involvement of support materials in the catalytic functions.

Figure 3.11 shows the rates of CO oxidation over model catalytic systems, Au/TiO$_2$(110) and well-ordered Au mono- and bilayer structures. The maximum

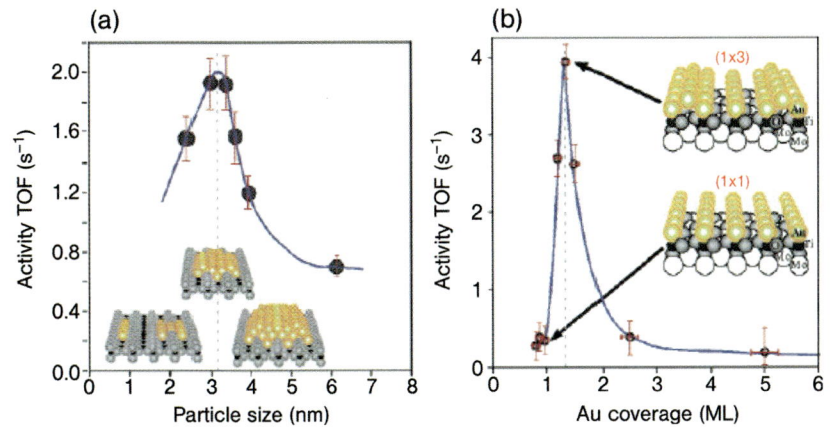

Figure 3.11 Catalytic activity for CO oxidation as a function of (a) particle size on TiO$_2$(110) at 353 K [88] and (b) Au coverage on the Mo-(112)-(8 × 2)-TiOx at room temperature [89].

reaction rate occurs at a Au coverage of ∼1.3 monolayers, where the particle morphology corresponds to the Au bilayer structure. The rate, a TOF of 4, is higher than any previously reported rate for supported Au NPs. These studies show that ultrathin Au films on a reduced titania surface exhibit catalytic activity comparable to or higher than the most active Au NPs supported on reducible metal oxides, that is, the thickness of the particle rather than the particle diameter is the critical structural feature with respect to catalytic activity.

Moreover, in the ordered Au monolayer and bilayer structures described above, the Ti^{4+} of the support titania is not accessible to the reactants, since each surface Ti site binds directly to a Au atom located at the topmost surface. The high catalytic activities for CO oxidation observed on ordered bilayer Au thus strongly suggest a Au-only CO oxidation pathway. The electronic nature of very small Au NPs and thin layers can be assumed to be significantly influenced by the nature and direct involvement of the TiO$_2$ support and the Mo metal substrate, especially the availability of defect sites [26, 27, 88, 89].

Alkaline earth metal hydroxides, Be(OH)$_2$ and Mg(OH)$_2$, require a specific size and structure of Au clusters [40]. Only 13-Au atom clusters with an icosahedral structure can lead to high catalytic activity, at a temperature as low as 203 K, even though it has been reported that the icosahedron is not the ground state structure of Au [90, 91]. As for single crystal of MgO supported Au anion cluster, it was reported, that there was appreciable size dependency of CO adsorption and that the highest reactivity to CO was observed for anion cluster consisting of 11 atoms [47]. It was also reported that 8 and 11 are the smallest and the second smallest number of atoms to give the catalytic activity for CO oxidation over MgO support. DFT studies performed by Molina and Hammer [23, 24] also give an evidence for NPs reaching a maximum reactivity per site for CO oxidation at a critical particle size, by providing the optimal degree of low-coordinated Au atoms in the neighborhood of the support.

3.2.3.3 Mechanisms Involving Cationic Gold

Most papers propose that only reduced gold (Au^0) is responsible for CO oxidation [21, 22, 26, 92]. However, some papers propose that oxidized Au is mainly responsible for CO oxidation. Hutchings and coworkers proposed that Au^{3+} in Au/Fe_2O_3 prepared by co-precipitation was an important component of active catalysts for the oxidation of CO [93, 94]. Notably, the catalyst was prepared by adding alkaline solution into acidic $HAuCl_4$ and $Fe(NO_3)_2$ aqueous mixtures. This procedure for co-precipitation was opposite to that employed by Haruta [15] and might produce heterogeneous microstructure because of precipitation in a wide range of pH from 4 to 9.

The results of X-ray photoelectron spectroscopy (XPS) and X-ray absorption fine structure (XAFS) proved that the oxidation state of Au was important for low temperature CO oxidation over Au catalysts supported on Fe_2O_3, TiO_2 and Al_2O_3. It was proposed that oxidized Au is more active than metallic Au; however, this conclusion was not directly evidenced [95].

Concerning Au/MgO prepared by gas-phase grafting using dimethyl Au(III) acetylacetonate [96], the TOF of which is lower than these reported for supported metallic Au NPs, EXAFS and XANES were used to identify the oxidation states of Au in the functioning catalysts. Temperature-programmed reduction (TPR) and temperature-programmed oxidation (TPO) were also performed to characterize the catalysts as they had been working at steady state in a flow reactor to characterize quantitatively the Au oxidation states. It was shown that both Au(I) and Au(0) are present in the working catalysts, and their relative amounts depend on the composition of the reacting atmosphere. The catalytic sites are assumed to be Au(I). Carbon monoxide plays a dual role as a reactant and a reducing agent that converts Au(I) into Au(0), thereby diminishing the catalytic activity.

Recently, an IR spectroscopic characterization of CO adsorption on Au supported on nanocrystalline CeO_2 was performed by Corma et al. [67, 86]. The presence of Au^{3+}, Au^+ and Au^0 species was indicated by their characteristic Au^x-CO frequency. Furthermore, it was proposed that there was a direct correlation between the concentration of Au^{3+} species and catalytic activity for CO oxidation over nanocrystalline CeO_2 supported Au. No correlation was found between catalytic activity and the concentration of Au^+ or Au^0. These results can also be interpreted as showing that the junction perimeter interface between Au NPs and the CeO_2 support is composed of Au^{3+} and acts as the reaction site, the length of which defines the catalytic activity.

3.3
Complete Oxidation of Volatile Organic Compounds

Volatile organic compounds (VOCs) are emitted from chemical plants, vehicles, electroplating, spray-painting, laundries and so forth. They are not only hazardous to human health and malodorous but also cause the formation of ozone and smog [97, 98]. Catalytic complete oxidation of VOCs is one of the most promising technologies for the abatement of VOCs emission because it can oxidize VOCs to CO_2 at much lower temperatures than thermal oxidation [99]. For the complete oxidation of hydrocarbons, in principle, Pt group metals are more active than Au. However, since highly dispersed Au NPs on base metal oxides exhibit surprisingly high catalytic activity for CO oxidation at low temperature [14, 42], and the catalytic activity can be remarkably tuned by the selection of the support metal oxides, a few attempts have been made to develop active Au catalysts for the complete oxidation of VOCs.

Formaldehyde (HCHO) is regarded as a major indoor pollutant emitted from the polymer-based floor, wall and their bonding materials in airtight buildings, and as a result it is one of the dominant VOCs [100–103]. Long-term exposure to indoor air containing even a few ppm of HCHO may cause adverse effects on human health [104]. Accordingly, great efforts have been made to reduce the indoor concentrations of HCHO to meet stringent environmental regulations.

Table 3.2 shows the catalytic activities of supported Au catalysts for the complete oxidation of HCHO, C_3H_6, C_3H_8 and CH_4. For complete oxidation of HCHO, Au/CeO_2 [105, 106], Au/FeO_x [107, 108] and Au/TiO_2 [102] have been tested so far. Jia et al. prepared different kinds of Au catalysts by the deposition-precipitation method [105]. In HCHO oxidation 2 wt% Au/CeO_2 gives higher activity than Au/CeO_2-TiO_2 and Au/TiO_2. A HCHO conversion of 100% was obtained at 353 K over Au/CeO_2. More recently, they optimized Au loading and calcination temperature and found that 0.85 wt% Au/CeO_2 calcined at 573 K gave 100% HCHO conversion at 376 K [106]. Taking into consideration the results obtained by HRTEM, XRD and XPS, it is proposed that the catalytic performance of Au/CeO_2 is closely related to the crystal structure of Au in the catalysts. The highly dispersed and poorly crystallized metallic gold and small amount of oxidized gold in the catalysts exhibit superior catalytic activity for HCHO oxidation. A series of Au/FeO_x catalysts were prepared by Zhu and coworker [108]. The catalyst containing 7.1 wt% of Au exhibited the highest catalytic activity and gave 100% HCHO conversion at 353 K. As identified by XPS, Au atoms with fractional positive charge ($Au^{\delta+}$) were found to exist in the catalyst and play an important role.

A comparative study of TiO_2-supported noble metal catalysts, which were prepared by impregnation, was carried out for the oxidation of a low concentration of HCHO (100 ppm) by Zhang and He [102]. As far as impregnation method is concerned, Au/TiO_2 is less active than Pt/TiO_2, giving 90% HCHO conversion at 393 K and 100% conversion at room temperature, respectively.

As for the complete oxidation of propene, propane and methane, Nieuwenhuys and coworkers studied the influence of metal oxides additives on the catalytic activity of Au/Al_2O_3 [109–115]. The addition of 3d transition metal oxides (MnO_x, CoO_x or FeO_x), which were active by themselves, or ceria that was poorly active by itself promoted the catalytic activity of Au/Al_2O_3 in the total oxidation of propene [112]. The most active catalyst was Au/CeO_x/Al_2O_3, with a T_{95} at 497 K and with a high stability. In these cases, ceria and the transition metal oxides may act as co-catalysts and the role is twofold: it stabilizes the Au NPs against sintering (ceria)

Table 3.2 Performance of supported Au catalysts for the oxidation of VOCs and hydrocarbons.

Catalyst	Cal (K)	Au (wt%)	d_{Au} (nm)[a]	Feed gas VOC	O_2(%)	Bal	T_{100} (K)	SV (h^{-1})	Reference
Au/TiO$_2$	673	1	<1	HCHO 100 ppm	20	He	393 (T_{90})	5 000	[102]
Pt/TiO$_2$	673	1 (Pt)	<1	HCHO 100 ppm	20	He	293	5 000	[102]
Au/FeOx	473	7.1	10–15[b]	HCHO 6.25 mg m^{-3}	21	N$_2$	353	54 000	[108]
Au/α-Fe$_2$O$_3$	673	3.5	3.6 ± 0.7	HCHO 0.5%	20	N$_2$	373 (T_{50})	20 000	[107]
Au/CeO$_2$	473	2	—	HCHO[c]	21	N$_2$	353	—	[105]
Au/CeO$_2$	573	0.85	—	HCHO 600 ppm	21	N$_2$	376	32 000	[106]
Au/CuO/Al$_2$O$_3$	623	7.4	6.8 ± 0.2	C$_3$H$_6$/O$_2$ = 1:9		He	<448 (T_{50})	9 000[d]	[114]
Au/CuO/Al$_2$O$_3$	623	4	3.0 ± 0.1	C$_3$H$_6$/O$_2$ = 1:9		He	528 (T_{50})	9 000[d]	[114]
Au/BaO/Al$_2$O$_3$	623	3.6 ± 0.2	1.5 ± 0.2[b]	C$_3$H$_6$/O$_2$ = 1:9		—	563 (T_{50})	9 000[d]	[113]
Au/CeOx/Al$_2$O$_3$	573	4.5	<3.0	C$_3$H$_6$ 0.4%	3.6%		497 (T_{95})	9 000[d]	[112]
Au/MnOx/Al$_2$O$_3$	573	4.3 ± 0.2	8.0 ± 0.1	C$_3$H$_8$/air = 0.5 : 99.5		He	598 (T_{50})	1800	[115]
Au/CoOx	673	3	N.d[e]	C$_3$H$_8$/air = 0.5 : 99.5		He	438 (T_{50})	15 000	[116]
Au/CuO/Al$_2$O$_3$	623	4	3.0 ± 0.1	C$_3$H$_8$/O$_2$ = 1:9		He	~613 (T_{50})	9 000[d]	[114]
Au/CuO/Al$_2$O$_3$	623	7.4	6.8 ± 0.2	C$_3$H$_6$/O$_2$ = 1:9		He	~653 (T_{50})	9 000[d]	[114, 115]
Au/CoOx	673	3	N.d[e]	C$_2$H$_4$/air = 0.5 : 99.5			~463 (T_{50})	15 000	[116]
Au/CoOx	673	3	N.d[e]	CH$_4$/air = 0.5 : 99.5			~553 (T_{50})	15 000	[116]
Au/FeOx/Al$_2$O$_3$	573	4.2 ± 0.2	<3.0	CH$_4$/O$_2$ = 1:4		He	816 (T_{50})	1800	[115]
Au/TiO$_2$	773	2.1(3.29)[b,f]	<4	Benzene: 250 ppm		Air	~608 (T_{50})	—	[118]
Au/CeO$_2$-HS[h]	573	1.5	4.5	Benzene: 0.5%		Air	480 (T_{50})	4.0[g]	[119]
Au/TiO$_x$N$_y$	573	1.68	>8	2-Propanol: 500 ppmv; O$_2$: 21%		He	~450 (T_{50})[i]	60 000[d]	[118]
Au/CeO$_2$-monolith	373	0.06	N.d[e]	2-Propanol: 12 000 ppm; O$_2$: 21%		He	470 (T_{50})[i]	9700	[120]
								175	

[a] Mean diameter of Au particles, as determined by XRD.
[b] Mean diameter of Au particles, determined by TEM.
[c] Air was passed through HCHO liquid at 0 °C.
[d] mL h^{-1} g-cat.$^{-1}$.
[e] Not determined.
[f] Bulk and surface Au content, respectively.
[g] g-cat min mL^{-1} VOC NPT.
[h] High surface area (HS).
[i] Acetone is major product.
[j] Based on CO$_2$ formation.

and provides active oxygen for the reaction. The addition of alkali (earth) metal oxides (MOx, with M = Li, Rb, Mg or Ba) also enhances the performance of Au/Al_2O_3 catalysts for propene oxidation [113]. The best effect is obtained when BaO is used. It is found that MOx induces a decrease in the size of Au NPs and stabilizes them against sintering. In turn, a direct dependence of the catalytic performance on the size of Au particles is found. The role of the additives is that of a structural promoter and not a chemical promoter, which is different to the addition of transition metal oxide or ceria.

Nieuwenhuys and coworkers [115] have also demonstrated that the addition of various MOx [M: alkali (earth), transition metal and cerium] to Au/Al_2O_3 improves the catalytic activity in both propane and methane oxidation. As expected, higher temperatures are required to oxidize CH_4 (above 673 K), which is a more stable hydrocarbon than C_3H_8 (above 523 K). Figure 3.12 shows that for methane oxidation the most efficient promoters for Au/Al_2O_3 are FeOx and MnOx, whereas for C_3H_8 oxidation MnOx and CoOx are the most efficient promoters for Au/Al_2O_3. The effect of Au particle size becomes less important for additives of transition metal oxides and ceria. The results suggest that the role of the alkali (earth) metal oxides is related to the stabilization of the Au NPs, whereas transition metal oxide and ceria additives may be involved in oxygen activation.

In the oxidation of C_3H_6 over Au/CuO/Al_2O_3 catalysts, a high Au loading of 7.4 wt% gives a half conversion for C_3H_6 at below 448 K, showing higher activity than that of 4 wt% (with T_{50} at ∼613 K). In contrast, in C_3H_8 oxidation, a lower loading of 4 wt% gives higher activity, indicating that the particle size effect prevails over the effect of Au loading. The authors also proposed that the different order in catalysis in C_3H_6 and C_3H_8 oxidation is not yet understand [114].

Carley et al. have studied intensively the total oxidation of methane, ethane and propane over Au catalysts [116]. Different kinds of metal oxides, such as TiO_2, MnOx, Fe_2O_3, CoOx, CuO and CeO_2, were used as supports. They were active for alkane activation but at relatively high temperatures (e.g., MnOx and CoOx). The addition of Au to these active metal oxides led to a marked increase in catalytic activity for the alkane combustion in all cases investigated. The most effective catalyst for alkane oxidation was Au/CoOx prepared by co-precipitation, which gave T_{50}s at 438, ∼463 and ∼533 K for propane, ethane and methane conversion, respectively. The correlation between alkane oxidation and CO oxidation was also investigated. Figure 3.13 shows a plot of CO conversion at room temperature against T_{10} calculated for propane oxidation over the same catalyst. For CO oxidation the only active catalysts at room temperature are those with Au co-precipitated and, interestingly, there is no correlation between the catalytic activity for CO oxidation and the catalytic activity for propane combustion. This fact indicates a difference in the active sites for CO oxidation and propane combustion or in the working mechanisms. It is most likely that the redox properties of the support materials and, mainly, Au define the catalytic activity for propane combustion, while the mean diameter of Au particles mainly defines the activity for CO oxidation.

To study the possibility of application in the reduction of cold start emission, Pitchon et al. tested the reaction of total oxidation of a mixture of light hydrocarbons

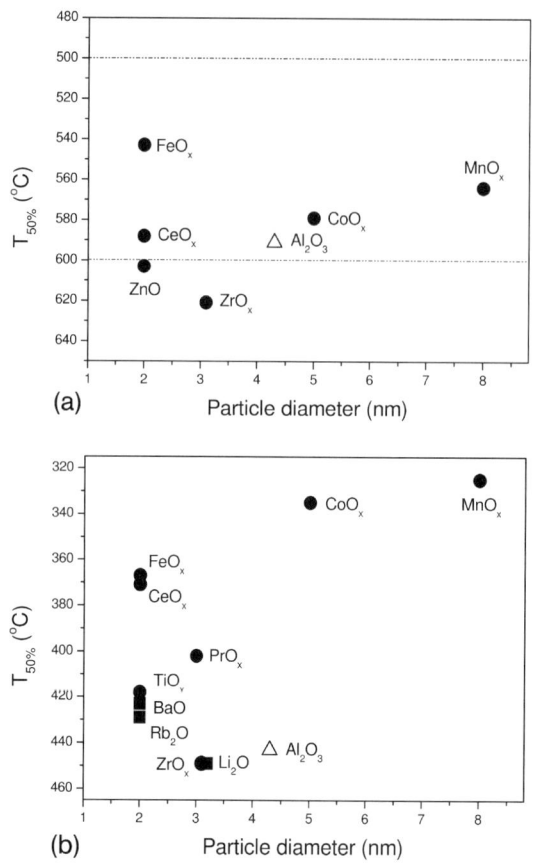

Figure 3.12 Temperature for 50% conversion (oxidation) for methane (a) and propane (b) over supported Au catalysts as a function of the mean diameter of Au particles measured by XRD. Au particles below 3.0 nm are positioned at 2.0 nm. Reaction conditions: $CH_4/O_2 = 1:4$ (or $C_3H_8/O_2 = 1:16$); all the gases were 4 vol.%; the total flow was 30 mL min^{-1} (GHSV ~ 1800 h^{-1}) [115].

and CO over Au/Al$_2$O$_3$ [117]. It was found that CO oxidation is surprisingly inhibited by the presence of acetylene (C$_2$H$_2$) but the presence of alkenes and alkanes such as C$_2$H$_4$ and C$_2$H$_6$ hardly changes CO oxidation. The adsorption of C$_2$H$_2$ is much stronger than that of CO. Both C$_2$H$_4$ and C$_2$H$_6$ adsorb moderately on the surface of Au NPs, not inhibiting the adsorption of CO.

In addition, complete oxidation of benzene [118, 119] and 2-propanol [118, 120] over Au catalysts was reported. In this case, Au/CeOx was often used and showed high activity [119, 120], indicating that the redox properties of ceria phase may play an important role in the formation of surface oxygen species, which results in the complete oxidation of benzene and 2-propanol.

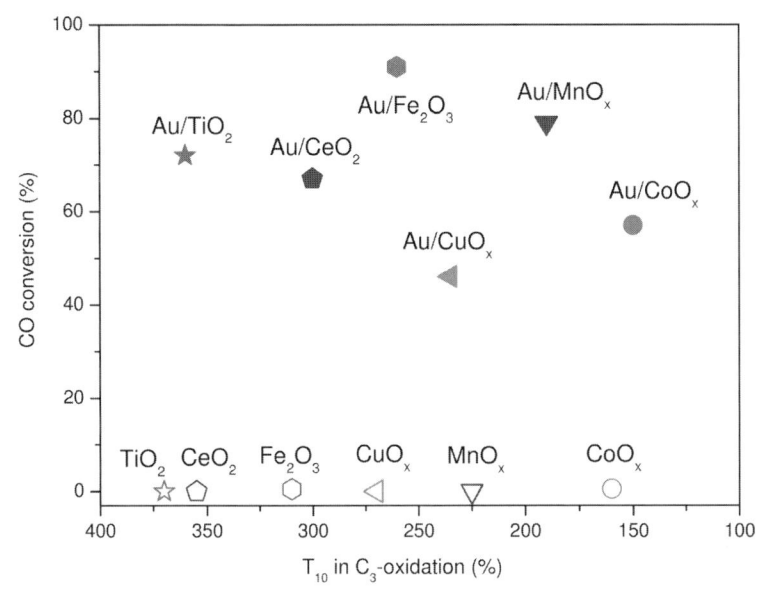

Figure 3.13 Relationship between conversion in CO oxidation at room temperature and the temperature for 10% conversion (T_{10}) in propane oxidation. Symbols: Supports (open symbols), gold catalyst prepared by co-precipitation (filled symbols). Reaction conditions for CO oxidation: 0.5% CO in air, 50 mg of catalyst, 22.5 mL min^{-1}, reaction temperature = 298 K and reaction time = 2h. Reaction conditions for C_3H_8 oxidation: 0.5% C_3H_8 in air, 50 mg of catalyst and 50 mL min^{-1} [116].

3.4
Gas-Phase Selective Oxidation of Organic Compounds

Because the petrochemical industry is based on hydrocarbons, especially alkenes, the selective oxidation of hydrocarbons to produce organic oxygenates occupies about 20% of total sales of current chemical industries. This is the second largest market after polymerization, which occupies about a 45% share. Selectively oxidized products, such as epoxides, ketones, aldehydes, alcohols and acids, are widely used to produce plastics, detergents, paints, cosmetics, and so on. Since it was found that supported Au catalysts can effectively catalyze gas-phase propylene epoxidation [121], the catalytic performance of Au catalysts in various selective oxidation reactions has been investigated extensively. In this section we focus mainly on the gas-phase selective oxidation of organic compounds.

3.4.1
Gas-Phase Selective Oxidation of Aliphatic Alkanes

Gas-phase selective oxidation of aliphatic alkanes with a $O_2 + H_2$ mixture over supported Au catalysts has been reported by Haruta [122], and related results are

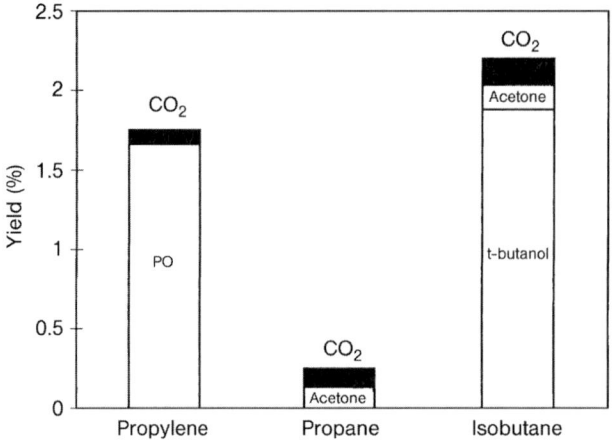

Figure 3.14 Yield of products in the oxidation of propylene at 373 K, propane at 393 K and isobutane at 393 K on 1.20 wt% Au/Ti- MCM-41(Ti/Si = 2.8 : 100) in the presence of oxygen and hydrogen. Reactant gas, C_3H_6 (or alkane)/O_2/H_2/Ar = 1 : 1 : 1 : 7; space velocity, 4000 mL h^{-1} g$_{cat}^{-1}$ [122].

shown in Figure 3.14. At 393 K, propane conversion and isobutane conversion are 0.3% and 2.2% and the selectivities to acetone and t-butanol are 48% and 85%, respectively. A steady STY (space–time yield) for t-butanol of 23 g-t-butanol kg-cat.$^{-1}$ h^{-1} on Au/Ti-MCM-41 was obtained.

3.4.2
Gas-Phase Selective Oxidation of Alcohols

Oxidation of alcohols to the corresponding aldehydes and ketones is very important in organic synthesis. However, less attention has been paid to the gas-phase reaction. Rossi first reported that on Au catalysts ethanol can be selectively oxidized in the gas phase to acetaldehyde [123]. Later, the same author found that primary and secondary aliphatic alcohols can also be selectively oxidized by air to the corresponding aldehydes and ketones on 1.0 wt% Au/SiO_2 [124]. As shown in Table 3.3, at 573 K, the conversions of 1-propanol, 1-butanol, 1-pentanol are 44%, 63% and 29%, respectively, and the selectivities to the corresponding aldehydes are 100%, 94% and 100%. At 423 K, the conversions of 2-propanol, 2-butanol and 2-pentanol are 100%, 64% and 85%, and the selectivities to the corresponding ketones are 100%, 100% and 84%, respectively. For 2-propanol, at 423 K, a very high steady STY of 150 mol-acetone kg-cat.$^{-1}$ h^{-1} is achieved. Therefore, 1.0 wt% Au/SiO_2 is an innovative catalyst suitable for gas-phase selective oxidation of primary and secondary alcohols under mild conditions.

Table 3.3 Gas-phase selective oxidation of primary and secondary aliphatic alcohols by air to the corresponding aldehydes or ketones on Au catalysts [124].

Entry	Reagent	T (K)	Conversion (%)	Selectivity (%)	STY (mol kg-cat.$^{-1}$ h^{-1})[a]
1[b]	1-Propanol	523	27	100[c]	40.5[c]
		573	44	100[c]	66.0[c]
2[b]	1-Butanol	523	51	100[c]	76.5[c]
		573	63	94[c]	88.8[c]
3[b]	1-Pentanol	523	23	100[c]	34.5[c]
		573	29	100[c]	43.5[c]
4[b]	Phenylcarbinol	523	50	100[c]	75.0[c]
		553	75	98[c]	110[c]
5[b]	Prop-2-en-1-ol	523	42	97[c]	61.1[c]
6[b]	2-Propanol	373	69	100[c]	104[c]
		423	100	100[c]	150[c]
7[b]	2-Butanol	393	34	100[c]	51.0[c]
		423	64	100[c]	96.0[c]
8[b]	2-Pentanol	393	72	87[c]	94.0[c]
		423	85	84[c]	107[c]
9[b]	3-Pentanol	393	85	100[c]	128[c]
		423	97	100[c]	146[c]
10[b]	1-Propanal	523	14	4[d]	0.84[d]
		573	16	3[d]	0.72[d]
11[e]	1-Propanol	523	20	4[b]	1.20[c]
12[e]	2-Propanol	413	4	0[b]	0[c]

Reaction conditions: catalyst of 0.2 g; air stream of 1.2 mmol min^{-1}; liquid reagent of 0.5 mmol min^{-1}. Liquid reagent was supplied through a syringe pump, and vaporized on the reactor wall prior to the reaction.
[a] STY, space–time yield (mol-product kg-cat.$^{-1}$ h^{-1}).
[b] Catalyst, 1.0 wt% Au/SiO$_2$.
[c] To carbonyl derivates.
[d] To propanoic acid.
[e] Catalyst, SiO$_2$.

3.4.3
Gas-Phase Propylene Epoxidation

3.4.3.1 Introduction

Propylene oxide (PO) is a major bulk chemical that is useful as an intermediate for the production of glycols, polyethers and polyols. World PO production is about 7 million tons per year, with the market growing by approximately 5% annually. PO comes from two current industrial processes: the chlorohydrin process and the organic hydroperoxide process (Scheme 3.1). These processes require multiple steps and suffer from the additional drawback of not producing the desired PO alone.

Scheme 3.2 shows recently established industrial processes for PO production. The cumene recycling process uses an organic hydroperoxide process combined with hydrogenation of cumyl alcohol to cumene. This process consumes only hydrogen

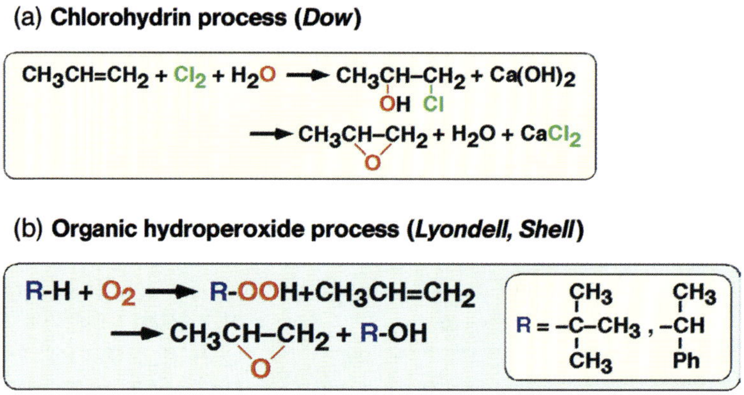

(a) Chlorohydrin process (*Dow*)

$$CH_3CH=CH_2 + Cl_2 + H_2O \longrightarrow CH_3\underset{OH}{C}H-\underset{Cl}{C}H_2 + Ca(OH)_2$$
$$\longrightarrow CH_3CH-CH_2 + H_2O + CaCl_2 \quad (\text{epoxide})$$

(b) Organic hydroperoxide process (*Lyondell, Shell*)

$$R-H + O_2 \longrightarrow R-OOH + CH_3CH=CH_2$$
$$\longrightarrow CH_3CH-CH_2 + R-OH \quad (\text{epoxide})$$

$$R = -\underset{CH_3}{\overset{CH_3}{\underset{|}{C}}}-CH_3, \; -\underset{Ph}{\overset{CH_3}{\underset{|}{C}}}H$$

Scheme 3.1 The two main industrial processes for propylene epoxidation production.

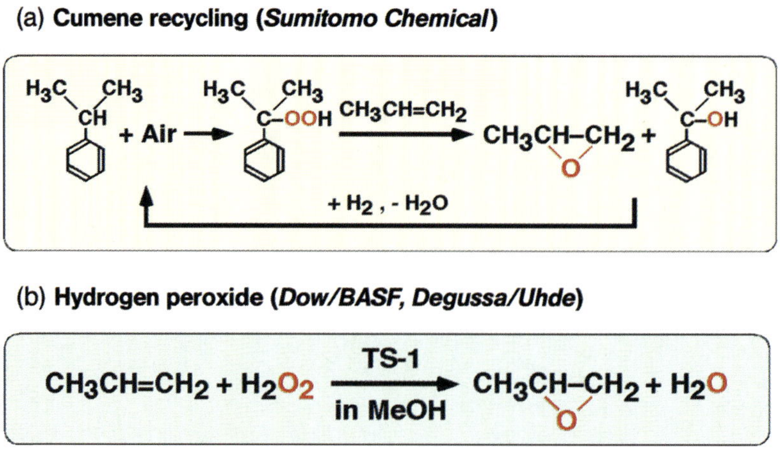

(a) Cumene recycling (*Sumitomo Chemical*)

cumene + Air → cumene hydroperoxide + $CH_3CH=CH_2$ → CH_3CH-CH_2 (epoxide) + cumyl alcohol
+ H_2, − H_2O (recycle)

(b) Hydrogen peroxide (*Dow/BASF, Degussa/Uhde*)

$$CH_3CH=CH_2 + H_2O_2 \xrightarrow[\text{in MeOH}]{\text{TS-1}} CH_3CH-CH_2 + H_2O \quad (\text{epoxide})$$

Scheme 3.2 Recently established industrial processes for PO.

and oxygen, and no co-product other than water is formed. For hydrogen peroxide process, propylene epoxidation on titanosilicalite (TS-1) is achieved by utilizing hydrogen peroxide as the oxidant. This process is more environmentally benign than the chlorohydrin and organic hydroperoxide processes; however, H_2O_2 is usually expensive. Therefore, the deficiencies of the above processes have spurred research into the production of PO through the epoxidation of propylene with oxygen (O_2) alone or with O_2 in the presence of reductive H_2.

The direct epoxidation of propylene is one of the top targets in industrial chemistry. Although the epoxidation of ethylene with O_2 is a commercial process that operates with selectivity of around 90% by using a supported Ag catalyst [125], the epoxidation of propylene with O_2 over many catalysts, such as Ag- and Cu-based catalysts, has not yet been successful [126–128]. Even at a very low conversion of propylene, PO selectivity is less than 60%. Recently, Hutchings and colleagues showed that, on Au NPs, liquid-phase selective oxidation of cyclohexene, styrene, cis-stilbene and cyclooctene with molecular oxygen can be achieved with a high selectivity to the corresponding epoxides in the presence of a catalytic amount of peroxides [2]. This may provide us with a new route for gas-phase alkene epoxidation by utilizing Au NPs to activate molecular oxygen.

3.4.3.2 Gas-Phase Propylene Epoxidation with Hydrogen–Oxygen Mixtures on Au/TiO₂

While direct activation of O_2 is very difficult, the co-presence of H_2 can activate O_2 at relatively mild conditions with an energy input of less than $10\,kJ\,mol^{-1}$. This leads to a feasible alternative method of controlling the reactivity of oxygen species so as to produce valuable oxygenated organic compounds [3]. Hayashi and Haruta first found that, over finely dispersed Au deposited on TiO_2, gas-phase epoxidation of propylene by O_2 can be achieved in the presence of reductive H_2 below 373 K, with selectivity to PO above 90% [121]. This finding spurred research into production of PO through the epoxidation of propylene with H_2–O_2 mixtures. Notably, in this reaction system, a Ti-based support is indispensable for obtaining high selectivity to PO.

Effect of preparation methods The preparation methods are crucial to the catalytic performance of Au/TiO₂ [121, 129, 130]. As shown in Figure 3.15, the impregnation method produces large spherical Au particles simply loaded on TiO_2, resulting in the combustion of C_3H_6 and H_2 to a large amount of H_2O and a small amount of CO_2 at a relatively high temperature. The Deposition-precipitation (DP) method produces hemispherical Au particles attached to the TiO_2 surfaces at their basal plane and forms the most efficient catalysts for PO production. Interestingly, among various metal oxide supports, TiO_2 with the anatase structure is effective but both rutile and amorphous titania are not.

Effect of the size of Au nanoparticles In many reactions catalyzed by Au catalysts, the size of Au NPs markedly affects the catalytic activities and product selectivities [42, 131]. Figure 3.16 shows that a diameter of 2–5 nm is optimum for the production of PO whereas smaller Au clusters, below 2 nm, produce almost exclusively

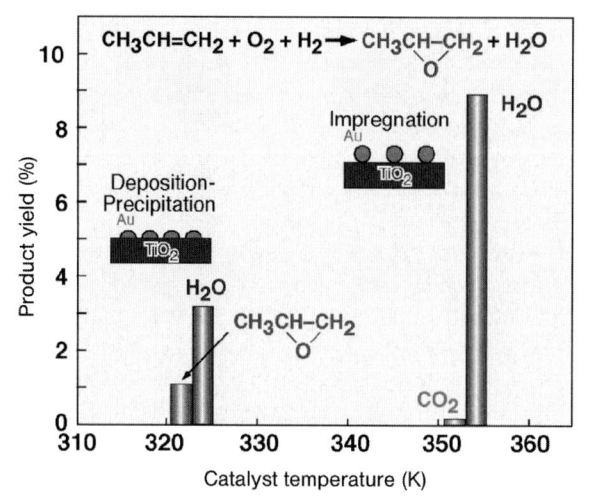

Figure 3.15 Product yields in the reaction of propylene over Au/TiO$_2$ catalysts prepared by different methods: catalyst, 0.5 g; feed gas, C$_3$H$_6$/O$_2$/H$_2$/Ar = 10 : 10 : 10 : 70; space velocity, 4000 h^{-1} mL g$_{cat}^{-1}$ [121].

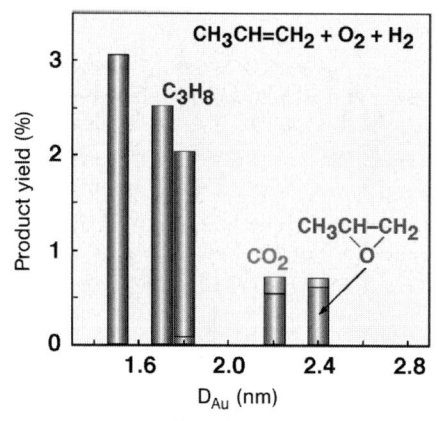

Figure 3.16 Size effect of Au nanoparticles on the reaction of propylene with O$_2$ and H$_2$; catalyst, Au/TiO$_2$ (P25) 0.5 g; catalyst bed temperature, 353 K; feed gas, C$_3$H$_6$/O$_2$/H$_2$/Ar = 10 : 10 : 10 : 70; space velocity, 4000 h^{-1} mL g$_{cat}^{-1}$ [121].

propane [21, 22, 121]. This phenomenon suggests that small Au clusters behave like Pd and Pt in the presence of O$_2$, namely Au clusters can dissociate H$_2$ molecules at low temperatures.

Reaction pathways for propylene epoxidation over Au/TiO$_2$ Sellers predicted that theoretically, on gold, H$_2$O$_2$ can be produced from O$_2$ and H$_2$ [132]. Later, Haruta and

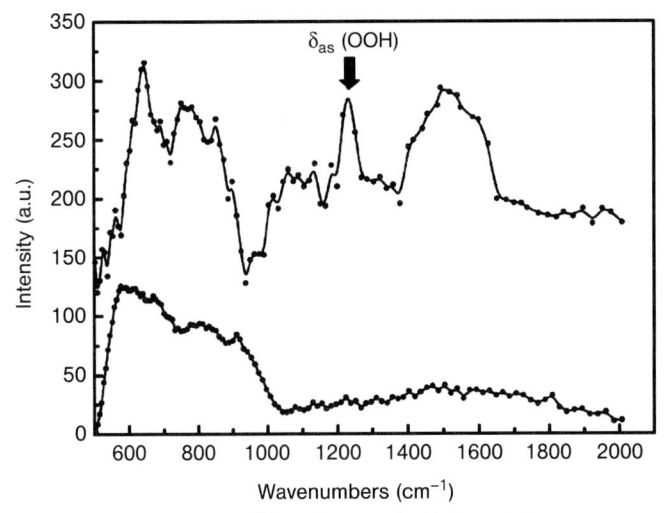

3.4 Gas-Phase Selective Oxidation of Organic Compounds | 103

Figure 3.17 INS spectrum of Au/TiO$_2$ reacted with H$_2$ and O$_2$ at 523 K for 4 h in flowing H$_2$/O$_2$/He (1:1:7) (top line). The INS spectrum of water at 523 K adsorbed on Au/TiO$_2$ is shown for comparison (bottom line) [135].

Hutchings found that on Au NPs H$_2$O$_2$ can be synthesized from H$_2$ and O$_2$ in an autoclave [133, 134]. Goodman and coworkers also reported inelastic neutron scattering (INS) evidence for the formation of OOH and H$_2$O$_2$ species from O$_2$ and H$_2$ on a Au/TiO$_2$ catalyst (Figure 3.17) [135]. Based on the above knowledge and detailed FT-IR spectroscopic investigations [136–138], Nijhuis proposed a possible reaction pathway of propylene epoxidation: (1) C$_3$H$_6$ is adsorbed at the perimeter interfaces between Au NPs and the TiO$_2$ to form bidentate propoxy species; (2) peroxide species are produced from H$_2$ and O$_2$ on Au NPs, which is the rate-determining step; and (3) with the assistance of the peroxide species, bidentate propoxy species desorb from the catalyst to produce propylene oxide.

3.4.3.3 Gas-Phase Propylene Epoxidation with Hydrogen–Oxygen Mixtures on Au/Ti-SiO$_2$

Although Au NPs deposited on TS-1 can catalyze gas-phase propylene epoxidation with oxygen in the presence of hydrogen [139–142], their catalytic performances have been inferior to Au supported on 3D mesoporous titanosilicates in terms of propylene conversion, PO selectivity and H$_2$ utilization efficiency. However, recently Delgass reported that a 0.081 wt% Au/TS-1 catalyst displayed a high propylene conversion of 10% with a PO selectivity of 76% [143]. A high PO STY of 134 g-PO kg-cat^{-1} h^{-1} was achieved, which is comparable to that of ethylene oxide in commercial plants.

Although Au particles with mean diameters around 5–6 nm were observed by transmission electron microscopy (TEM), Delgass suggested, based on DFT calculations, that small Au clusters such as Au$_3$ rather than the large particles are responsible

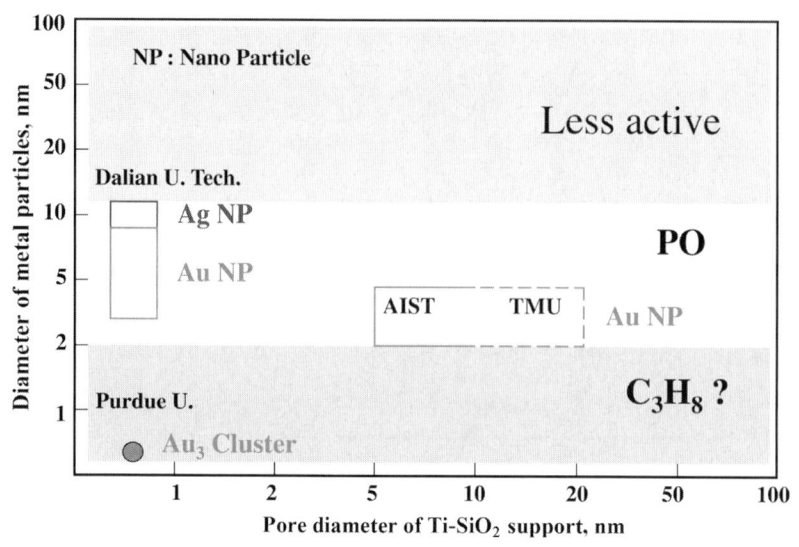

Figure 3.18 Different research strategies of Delgass's group (Purdue University) and Haruta's group (TMU) for propylene epoxidation. Also shown are the catalysts used by Guo's group (Dalian U. Tech.).

for the epoxidation of propylene. Figure 3.18 shows a sharp contrast in research strategy between Delgass's group and that of Haruta in terms of the size of Au particles and the pore diameter of Ti-SiO$_2$ supports. The figure also shows the position of Ag/TS-1 catalysts studied by Guo's group, who reported that Ag nanoparticles (around 8 nm) deposited on TS-1 were active in propylene epoxidation with H$_2$–O$_2$ mixtures, with a propylene conversion of 1.4% and a PO selectivity of 93.5% [144].

Effect of pore diameters of titanium silicate supports Titanium silicate support materials with different pore diameters (TS-1, Ti-MCM-41/48, 3D-Ti-SiO$_2$) have been used to support Au nanoparticles for propylene epoxidation (Figure 3.19) [122, 145, 146]. With increasing pore diameter of the supports, the PO yield increases; Au/3D (dimension)-Ti-SiO$_2$ displays the highest catalytic performance. This suggests that the benefits of large pore diameters for the effective diffusion and transport of reactants and products results in higher catalytic performance. In addition, Au/Ti-MCM-48 displays a slightly higher PO yield than Au/Ti-MCM-41, which indicates that the 3D structure of Ti-MCM-48 has an advantage in the diffusion of reactants and products over the 1D structure of Ti-MCM-41.

Improvement of Au/3-D mesoporous Ti-SiO$_2$ As for propylene epoxidation, alkaline or alkaline earth metal chlorides were efficient solid-phase promoters; however, chloride anion markedly enhanced the coagulation of Au particles [147]. For Au/3D Ti-SiO$_2$, Ba(NO$_3$)$_2$ was the best solid-phase promoter, which might be transformed into BaO after calcination at 573 K and might function by neutralizing acid sites on

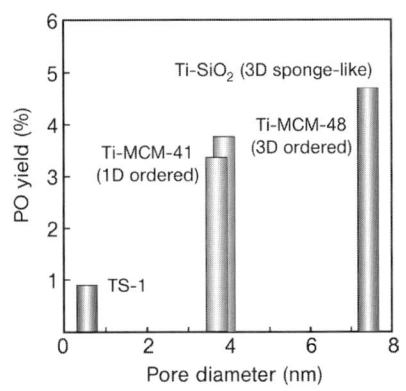

Figure 3.19 Propylene oxide (PO) yield as a function of the mean pore diameter of supports: catalyst, 0.15 g; catalyst bed temperature, 423 K; feed gas, $C_3H_6/O_2/H_2/Ar = 10:10:10:70$; space velocity, 4000 h^{-1} mL g_{cat}^{-1}. Actual Au loading, about 0.3 wt% [122, 145, 146].

the catalyst surfaces [148]. After further modifying the surface of 3D Ti-SiO$_2$ with methoxytrimethylsilane to give a hydrophobic character, Au-Ba(NO$_3$)$_2$/Ti-SiO$_2$ displayed a high STY. At an SV of 4000 h^{-1} mL g-cat.$^{-1}$, the catalyst gave a propylene conversion above 8%, a PO selectivity of 91%, and a steady STY of 80 g-PO kg-cat.$^{-1}$ h^{-1} [148].

In propylene epoxidation catalyzed by trimethylsilylated Au-Ba(NO$_3$)$_2$/Ti-SiO$_2$, the introduction of the gas promoter trimethylamine (13–15 ppm) to the reactant gas stream appreciably improved catalytic performance [149] (Figure 3.20). Surprisingly, trimethylamine made the used catalysts better than fresh catalysts in catalytic performance. Possible reasons are (i) trimethylamine might destroy the mobile acid sites that appear and disappear intermittently, suppressing by-product formation from PO; (ii) trimethylamine can also adsorb on the surfaces of Au and depress the combustion of H$_2$ to form H$_2$O, thus leading to improved H$_2$ utilization efficiency.

Reaction pathways for propylene epoxidation In the catalytic system shown in Figure 3.21, a most probable pathway is (1) O$_2$ and H$_2$ react over the Au surfaces to form H$_2$O$_2$; (2) H$_2$O$_2$ moves to isolated sites of Ti cations to form Ti-OOH species [150]; (3) Ti-OOH species react with propylene adsorbed on the support surfaces to form PO. This is a "sequential" mechanism. It has recently been verified that the Ti-OOH species is a true reaction intermediate and that a bidentate propoxy species is probably a spectator on the surface [151]. The coverage (θ) of the Ti-hydroperoxo species was determined from the area of the pre-edge peak in the Ti K-edge XANES spectra at reaction conditions. Measurement of the changes in Ti-hydroperoxo coverage, dθ/dt, under transient experiments at reaction conditions with H$_2$/O$_2$/Ar (1:1:8) and C$_3$H$_6$/H$_2$/O$_2$/Ar (1:1:1:7) gas mixtures, allowed the estimation of the initial net rate of propylene epoxidation (3.4×10^{-4} s^{-1}), which closely matched the TOF (2.5×10^{-4} s^{-1}) obtained for the same catalyst at steady-state conditions.

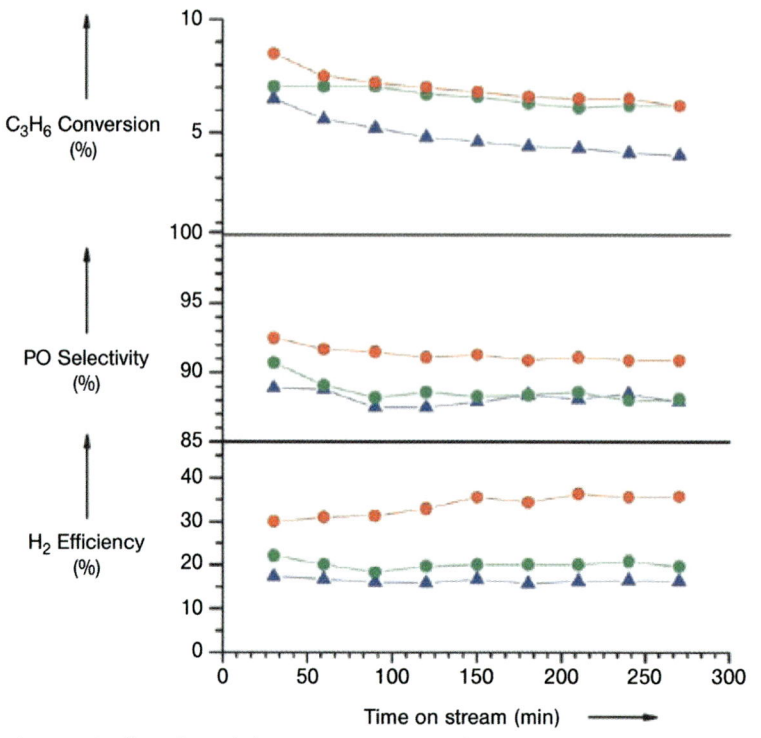

Figure 3.20 Effect of trimethylamine (13–15 ppm) on the activity of a trimethylsilylated Au-Ba(NO$_3$)$_2$/Ti-SiO$_2$ (Ti/Si = 3 : 100) catalyst for propylene epoxidation with O$_2$ and H$_2$ at 423 K. Reactant gas, C$_3$H$_6$/O$_2$/H$_2$/Ar = 1 : 1 : 1 : 7; space velocity, 4000 h^{-1} mL g$_{cat}^{-1}$; (blue ▲) fresh catalyst in the absence of TMA; (green ●) fresh catalyst in the presence of TMA; (red ●) regenerated catalyst in the presence of TMA [149].

Based on Au/TS-1, Delgass envisioned that both the "simultaneous" mechanism and "sequential" mechanism operated in parallel [152]. For the combination of Au sites and bare Ti-defect sites, propylene epoxidation proceeded by the "sequential" mechanism described above. However, the "simultaneous" mechanism dominated on Au–Ti interface sites (at least on the Au/TS-1 with a low Au loading of 0.01–0.06 wt%), where propylene epoxidation was accomplished by the attack of propylene adsorbed on H-Au-OOH species [153].

3.5
Liquid-Phase Selective Oxidation of Organic Compounds

Liquid-phase selective oxidations are of industrial importance for fine chemical syntheses. However, stoichiometric oxidizing agents have been mostly used and the generation of unwanted by-products exceeds that of desired products in quantity. To

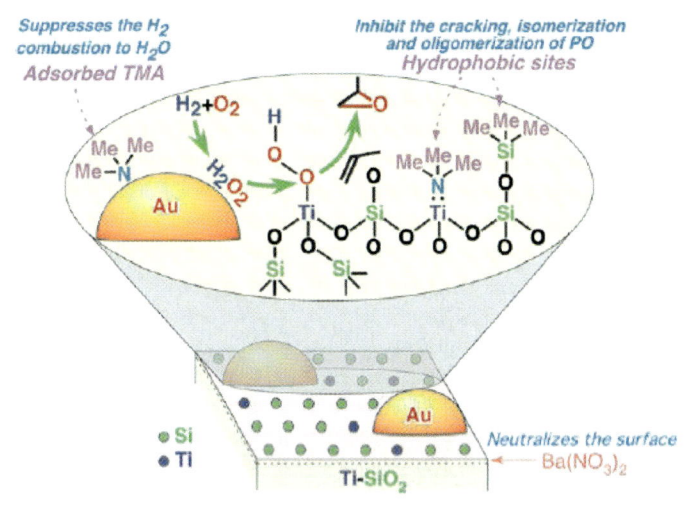

Figure 3.21 Probable reaction pathways for propylene epoxidation with O_2 and H_2 in the presence of trimethylamine on trimethylsilylated Au-Ba$(NO_3)_2$/Ti-SiO$_2$ catalyst [149].

replace current industrial processes with green, sustainable chemical processes, heterogeneous catalytic reactions should be exploited, which proceed under moderate conditions such as atmospheric pressure and at room temperature, in aqueous media or under solvent-free conditions using molecular oxygen or air as an oxidizing agent. Another important area of research is the direct transformation of biomass derived feedstock such as glucose, glycerol as a by-product of biodiesel production, and bioethanol into valuable compounds such as acetic acid by oxidation with air. Since these raw materials are usually obtained as aqueous solutions, their transformations are preferably carried out with highly active, selective, stable catalysts in water.

Since Prati and Rossi reported the aerobic oxidation of ethylene glycol to produce glycolic acid with high selectivity over supported Au catalysts in aqueous media [154], much attention has been paid to Au-catalyzed liquid-phase oxidation reactions during the last decade. The heterogeneous catalysis by Au is enhanced by water, so that it will have great potential and capability for providing environmentally benign processes using water as a solvent. Supported Au NPs appear to be advantageous over Pd or Pt catalysts, which sometimes cause overoxidation of substrates and deactivation by surface oxidation in water.

3.5.1
Oxidation of Mono-Alcohols

In the oxidation of mono-alcohols, in particular aliphatic and aromatic alcohols, Au/metal oxides have attracted more attention than Au NPs supported on activated carbon (AC) owing to the relatively higher catalytic activity and because of the adsorption of substrates or products on AC. Whereas base is necessary for the oxidation of alcohols to proceed over Au/AC, when Au/metal oxide systems can be used for the oxidation reactions even without base to afford the corresponding aldehydes, while alcohol was fully oxidized to form carboxylic acids in the presence of base in water. In addition, Au on metal oxides can be used under solvent-free conditions, while Au/AC can not.

A recent advance in Au-catalyzed aerobic oxidation of mono-alcohols reported by Corma and his coworkers was the use of Au NPs supported on nanoparticulate (<5 nm) CeO_2, which showed high catalytic activity and selectivity for allylic alcohols [155–158]. For the oxidation of allylic alcohols, Au/CeO_2 gave allylic ketones or acids with high selectivity at high conversions, whereas Pd catalysts showed lower selectivity due to the formation of isomerized and hydrogenated molecules. The attempted oxidation of 1-octen-3-ol under solvent-free conditions yielded 1-octen-3-one over Au/CeO_2 with a selectivity of 90% at a conversion of >99% [158]. Although Pd/hydroxyapatite exhibited high selectivity to cinnamic acid from the oxidation of cinnamyl alcohol in trifluorotoluene [159], the selectivity to 1-octen-3-one from 1-octen-3-ol was only 49% [158].

The aerobic oxidation of alcohol under neutral or acidic conditions to produce the corresponding acids, which can avoid the neutralization of the carboxylate salts, is also an important R&D issue. In Au-catalyzed alcohol oxidation in methanol, the corresponding methyl esters are obtained with high selectivity instead of carboxylic acids by using metal oxide supported Au NPs [157, 160]. In this case, base is not necessary, or only a catalytic amount of base is required to promote the reaction. However, in water, it was demonstrated that alcohols were not oxidized under acidic conditions [161] and only aldehydes were oxidized to carboxylic acids [162]. Even under solvent-free conditions or in organic solvents, alcohols were converted into aldehydes without base; however, the alcohols were not fully oxidized to carboxylic acid under acidic conditions [163–166].

Recently, Christensen *et al.* succeeded in oxidizing ethanol to obtain directly acetic acid under acidic conditions in water over Au/$MgAl_2O_4$ [167] and Au/TiO_2 [168]. As shown in Figure 3.22, the selectivity to acetic acid was 83% [167] and 92% [168] over Au/$MgAl_2O_4$ and Au/TiO_2 at conversions over 90%, respectively, in the absence of base at 453 K with 3.5 MPa O_2 after 8 h. The catalytic performances of Au/$MgAl_2O_4$ and Au/TiO_2 were almost the same, but Au/$MgAl_2O_4$ was superior in terms of long-term stability owing to the stable spinel structure of $MgAl_2O_4$ under acidic conditions.

Bioethanol can be regarded as a potential renewable feedstock and is produced as 3–15 vol.% aqueous solution by fermentation. Therefore, the direct transformation of dilute bioethanol to valuable compounds with air can be an environmentally friendly process. Gold catalysts appear to be advantageous in this process over Pd and Pt catalysts. The latter showed inferior selectivity to acetic acid of 60% and 16% for Pd and Pt, respectively, under the same conditions [167]. Furthermore, CO_2, which can be easily removed from the product solution, was formed as the major by-product over Au catalysts, whereas acetaldehyde was also co-produced over Pd and Pt catalysts.

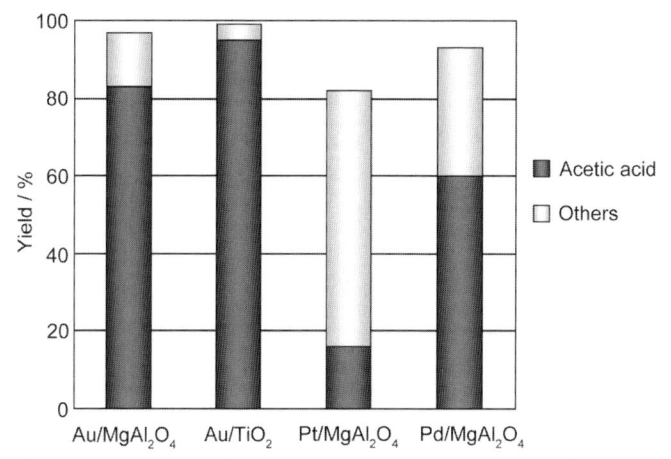

Figure 3.22 Oxidation of ethanol catalyzed by supported Au catalysts [167, 168].

Another growing research interest is the scope of support materials for Au NPs, since catalytic performance markedly depends on the size of Au particles and the interaction between Au and the supports. In particular, organic polymers have received attention as a new kind of support for Au NPs, because polymers are expected to not only act as a support to stabilize small Au NPs and clusters but also to provide a suitable reaction environment through the design of polymer structures and surface modifications [11].

Miyamura et al. [170] and Kanaoka et al. [171] have succeeded in stabilizing Au clusters on polymer supports for aerobic oxidation at room temperature in the mixed solvent of water–benzotrifluoride and in water, respectively. Polymer supports could also offer new functions, such as a recycling system by using a thermoresponsive polymer-supported Au catalyst [171].

Biffis et al. compared the catalytic performance of polymer gel immobilized Au NPs and Au/AC, with similar sized Au NPs, for alcohol oxidations in water [172]. Gold stabilized by polymer gel is advantageous over Au/AC for the use of hydrophobic substrates such as 1-octanol and 1-phenylethanol in aqueous media, although lower selectivity was obtained in some cases. For instance, polymer microgel supported Au NPs gave 1-octanoic acid by the oxidation of 1-octanol with a selectivity of 84% at 59% conversion, whereas Au/AC gave a selectivity of 93% at 65% conversion.

When the catalytic activities for benzyl alcohol oxidation were compared at moderate temperature, in the range of room temperature to 100 °C, higher catalytic activity and selectivity to benzaldehyde were obtained in organic solvents than in aqueous solution and under solvent-free conditions. Accordingly, TOFs ranged over 500 h^{-1} under solvent-free conditions with low selectivity due to the formation of benzyl benzoate [173], 100–400 h^{-1} in organic solvents with high selectivity [164, 166, 174], 11 h^{-1} at room temperature with high selectivity [171] in water, and 960 h^{-1} at 60 °C with low selectivity in water [172].

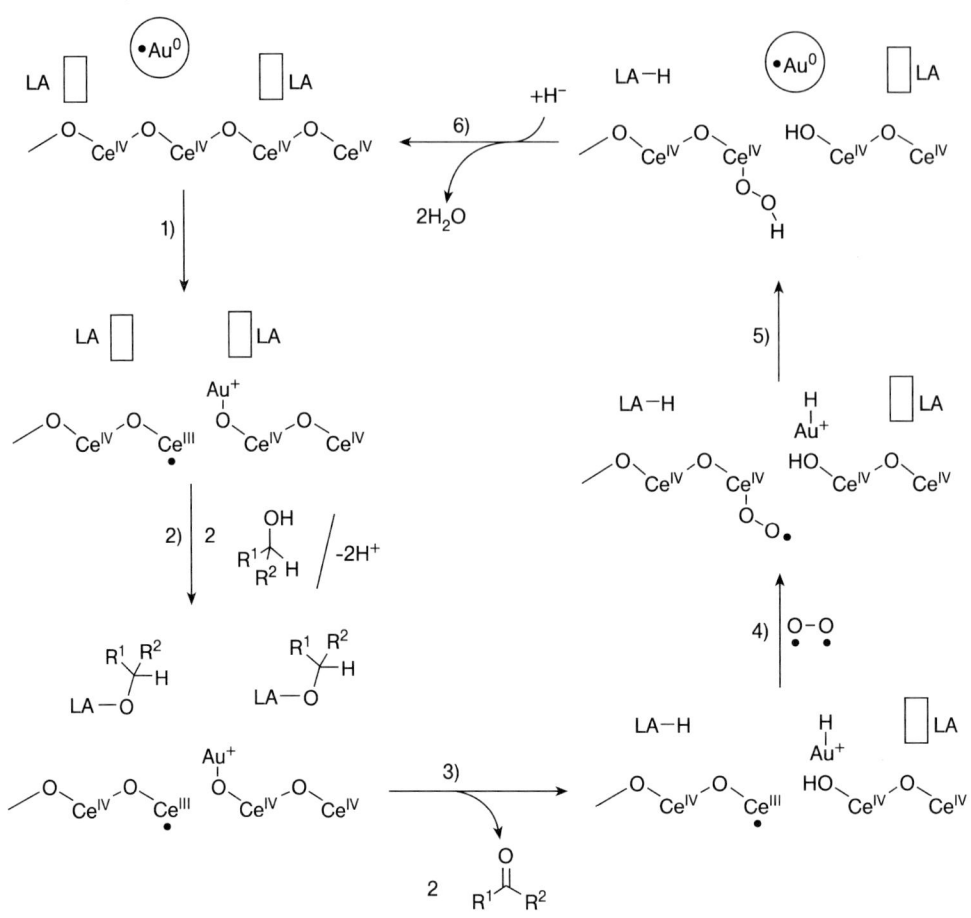

Figure 3.23 Proposed mechanism for the oxidation of alcohols in the presence of Au/CeO$_2$ as catalyst; LA = Lewis acid [155].

The reaction mechanism of alcohol oxidation has been studied independently by Abad et al. and Comotti et al. for Au/CeO$_2$ [155, 156] and Au/C [175], respectively. With Au/CeO$_2$, as shown in Figure 3.23, cationic Au species that exist at the perimeter interface with the CeO$_2$ support act as Lewis acidic sites to adsorb alcohols. Subsequently, hydride transfer might occur from alcoholate to form Au-H species and the corresponding aldehyde or ketone. After the desorption of aldehyde or ketone into the liquid phase, the oxygen vacancy sites at the surface of CeO$_2$ near the Au particles adsorb molecular oxygen. The adsorbed oxygen is used to form water by interacting with Au-H. By comparing metal-H species, the instability of Au-H might lead to high selectivity to allylic ketones or aldehydes whereas the more stable Pd-H might cause the hydrogenation of a C=C double bond to form aliphatic oxygenates. In this proposed mechanism, dehydrogenation of adsorbed alcoholate on Au catalytic sites is regarded as the rate-determining step and the redox properties of the metal oxide support play an important role in determining the catalytic performance.

Christensen and coworkers also reported that dehydrogenation of ethanol was supposed to be the rate-determining step in ethanol oxidation using Au/MgAl$_2$O$_4$ and Au/TiO$_2$ [168]. In contrast to CeO$_2$, MgAl$_2$O$_4$ does not have redox properties. Therefore, the redox property is an advantageous factor but it may not be essential or indispensable for Au-catalyzed oxidation in the liquid phase.

In contrast, a different reaction mechanism has been proposed by Comotti et al. for the oxidation of glucose over Au/AC (Figure 3.24) [175]. In their mechanism, the first step is the adsorption of hydrated alcoholate, which is formed in advance by alkali, onto Au catalytic sites. The electron-rich Au species might be formed by Au-alcoholate intermediates and then oxygen could be adsorbed onto Au by nucleophilic attack to form Au$^+$-O$_2^-$ or Au^{2+}-O$_2^{2-}$. Then, carboxylic acids are formed by two-electron reduction of oxygen. In this reaction, H$_2$O$_2$ must be formed as a by-product and was in fact detected experimentally.

For Au/C, gold was completely reduced to metallic NPs, where the fraction of edges and corners is not negligible. Alcoholate might be adsorbed by the Au catalytic sites, although such species can not adsorb alcohols themselves. This is supposed to be the crucial role of base. The rate-determining step would be the adsorption of oxygen or the two-electron transfer to form carboxylic acid, which was supposed to give rise to a first-order dependence of the reaction rate on pO_2. Different behaviors between Au/CeO$_2$ and Au/C by changing the reaction conditions such as pH and solvent can be accounted for by the different mechanisms.

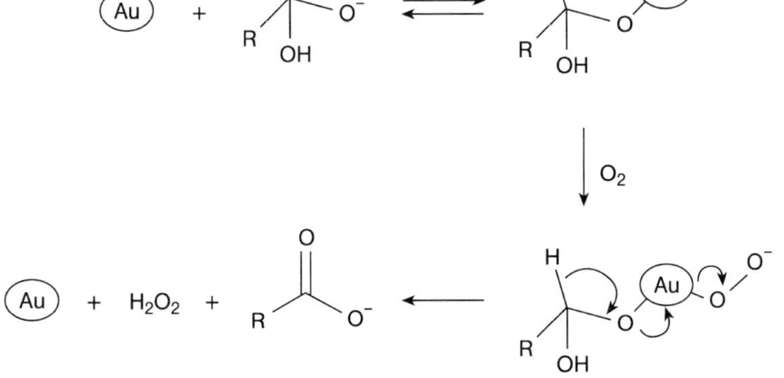

Figure 3.24 Molecular mechanism of glucose oxidation with gold nanoparticles (Au/C) [175].

3.5.2
Oxidation of Diols

The first prominent work on Au-catalyzed liquid-phase oxidation was reported by Rossi and Prati in 1998 (Scheme 3.3) [154]. The aerobic oxidation of ethylene glycol produces sodium glycolate with high selectivity (98%) by oxidizing only one OH group in water in the presence of an equimolar amount of NaOH. Conversely, Pd and Pt catalysts gave lower selectivity to glycolate, such as 77 and 71%, respectively, owing to the overoxidation and C–C bond cleavage (Scheme 3.3) (Table 3.4) [176].

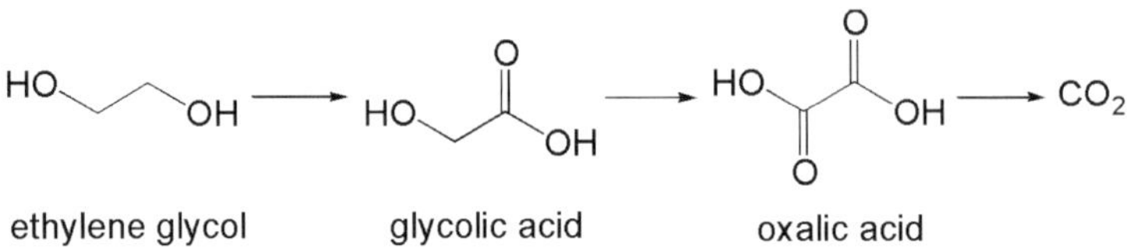

ethylene glycol → glycolic acid → oxalic acid → CO_2

Scheme 3.3 Oxidation of ethylene glycol.

Since the promoting effect of both water and basic conditions in Au catalysts was also observed in the liquid phase. Au-catalyzed oxidations of diol, glycerol and glucose that are miscible in water were examined in the presence of a strong base such as NaOH. Gold NPs supported on activated carbon were mainly used for these oxidations of polyols because the selectivities to glycolic acid from ethylene glycol and to glyceric acid from glycerol over Au/AC were superior to those of Au on metal oxides such as Al_2O_3 and TiO_2 [154, 176].

For the aerobic oxidation of ethylene glycol, Au/AC exhibited high selectivities to sodium glycolate of over 90% in a wide range of conversion, indicating that the glycolate was stable in the presence of Au/C, whereas Pd/C and Pt/C catalyzed further oxidation of glycolate to form oxalate or to cause C–C bond cleavage resulting in the formation of CO_2 (Table 3.4). The TOF for Au/AC reached $1000\,h^{-1}$ with a selectivity of 98% to glycolic acid at 343 K with 0.3 MPa O_2 in basic aqueous solution. Other diols were also converted into the corresponding mono-acids, esters or

Table 3.4 Oxidation of diols catalyzed by supported Au catalysts.

Substrate	Catalyst	Product	Solvent	Temperature (°C)	TOF (h^{-1})	Selectivity (%)	Ref.
Ethylene glycol	Au/C	Glycolate	H_2O	70	1000	98	[176]
Ethylene glycol	Pd/C	Glycolate	H_2O	70	500	77	[176]
Ethylene glycol	Pt/C	Glycolate	H_2O	70	475	71	[176]
Phenylethan-1,2-diol	Au/C	Mandelate	H_2O	70	800	83	[176]
1,2-Propylene glycol	Au/C	Lactate	H_2O	90	780	100	[154]
1,3-Propylene glycol	Au/TiO_2	a)	MeOH	100	53	90	[177]
1,4-Butanediol	Au/TiO_2	γ-Butyrolactone	PBu_3	300	64	99	[178]

a) Methyl 3-hydroxypropionate.

lactones. In the case of 1,4-butanediol, oxidative cyclization takes pace with high selectivity to γ-butyrolactone [176–178].

Glycolic acid is used as a detergent for electronic devices, raw material for cosmetics and co-monomer for biodegradable polymers. Nippon Shokubai Co. Ltd. demonstrated the direct synthesis of methyl glycolate by the aerobic oxidation of ethylene glycol and esterification in MeOH over Au catalyst in a pilot-plant scale in 2004. The catalytic activity and selectivity depends on the kind of support and reaction conditions, particularly pH, to avoid overoxidation or the Cannizzaro disproportionation reaction of keto aldehyde that was produced by the second oxidation [176]. Gold on AC always shows high selectivities towards mono-oxidation under various conditions, while the second oxidation took place over Au/TiO_2 with an increase in the NaOH/substrate ratio [176].

3.5.3
Oxidation of Glycerol

Glycerol is a raw material that is readily available through the production of biodiesel fuels from vegetable oils as a by-product. Transformation of cheap glycerol into valuable organic compounds has become an increasingly important R&D target with increasing production of biodiesel fuels year by year. Since glycerol is co-produced in a basic methanol solution in the biodiesel fuel synthetic process, it is desirable that the catalytic oxidation can be carried out in this waste solution.

As shown in Scheme 3.4, there are several reaction pathways of glycerol oxidation to form dihydroxyacetone, glyceric acid, hydroxypyruvic acid, mesoxalic acid, and so on. Dihydroxyacetone is formed by the oxidation of a secondary hydroxy group under

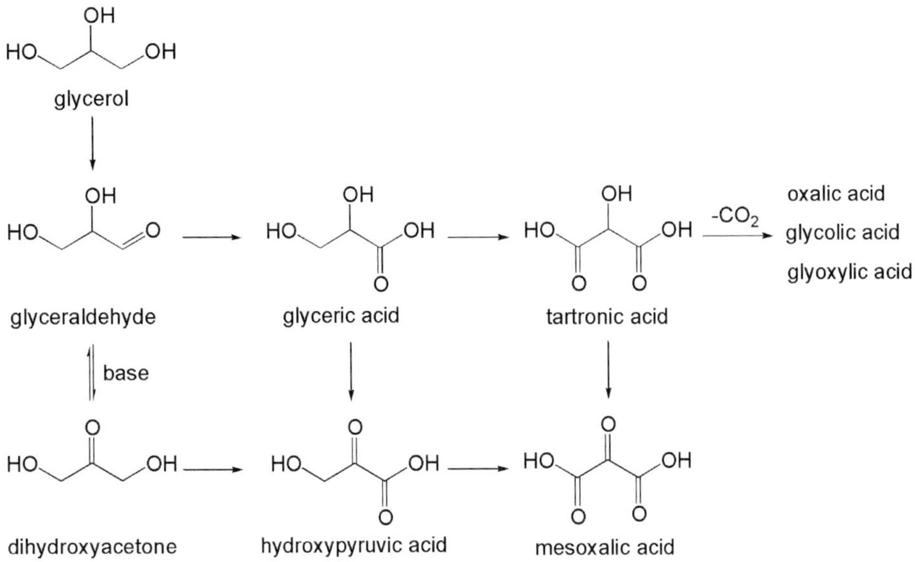

Scheme 3.4 Reaction pathways of glycerol oxidation.

basic condition, while glyceric acid is easily formed under basic conditions. However, glyceric acid can be converted into tartronic acid and mesoxalic acid, by successive oxidation, and into glycolic acid and oxalic acid by C—C bond cleavage. The selectivity is influenced by the kinds of catalytic metals. Gold and Pd are selective to glyceric acid while Pt is selective to dihydroxyacetone under basic conditions. Thus, the reaction conditions and the metallic species (and their ratio for bimetallic catalysts) should be carefully chosen to obtain the desired product with high selectivity.

Hutchings and coworkers demonstrated, that in the selective oxidation of glycerol the selectivity to glyceric acid was 100% at a conversion of 50% but gradually decreased to 86% at 72% conversion under 0.3 MPa of O_2 at 333 K over Au/AC in the presence of an equimolar amount of NaOH [169, 179–181]. More than an equimolar amount of NaOH was necessary to promote the reaction; the reaction did not occur without base.

The Au-catalyzed glycerol oxidation was influenced by the kind of support, the size of Au particles and the reaction conditions such as concentration of glycerol, pO_2 and molar ratio of NaOH to glycerol. As metal oxide supports showed inferior selectivity to glyceric acid compared to carbons, due to successive oxidation and C—C bond cleavage to form di-acids such as tartronic acid and glycolic acid, research has focused on Au NPs supported on carbon, as in the case of ethylene glycol oxidation [182]. Indeed, the catalytic activity was influenced by the kind of carbon support in terms of porous texture [183].

For the glycerol oxidation, smaller Au NPs showed higher catalytic activity than those with a diameter larger than 20 nm. However, the selectivity to glyceric acid decreased with an increase in conversion. In contrast, larger Au NPs could maintain the selectivity even at full conversion owing to the inhibition of overoxidation [184, 185]. More than an equimolar amount of strong base is essential for the glycerol oxidation over Au catalysts and the reaction rates increased with the NaOH/glycerol ratio. However, a molar ratio of 4 resulted in overoxidation at the second primary alcoholic group to form tartronic acid [186].

Porta et al. have investigated the effect of reaction conditions and concluded that the following conditions are optimum: glycerol/Au = 500 : 1 (mol/mol), glycerol/NaOH = 1 : 4, glycerol 0.3 M, pO_2 0.3 MPa, 303 K, 6 h, using Au/AC prepared from citrate-protected Au sols. They obtained the highest selectivity of 92% for glyceric acid at a full conversion with a TOF of 83 h^{-1} [186]. The highest TOF after 0.25 h of the reaction was reported as 1090 h^{-1} while the selectivity to glyceric acid was 65% at 50% conversion at 323 K [187].

Interestingly, Pd was glyceric acid selective whereas Pt showed selectivity for dihydroxyacetone, which was formed by oxidation at the secondary alcoholic group. A promoting effect in catalytic activity was observed by the introduction of Pd metal to Au/AC [184, 187, 188]. Dihydroxyacetone was preferentially formed over Au-Pt/C; however, the selectivity was lower than that of Bi-modified Pt catalyst [189].

Most recently, Taarning et al. reported that the fully oxidized ester dimethyl mesoxalate, which can be used as a monomer to yield poly(ketomalonate) [190], was produced by glycerol oxidation over Au/TiO_2 and Au/Fe_2O_3 with 10% $NaOCH_3$ in methanol with a selectivity of 89% at a full conversion [177]. Alcoholic media facilitated the oxidation with respect to aqueous media; thus dimethyl mesoxalate was obtained with high yields.

3.5.4
Aerobic Oxidation of Glucose

The conversion of carbohydrates into sugar acids is of great interest to industrial applications. Gluconic acid, which is used as a detergent for beverages, food additives and chelating agent for metal ions, has been produced commercially by fermentation. Since Rossi and coworkers demonstrated the aerobic oxidation of glucose in water by Au NPs supported on activated carbon (AC) with excellent selectivity to gluconic acid in 2002 (Scheme 3.5) [191], the catalytic oxidation of glucose has been a very active research area.

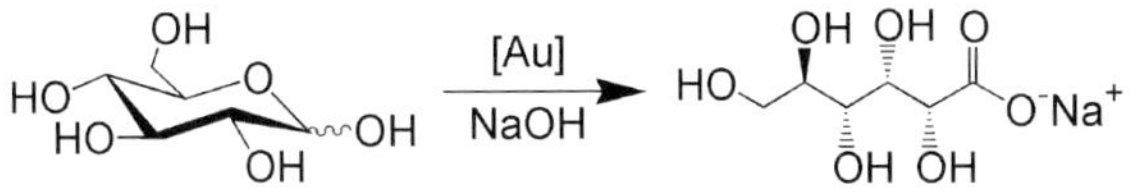

Scheme 3.5 Oxidation of glucose.

For glucose oxidation over Au/AC, Comotti et al. showed that the catalytic activity per unit weight of Au is inversely proportional to the mean diameter of Au NPs in the range 3–6 nm, whereas Au NPs larger than 6 nm deviate from linearity, and Au NPs larger than 10 nm are almost inactive [192]. The same workers have studied the large-scale production of Au catalysts [193], recycling tests [194] and optimization of the reaction conditions for industrial applications: 3 M glucose, glucose/Au = 40 000 mol mol^{-1}, pH 9.5, stirring rate of 39 000 rpm, O_2 100 mL min^{-1}, at 323 K by using AC supported Au NPs with a diameter of 3.6 nm [194]. In the first run, the TOF was reported as 1.5×10^5 h^{-1}, which was calculated for surface exposed Au atoms. As there are 36% of the Au atoms at the external surface of Au NPs with a diameter of 3.6 nm, the TOF for the total amount of Au loaded was estimated to be 54 000 h^{-1}. However, catalytic activity decreased with reaction cycles and became almost one-third of the original activity at the fifth run.

Thielecke et al. have recently developed a flow reactor system for glucose oxidation by using Au/Al$_2$O$_3$ and Au/TiO$_2$ as catalysts and have examined the long-term stability of the catalysts for industrial use [195–197]. The productivity of sodium gluconate was estimated to be 4.2 t g$_{Au}^{-1}$ and Au/Al$_2$O$_3$ showed no loss in catalytic activity and selectivity after 70 days operation [195].

Direct oxidation of glucose to obtain free gluconic acid is another process strongly demanded by industry. However, a lower pH slows down the reaction because of deactivation of Pd, Pt and Au catalysts. Comotti et al. reported that a Au-Pt (2 : 1) bimetallic catalyst presented a significantly improved TOF of 924 h^{-1} in the absence of base with respect to that of Au/C (51 h^{-1}) at 343 K under 0.3 MPa O_2, while the catalytic activity of Au/C was only slightly improved by alloying with Pt under pH 9.5 at 323 K [198]. The TOF was one-order of magnitude lower at uncontrolled pH than that at pH 9.5; however, the choice of bimetallic catalyst containing Au and other noble metals would be one possibility to obtain free acids.

3.5.5
Oxidation of Alkanes and Alkenes

Gold-catalyzed oxidation of styrene was firstly reported by Choudhary and coworkers for Au NPs supported on metal oxides in the presence of an excess amount of radical initiator, *t*-butyl hydroperoxide (TBHP), to afford styrene oxide, while benzaldehyde and benzoic acid were formed in the presence of supports without Au NPs [199]. Subsequently, Hutchings and coworkers demonstrated the selective oxidation of cyclohexene over Au/C with a catalytic amount of TBHP to yield cyclohexene oxide with a selectivity of 50% and cyclohexenone (26%) as a by-product [2]. Product selectivity was significantly changed by solvents. Cyclohexene oxide was obtained as a major product with a selectivity of 50% in 1,2,3,5-tetramethylbenzene while cyclohexenone and cyclohexenol were formed with selectivities of 35 and 25%, respectively, in toluene. A promoting effect of Bi addition to Au was also reported for the epoxidation of cyclooctene under solvent-free conditions.

Caps and coworkers studied the solvent effect in the epoxidation of stilbene by varying solvents and the supports [200]. In methylcyclohexane (MCH), the activated radical species proposed were MCH peroxy radicals, which were formed by the radical transfer from TBHP and reaction with molecular oxygen. Except for MCH, the solvent effect is not fully understood; however, the choice of solvent and supports that can trap or stabilize the radical species affected the catalytic performance of Au.

Aerobic oxidation of cyclohexane to cyclohexanol or cyclohexanone over Au catalysts has also been studied under solvent-free conditions and without radical initiators [201–203]. Gold on Al_2O_3 was relatively cyclohexanol selective while Au supported on more acidic supports such as silica exhibited cyclohexanone selective [201, 202]. Gold on MCM-41 showed a good selectivity to cyclohexanone of 76% accompanied by the formation of cyclohexanol (21%) at 19% conversion [202].

3.6
Conclusions

Although bulk gold is poorly active as a catalyst, Au nanoparticles (NPs) attached to various support materials exhibit unique catalytic properties in oxidation. General features can be summarized as follows:

1. Highly active Au catalysts can be prepared by an appropriate selection of preparation methods such as co-precipitation (CP), deposition-precipitation (DP), deposition-reduction (DR) and solid grinding (SG) with dimethyl Au(III) acetylacetonate, depending on the kind of support materials and reactions targeted.

2. There are four important conditions for Au to be active as a catalyst in oxidation: (1) strong junction with reducible metal oxides, (2) H_2O, (3) OH^- and (4) the size of Au particles and films. At least two conditions should be fulfilled for Au to exhibit high catalytic activity. Catalysis by Au is often promoted by water and alkaline but destroyed by acids, which presents an interesting research topic in catalysis science.

3. Although Au NPs supported on carbons and polymers are catalytically almost inactive in gas-phase oxidation, they are active in liquid-phase oxidation.

4. For CO oxidation in the gas phase at room temperature, Au NPs supported on reducible metal oxides such as TiO_2, Fe_2O_3, Co_3O_4, NiO and CeO_2 are much more active than Pt group metal catalysts. They are active even at a temperature as low as 200 K when Au loadings are higher than 3 wt%.

5. In the co-presence of H_2O, Au NPs supported on insulating metal oxides such as Al_2O_3 and SiO_2 can also exhibit catalytic activity at room temperature for CO oxidation.

6. Over most reported Au catalysts, CO oxidation takes place at the junction perimeter between Au NPs and the metal oxide supports. Carbon monoxide is adsorbed on the edges, corners and steps of Au NPs. Molecular oxygen is adsorbed on the support surfaces and may be activated at the oxygen defect sites at the perimeter interfaces, where the two adsorbates react to form CO_2 in the gas phase. At the perimeter interfaces Au is assumed to exist as Au^{3+}, which might be stabilized through bonding with OH^-.

7. Supported Au catalysts are, in general, less active than Pt group metal catalysts in the complete oxidation of hydrocarbons. However, by choosing appropriate metal oxide supports Au catalysts are also applicable to the removal of HCHO at room temperature.

8. Direct one-step propylene epoxidation can be achieved in the gas phase containing O_2 and H_2 with Au NPs deposited on TiO_2 or Ti-SiO_2 by the deposition-precipitation method.

9. Representative performances in steady states obtained to date are (i) a conversion of 5%, a selectivity above 90%, and hydrogen utilization efficiency of 35% and a STY (space–time yield) of 80–100 h^{-1} g-PO kg-cat^{-1} over Au/mesoporous Ti-SiO_2 and (ii) a conversion of 10%, a selectivity of 76% and a STY of 134 h^{-1} g-PO kg-cat^{-1} over Au/microporous TS-1.

10. In the propylene reaction with a O_2–H_2 mixture, the contact structure of Au NPs and the selection of support metal oxides are critical for producing propylene oxide (PO).

11. The diameter of Au NPs defines the products: PO is predominantly produced with Au NPs larger than 2 nm, whereas propane (C_3H_8) is formed with smaller NPs. This dramatic shift of products can be correlated with a change in the work function and energy gap of Au in contact with TiO_2.

12. In liquid-phase oxidation, supported Au NPs are in principle advantageous over Pt group metal catalysts in the applicability of water as a solvent and the avoidance of overoxidation of reactants and catalyst deactivation by surface oxidation. Gold clusters deposited on mesoporous carbon or Al_2O_3 exhibit extraordinarily high catalytic activity for glucose transformation into sodium gluconate. Their performances are excellent in terms of activity(TOF > 20 h^{-1} molGl Au-mol^{-1}) and selectivity(>98%) so that current industrial fermentation processes can be replaced with a Au catalyst process.

13. Gold NPs deposited on carbons are active and selective for mild oxidations in liquid phase although they exhibit almost no catalytic activity in the gas phase. Examples are aerobic oxidation of mono-alcohols, diols, glycerol, glucose, alkenes and alkanes.

14. The catalytic capability of Au in liquid-phase oxidation can be tuned to a wider scope by choosing the support, size of Au, alkaline(with or without) and solvents. The products changed dramatically, depending on the solvents: reactant alone without solvent, water, nonpolar and polar organic solvents.

References

1 Thayer, A.M. (1992) *Chemical & Engineering News*, March 9, **1992**, 27–49.
2 Hughes, M.D., Xu, Y.-J., Jenkins, P. et al. (2005) *Nature*, **437**, 1132–1135.
3 Haruta, M. (2005) *Nature*, **437**, 1098–1099.
4 Hashmi, A.S.K. and Hutchings, G.J. (2006) *Angewandte Chemie – International Edition*, **45**, 7896–7936.
5 (a) Cho, A. (2003) *Science*, **299**, 1684–1685; (b) Campbell, C.T. (2004) *Science*, **306**, 234–235.
6 Sanchez-Castillo, M.A., Couto, C., Kim, W.B. and Dumesic, J.A. (2004) *Angewandte Chemie – International Edition*, **43**, 1140–1142.
7 Zielasek, V., Jurgens, B., Schulz, C. et al. (2006) *Angewandte Chemie – International Edition*, **45**, 8241–8244.
8 Xu, C., Su, J., Xu, X. et al. (2007) *Journal of the American Chemical Society*, **129**, 42–43.
9 Tsunomiya, H., Sakurai, H. and Tsukuda, T. (2006) *Chemical Physics Letters*, **429**, 528–532.
10 Shi, F., Zhang, Q., Ma, Y. et al. (2005) *Journal of the American Chemical Society*, **127**, 4182–4183.
11 Ishida, T. and Haruta, M. (2007) *Angewandte Chemie – International Edition*, **46**, 7154–7156.
12 Haruta, M. (2007) *ChemPhysChem*, **8**, 1911–1913.
13 Langmuir, I. (1918) *Journal of the American Chemical Society*, **40**, 1361–1403.
14 Haruta, M., Kobayashi, T., Sano, H. and Yamada, N. (1987) *Chemistry Letters*, **16**, 405–408.
15 Haruta, M., Yamada, N., Kobayashi, T. and Iijima, S. (1989) *Journal of Catalysis*, **115**, 301–309.
16 Cunningham, D.A.H., Kobayashi, T., Kamijo, N. and Haruta, M. (1994) *Catalysis Letters*, **25**, 257–264.
17 Huber, H., McIntosh, D. and Ozin, G.A. (1977) *Inorganic Chemistry*, **16**, 975–979.
18 Lahr, D.L. and Ceyer, S.T. (2006) *Journal of the American Chemical Society*, **128**, 1800–1801.
19 Bond, G.C. and Thompson, D.T. (2000) *Gold Bulletin*, **33**, 41–51.
20 Haruta, M. and Daté, M. (2001) *Applied Catalysis A – General*, **222**, 427–437.

21 Haruta, M. (2002) *Cattech*, **6**, 102–115.
22 Haruta, M. (2003) *Chemical Record*, **3**, 75–87.
23 Molina, L.M. and Hammer, B. (2004) *Physical Review B – Condensed Matter*, **69**, 155424.
24 Molina, L.M. and Hammer, B. (2005) *Applied Catalysis A – General*, **291**, 21–31.
25 Hutchings, G.y. (2005) *Catalysis Today*, **100**, 55–66.
26 Chen, M.S. and Goodman, D.W. (2006) *Accounts of Chemical Research*, **39**, 739–746.
27 Chen, M.S. and Goodman, D.W. (2006) *Catalysis Today*, **111**, 22–33.
28 Kuny, M.C., Davis, R.J. and Kuny, H.H. (2007) *The Journal of Physical Chemistry C*, **111**, 11767–11775.
29 Cho, A. (2003) *Science*, **299**, 1684–1685.
30 Campbell, C.T. (2004) *Science*, **306**, 234–235.
31 Akita, T., Tanaka, K., Tsubota, S. and Haruta, M. (2000) *Journal of Electron Microscopy*, **49**, 657–662.
32 Ichikawa, S., Akita, T., Okumura, M. et al. (2003) *Journal of Electron Microscopy*, **52**, 21–26.
33 Akita, T., Lu, P., Ichikawa, S. et al. (2001) *Surface and Interface Analysis*, **31**, 106–113.
34 Bamwenda, G.R., Tsubota, S., Nakamura, T. and Haruta, M. (1997) *Catalysis Letters*, **44**, 83–87.
35 Liu, Z.M. and Vannice, M.A. (1997) *Catalysis Letters*, **43**, 51–54.
36 Tsubota, S., Nakamura, T., Tanaka, K. and Haruta, M. (1998) *Catalysis Letters*, **56**, 131–135.
37 Okumura, M., Nakamura, S., Tsubota, S. et al. (1998) *Catalysis Letters*, **51**, 53–58.
38 Daté, M. and Haruta, M. (2001) *Journal of Catalysis*, **201**, 221–224.
39 Haruta, M., Kobayashi, T., Iijima, S. and Delannay, F. (1988) Proceedings 9th International Congress Catal, Calgary, Canada, July 1988, p. 206.
40 Cunningham, D.A.H., Vogel, W., Kageyama, H. et al. (1998) *Journal of Catalysis*, **177**, 1–10.
41 Ishida, T., Nagaoka, M., Akita, T. and Haruta, M. (2008) *Chemistry A – European Journal*, **14**, 8456–8460.
42 Haruta, M., Tsubota, S., Kobayashi, T. et al. (1993) *Journal of Catalysis*, **144**, 175–192.
43 Haruta, M. (2003) *Catalysis and Electrocatalysis on Nanoparticles* (eds A. Wieckowski, E.R. Savinova and C.G. Vayenas), **240**, Marcel Dekker, New York, p. 243.
44 Iizuka, Y., Tode, T., Takao, T. et al. (1999) *Journal of Catalysis*, **187**, 50–58.
45 Mavrikakis, M., Stoltze, P. and Nørskov, J.K. (2000) *Catalysis Letters*, **64**, 101–106.
46 Cunningham, D.A.H., Vogel, W. and Haruta, M. (1999) *Catalysis Letters*, **63**, 43–47.
47 Heiz, U. and Schneider, W.-D. (2000) *Journal of Physics D – Applied Physics*, **33**, R85–R102.
48 Abbet, S., Heiz, U., Häkkinen, H. and Landman, U. (2001) *Physical Review Letters*, **86**, 5950–5953.
49 Heiz, U., Sanchez, A., Abbet, S. and Schneider, W.-D. (1999) *Journal of the American Chemical Society*, **121**, 3214–3217.
50 Cameron, D., Holliday, R. and Thompson, D. (2003) *Journal of Power Sources*, **118**, 298–303.
51 Kahlich, M.J., Gasteiger, H.A. and Behm, R.J. (1997) *Journal of Catalysis*, **171**, 93–105.
52 Ozkara, S. and Aksoylu, A.E. (2003) *Applied Catalysis A – General*, **251**, 75–83.
53 Snytnikov, P.V., Sobyanin, V.A., Belyaev, V.D. et al. (2003) *Applied Catalysis A – General*, **239**, 149–156.
54 Landon, P., Ferguson, J., Solsona, B.E. et al. (2005) *Chemical Communications*, 3385–3387.
55 Watanabe, M., Uchida, H., Igarashi, H. and Suzuki, M. (1995) *Chemistry Letters*, **24**, 21.
56 Fukuoka, A., Kimura, J., Oshio, T. et al. (2007) *Journal of the American Chemical Society*, **129**, 10120–10125.
57 Echigo, M. and Tabata, T. (2004) *Catalysis Letters*, **98**, 37–42.

58 Sanchez, R.M.T., Ueda, A., Tanaka, K. and Haruta, M. (1997) *Journal of Catalysis*, **168**, 125–127.
59 Okumura, M. and Haruta, M. (2000) *Chemistry Letters*, 396–397.
60 Landon, P., Ferguson, J., Solsona, B.E. et al. (2006) *Journal of Materials Chemistry*, **16**, 199–208.
61 Schubert, M.M., Plzak, V., Garche, J. and Behm, R.J. (2001) *Catalysis Letters*, **76**, 143–150.
62 Ko, E.-Y., Park, E.D., Seo, K.W. et al. (2006) *Catalysis Today*, **116**, 377–383.
63 Quinet, E., Morfin, F., Diehl, F. et al. (2008) *Applied Catalysis B*, **80**, 195–201.
64 Qiao, B.T. and Deng, Y.Q. (2003) *Chemical Communications*, 2192–2193.
65 Schubert, M.M., Venugopal, A., Kahlich, M.J. et al. (2004) *Journal of Catalysis*, **222**, 32–40.
66 Avgouropoulos, G., Ioannides, T., Papadopoulou, Ch. et al. (2002) *Catalysis Today*, **75**, 157–167.
67 Carrettin, S., Concepcion, P., Corma, A. et al. (2004) *Angewandte Chemie – International Edition*, **43**, 2538–2540.
68 Avgouropoulos, G., Papavasiliou, J., Tabakova, T. et al. (2006) *Chemical Engineering Journal*, **124**, 41–45.
69 Luengnaruemitchai, A., Osuwan, S. and Gulari, E. (2004) *International Journal of Hydrogen Energy*, **29**, 429–435.
70 Panzera, G., Modafferi, V., Candamano, S. et al. (2004) *Journal of Power Sources*, **135**, 177–183.
71 Fu, Q., Saltsburg, H. and Flytzani-Stephanopoulos, M. (2003) *Science*, **301**, 935–938.
72 Tibiletti, D., Amieiro-Fonseca, A., Burch, R. et al. (2005) *The Journal of Physical Chemistry. B*, **109**, 22553–22559.
73 Grigorova, B. Mellor J. Palazov, A. and Greyling, F.WO Pat., 00/59631/2000.
74 Wang, H., Zhu, H., Qin, Z. et al. (2008) *Catalysis Communications*, **9**, 1487–1492.
75 Chang, L.-H., Sasirekha, N.N. and Chen, Y.-W. (2007) *Catalysis Communications*, **8**, 1702–1710.
76 Naknama, P., Luengnaruemitchai, A., Wongkasemjit, S. and Osuwan, S. (2007) *Journal of Power Sources*, **65**, 353–358.
77 Avgouropoulos, G. and Ioannides, T. (2003) *Applied Catalysis A*, **244**, 155–167.
78 Monyanon, S., Pongstabodee, S. and Luengnaruemitchai, A. (2006) *Journal of Power Sources*, **163**, 547–554.
79 Louis, C. (2008) *Nanoparticles and Catalysis* (ed. Didier Astruc), Wiley-VCH, Weinheim, pp. 475–502.
80 Boccuzzi, F., Chiorino, A., Manzoli, M. et al. (2001) *Journal of Catalysis*, **202**, 256–267.
81 Ruggiero, C. and Hollins, P. (1997) *Surface Science*, **377–379**, 583–586.
82 Olea, M., Kunitake, M., Shido, T. and Iwasawa, Y. (2001) *Physical Chemistry Chemical Physics*, **3**, 627–631.
83 Schubert, M.M., Hackenberg, S., van Veen, A.C. et al. (2001) *Journal of Catalysis*, **197**, 113–112.
84 Liu, H., Kozlov, A.I., Kozlova, A.P. et al. (1999) *Journal of Catalysis*, **185**, 252–264.
85 Okumura, M., Coronado, J.M., Soria, J. et al. (2001) *Journal of Catalysis*, **203**, 168–174.
86 Guzman, J., Carrettin, S. and Corma, A. (2005) *Journal of the American Chemical Society*, **127**, 3286–3287.
87 Guzman, J., Carrettin, S., Fierro-Gonzalez, J.C. et al. (2005) *Angewandte Chemie – International Edition*, **44**, 4778–4781.
88 Valden, M., Lai, X. and Goodman, D.W. (1998) *Science*, **281**, 1647–1650.
89 Chen, M.S. and Goodman, D.W. (2004) *Science*, **306**, 252–255.
90 Gilb, S., Weis, P., Furche, F. et al. (2002) *Journal of Chemical Physics*, **116**, 4094–4101.
91 Cleveland, C.L., Landman, U., Schaaff, T.G. et al. (1997) *Physical Review Letters*, **79**, 1873–1876.
92 Comotti, M., Li, W.-C., Spliethoff, B. and Schüth, F. (2006) *Journal of the American Chemical Society*, **128**, 917–924.
93 Finch, R.M., Hodge, N.A., Hutchings, G.J. et al. (1999) *Physical Chemistry – Chemical Physics*, **1**, 485–489.

94 Golunski, S., Rajaram, R., Hodge, N. et al. (2002) *Catalysis Today*, **72**, 107–113.
95 Park, E.D. and Lee, J.S. (1999) *Journal of Catalysis*, **186**, 1–11.
96 Guzman, J. and Gates, B.C. (2004) *Journal of the American Chemical Society*, **126**, 2672–2673.
97 Armor, J.N. (1992) *Applied Catalysis B – Environmental*, **1**, 221–256.
98 Crump, D.R. (1995) *Volatile Organic Compounds in Indoor Air*, in Volatile Organic Compounds in the Atmosphere (eds R.E. Hester and R.M. Harrison), The Royal Society of Chemistry, p. 118.
99 Spivey, J.J. (1987) *Industrial & Engineering Chemistry Research*, **26**, 2165–2180.
100 Zhang, C.B., He, H. and Tanaka, K. (2005) *Catalysis Communications*, **6**, 211–214.
101 Zhang, C.B., He, H. and Tanaka, K. (2006) *Applied Catalysis B – Environmental*, **65**, 37–43.
102 Zhang, C.B. and He, H. (2007) *Catalysis Today*, **126**, 345–350.
103 Peng, J. and Wang, S. (2007) *Applied Catalysis B – Environmental*, **73**, 282–291.
104 Sekine, Y. (2002) *Atmospheric Environment*, **36**, 5543–5547.
105 Jia, M., Shen, Y., Li, C. et al. (2005) *Catalysis Letters*, **99**, 235–239.
106 Shen, Y., Yang, X., Wang, Y. et al. (2008) *Applied Catalysis B – Environmental*, **79**, 142–148.
107 Haruta, M., Ueda, A., Tsubota, S. and Torres Sanchez, R.M. (1996) *Catalysis Today*, **29**, 443–447.
108 Li, C., Shen, Y., Jia, M. et al. (2008) *Catalysis Communications*, **9**, 355–361.
109 Grisel, R.J.H., Slyconish, J.J. and Nieuwenhuys, B.E. (2001) *Topics in Catalysis*, **16–17**, 425–431.
110 Grisel, R.J.H. and Nieuwenhuys, B.E. (2001) *Catalysis Today*, **64**, 69–81.
111 Gluhoi, A.C., Lin, S.D. and Nieuwenhuys, B.E. (2004) *Catalysis Today*, **90**, 175–181.
112 Gluhoi, A.C., Bogdanchikova, N. and Nieuwenhuys, B.E. (2005) *Journal of Catalysis*, **229**, 154–162.
113 Gluhoi, A.C., Bogdanchikova, N. and Nieuwenhuys, B.E. (2005) *Journal of Catalysis*, **232**, 96–101.
114 Gluhoi, A.C., Bogdanchikova, N. and Nieuwenhuys, B.E. (2006) *Catalysis Today*, **113**, 178–181.
115 Gluhoi, A.C., Nieuwenhuys, B.E., Gluhoi, A.C. and Nieuwenhuys, B.E. (2007) *Catalysis Today*, **119**, 305–310.
116 Solsona, B.E., Garcia, T., Jones, C. et al. (2006) *Applied Catalysis A – General*, **312**, 67–76.
117 Ivanova, S., Petit, C. and Pitchon, V. (2006) *Catalysis Today*, **113**, 182–186.
118 Centeno, M.A., Paulis b, M., Montes, M. and Odriozola, J.A. (2005) *Applied Catalysis B – Environmental*, **61**, 177–183.
119 Lai, S.Y., Qiu, Y. and Wang, S. (2006) *Journal of Catalysis*, **237**, 303–313.
120 Domínguez, M.I., Sánchez, M., Centeno, M.A. et al. (2007) *Journal of Molecular Catalysis A – Chemical*, **277**, 145–154.
121 Hayashi, T., Tanaka, K. and Haruta, M. (1998) *Journal of Catalysis*, **178**, 566–575.
122 Kalvachev, Y.A., Hayashi, T., Tsubota, S. and Haruta, M. (1999) *Journal of Catalysis*, **186**, 228–233.
123 Biella, S., Prati, L. and Rossi, M. (2001) IV World Congress on Oxidation Catalysis, Berlin, I, p. 371.
124 Biella, S. and Rossi, M. (2003) *Chemical Communications*, 378–379.
125 Boxhoorn, G. (1988) Shell Internationale Research Maatschappij B V, Netherlands, EP 255975, 8.
126 Lu, J., Bravo-Surez, J.J., Haruta, M. and Oyama, S.T. (2006) *Applied Catalysis A – General*, **302**, 283–295.
127 Vaughan, O.P.H., Kyriakou, G., Macleod, N. et al. (2005) *Journal of Catalysis*, **236**, 401–404.
128 Chu, H., Yang, L., Zhang, Q. and Wang, Y. (2006) *Journal of Catalysis*, **241**, 225–228.
129 Stangland, E.E., Taylor, B., Andres, R.P. and Delgass, W.N. (2005) *The Journal of Physical Chemistry B*, **109**, 2321–2330.
130 Haruta, M., Uphade, B.S., Tsubota, S. and Miyamoto, A. (1998) *Research on Chemical Intermediates*, **24** (3), 329–336.

131 Bianchi, C., Porta, F., Prati, L. and Rossi, M. (2000) *Topics in Catalysis*, **13**, 231–236.
132 Olivera, P.P., Patrito, E.M. and Sellers, H. (1994) *Surface Science*, **313** (1–2), 25–40.
133 Okumura, M., Kitagawa, Y., Yamagcuhi, K. et al. (2003) *Chemistry Letters*, **32**, 822–823.
134 Landon, P., Collier, P.J., Carley, A.F. et al. (2003) *Physical Chemistry Chemical Physics*, **5**, 1917–1923.
135 Sivadinarayana, C., Choudhary, T.V., Daemen, L.L. et al. (2004) *Journal of the American Chemical Society*, **126**, 38–39.
136 Nijhuis, T.A., Visser, T. and Weckhuysen, B.M. (2005) *Angewandte Chemie – International Edition*, **44**, 1115–1118.
137 Nijhuis, T.A., Visser, T. and Weckhuysen, B.M. (2005) *The Journal of Physical Chemistry B*, **109**, 19309–19319.
138 Nijhuis, T.A., Gardner, T.Q. and Weckhuysen, B.M. (2005) *Journal of Catalysis*, **236**, 153–163.
139 Uphade, B.S., Tsubota, S., Hayashi, T. and Haruta, M. (1998) *Chemistry Letters*, **27**, 1277–1278.
140 Nijhuis, T.A., Huizinga, B.J., Makkee, M. and Moulijn, J.A. (1999) *Industrial & Engineering Chemistry Research*, **38**, 884–891.
141 Yap, N., Andres, R.P. and Delgass, W.N. (2004) *Journal of Catalysis*, **226**, 156–170.
142 Taylor, B., Lauterbach, J. and Delgass, W.N. (2005) *Applied Catalysis A – General*, **291**, 188–198.
143 Cumaranatunge, L. and Delgass, W.N. (2005) *Journal of Catalysis*, **232**, 38–42.
144 Wang, R., Guo, X., Wang, X. et al. (2004) *Applied Catalysis A – General*, **261**, 7–13.
145 Sinha, A.K., Seelan, S., Tsubota, S. and Haruta, M. (2004) *Topics in Catalysis*, **29**, 95–102.
146 Uphade, B.S., Akita, T., Nakamura, T. and Haruta, M. (2002) *Journal of Catalysis*, **209**, 331–340.
147 Uphade, B.S., Okumura, M., Tsubota, S. and Haruta, M. (2000) *Applied Catalysis A – General*, **190**, 43–50.
148 Sinha, A.K., Seelan, S., Tsubota, S. and Haruta, M. (2004) *Angewandte Chemie – International Edition*, **43**, 1546–1548.
149 Chowdhury, B., Bravo-Suarez, J.J., Date, M. et al. (2006) *Angewandte Chemie – International Edition*, **45**, 412–415.
150 Chowdhury, B., Bravo-Suarez, J.J., Mimura, N. et al. (2006) *The Journal of Physical Chemistry B*, **110**, 22995–22999.
151 Bravo-Suarez, J.J., Bando, K.K., Lu, J. et al. (2008) *The Journal of Physical Chemistry C*, **112**, 1115–1123.
152 Joshi, A.M., Delgass, W.N. and Thomson, K.T. (2007) *The Journal of Physical Chemistry C*, **111**, 7841–7844.
153 Taylor, B., Lauterbach, J., Blau, G.E. and Delgass, W.N. (2006) *Journal of Catalysis*, **242**, 142–152.
154 Prati, L. and Rossi, M. (1998) *Journal of Catalysis*, **176**, 552–560.
155 Abad, A., Concepción, P., Corma, A. and García, H. (2005) *Angewandte Chemie – International Edition*, **44**, 4066–4069.
156 Abad, A., Almela, C., Corma, A. and García, H. (2006) *Tetrahedron*, **62**, 6666–6672.
157 Abad, A., Corma, A. and García, H. (2007) *Pure and Applied Chemistry*, **79**, 1847–1854.
158 Abad, A., Almela, C., Corma, A. and García, H. (2006) *Chemical Communications*, 3178–3180.
159 Mori, K., Hara, T., Mizugaki, T. et al. (2004) *Journal of the American Chemical Society*, **126**, 10657–10666.
160 Nielsen, I.S., Taarning, E., Egeblad, K. et al. (2007) *Catalysis Letters*, **116**, 35–40.
161 Dimitratos, N., Villa, A., Wang, D. et al. (2006) *Journal of Catalysis*, **244**, 113–121.
162 Biella, S., Prati, L. and Rossi, M. (2003) *Journal of Molecular Catalysis A – Chemical*, **197**, 207–212.
163 Dimitratos, N., Lopez-Sanchez, J.A., Morgan, D. et al. (2007) *Catalysis Today*, **122**, 317–324.
164 Su, F.-Z., Liu, Y.-M., Wang, L.-C. et al. (2008) *Angewandte Chemie – International Edition*, **47**, 334–337.

165 Kimmerle, B., Grunwaldt, J.-D. and Baiker, A. (2007) *Topics in Catalysis*, **44**, 285–292.
166 Hu, J., Chen, L., Zhu, K. et al. (2007) *Catalysis Today*, **122**, 277–283.
167 Christensen, C.H., Jørgensen, B., Rass-Hansen, J. et al. (2006) *Angewandte Chemie – International Edition*, **45**, 4648–4651.
168 Jørgensen, B., Christensen, S.E., Thomsen, M.L.D. and Christensen, C.H. (2007) *Journal of Catalysis*, **251**, 332–337.
169 Carrettin, S., McMorn, P., Johnston, P. et al. (2002) *Chemical Communications*, 696–697.
170 Miyamura, H., Matsubara, R., Miyazaki, Y. and Kobayashi, S. (2007) *Angewandte Chemie – International Edition*, **46**, 4151–4154.
171 Kanaoka, S., Yagi, N., Fukuyama, Y. et al. (2007) *Journal of the American Chemical Society*, **129**, 12060–12061.
172 Biffis, A., Cunial, S., Spontoni, P. and Prati, L. (2007) *Journal of Catalysis*, **251**, 1–6.
173 Zheng, N. and Stucky, G.D. (2007) *Chemical Communications*, 3862–3864.
174 Haider, P. and Baiker, A. (2007) *Journal of Catalysis*, **248**, 175–187.
175 Della Pina, C., Falletta, E. and Rossi, M. (2008) *Nanoparticles and Catalysis* (ed. Didier Astruc), Wiley-VCH, Weinheim, pp. 427–455.
176 Biella, S., Castiglioni, G.L., Fumagalli, C. et al. (2002) *Catalysis Today*, **72**, 43–49.
177 Taarning, E., Madsen, A.T., Marchetti, J.M. et al. (2008) *Green Chemistry*, **10**, 408–414.
178 Huang, J., Dai, W.-L., Li, H. and Fan, K. (2007) *Journal of Catalysis*, **252**, 69–76.
179 Carrettin, S., McMorn, P., Johnston, P. et al. (2004) *Topics in Catalysis*, **27**, 131–136.
180 Carrettin, S., McMorn, P., Johnston, P. et al. (2003) *Physical Chemistry – Chemical Physics*, **5**, 1329–1336.
181 Hutchings, G.J., Sarrettin, S., Landon, P. et al. (2006) *Topics in Catalysis*, **38**, 223–230.
182 Damirel, S., Kern, P., Lucas, M. and Claus, P. (2007) *Catalysis Today*, **122**, 292–300.
183 Prati, L. and Porta, F. (2005) *Applied Catalysis A – General*, **291**, 199–203.
184 Dimitratos, N., Lopez-Sanchez, J.A., Lennon, D. et al. (2006) *Catalysis Letters*, **108**, 147–153.
185 Ketchie, W.C., Fang, Y., Wong, M.S. et al. (2007) *Journal of Catalysis*, **250**, 94–101.
186 Porta, F. and Prati, L. (2004) *Journal of Catalysis*, **224**, 397–403.
187 Bianchi, C.L., Canton, P., Dimitratos, N. et al. (2005) *Catalysis Today*, **102–103**, 203–212.
188 Dimitratos, N., Porta, F. and Prati, L. (2005) *Applied Catalysis A – General*, **291**, 210–214.
189 Damirel, S., Lehnert, K., Lucas, M. and Claus, P. (2007) *Applied Catalysis B – Environmental*, **70**, 637–643.
190 Kimura, H. (2001) *Polymers for Advanced Technologies*, **12**, 697–710.
191 Biella, S., Prati, L. and Rossi, M. (2002) *Journal of Catalysis*, **206**, 242–247.
192 Comotti, M., Della Pina, C., Matarrese, R. and Rossi, M. (2004) *Angewandte Chemie – International Edition*, **43**, 5812–5815.
193 Comotti, M., Della Pina, C., Matarrese, R. et al. (2005) *Applied Catalysis A – General*, **291**, 204–209.
194 Comotti, M., Della Pina, C., Falletta, E. and Rossi, M. (2006) *Journal of Catalysis*, **244**, 122–125.
195 Thielecke, N., Aytemir, M. and Prüße, U. (2007) *Catalysis Today*, **121**, 115–120.
196 Thielecke, N., Vorlop, K.-D. and Prüße, U. (2007) *Catalysis Today*, **122**, 266–269.
197 Mirescu, A., Berndt, H., Martin, A. and Prüße, U. (2007) *Applied Catalysis A – General*, **317**, 204–209.
198 Comotti, M., Della Pina, C. and Rossi, M. (2006) *Journal of Molecular Catalysis A – Chemical*, **251**, 89–92.
199 Patil, N.S., Uphade, B.S., Jana, P. et al. (2004) *Journal of Catalysis*, **223**, 236–239.

200 Lignier, P., Morfin, F., Piccolo, L. et al. (2007) *Catalysis Today*, **122**, 284–291.

201 Xu, L.-X., He, C.-H., Zhu, M.-Q. and Fang, S. (2007) *Catalysis Letters*, **114**, 202–205.

202 Lü, G., Zhao, R., Qian, G. et al. (2004) *Catalysis Letters*, **97**, 115–118.

203 Xu, Y.-J., Landon, P., Enache, D. et al. (2005) *Catalysis Letters*, **101**, 175–179.

4
Metal-Substituted Zeolites as Heterogeneous Oxidation Catalysts

Takashi Tatsumi

4.1
Introduction – Two Ways to Introduce Hetero-Metals into Zeolites

There are two ways to introduce heteroatoms into zeolites. One is the introduction of heteroatoms as extraframework species and the other as framework species.

Most zeolites have an intrinsic ability to exchange cations [1]. This exchange ability is a result of isomorphous substitution of a cation of trivalent (mostly Al) or lower charges for Si as a tetravalent framework cation. As a consequence of this substitution, a net negative charge develops on the framework of the zeolite, which is to be neutralized by cations present within the channels or cages that constitute the microporous part of the crystalline zeolite. These cations may be any of the metals, metal complexes or alkylammonium cations. If these cations are transition metals with redox properties they can act as active sites for oxidation reactions.

Research on coordination chemistry in zeolites started in the 1970s [2]. A metal complex of appropriate dimensions can be encapsulated in a zeolite, being viewed as a bridge between homogeneous and heterogeneous systems. Complexes that are smaller than the free diameters of the channels and windows have access to the cavities. Conversely, complexes that are larger than the diameters of the windows must be synthesized *in situ*, namely, by adsorption of the ligands into the zeolites containing transition metal ions or by synthesis of the ligands in those zeolites [3–5]. Herron *et al.* first referred to such zeolite guest molecules as ship-in-a-bottle complexes [6]. Cationic complexes can be tethered to zeolites through the electrostatic interaction. However, notably, ship-in-a-bottle complexes, even if they are neutral, can be stabilized in zeolite pores. Since the first report on the synthesis of a metal phthalocyanine inside zeolite Na-Y in 1977, numerous examples of encapsulation of metal phthalocyanine complexes have been provided. Related porphyrin and N,N'-bis(salicylidene)ethylenediimine (Salen) complexes have also been trapped in a zeolite cavity that has restricted apertures. These are typical examples of ship-in-a-bottle complexes and are given name zeozymes [7] and inorganic protein [8] in regard to biomimetic chemistry such as models for dioxygen binding and oxygenase.

The other way to introduce heterometals is their isomorphous substitution for Si in the framework, in a similar manner to the isomorphous substitution of Al. The heteroatoms should be tetrahedral (T) atoms. In hydrothermal synthesis, the type and amount of T atom, other than Si, that may be incorporated into the zeolite framework are restricted due to solubility and specific chemical behavior of the T-atom precursors in the synthesis mixture. Breck has reviewed the early literature where Ga, P and Ge ions were potentially incorporated into a few zeolite structures via a primary synthesis route [9]. However, until the late 1970s, exchangeable cations and other extraframework species had been the primary focus of researchers.

Isomorphous substitution of Ti for Si was claimed by Taramasso et al. in 1983 [10]. The resulting material has the structure of silicalite-1 (pure silica MFI) with Ti in the framework positions and is named TS-1 or titanium silicalite-1. The new findings, including the claim that other metals can be inserted into the zeolite lattice, met with skepticism. Ione et al. predicted the probability of isomorphous substitution of metal ion (M^{n+}) and the stability of the M^{n+} position in the tetrahedrally surrounding oxygen atoms by using the Pauling criterion [11]. Based on the ratio of ionic radii ρ of the cation and anion, the value for Ti and O ($\rho = 0.515$) falls out of the range ($\rho = 0.225$–0.414) for which tetrahedral coordination is expected [12]. The allowed cations would include only Al^{3+}, Mn^{4+}, Ge^{4+}, V^{5+}, Cr^{6+}, Si^{4+}, P^{5+}, Se^{6+} and Be^{2+}. Presumably this type of estimate is effective, which can explain the preference of B^{3+} for trigonal coordination and the resultant instability of B^{3+} in the zeolite matrix. However, it is a very rough approximation since the T–O bond is not completely ionic and the model also assumes the atoms have a perfect round shape.

TS-1 proved to be a very good catalyst for liquid-phase oxidation of various organic compounds using H_2O_2 as oxidant. This chapter deals with oxidation catalysis exhibited mainly by isolated transition metals incorporated into the framework of zeolites. Although the metals include titanium, vanadium, chromium and cobalt, most of this chapter is devoted to titanium-containing zeolites. The success of TS-1 has encouraged researchers to synthesize other titanosilicates with different zeolite structures, especially those with larger pores, since TS-1 was inapplicable to bulky molecules owing to the medium-size pores of the ten-membered rings (MR). Table 4.1 lists representative titanosilicates prepared by various techniques.

4.2
Titanium-Containing Zeolites

4.2.1
TS-1

A very comprehensive review has been published on TS-1 and other titanium-containing molecular sieves [26]. TS-1 was synthesized by the hydrothermal crystallization of a gel obtained from $Si(OC_2H_5)_4$ and $Ti(OC_2H_5)_4$ [10, 27] (Enichem method, hereafter named method A). The incorporation of Ti into the framework of

4.2 Titanium-Containing Zeolites

Table 4.1 Representative titanosilicates.

Material	Structure code	Channel system	Preparation methods[a]	Reference
TS-1	MFI	10-10	HTS	[10]
TS-2	MEL	10-10	HTS	[13]
Ti-ZSM-48	N.A.[b]	10	HTS	[14]
Ti-beta	*BEA	12-12-12	HTS, F$^-$mm, DGC	[15–17]
TAPSO-5	AFI	12	HTS	[18]
Ti-ZSM-12	MTW	12	HTS	[19]
Ti-MOR	MOR	12-8-8	PS	[20]
Ti-ITQ-7	ISV	12-12-12	HTS	[21, 22]
Ti-MWW	MWW	10, 10	HTS, PS	[23, 24]
T-YNU-2	MSE	12-10-10	PS	[25]

[a] HTS, hydrothermal synthesis in alkali media; DGC, dry gel conversion; PS, post-synthesis; F$^-$mm, fluoride media method.
[b] Not assigned.

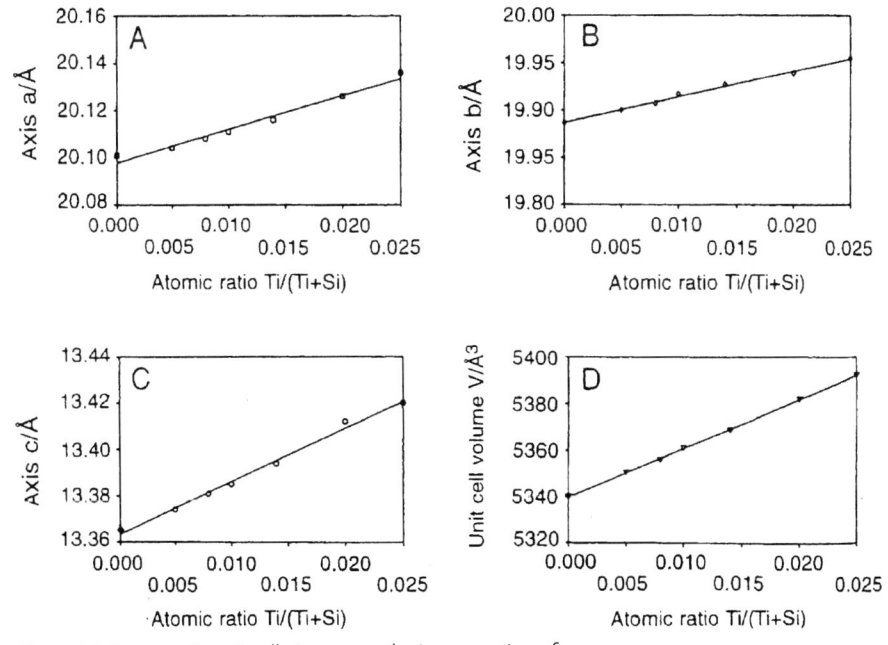

Figure 4.1 Increase in unit cell size versus the incorporation of Ti into the framework of the MFI structure [28].

MFI structure was demonstrated by the increase in unit cell size in XRD pattern (Figure 4.1) [28], and the appearance of tetrahedral Ti species in UV/Vis spectra. The maximum amount of Ti that can be accommodated in the framework positions is claimed to be limited to a value of $x = \text{Ti}/(\text{Ti} + \text{Si})$ of 0.025.

TS-1 can serve as a highly efficient catalyst for the oxidation of various organic substrates, for example, alkanes, alkenes, alcohols and aromatics, with H_2O_2 as an oxidant under mild conditions [13, 29–32]. This discovery was a major breakthrough in research into the oxidation of organic substrates. For example, in current epoxidation processes, chlorine, hydroperoxides and peracids are the most commonly used oxidants. Organic and inorganic by-products are co-produced in the reaction, and need to be recycled or disposed of. In the chlorohydrin route, generally preferred in the epoxidation of C_3 and higher olefins, stoichiometric amounts of sodium or calcium chlorides are produced by the dehydrohalogenation of intermediate chlorohydrins. Chlorinated organic by-products, such as halogen ethers and dichlorides, are formed as well in the process, further increasing the quantity of waste. In contrast, the epoxidation reaction catalyzed by TS-1 may be performed under mild conditions in dilute aqueous or methanolic solution. The active oxygen content H_2O_2, 47 wt% (16/34), is much higher than that of organic peracids and hydroperoxides; water is an only co-product. Besides epoxidation, TS-1 catalyzes a broad range of oxidation reactions with hydrogen peroxide as oxidant (Table 4.2).

Notably, in the presence of alkalis, extraframework Ti species are formed, giving rise to inferior catalytic properties [13]. It was widely believed that the presence of alkali metal, even in very small amounts, cramps the activity of TS-1 for the oxidations with H_2O_2 by preventing the insertion of titanium into the silicalite framework. However, Khouw and Davis reported that the presence of alkali metal ion in the preformed TS-1 has no significant effect on the activity [33]; although neither sodium-exchanged TS-1 nor TS-1 synthesized in the presence of high alkali metal concentrations (Si/Na < 20) is active for alkane oxidation, the catalytic activity can be restored by washing the solid with acid solution. The restoration of the activity may be ascribed to the conversion of the Na-exchanged TS-1 into its original form (Scheme 4.1).

No satisfactory explanation has yet been offered for the lack of activity of sodium-exchanged Ti species. If this acid treatment is generally applicable, it will be extremely

Table 4.2 Catalytic chemistry with TS-1.

Substrate	Product
Olefins	Epoxides
Olefins and methanol	Glycol monomethyl ethers
Diolefins	Monoepoxides
Phenol	Hydroquinone and catechol
Benzene	Phenol
Paraffins	Alcohols and ketones
Primary alcohols	Aldehydes
Secondary alcohols	Ketones
Ammoximation of cyclohexanone	Oximes
N,N-Dialkylamines	N,N-Dialkylhydroxylamines
Thioethers	Sulfoxides

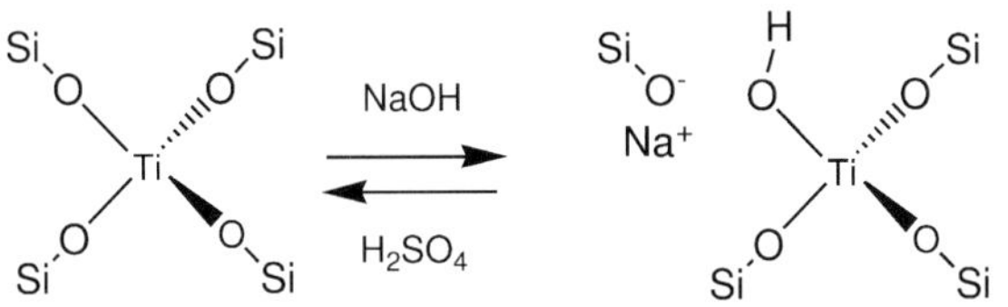

Scheme 4.1 Possible interconversion between TS-1 and Na-exchanged TS-1.

useful for synthesizing various titanium silicate structures that require the presence of alkali metal ions for their crystallization.

The catalytic properties of TS-1 depend on the lattice Ti content, which is usually less than 2 wt% [34, 35]. Finding an effective way to increase the Ti content in the framework of TS-1 is still a huge challenge. Thangaraj and Sivasanker reported that eight Ti ions could be incorporated in the lattice sites per unit cell (Si/Ti = approx. 10) by an improved method (method B) in which titanium tetra-n-butoxide was first dissolved in isopropyl alcohol before addition to the hydrolyzed tetraethyl orthosilicate aqueous solution for the purpose of avoiding formation of TiO_2 precipitate by reducing the hydrolysis rate of the alkoxide [36]. However, Schuchardt and his coworkers could not reproduce this work, and found no difference in the framework Ti content between the samples synthesized by methods A and B [37].

To synthesize Ti-rich TS-1 it is helpful to make its crystallization mechanism clear. However, very few reports have been devoted to this subject [38]. The crystallization process of titanosilicates is much more complex than that of aluminosilicates because Ti^{4+} has a weak structure-directing role and is much more difficult to incorporate into the framework than Al^{3+}. Isomorphous substitution of metal atoms for Si in zeolites is not only related to zeolite structures/framework composition flexibility and the chemical nature of metals but also strongly influenced by the crystallization mechanism. The framework composition flexibility of zeolites is chemically important. Ti K-edge EXAFS studies have shown that the Ti–O bond length of tetrahedral $Ti(OSi)_4$ species is about 1.80 Å, in contrast to 1.61 Å for the Si–O bond [39, 40]. The Ti–O bond is much longer than the Si–O bond, probably making the local structure around Ti seriously distorted. This results in the slow inclusion of Ti into the framework, compared to Si ions. If crystallization proceeded too fast, Ti ions would not have enough time to be incorporated into the lattice. However, too slow a crystallization would possibly cause the formation of transition metal oxides, preventing metal cations from being incorporated into the framework. In addition, the difficult crystallization may also result from the strong competition between the interaction of soluble silicate ions and mother liquor and the condensation of silicate ions. Furthermore, a mismatch among hydrolysis of Ti and Si alkoxides, polymerization of Ti^{4+} and/or Si^{4+} ions, nucleation and crystal growth would lead to much difficulty in the inclusion of Ti in the framework. Since the chemical nature of Ti and the rigidity of the framework of TS-1 cannot be altered, finding an effective crystallization-mediating agent would be the sole way to increase the lattice Ti content in TS-1 by harmonizing the hydrolysis rate of Ti alkoxide with that of silicate species as well as the nucleation and crystal growth rates.

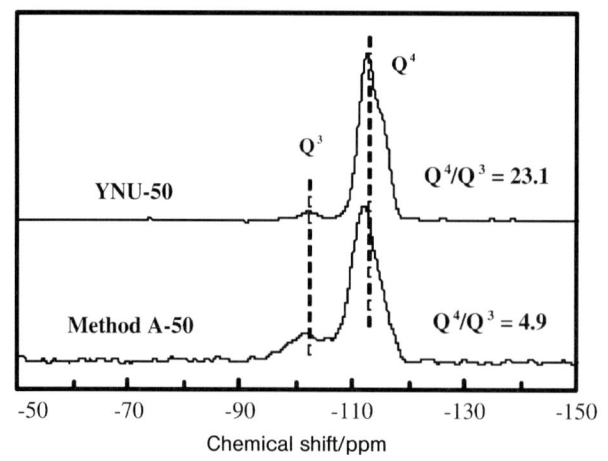

Figure 4.2 ^{29}Si MAS NMR spectra of the prepared TS-1 catalysts (the Si/Ti ratios are 58.3 and 57.2 for the A-50 and YNU-50 samples, respectively).

In this respect, a new route to the synthesis of TS-1 has been developed by using $(NH_4)_2CO_3$ as a crystallization-mediating agent (the YNU method) [41]. By this method, the framework Ti content can be significantly increased without forming extraframework Ti species. The prepared catalyst has a Si/Ti ratio as low as 34, although under the same synthesis conditions we only achieved the ratio of 58 by the methods A and B established by the Enichem group [27] and Thangaraj and Sivasanker [36], respectively. The YNU samples are well crystalline, containing less defect sites than samples synthesized by the other two methods. As shown in Figure 4.2, the ^{29}Si NMR spectra indicate that the YNU-50 sample has a higher Q^4/Q^3 ratio than the method A-50 sample (Enichem method), which indicates that more defective sites are present in the A samples. This is consistent with the finding that the silanol groups or the defect sites decreased with increasing Ti content in the framework [34, 39], since Ti has a mineralizing effect and more Ti cations are incorporated into the YNU samples than in the A samples. Therefore, compared with the YNU samples, the A samples are not only relatively hydrophilic but might contain more $(SiO)_3TiOH$ species. The higher hydrophilicity of the A samples is further corroborated by the fact that the weight loss below 150 °C due to the desorption of water was about 1.0 wt% for as-synthesized A-50 in contrast to 0.3 wt% for as-synthesized YNU-50.

Although the catalytic properties of titanosilicates are related to their crystalline structures and/or characters, the hydrophilicity caused by the presence of more defect sites in titanosilicates would be unfavorable for the oxidation of hydrocarbon reactants, as suggested by the following facts: (1) compared with TS-1, Ti-MWW and Ti-beta with more defect sites showed very low activity for the oxidation of hexane, styrene and benzene; (2) Ti-MCM-41, with a low Q^4/Q^3 ratio, showed very low activity in the oxidation of hexane and 1-hexene; however, silylation resulted in a remarkable

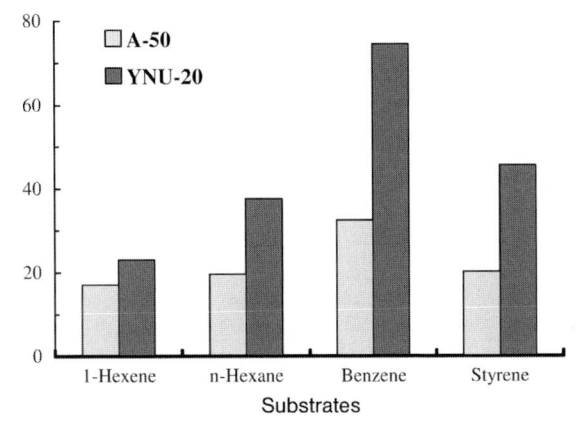

Figure 4.3 Catalytic results of the oxidation of various organic substrates over the YNU-20 and A-50 catalysts. H_2O_2 based conv. (%): the percent used for the oxidation of substrates. Reaction conditions: for 1-hexene, 0.05 g catalyst, 10 mL methanol, 10 mmol substrate, 10 mmol H_2O_2, 60 °C, 2 h; for n-hexane, 0.1 g catalyst, 10 mL methanol, 10 mmol substrate, 20 mmol H_2O_2, 60 °C, 4 h; for benzene, 0.1 g catalyst, 10 mL sulfolane, 10 mmol substrate, 1 mmol H_2O_2, 100 °C, 2 h; for styrene, 0.1 g catalyst, 10 mL acetone, 5 mmol substrate, 2.5 mmol H_2O_2, 60 °C, 4 h.

improvement in the oxidation activity [42]. Furthermore, it has been proven that $(SiO)_3TiOH$ species are much less active than $(SiO)_4Ti$ species for the liquid-phase oxidation of organics over TS-1 [43, 44]. As a result, the YNU sample showed much higher activity for the oxidation of various organic substrates, such as linear alkanes/alkenes and alcohols, styrene and benzene, than the sample A because of their higher hydrophobicity (Figure 4.3).

Figure 4.4 shows the crystallinity of the TS-1 and silicalite-1 samples synthesized under different conditions in the YNU and method A systems (the Si/Ti ratio in the synthesis gel was 50 for both series of samples). Clearly, the crystallization rate of TS-1 was much slower in the YNU system than that in the method A system and the presence of Ti in the gel not only severely retarded the crystallization process by prolonging the induction period but also reduced the crystallinity of most of the samples. Nevertheless, higher crystallinity of TS-1 was finally achieved in the YNU system than in the method A system.

Compared to the sample crystallized in the method A system, TS-1 synthesized in the YNU system was more stable against hydrolysis under the crystallization conditions after the crystallization was completed. This may be due to the incorporation of more Ti cations into the framework associated with a reduction in defect sites, resulting in the higher hydrophobicity. Lamberti et al. found that the intensity of the OH band due to internal defective Si–OH groups progressively decreased with increasing Ti amount incorporated in the framework [34, 35]. This shows that the insertion of Ti into the lattice of TS-1 has a mineralizing effect, reducing the framework defects, and thus increasing the hydrophobicity. This is also evidenced by the ^1H and ^{29}Si MAS NMR spectroscopic findings that silanol protons of H-bonded

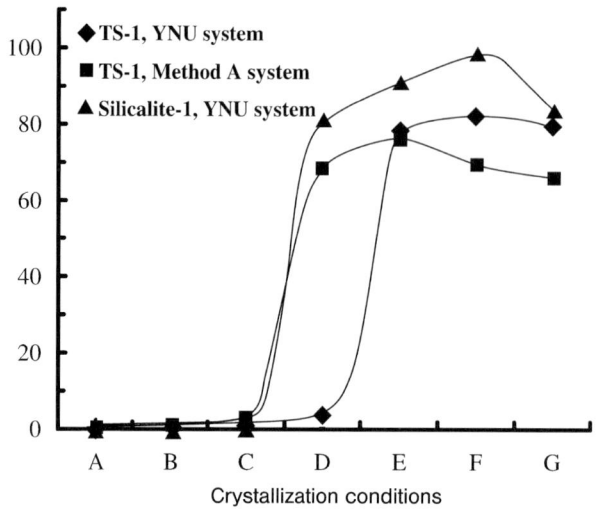

Figure 4.4 Crystallization curves of TS-1 and silicalite-1 synthesized by the YNU system and of TS-1 synthesized by method A. A: 30 °C, 1 day; B: 60 °C, 1 day; C: 80 °C, 1 day; D: 100 °C, 1 day; E: 140 °C, 1 day; F: 170 °C, 1 day; G: 170 °C, 3 days.

siloxy oxygens at defect sites proportionally decreased with the Ti atoms per unit cell [45].

An increase in hydrophobicity by inserting Ti into the framework is further substantiated by the crystallization curves; the crystallinity of silicalite-1 decreased with increasing crystallization time after reaching the maximum. The presence of defect sites would make the material hydrophilic, resulting in an easy attack by adsorbed H_2O, and thus a collapse of the framework (4.1):

$$\equiv Si-O-Si \equiv + H_2O \rightarrow 2 \equiv Si-OH \tag{4.1}$$

Figure 4.5 shows the Si/Ti ratio of the solid samples collected in the whole crystallization process. The Si/Ti ratio in the solid samples synthesized in the YNU system was kept almost constant, being in the range 48–50.5. This constant Si/Ti ratio is similar to that found in the synthesis of TS-1 from amorphous wetness impregnated SiO_2–TiO_2 xerogels and the non-aqueous synthesis of zeolites with solid reaction mixtures where a solid-phase transformation mechanism predominated [38, 46]. In addition, the solid yields recovered after calcination of all the YNU samples were always higher than 95% based on the added SiO_2 and TiO_2. This is also consistent with the occurrence of a solid-phase transformation mechanism, which would be favorable for the incorporation of Ti cations into the framework since dissociation, coalescence and reorganization are the main processes during nucleation and crystal growth.

In contrast, as shown in Figure 4.5, the Si/Ti ratio of the samples synthesized by the method A drastically increased during the period of crystal growth; the solid samples in the induction period were obtained directly by drying the liquid at 100 °C

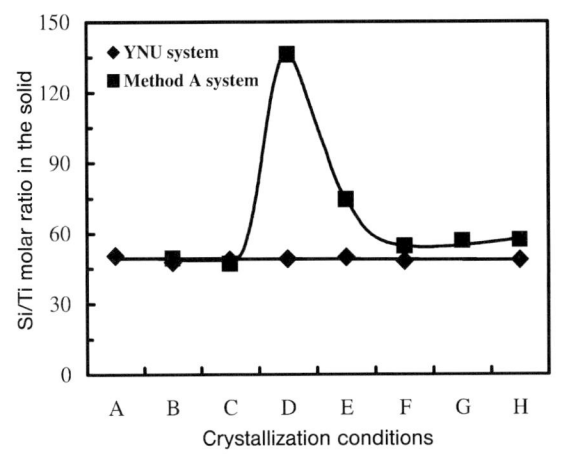

Figure 4.5 Dependence of the Si/Ti molar ratio of the solid fraction on the crystallization conditions in the YNU and method A systems. A: 30 °C, 1 day; B: 60 °C, 1 day; C: 80 °C, 1 day; D: 100 °C, 1 day; E: 140 °C, 1 day; F: 170 °C, 1 day; G: 170 °C, 2 days; H: 170 °C, 3 days.

since no solid was formed during this period. This suggests that during the period of rapid crystal growth in the method A system, silicic and/or silicate species ($\equiv$Si–OH, $\equiv$Si–O$^-$) were polymerized with each other at a much higher rate than the condensation of silicic or silicate species and titanic or titanate ($\equiv$Ti–OH, $\equiv$Ti–O$^-$) species to form TS-1 crystals. The Ti content in the solid increased remarkably after the sample reached the highest crystallinity (Figure 4.4). The crystallization in the method A system occurred via a homogeneous nucleation mechanism although precursor aggregates might be formed before nucleation. It can be also seen from Figure 4.5 that in this stage the Ti content in the as-synthesized YNU products was higher than in the A products, suggesting the incorporation of more Ti cations into the framework.

In contrast, the evolution of the Ti/Si ratio in the mother liquid during the whole crystallization process is in agreement with that of the solid samples. In the method A system, the Ti/Si ratio of the liquid phase increased sharply from the point after the nuclei formed, and reached the highest value when crystallization was almost completed. This again shows that in the method A system Ti cations started to be inserted into the framework after the zeolitic architecture was almost built. In contrast, in the YNU system, the Ti/Si ratio in the liquid was almost constant (about 0.02) during the induction period, being the same as that in the solid fraction. After nuclei formed, a small amount of titanic and/or titanate species (>95% of silicon and titanium was kept in the solid fraction through the crystallization process) in the mother liquid quickly transferred into the solid, and the transfer rate was consistent with the crystal growth speed; the Ti content in the mother liquid decreased to 0 when the crystallization conditions were changed from 100 °C for 1 day (the formation of the MFI structure was just detected by XRD) to 140 °C for

1 day (the crystallization was nearly completed). This shows that the crystallization mechanism in the YNU system is completely different from that working in the method A system. Nevertheless, the decrease in the Ti/Si ratio in the liquid during the crystal growth period is indicative of the occurrence of a liquid-to-solid mass transfer of Ti as well as the solid-phase transformation that probably dominated the crystallization process. The complete depletion of Ti species in the liquid in the YNU system is in contrast to a fraction of Ti species left in the liquid in method A, further evidencing that more Ti cations can be incorporated in the framework by the YNU method than by method A.

In the synthesis of TS-1 in the method A system, the hydroxide ion of TPAOH plays a major role in accelerating the hydrolysis of silica source and Ti alkoxide, and the oligomerization of Si–OH and Ti–OH species (4.2). These processes would lower the pH. Therefore, a high pH value could promote these processes, and thus decrease the time taken to reach the critical concentration of soluble silica and titanium species from which zeolites are ultimately precipitated. However, a very high alkaline condition would make the synthesis gel unreactive, but dissolve the silicate/titanosilicate species [(4.3), from right to left], resulting in a lower crystallinity of the products crystallized by method A and the presence of more defect sites (Figure 4.4). In contrast, for the YNU system the synthesis gel was quickly solidified after the addition of $(NH_4)_2CO_3$. Thus, most TPA^+ species were embedded in the solid with the amount of free OH^- species drastically decreasing, such that the pH of the liquid decreases. The lower the pH, the higher the yield of crystalline material, or the more well-crystallized the product because the condensation reaction [(4.3), from left to right] proceeds to a greater extent. This accounts for the higher crystallinity of the TS-1 samples achieved by the YNU method than by method A:

$$x(TPA^+OH^-) + T(OR)_4 + (4-x)H_2O \rightleftharpoons (TPA^+)_x(^-O)_x - T(OR)_{4-x}$$

$$+ xROH + (4-x)H_2O \rightleftharpoons (TPA^+)_x(^-O)_x - T(OH)_{4-x}$$

$$+ (4-x)ROH \quad (T = Si, Ti) \tag{4.2}$$

$$\equiv Si - O^- + \equiv TOH \rightleftharpoons \equiv Si - O - T \equiv + OH^- (T = Si, Ti) \tag{4.3}$$

As expected, an increase in pH during the crystal growth period was observed for both synthesis systems, but the degree was much more appreciable for method A than for the YNU system. The pH increased from 11.9 to 12.8 with method A, while it increased only slightly, from 10.4 to 10.7, in the YNU system. The increase in pH is attributed to the incorporation of silicate species into the framework of TS-1, which releases free OH^- (4.3). This increase continued till completion of crystal growth. The higher OH^- concentration in the method A system also accelerates the crystal growth. Such a high crystallization rate is not beneficial for the incorporation of Ti into the framework since this process would make the local structure around Ti distorted. In addition, the high alkalinity of the liquid is unfavorable to the condensation of Ti–OH and silicate species. In contrast, the presence of $(NH_4)_2CO_3$ appropriately slows down the crystallization rate by significantly decreasing the pH value, buffering the synthesis gel and introducing NH_4^+, which is a structure-breaking cation in water, and consequently reducing the polymerization rate of silicic/silicate/titanic/titanate species. This would provide enough time for Ti species to be inserted into the lattice during the crystallization process, as indicated by the much lower Si/Ti ratio in the solid samples obtained during the crystal growth period in the YNU system than in the method A system.

Thus it is concluded that the presence of $(NH_4)_2CO_3$ not only drastically lowered the pH, slowing down the crystallization process and making the incorporation of Ti into the framework match well with nucleation and crystal growth, but it also modified the crystallization mechanism. It seems that the solid-phase transformation mechanism predominated in the crystallization process initiated by dissociation, reorganization and re-coalescence of the solidified gel, although a small amount of non-gelatinated Ti shifted to the solid during the crystal growth period. In contrast, a typical homogeneous nucleation mechanism occurred with method A.

The most widely accepted mechanism for TS-1 catalyzed epoxidation is the peracid-like mechanism, which involves a hydroperoxo rather than a peroxo species, and coordination of an alcohol or water molecule to the site (Scheme 4.2) [26, 31]. Titanium isomorphously substituted for Si in the zeolite framework in tetrahedral coordination is much more resistant to hydrolysis than titanium species on amorphous silica. Thus Ti species in TS-1 are negligibly leached out from the framework during the oxidation reactions in the liquid phase, in contrast to those on amorphous silica. A key characteristic of these materials is their relatively high hydrophobicity, resulting in the favorable adsorption of alkanes and other hydrocarbons, as described above. Thus the low concentration of hydrogen peroxide present at all times in the catalyst results in its efficient use. The strong hydrophobicity also enables fast desorption of the oxygenated products. Therefore, oxidations can occur up to high conversions with high H_2O_2 selectivities and high efficiencies.

TS-1 has the MFI structure and a medium pore (10-MR) zeolite. The structure of TS-1 prevents all compounds having a cross section larger than about 0.55 nm

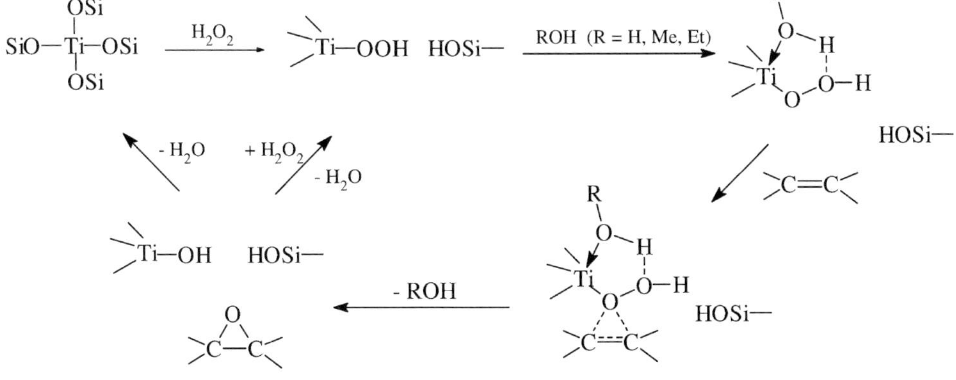

Scheme 4.2 Mechanisms of epoxidation on titanosilicates.

from diffusing inside its channels and, therefore, from reacting or interfering with reactions occurring at Ti-sites. Thus TS-1 shows remarkable shape selectivity. The rate of oxidation of the branched or cyclic hydrocarbons is much lower than that of the straight one [29, 47]. Hayashi et al. have reported that oxidation of 2-propanol over TS-1 is strongly retarded in the presence of 1-propanol [48]; 1-propanol coordinates strongly on the active site followed by slow oxidation, while 2-propanol coordinated weakly but is oxidized quickly. The importance of coordination of the group to be oxidized to the active site present in the sterically constrained environment in the zeolite pores has been revealed by intra- and intermolecular competitive oxidation [49]. Since oxidations are irreversible in principle, and usually not accompanied by reactions that can carry out the interconversion of products, only reactant selectivity or restricted transition state selectivity can be practically important.

4.2.2
Ti-Beta

Although the discovery of TS-1 is a milestone in zeolite catalysis, the smaller pore size of the zeolite-type framework restricts its use, even for small molecules such as simple cyclic alkenes. Thus following the success of TS-1 as a liquid-phase oxidation catalyst great efforts have been devoted to the synthesis of titanium silicalites with different zeolitic structures. One of the interesting titanium silicalites is Ti-beta (BEA*), a large-pore Ti-containing zeolite, which has been synthesized hydrothermally from gels containing tetraethylammonium hydroxide (TEAOH) as a structure-directing agent (SDA) [50]. Owing to its large pore size, Ti-beta was shown to be more active than TS-1 for the oxidation of bulky substrates such as cyclic and branched molecules [51].

Ti-beta was usually obtained in very low yield [51], in contrast to TS-1. Moreover, an additional factor, viz., the presence of aluminium in the framework of Ti-beta, can contribute to the different catalytic behavior observed between Ti-beta and TS-1. In contrast to the hydrophobic characteristics of TS-1, the presence of Al and the large concentration of internal and external silanol groups confer a rather hydrophilic character on Ti-beta. There is, therefore, a strong incentive for the preparation of Ti-beta with low Al content in a better zeolite yield by using new methods. Improved methods for the synthesis of Ti-beta, for example, use of special SDA [52], the fluoride method [53] and the dry gel conversion method [54], have been developed to obtain Ti-beta zeolites active for H_2O_2 oxidation in high yields. The improvement of epoxide selectivity by selectively poisoning the acid function without spoiling the oxidation activity is attained by modification by ion exchange with quaternary ammonium ions [55]. Notably, cleavage of the oxirane ring is promoted by the presence of H_2O_2. This indicates that the acidity of titanosilicates is generated by the contact of Ti sites with H_2O_2 [55].

The fluoride method gives rise to a very hydrophobic Ti-beta free of framework Al and silanol defects. Although this material was expected to have high catalytic activity, in the epoxidation of alkenes, the turnover number of such Ti-beta was only comparable to hydrothermally synthesized Ti-beta. One reason might be the larger crystal size obtained by the fluoride method, which decreases the effectiveness of the solid catalyst. Koller *et al.* have reported the existence of five-coordinated Si species, $(SiO)_4SiF^-$, in high-silica zeolites (MFI, BEA, MTW, etc.) synthesized in the fluoride media, as confirmed by ^{19}F MAS NMR and $^{29}Si\{^{19}F\}$ CP MAS NMR [56]. It has been proved that the presence of the fluorine attached to the zeolite framework would be harmful to the catalytic activity of Ti-beta [57]. Treatment of Ti-beta synthesized by the fluoride method with basic quaternary ammonium solution followed by calcination enhanced the epoxidation activity with hydrogen peroxide, which seems to be due to the reduction of the amount of fluorine contained in the catalyst. FT-IR and UV/Vis spectra revealed the increase in Si−OH and Si−O−Ti groups and the decrease in the Ti species with high coordination number after the treatment followed by calcination.

A different approach to the substitution of metal atoms into the framework is the secondary synthesis or post-synthesis method. This is particularly effective in synthesizing metallosilicates that are difficult to crystallize from the gels containing other metal atoms or hardly incorporate metal atoms by the direct synthesis method. Substitution of Ti for Al goes back to the 1980s. The reaction of zeolites with an aqueous solution of ammonium fluoride salts of Ti or Fe under relatively mild conditions yields materials that are dealuminated and contain substantial amounts of either iron or titanium and are essentially free of defects [58]. However, no sufficient evidence for the Ti incorporation has been provided.

The post-synthetic incorporation of titanium into the framework of zeolite beta has been achieved by liquid-phase treatment employing ammonium titanyl oxalate [59]. Ti-beta has been also prepared by treating Al-containing beta with a concentrated solution of perchloric or nitric acid in the presence of dissolved titanium [60]. Although $Ti(OBu)_4$ and TiF_4 were efficient sources for the incorporation of tetrahedral Ti, the use of TiO_2 as the Ti source gives rise to octahedral Ti as well as tetrahedral Ti. In this case, extraction of Al from the framework occurred simultaneously, giving almost Al-free Ti-beta.

Post-synthesis gas–solid isomorphous substitution methods are also known [61]. Ti-beta essentially free of trivalent metals can be prepared from boron-beta. However, the gas-phase method is not efficient for Ti incorporation and could have some disadvantages such as the deposition of TiO_2 [62].

4.2.3
Ti-MWW

MWW aluminosilicate (generally known as MCM-22) is hydrothermally synthesized without difficulty; however, the synthesis of MWW titanosilicate (Ti-MWW) was a challenge until it was first shown that Ti is effectively incorporated into the MWW framework when boric acid coexists in the synthesis media [23a].

Since titanosilicates generally require specific synthesis conditions in comparison to silicalites and aluminosilicates, many efforts made to synthesize those of numerous zeolite structures have led to a very limited success. This has also been the case with the MWW zeolite. Although it is possible to hydrothermally synthesize MWW silicalite in alkali-free media using a specific organic SDA of trimethyladamantylammonium hydroxide, the addition of other metal cations such as Al and Ti into the synthesis gel results in failure [63]. When boric acid, in an amount even greater than that of silicon, and a Ti source were co-existent in the synthesis gel composed of fumed silica and cyclic amine SDA such as hexamethyleneimine (HM) or piperidine (PI), Ti-MWW was crystallized readily by autoclaving the gel at 403–443 K [23a, 23b]. These are designated as Ti-MWW-HM or Ti-MWW-PI.

All the as-synthesized samples of Ti-MWW-PI and Ti-MWW-HM showed the XRD patterns totally consistent with those of the lamellar precursor of MWW topology, generally designated as MCM-22(P) [64, 65]. Upon calcination at 803 K, all the samples were converted into the porous three-dimensional (3D) MWW structure with good quality. The amount of B incorporated into the products was in the Si/B range 11–13, which is far lower than that in the gel with a Si/B ratio of 0.75. In contrast, there was little difference in the Si/Ti ratios between the gel and the solid product, except for the gel of Si/Ti = 100, indicating that the synthesis system is very effective for Ti incorporation.

Figure 4.6a shows the UV/Vis spectra of Ti-MWW-PI without calcination. No obvious band around 330 nm was observed even at a very high Ti level, indicating that the anatase-like Ti phase was hardly formed during crystallization. However, the spectra are quite different from those reported for TS-1 and Ti-beta that generally show only a narrow band around 210 nm. Irrespective of the Si/Ti ratio and the SDA used, all as-synthesized Ti-MWW samples exhibited a main band at 260 nm together with a weak shoulder around 220 nm. The 220 nm band, resulting from the charge transfer from O^{2-} to Ti^{4+}, has been widely observed for Ti-substituted zeolites and is characteristic of tetrahedrally coordinated Ti highly dispersed in the framework [30]. The 260 nm band has been attributed to octahedral Ti species, related to a kind of extraframework Ti species, probably with Ti–O–Ti bonds in the case of Ti-beta [15].

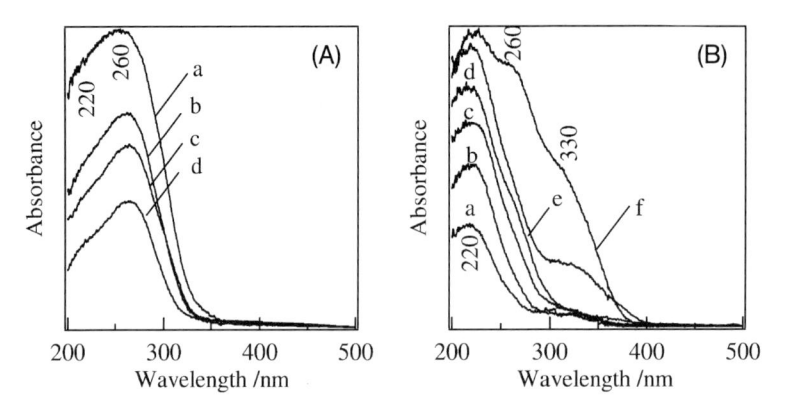

Figure 4.6 UV/Vis spectra. (A) As-synthesized Ti-MWW-PI with a Si/Ti ratio of a: 100, b: 50, c: 30 and d: 10. (B) Acid-treated and further calcined Ti-MWW-PI with a Si/Ti ratio of a: 170, b: 116, c: 72, d: 59, e: 38 and f: 17.

Upon calcination, dehydroxylation between the lamellar sheets occurred to form the MWW structure. This recrystallization process also led to a change in the nature of the Ti species; a new band around 330 nm (Figure 4.6b) is ascribed to anatase. The anatase-type Ti species are not active and may cause unproductive decomposition of the oxidant H_2O_2 when employed as an oxidation catalyst. Once octahedral Ti species are converted into anatase they could not be removed by washing with a HNO_3 or H_2SO_4 solution under refluxing conditions. However, when as-synthesized Ti-MWW was first refluxed with an acid solution and then calcined, the octahedral Ti species were eliminated selectively (Figure 4.6b); only the narrow band at 220 nm due to tetrahedral Ti species was observed for the samples prepared by acid-treating the precursors with the Si/Ti ratio of 100 : 30. Extraframework Ti species, both octahedral and anatase-like, still remained to a certain level for the samples obtained from the precursors with Si/Ti of 20 and 10 because they contained too a high concentration of octahedral Ti. Thus it should be emphasized that the pretreatment sequences are essential for obtaining Ti-MWW with tetrahedrally substituted Ti species. Together with extraframework Ti, some of the framework boron was also extracted, to a level corresponding to an Si/B ratio of about 30.

Large-pore titanosilicates developed after TS-1, for example, Ti-beta, Ti-ITQ-7, Ti-MCM-41 and Ti-MCM-48, have been claimed to have advantages for the oxidation of bulky alkenes because of their pore size [15–17, 66, 67]. However, none of them is intrinsically more active than TS-1 in the reactions of small substrates that have no obvious diffusion problem for the medium pores. Therefore, in parallel with developing large pore titanosilicates, the search for more intrinsically active ones than TS-1 is also an important research subject. The catalytic performance of Ti-MWW is compared in the oxidation of 1-hexene with H_2O_2 with that of TS-1 and Ti-beta. Consistent with the results reported elsewhere [15–17], TS-1 showed higher conversion than Ti-beta with a similar Ti content. However, Ti-MWW exhibited activity about three times as high as TS-1 based on the specific conversion per Ti site (TON).

A more special property exhibited only by Ti-MWW is that it shows unique shape selectivity in the epoxidation of geometric alkene isomers. Table 4.3 shows the results for various titanosilicates in the epoxidation of 2-hexenes with a cis/trans ratio of 41 : 59. The products were 2,3-epoxyhexanes with both cis and trans-configurations, and diols formed by successive hydrolysis of epoxides over acid sites. Although the Ti content varied greatly with the titanosilicates, Ti-MWW obviously exhibited the highest specific activity in the conversion of 2-hexenes with high H_2O_2 efficiency. Interestingly, Ti-MWW showed totally different behavior in the epoxide distribution, that is, only Ti-MWW exhibited a selectivity as high as 81% for the trans-epoxide. In contrast, the other titanosilicates selectively promoted epoxidation of the cis-isomer to give a selectivity for the corresponding epoxide higher than the percentage in the starting substrates. TS-1 is generally a cis-selective catalyst for alkene stereoisomers, selectively producing the cis-epoxide in the epoxidation of cis/trans 2-butenes or cis/trans 2-hexenes with H_2O_2 [68]. Thus there was an essential difference in the geometric selectivity between Ti-MWW and conventional titanosilicates in the epoxidation of cis/trans alkenes.

Table 4.3 Epoxidation of hex-2-ene isomers with hydrogen peroxide over various titanosilicates[a].

Catalyst	Si/Ti	Conversion (mol.%)	Product selectivity (mol.%)		Epoxide selectivity (mol.%)		H_2O_2 (mol.%)	
			Epoxides	Diols	Cis	Trans	Conv.	Eff.
Ti-MWW	46	50.8	99	1	19	81	55.1	92
TS-1	42	29.1	96	4	66	34	32.5	89
TS-2	95	13.6	96	4	67	33	18.0	77
Ti-beta	40	15.9	91	9	73	27	35.8	45
Ti-MOR	79	2.6	99	1	52	48	3.9	66
Ti-Y	43	3.8	40	60	55	45	8.4	46
Ti-MCM-41	50	3.1	36	64	62	38	21.0	15
SiO_2-TiO_2	85	0.8	37	63	61	39	7.6	10

[a] Cat., 0.05 g; hex-2-enes (cis/trans = 41 : 59), 10 mmol; H_2O_2, 10 mmol; MeCN, 10 mL; temp., 333 K; time, 2 h.

As described above, despite a relatively high content of boron (generally corresponding to a Si/B ratio of 30) in the framework, hydrothermally synthesized Ti-MWW proved to be an extremely active catalyst for alkene epoxidation. Therefore, B-free Ti-MWW is expected to exhibit excellent activity, much higher than that of B-containing Ti-MWW. Since the direct synthesis of Ti-MWW without using the crystallization-support agent of boric acid is still a challenge, post-synthesis is an alternative choice for the preparation of B-free catalysts. Treatment with $TiCl_4$ vapor at elevated temperatures is the usual method employed for modifying MOR and BEA zeolites of 12-MR channels [20, 61]. In fact, the preparation of Ti-MCM-22 by the reaction of dealuminated MCM-22 with $TiCl_4$ vapor has been patented [69]. However, it is suspected that the $TiCl_4$ treatment method is actually ineffective for the MWW zeolite because $TiCl_4$ (6.7 × 6.7 Å) is expected to suffer serious steric restriction when penetrating the 10-MR pores (4.0 × 5.9 Å, 4.0 × 5.4 Å) of MWW and thus might give rise to uneven Ti distribution.

As a totally different post-synthesis method that is firmly based on the structural characteristics of MWW, reversible structural conversion between 3D MWW silicate and its corresponding 2D lamellar precursor MWW (P) has been developed to construct more active Ti species within the framework [24, 70]. Figure 4.7 illustrates the strategy of this post-synthesis method of "reversible structural conversion." First, highly siliceous MWW is prepared from hydrothermally synthesized MWW borosilicate by the combination of calcination and acid treatment. Second, the MWW silicalite is treated with an aqueous solution of HM or PI and a Ti source. A reversible structure conversion from MWW into the corresponding lamellar precursor occurs as a result of Si–O–Si bond hydrolysis catalyzed by OH^-, which is supplied by basic amine molecules. This is accompanied by the intercalation of the amine molecules.

Calcination and acid treatment of the as-synthesized borosilicate MWW not only removed the framework boron but also converted the lamellar precursor into a

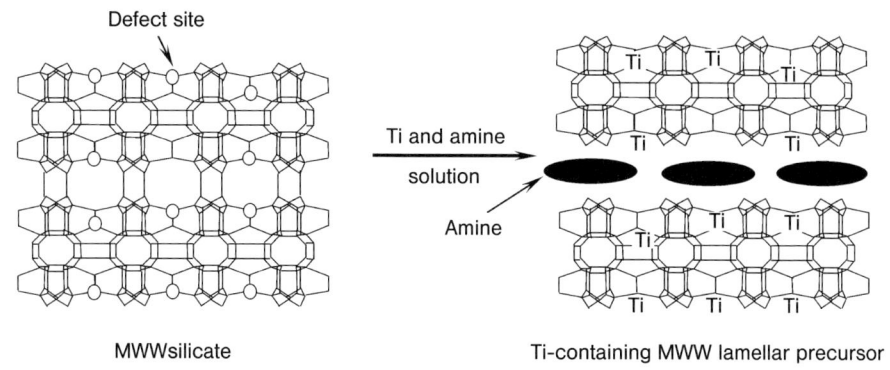

Figure 4.7 Reversible structural conversion from MWW into MWW(P) lamellar precursor as the method of post-synthesizing Ti-MWW.

MWW silicate (Si/B > 500). When this deboronated MWW was treated with Ti(OBu)$_4$ (TBOT) in the presence of HM or PI, Ti was incorporated and, more interestingly, the lamellar structure was simultaneously restored. The Ti species entered the interlayer space freely through the pore entrance of expanded layers to fill up the defect sites such as hydroxyl nests. Extraction of the extraframework Ti species by acid treatment (see next section) followed by calcination caused the layers to dehydroxylate, resulting in B-free Ti-MWW. For the Si/Ti ratio range 20–100 in the starting gel, nearly all the Ti was incorporated into the solid product, indicating that this post-synthesis method was highly effective in introducing Ti. Notably, this structural conversion occurred only in the presence of HM or PI, the two typical SDAs for the crystallization of MWW zeolites, but was never caused by pyridine or piperazine although these cyclic amines have similar molecular shapes. This means that there is a "molecular recognition" of amine molecules by the layered MWW sheets and that HM or PI molecules stabilize the lamellar MWW (P) structure.

The catalytic properties of post-synthesized PS-Ti-MWW were compared with directly hydrothermally synthesized HTS-Ti-MWW and TS-1 in the epoxidation of 1-hexene with H$_2$O$_2$ (Figure 4.8). For reasonable comparison, the reactions were carried out in the most suitable solvents for the two titanosilicates – in acetonitrile for Ti-MWW and in methanol for TS-1. HTS-Ti-MWW showed much higher intrinsic activity than TS-1 for 1-hexene. PS-Ti-MWW further proved to be about twice as active as HTS-Ti-MWW. The efficiency of H$_2$O$_2$ utilization was also very high on PS-Ti-MWW. Thus, in terms of the activity, epoxide selectivity and H$_2$O$_2$ efficiency, PS-Ti-MWW has so far been a most efficient heterogeneous catalyst for liquid-phase epoxidation of linear alkenes.

At first, the main difference between PS-Ti-MWW and HTS-Ti-MWW seemed to be the boron content. A further treatment with refluxing 2 M HNO$_3$ was able to deboronate HTS-Ti-MWW to produce a sample nearly free of boron (Si/B > 500), while removing the Ti species only slightly. However, this did not result in any substantial increase in TON. Consequently, a reason other than the boron content may account for the above results. Since HTS-Ti-MWW is prepared by using boric

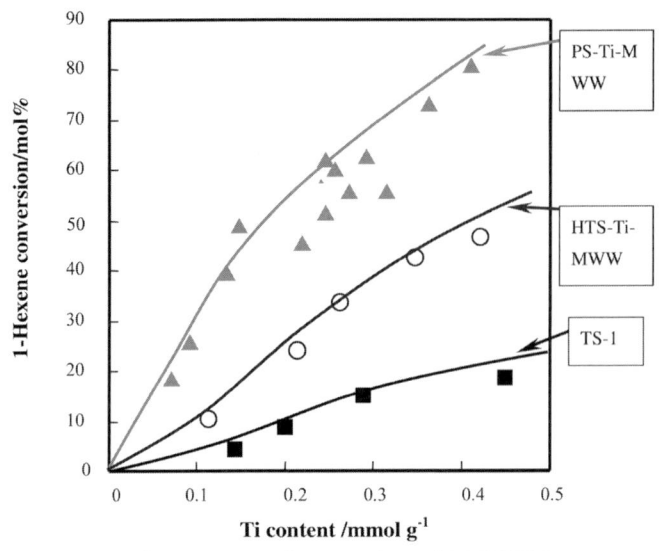

Figure 4.8 Catalytic properties of post-synthesized PS-Ti-MWW, directly hydrothermally synthesized HTS-Ti-MWW and TS-1 in the epoxidation of 1-hexene with H_2O_2.

acid as a structure-supporting agent, the coexisting boron would preferentially occupy the specific framework position, which would hinder the uniform incorporation of titanium. In contrast, PS-Ti-MWW was prepared through inserting the Ti species mainly into defect sites formed by elimination of boron atoms. Thus the Ti species occupying crystallographically different tetrahedral sites may account for the different catalytic behavior between PS-Ti-MWW and HTS-Ti-MWW.

Ti-MWW catalysts have also been found to exhibit catalytic performance superior to conventional titanosilicates such as TS-1 and Ti-beta in other oxidation reactions. For example, epoxidation of allyl alcohol to glycidol [71], epoxidation of diallyl ether to allyl glycidyl ether [72], epoxidation of allyl chloride to epichlorohydrin [73], epoxidation of 2,5-dihydrofuran to 3,4-epoxytetrahydrofuran [74], hydroxylation of 1,4-dioxane to 1,4-dioxane-2-ol [75] and ammoximation of cyclohexanone to cyclohexanone oxime [76]. The catalytic performance of Ti-MWW in the liquid-phase ammoximation of cyclohexanone depends greatly on the operating conditions of the reaction, especially the method of adding H_2O_2. Only when H_2O_2 is added slowly into the reaction system is a high yield of cyclohexanone oxime. The process consists of reaction of NH_3 with H_2O_2 to produce hydroxylamine and subsequent non-catalytic reaction of hydroxylamine with cyclohexanone to afford cyclohexanone oxime. In the presence of excess H_2O_2, extensive oxidation of hydroxylamine occurs on the Ti species of Ti-MWW with extremely high oxidation ability. Thus, under optimized conditions, Ti-MWW can catalyze the ammoximation of cyclohexanone to cyclohexanone oxime at a conversion and selectivity of >99%, thereby proving to be a superior catalyst to TS-1, the currently used industrial catalyst.

Reversible structural conversion did not occur when as-synthesized Ti-MWW (P) with a Si/Ti ratio of >100 was calcined subsequent to washing with 2 M HNO$_3$. The so-obtained novel titanosilicate with structure analogues to the MWW precursor, designated as Ti-YNU-1, shows much higher oxidation ability, epoxide selectivity and stability than Ti-beta in the oxidation of bulky cycloalkenes [77, 78]. Ti-YNU-1 has proved to have a large interlayer pore space corresponding to 12-MR zeolites [79]. Whereas direct condensation of the layers results in the formation of the MWW structure having a 10-MR interlayer pore, apparently monomeric Si species have been inserted into the interlayer spaces followed by condensation to provide a 12-MR pore. Since no Si source has been added, it is assumed that silica "debris" formed by decomposition of part of the MWW layer acted as the source.

This assumption has been tested by deliberately adding a Si source to the as-synthesized MWW(P); by using silylating agents such as SiMe$_2$(OR)$_2$ and SiMe$_2$Cl$_2$ silylene units can be inserted between the layers, which was followed by removal of the organic moieties to give the material that is very similar to Ti-YNU-1. Furthermore, this method of inserting monomeric Si sources into the interlayer spaces can be widely applied to the conversion of various 2D lamellar precursors into novel 3D crystalline metallosilicates with expanded pore apertures between the layers [80, 81].

Based on well-established knowledge about ITQ-2 [65], delaminated titanosilicates, Del-Ti-MWW, which essentially consist of thin sheets, have been prepared [82]. Subsequent treatment of the acid-treated Ti-MWW precursor in a basic solution of tetrapropylammonium hydroxide (TPAOH) and cetyltrimethylammonium bromide (CTMABr) cleaved the interlayer linkages and so allowed easy entry of the surfactant molecules to be intercalated between the layers. This afforded a swollen material with expanded interlayer space, exhibiting a diffraction peak at lower 2° with a corresponding d-spacing of 3.9 nm. After the swollen material was treated in an ultrasound bath and further calcined at 823 K, the expanded layered structure substantially collapsed, leading to a sample with extensively weakened diffractions due to the MWW structure. Nevertheless, this sample exhibited an enlarged surface area, over 1000 m^2 g^{-1}, compared to Ti-MWW, and contained a large amount of silanol groups, as revealed by a strong band at 3742 cm^{-1} in the IR spectra. ^{29}Si MAS NMR spectra also verified the increase in silanol groups.

The catalytic properties of Del-Ti-MWW have been compared with those of other titanosilicates in the epoxidation of cyclic alkenes (Table 4.4). The TON decreased sharply for TS-1, Ti-beta and 3D Ti-MWW with increasing molecular size of cyclic alkenes. Ti-MCM-41 with mesopores, however, showed higher TONs for cyclooctene and cyclododecene. This implies that the reaction space is extremely important for the reactions of bulky molecules. The delamination of Ti-MWW increased the TON greatly for not only cyclopentene but also bulkier cycloalkenes. Especially, the catalytic activity of Del-Ti-MWW was about 6× higher than that of Ti-MWW for cyclooctene and cyclododecene. Del-Ti-MWW even turned out to be superior to Ti-MCM-41 in the epoxidation of bulky substrates. This should be due to the high accessibility of Ti active sites in Del-Ti-MWW. Thus the delamination was able to change Ti-MWW into an effective catalyst applicable to reactions of bulky substrates.

Table 4.4 Epoxidation of cycloalkenes with H_2O_2 over various titanosilicates[a].

Catalyst	Si/Ti	Surface area $(m^2 g^{-1})$	Cyclopentene		Alkene epoxidation[b] Cyclooctene		Cyclododecenes	
			Conversion	TON	Conversion	TON	Conversion	TON
Del-Ti-MWW	42	1075	58.9	306	28.2	147	20.7	57
3D PS-Ti-MWW	46	520	15.7	89	4.3	24	3.3	9
TS-1	34	525	16.3	69	1.6	7	1.2	3
Ti-beta	35	621	9.9	43	4.6	20	1.9	4
Ti-MCM-41	46	1144	3.5	20	5.1	29	4.1	12

[a] Reaction conditions: cat., 10–25 mg; alkene, 2.5–10 mmol; H_2O_2, equal to the alkene amount; CH_3CN, 5–10 mL; temp., 313 K for cyclopentene and 333 K for other substrates; time, 2 h.
[b] Conv. in mol.%; TON in mol (mol-Ti)$^{-1}$.

4.2.4
Other Titanium-Containing Zeolites

It is possible to directly incorporate Ti into the framework of other zeolites. The structure of UTD-1, which can be synthesized using bis(pentamethylcyclopentadienyl) cobalt(III) hydroxide (Cp*2CoOH) as a structure-directing agent, has monodimensional extra-large elliptically-shaped (7.5 × 10 Å) pores circumscribed by 14-MR [83]. Despite the existence of the extra-large pores, the thermal stability of the framework is remarkably high. The addition of titanium into the UTD-1 synthesis gel results in the isostructural titanosilicate Ti-UTD-1 [84]. The Ti-UTD-1 molecular sieves are active in the oxidation of cyclohexane at room temperature with *tert*-butyl hydroperoxide as oxidation agent to give cyclohexanone as the major product with lesser amounts of cyclohexanol and adipic acid. Cyclohexene oxidation with H_2O_2 results in allylic oxidation and epoxidation followed by hydrolysis [85]. Tuel has synthesized Ti-ZSM-12 using a diquaternary ammonium structure-directing agent $(Et_2MeN^+C_3H_6)_2$ [19]. Ti-ZSM-48 [14], TAPSO-5 [18] and Ti-ITQ-7 [21, 22] were also synthesized and used as oxidation catalysts.

Post-synthesis modifications have been successful in preparing titanium-containing molecular sieves active in oxidation. The method employed for the post-synthesis incorporation of Ti to the zeolite beta was also applied to the incorporation of Ti into MOR and FAU [60].

Yashima *et al.* proposed the "atom planting" method as the development of the alumination through the reaction of internal silanol groups with aluminium halides. By using halides of other metals, metallosilicates with the MFI structure (Ga, In, Sb, As and Ti) and with the MOR structure (Ga, Sb and Ti) have been prepared [86]. The incorporation of Ti into the MOR structure was confirmed by the appearance of the specific absorption band in the IR spectra [20]. Very recently, Kubota *et al.* have synthesized Ti-YNU-2 (MSE) [25] by the post-synthesis modification of YNU-2 (P) that has a large number of defect sites [87]. Ti-YNU-2 has proved to be a very active catalyst in liquid-phase oxidation using H_2O_2 as oxidant [25].

4.2.5
Solvent Effects and Reaction Intermediate

The role of the solvent is complex: polarity, solubility of reactants and products, diffusion and counter-diffusion effects, and also interaction with the active centers [88]. Using a triphase system (solid–liquid–liquid) in the absence of any cosolvent, a considerable increase in the conversion of various water-immiscible organic compounds (toluene, anisole, benzyl alcohol, etc.) can be achieved [89]. A similar retardation effect of cosolvent was reported previously for benzene oxidation [90, 91]. The solvent may compete with reactant for diffusion in the channels and adsorption at the active sites of TS-1 catalyst. The activity Ti-beta for 1-hexene and cyclohexanol oxidations is highest in acetonitrile, which is a polar, nonprotic solvent [92]. This is in contrast with the observed enhancement of the activity of TS-1 by methanol and protic solvents [68]. These differences have been

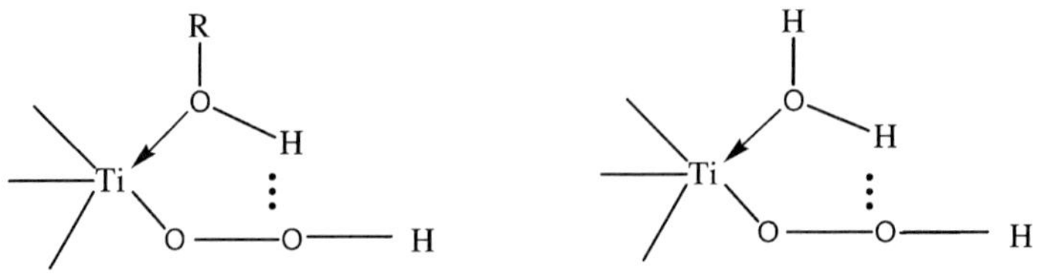

Scheme 4.3 Proposed intermediate Ti species.

ascribed to the hydrophilic character of Ti-beta compared to TS-1. The presence of Al and a large concentration of internal and external silanol groups (beta is composed of an intergrowth of polymorphs A and B) give Ti-beta a greater hydrophilicity. Corma et al. pointed out that Ti-beta is more active than TS-1 for 1-hexene oxidation when the reaction is carried out in acetonitrile as solvent, suggesting that the active species in Ti-beta is a cyclic species in which a water molecule coordinates the Ti atoms instead of an alcohol molecule [92].

Species I with a stable five-membered ring structure (Scheme 4.3), formed by the coordination of ROH to Ti centers and hydrogen bonding to Ti-peroxo complex, was believed to be the active intermediates in protic alcohol solvents, while species II was assumed to contribute to the oxidation of substrates in aprotic solvents. Recently, Lamberti et al. found that an end-on η^2 Ti hydroperoxo complex was generated, probably by the reversible rupture of Ti—O—Si bridge under anhydrous H_2O_2 conditions, and that this complex was reversibly transformed into a side-on η^2 Ti peroxo complex after addition of water, indicating that water molecules play an active role in determining the relative concentration of Ti peroxo to hydroperoxo species present on the working catalyst [93–95]. Despite such findings and interpretations, aprotic acetone was reported to be the solvent of choice in terms of both activity and selectivity for the epoxidation of styrene and allyl alcohol on TS-1 [96, 97].

A comprehensive investigation into the solvent effect on the catalytic performance of three types of representative titanosilicates, viz. TS-1, Ti-MWW and Ti-beta, revealed that the solvent effect was highly dependent on substrates [98]. Figure 4.9 shows the catalytic results for the oxidation of 1-hexene over TS-1. In agreement with the results reported in the literature [27, 23c, 92], methanol as solvent resulted in a substantial enhancement in catalytic activity, compared with acetonitrile. As expected, the difference in activity became small with decreasing Ti content since the number of active sites decreases. However, with respect to the oxidation of cyclohexene, a converse solvent effect was observed (Figure 4.10). When acetonitrile was used as solvent, the conversion was nearly 4× higher than that obtained with methanol as solvent. Almost no dependence of the conversion on the Ti content was observed when the Ti/(Ti + Si) molar ratio was larger than 0.0086. This is probably due to the occurrence of the reaction mainly on the external surface and near the pore mouth; the increase in Ti content incorporated into the internal surface does not significantly contribute to the enhancement in activity in the oxidation of cyclohexene.

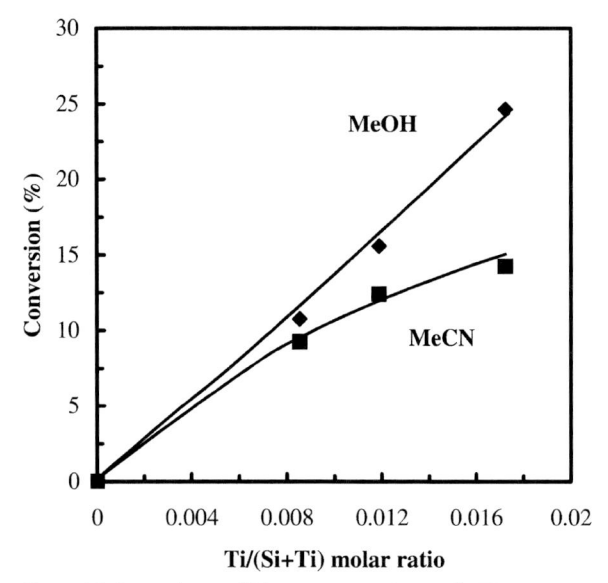

Figure 4.9 Dependence of 1-hexene conversion on the Ti content of TS-1 in the oxidation of 1-hexene in methanol and acetonitrile solvents. Reaction conditions: 60 °C, 2 h, 0.05 g catalyst, 10 mL solvent, 10 mmol substrate and 10 mmol H_2O_2 (31% aqueous solution).

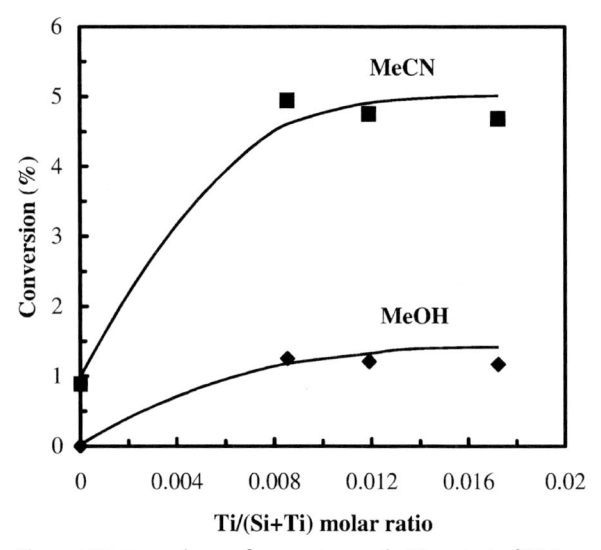

Figure 4.10 Dependence of conversion on the Ti content of TS-1 in the oxidation of cyclohexene in methanol and acetonitrile solvents. Reaction conditions: 60 °C, 4 h, 0.05 g catalyst, 10 mL solvent, 10 mmol substrate and 10 mmol H_2O_2 (31% aqueous solution).

Compared with species II (Scheme 4.3), species I is likely to form on TS-1. This is due to its hydrophobic character [99], which would make methanol approach Ti sites more easily than water, leading to the formation of a large amount of active sites of species I [99]. Therefore, when 1-hexene is oxidized on TS-1, methanol should be the solvent of choice. However, because cyclohexene is a large molecule, its oxidation would occur mainly on the exterior surface and/or near the pore mouth, where much more Si–OH and Ti–OH groups are present than inside the channels, making these areas relatively hydrophilic. To confirm this hypothesis, a TS-1 sample was poisoned by adding 2,4-dimethylquinoline to the reaction mixture. The conversion of cyclohexene drastically declined in acetonitrile solvent, while the conversion of 1-hexene was not significantly changed in methanol. This shows that the oxidation of cyclohexene was indeed primarily catalyzed by Ti sites on the exterior surface and/or near the pore mouth.

Species II (Scheme 4.3) should be formed more easily on the external surface and act as an active intermediate when acetonitrile is used as solvent [92]. Additionally, species II is more intrinsically active than species I due to its higher electrophilic character [92]. This makes acetonitrile a better solvent in the oxidation of cyclohexene over TS-1, which mainly occurs on the external surface of TS-1, in contrast to that of 1-hexene. This is also supported by the fact that the silylation slightly increased the conversion of cyclohexene over TS-1 when methanol was used as solvent. The selective silylation of the external surface of TS-1 by bulky 1,1,1,3,3,3-hexamethyldisilazane could increase the hydrophobicity, favoring the formation of species I. The silylation of Ti-MCM-41 led to a remarkable increase in the Q^4/Q^3 ratio, significantly increasing its hydrophobicity and enhancing its activity for the oxidation of 1-hexene and hexane [42]. Nevertheless, after silylation, considerable amounts of OH groups were still present on the exterior surface of TS-1 as a result of the steric limitation originating from the large molecular size of 1,1,1,3,3,3-hexamethyldisilazane. This makes the external surface of TS-1 still relatively hydrophilic even after being silylated. Methanol as solvent would strongly compete with water to adsorb on the Ti sites because of its preponderance, resulting in a lower activity than acetonitrile as solvent.

In contrast to TS-1, Ti-MWW exhibited much higher activity in acetonitrile solvent than in methanol for the oxidation of 1-hexene, whereas for the oxidation of cyclohexene, acetonitrile as solvent gave slightly higher conversion than methanol. Ti-MWW was much more active in acetonitrile than in methanol in the epoxidation of 1-hexene (Figure 4.11), because of its hydrophilicity [24, 78]. The silylation had no marked effect on the catalytic results due to the reaction mainly occurring inside the channels with distorted 10-MR pore openings; the silylation of internal Ti species by bulky 1,1,1,3,3,3-hexamethyldisilazane would be difficult. It is believable that the epoxidation of cyclohexene over Ti-MWW, in a similar manner to its epoxidation over TS-1, mainly took place on the external surface since the pore-openings of MWW-type materials are slightly smaller than those of the MFI-type [100]; in addition, poisoning by 2,4-dimethylquinoline drastically lowered the activity for oxidation of cyclohexene regardless of the employed solvent but did not significantly affect the activity in 1-hexene epoxidation.

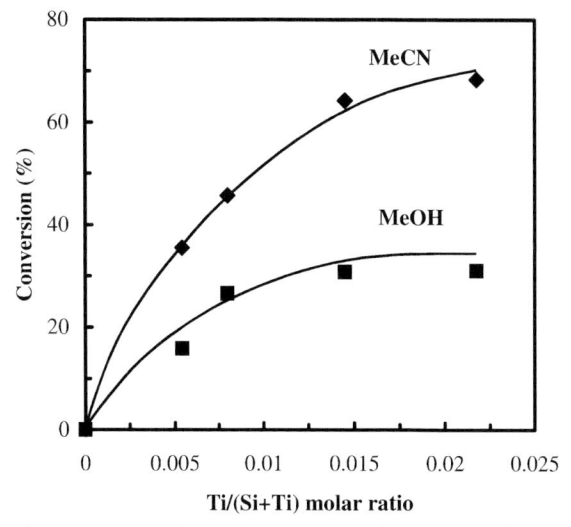

[Figure: plot of Conversion (%) vs Ti/(Si+Ti) molar ratio, showing MeCN and MeOH curves]

Figure 4.11 Dependence of conversion on the Ti content of Ti-MWW in the oxidation of 1-hexene in methanol and acetonitrile solvents. Reaction conditions: 60 °C, 2 h, 0.05 g catalyst, 10 mL solvent, 10 mmol substrate and 10 mmol H_2O_2 (31% aqueous solution).

The presence of a lot of 12-MR side-pockets on the external surface of Ti-MWW led to a moderate cyclohexene conversion, between those obtained with TS-1 and Ti-beta. Notably, the Ti species inside the side-pockets are located on the intracrystalline surface, and consequently have an environmental state similar to that of the Ti species in TS-1; these Ti species are relatively hydrophobic. As a result, methanol as solvent gave almost the same activity as acetonitrile, probably because both the hydrophilic Ti species on the external surface and hydrophobic ones in the side pockets could serve as active sites for the epoxidation of cyclohexene. Upon silylation with 1,1,1,3,3,3-hexamethyldisilazane, the external surface silanol groups would be silylated. Therefore, the sites active for the cyclohexene epoxidation becomes totally hydrophobic, resulting in methanol being the solvent of choice. However, silylation close to the entrance of side pockets with such a large-molecule agent would result in a considerable steric constraint on the diffusion of substrate and product molecules into and out of these side pockets, giving rise to a slight decrease in activity.

When 1-hexene was oxidized over Ti-beta, the conversion in acetonitrile solvent was about twice as high as that attained in methanol. However, for cyclohexene oxidation, methanol as solvent was much superior to acetonitrile, which conflicts with the results reported by Corma *et al.* [92], but is consistent with our previous findings [55].

It has long been known that acetonitrile should be the solvent of choice for the oxidation of 1-hexene over Ti-beta and Ti-MWW while methanol is preferred by TS-1. In contrast, for cyclohexene oxidation, methanol is favorable for Ti-beta, whereas acetonitrile is the best for TS-1 and Ti-MWW. The effect of hydrophilicity/hydrophobicity is demonstrated by a series of catalytic results, but this can not interpret the solvent effect on the oxidation of cyclohexene over Ti-beta.

It has also been shown that the catalytic activity can be enhanced when appropriate amounts of mixed solvents are present in the oxidation system, compared to a single solvent [98]. In particular, converse results are obtained when the same titanosilicate is used for catalyzing the oxidation of 1-hexene and cyclohexene. Thus, further insight into the solvent effect is required.

4.3
Other Metal-Containing Zeolites

Vanadium is claimed to be incorporated into the zeolite MEL framework to give VS-2 [101]. A conspicuous characteristic of VS-2 is that it can catalyze the oxidation of terminal methyl groups of linear alkanes in the presence of H_2O_2 as oxidant to produce 1-alkanol and aldehydes, whereas TS-1 and TS-2 catalyze only the oxidation of internal carbons (methylene groups) [29]. A comparative study of alkane oxidation on VS-2 and TS-2 has been conducted [102]. Spin-trapping experiments revealed that the VS-2–H_2O_2–hexane system generates primary hexyl radical species, which could not be observed in the TS-2 system. It is proposed that the oxidation of internal carbons and that of terminal carbons proceed by different mechanisms. Gallot et al. have provided another explanation for the complete absence of oxidation of terminal carbon based on the agostic interaction of Ti^{4+} and C–H bond of a terminal methyl [103].

A detailed study on the form of the vanadium species in the V-MEL samples has been carried out [104]. As-synthesized V-MEL samples contain two different V species, framework V^{5+} in a distorted tetrahedral environment and extraframework V^{4+} in an octahedral environment. Upon calcination the V^{4+} species transform into two types of V^{5+} species. However, these V^{5+} species are easily extracted by NH_4OAc and only the nonextractable vanadium species are active in the oxidation of toluene and phenol. V-beta can be prepared in a similar two-step manner. The method consists of first creating a vacant site by dealumination of the beta zeolite with nitric acid followed by contact with an NH_4VO_3 solution [105]. Reddy et al. have prepared V-NCL-1 using the diquaternary ammonium ion $(Et_3N^+C_3H_6)_2$ [106]. Notably, however, in terms of H_2O_2 utilization efficiency and resistance against leaching, the vanadosilicates synthesized so far are inferior to titanosilicates.

Tin-containing silicalite-2, Sn-Sil-2, has been synthesized [107]. From an ^{119}Sn NMR study, however, it has been suggested that Sn^{4+} ions are mostly in octahedral coordination. Sn-beta catalyzes the Baeyer–Villiger oxidation of cyclic ketones to lactones without using peracids but using H_2O_2 [108].

Chromium-containing silicalite-2, CrS-2, has been synthesized and shown to catalyze similar reactions using TBHP as an oxidant [109]. Notably, TBHP has been reported to be an ineffective oxidant in TS-1 catalyzed oxidations [26]. However, it has been claimed that a TS-1–TBHP combination exhibits activity in the oxidative cleavage of the C=C double bond of silyl enol ethers to produce dicarboxylic acids [110].

Lempers and Sheldon have reported that small amounts of chromium that are leached from CrAPO-5, CrAPO-11 and CrS-1 catalyze the liquid-phase oxidation of bulky alkenes with TBHP [111]. The leaching seems to be caused by TBHP that extracts chromium from the micropores. The authors emphasize that experiments demonstrating that heterogeneous catalysts can be recovered and recycled without apparent loss of activity is not definite proof of heterogeneity.

The Fe ion is easily incorporated into zeolites. Ferrosilicates and ferrisilcates are often used as acid catalysts. Direct oxidation of benzene to phenol over Fe-MFI zeolites has been shown to be practical when N_2O is used as an oxidant (AlphOx process) [112]. This process is suitable for the abatement of highly concentrated N_2O (20–40%) gas discharged from adipic acid plants to meet the tightened regulation of emission levels of N_2O.

4.4
Conclusion

Oxidation catalysis is exhibited by introducing isolated metals with redox properties into the framework of zeolites. So far, titanosilicate zeolites have been superior in catalytic performance in oxidation reactions to any other metallosilicate using H_2O_2 as oxidant. In addition to high activity and selectivity in oxidation, H_2O_2 utilization efficiency and resistance against leaching would be critically important to putting the metallosilicate zeolites to practical use. Ammoximation of cyclohexanone using TS-1 to produce cyclohexanone is already industrialized, and it is reported that propylene oxide manufacturing from propylene and H_2O_2 in the presence of TS-1 was put into operation in 2008. Ti-beta and Ti-MWW (MCM-22) as well as TS-1 proved to be promising catalysts for various oxidation reactions. Although H_2O_2 is an excellent oxidant that has high active oxygen content and produces water as by-product, molecular oxygen is ideal for oxidation reactions. One approach would be *in situ* preparation of H_2O_2 from O_2 and H_2, whereas the catalytic synthesis of H_2O_2 from O_2 and H_2 itself is extremely difficult.

Direct synthesis of metal-substituted zeolites has long been sought. However, since the post-synthesis modifications can be made under wide-ranging conditions (temperature, solvent, atmosphere, pH, etc.) far from those for the zeolite synthesis, the modifications of zeolites present us with powerful indirect methods for manipulating the properties of zeolites. Therefore, the fine-tuning of the properties of zeolites will continue to be achieved by developing various post-synthesis modification procedures as well as direct synthetic techniques.

References

1 Kühl, G.H. (1999) *Catalysis and Zeolites* (eds J. Weitkamp and L. Puppe), Springer, Berlin, p. 81.
2 Lunsford, J. (1977) *ACS Symposium Series*, **40**, 473.
3 Weitkamp, J. (1993) Proceedings from the Ninth International Zeolite Conference, Vol. 1 (eds R. von Ballmoos, J.B. Higgins and M.M.J. Treacy), Buttherworth-Heinemann, Boston, p. 13.

4 Balkus, K.J. Jr and Gabrielov, A.G. (1995) *Journal of Inclusion Phenomena and Molecular Recognition in Chemistry*, **21**, 159.

5 Schulz-Ekloff, G. and Ermst, S. (1999) *Preparation of Solid Catalysts* (eds G. Ertl, K. Knözinger and J. Weitkamp), Wiley-VCH, Weinheim, p. 405.

6 Herron, N., Stucky, G.D. and Tolman, C.A. (1985) *Inorganica Chimica Acta*, **100**, 135.

7 Parton, R., de Vos, D. and Jacobs, P.A. (1992) Proceedings of the NATO Advanced Study on Zeolite Microporous Solids: Synthesis, Structure and Reactivity (eds E.G. Derouane, F. Lemos, C. Naccache and F. Ribeiro), Kluwer, Dordrecht, p. 555.

8 Herron, N., Tolman, C.A. and Stucky, G.D. (1986) *Journal of the Chemical Society. Chemical Communications*, 1521.

9 Breck, D.W. (1974) *Zeolite Molecular Sieves*, John Wiley & Sons, Inc., New York, p. 320.

10 Taramasso, M., Perego, G. and Notari, B. (1983) U.S. Pat., 4,401,051.

11 Ione, K.G., Vostrikova, L.A. and Mastikhin, V.M. (1985) *J Mol Catal*, **31**, 355.

12 Tielen, M., Geelen, M. and Jacobs, P.A. (1985) *Acta Physica et Chemica*, **31**, 1.

13 Reddy, J.S., Kumar, R. and Ratnasamy, P. (1990) *Applied Catalysis*, **58**, L1.

14 Serrano, D.P., Li, H.-X. and Davis, M.E. (1992) *Journal of the Chemical Society. Chemical Communications*, 745.

15 (a) Camblor, M.A., Corma, A., Martínez, A. and Pérez-Pariente, J. (1992) *Chemical Communications*, 589; (b) Camblor, M.A., Constantini, M., Corma, A. et al. (1996) *Chemical Communications*, 1339.

16 (a) Blasco, T., Camblor, M.A., Corma, A. et al. (1996) *Chemical Communications*, 2367; (b) Blasco, T., Camblor, M.A., Corma, A. et al. (1998) *The Journal of Physical Chemistry B*, **102**, 75.

17 (a) Jappar, N., Xia, Q., Tatsumi, T. (1998) *Journal of Catalysis*, **180**, 132; (b) Tatsumi, T., Jappar, N. (1998) *The Journal of Physical Chemistry B*, **102**, 7126.

18 Tuel, A. (1995) *Zeolites*, **15**, 228.

19 Tuel, A. (1995) *Zeolites*, **15**, 236.

20 (a) Wu, P., Komatsu, T., Yashima, T. (1996) *The Journal of Physical Chemistry*, **100**, 10316; (b) Wu, P., Komatsu, T., Yashima, T. (1997) *Journal of Catalysis*, **168**, 400; (c) Wu, P., Komatsu, T. and Yashima, T. (1997) *Stud Surf Sci Catal*, **105**, 663; (d) Wu, P., Komatsu, T. and Yashima, T. (1998) *The Journal of Physical Chemistry B*, **102**, 9297.

21 Díñaz-Cabañas, M.J., Villaescusa, L.A. and Camblor, M.A. (2000) *Chemical Communications*, 761.

22 Corma, A., Díñaz-Cabañas, M.J., Domine, M.E. and Rey, F.Z. (2000) *Chemical Communications*, 1725.

23 (a) Wu, P., Tatsumi, T., Komatsu, T. and Yashima, T. (2000) *Chemistry Letters*, 774; (b) Wu, P., Tatsumi, T., Komatsu, T. and Yashima, T. (2001) *The Journal of Physical Chemistry. B*, **105**, 2897; (c) Wu, P., Tatsumi, T., Komatsu, T. and Yashima, T. (2001) *Journal of Catalysis*, **202**, 245; (d) Wu, P. and Tatsumi, T. (2001) *Chemical Communications*, 897; (e) Wu, P., Tatsumi, T. (2002) *The Journal of Physical Chemistry. B*, **106**, 748; (f) Wu, P., Tatsumi, T. (2003) *Journal of Catalysis*, **214**, 317.

24 Wu, P. and Tatsumi, T. (2002) *Chemical Communications*, 1026.

25 Kubota, Y., Koyama, Y., Yamada, T., Inaki, S. and Tatsumi, T. (2008) *Chemical Communications*, 6224.

26 Notari, B. (1996) *Advances in Catalysis*, Vol. 41 (eds D.D. Eley, W.O. Haag and B.C. Gates), Academic Press, San Diego, p. 253.

27 Clerici, M., Bellussi, G. and Romano, U. (1991) *Journal of Catalysis*, **129**, 159.

28 Millini, R., Previde-Massara, E., Perego, G. and Bellussi, G. (1992) *Journal of Catalysis*, **137**, 497.

29 Tatsumi, T., Nakamura, M., Negishi, S. and Tominaga, H. (1990) *Journal of the Chemical Society. Chemical Communications*, 476.

30 Bellussi, G. and Rigutto, M.S. (1994) *Stud Surf Sci Catal*, **105**, 177.

31 Ratnasamy, P. (2004) *Advances in Catalysis*, Vol. 48 (eds B.C. Gates and H. Knoezinger), Elsevier, Amsterdam, pp. 1–179.
32 Perego, G., Bellussi, G., Corno, C. *et al.* (1986) *Stud Surf Sci Catal*, **28**, 129.
33 Khouw, C.B. and Davis, M.E. (1995) *Journal of Catalysis*, **151**, 77.
34 Lamberti, C., Bordiga, S., Zecchina, A. *et al.* (2001) *Journal of the American Chemical Society*, **123**, 2204.
35 Bordiga, S., Damin, A., Bonino, F. *et al.* (2003) *Physical Chemistry Chemical Physics*, **5**, 4390.
36 Thangaraj, A. and Sivasanker, S. (1992) *Journal of the Chemical Society. Chemical Communications*, 123.
37 Schuchardt, U., Pastore, H.O. and Spinace, E.V. (1994) *Studies in Surface Science Catalysis*, **84**, 1877.
38 Serrano, D.P., Uguina, M.A., Ovejero, G. *et al.* (1996) *Microporous Mater*, **7**, 309.
39 Bordiga, S., Bonino, F., Damin, A. and Lamberti, C. (2007) *Physical Chemistry Chemical Physics*, **9**, 4854.
40 Thomas, J.M. and Sankar, G. (2001) *Accounts of Chemical Research*, **34**, 571.
41 Fan, W., Duan, R.-G., Yokoi, T. *et al.* (2008) *Journal of the American Chemical Society*, 130, 10150.
42 Tatsumi, T., Koyano, K.A. and Igarashi, N. (1998) *Chemical Communications*, 325.
43 Srinivas, D., Manikandan, P., Laha, S.C. *et al.* (2003) *Journal of Catalysis*, **217**, 160.
44 Zhuang, J., Ma, D., Yan, Z. *et al.* (2004) *Journal of Catalysis*, **221**, 670.
45 Parker, W.O. and Millini, R. (2006) *Journal of the American Chemical Society*, **128**, 1450.
46 Fan, W., Li, R., Ma, J. *et al.* (1997) *Microporous Mater*, **8**, 131.
47 Tatsumi, T., Nakamura, M., Yuasa, K. and Tominaga, H. (1990) *Chemistry Letters*, 297.
48 Hayashi, H., Kikawa, K., Murei, U., Shigemoto, N., Sugiyama, S. and Kawashivo, K. (1996) *Catalysis Letters*, **36**, 99.
49 Tatsumi, T., Yako, M., Nakamura, M., Yukora, H. and Tominaga, H. (1993) *Journal of Molecular Catalysis*, **78**, L41.
50 Camblor, M.A., Corma, A. and Perez-Pariente, J. (1993) *Zeolites*, **13**, 82.
51 Corma, A., Esteve, P., Martinez, A. and Valencia, S. (1995) *Journal of Catalysis*, **152**, 18.
52 vandel Waal, J.C., Lin, P., Rigutto, M.S. and van Bekkum, H. (1997) *Stud Surf Sci Catal*, **105**, 1093.
53 Camblor, M.A., Corma, A. and Valencia, S. (1998) *The Journal of Physical Chemistry B*, **102**, 75.
54 Tatsumi, T. and Jappar, N. (1998) *The Journal of Physical Chemistry. B*, **102**, 7126.
55 (a) Goa, Y., Wu, P. and Tatsumi, T. (2001) *Chemical Communications*, 1714; (b) Goa, Y., Wu, P. and Tatsumi, T. (2004) *The Journal of Physical Chemistry B*, **108**, 8401.
56 Koller, H., Woelker, A., Villaescusa, L.A. *et al.* (1999) *Journal of the American Chemical Society*, **121**, 3368.
57 Goa, Y., Wu, P. and Tatsumi, T. (2004) *The Journal of Physical Chemistry. B*, **108**, 4242.
58 Skeels, G.W. and Flanigen, E.M. (1989) *Synthesis of Zeolites, ACS Symposium Series*, **398**, 420.
59 Reddy, J.S. and Sayari, A. (1995) *Stud Surf Sci Catal*, **94**, 309.
60 Di Renzo, F., Gomez, S., Teissier, R. and Fajula, F. (2000) *Stud Surf Sci Catal*, **130**, 1631.
61 Rigutto, M.S., de Ruiter, R., Niederer, J.P.M. and van Bekkum, H. (1994) *Stud Surf Sci Catal*, **84**, 2245.
62 Tatsumi, T., Nakamura, M., Yuasa, K. and Tominaga, H. (1991) *Catalysis Letters*, **10**, 259.
63 Camblor, M.A., Corma, A., Díaz-Cabanas, M.J. and Baerlocher, C. (1998) *The Journal of Physical Chemistry. B*, **102**, 44.
64 Roth, W.J., Kresge, C.T., Vartuli, J.C. *et al.* (1995) *Stud Surf Sci Catal*, **94**, 301.
65 Corma, A., Fornés, V., Pergher, S.B. *et al.* (1998) *Nature*, **396**, 353.
66 Blasco, T., Navarro, M.T., Corma, A. and Pérez-Pariente, J. (1995) *Journal of Catalysis*, **156**, 65.
67 Koyano, A.K. and Tatsumi, T. (1997) *Chemical Communications*, 145.

68 Clerici, M.G. and Ingallina, P. (1993) *Journal of Catalysis*, **140**, 71.
69 Levin, D., Chang, A.D., Luo, S. et al. (2000) US Pat. 6,114,551.
70 Wu, P. and Tatsumi, T. (2004) *Catal Surveys Asia*, **8**, 137.
71 Wu, P. and Tatsumi, T. (2003) *Journal of Catalysis*, **214**, 317.
72 Wu, P., Liu, Y., He, M. and Tatsumi, T. (2004) *Journal of Catalysis*, **228**, 183.
73 Wang, L., Liu, Y., Xie, W. et al. (2007) *Journal of Catalysis*, **246**, 205.
74 Wu, H., Wang, L., Zhang, H. et al. (2006) *Green Chemistry*, **8**, 78.
75 Fan, W., Kubota, Y. and Tatsumi, T. (2008) *Chemistry & Sustainability: Energy & Materials*, **1**, 175.
76 (a) Song, F., Liu, Y., Wu, H. et al. (2005) *Chemistry Letters*, **34**, 1436; (b) Song, F., Liu, Y., Wu, H. et al. (2006) *Journal of Catalysis*, **237**, 359.
77 Fan, W., Wu, P., Namba, S. and Tatsumi, T. (2003) *Angewandte Chemie – International Edition*, **43**, 236.
78 Fan, W., Wu, P., Namba, S. and Tatsumi, T. (2006) *Journal of Catalysis*, **243**, 183.
79 Ruan, J., Wu, P., Slater, B. and Terasaki, O. (2005) *Angewandte Chemie – International Edition*, **44**, 6719.
80 Wu, P., Ruan, J., Wang, L. et al. (2008) *Journal of the American Chemical Society*, **130**, 8178.
81 Inagaki, S., Yokoi, T., Kubota, Y. and Tatsumi, T. (2007) *Chemical Communications*, 5188.
82 Wu, P., Nuntasri, D., Ruan, J. et al. (2004) *The Journal of Physical Chemistry B*, **108**, 19126.
83 Freyhardt, C.C., Tsapatsis, M., Lobo, R.F. et al. (1996) *Nature*, **381**, 295.
84 Balkus, K.J., Gavrielov, A.G. and Zones, S.I. (1995) *Zeolites: A Refined Tool for Designing Catalytic Sites* (eds L. Bonneviot and S. Kalliaguine), Elsevier, Amsterdam, pp. 519–525.
85 Balkus, K.J., Khanmamedova, A., Gavrielov, A.G. and Zones, S.I. (1996) *11th International Congress on Catalysis* (eds J.W. Hightower, W.N. Delgass and E. Iglesia), Elsevier, Amsterdam, pp. 1341–1348.
86 Yashima, T., Yamagishi, K. and Namba, S. (1991) *Stud Surf Sci Catal*, **60**, 171.
87 Koyama, Y., Ikeda, T., Tatsumi, T. and Kubota, Y. (2008) *Angewandte Chemie – International Edition*, **47**, 1042.
88 Clerici, M.G. (2001) *Topics in Catalysis*, **15**, 257.
89 Bhaumik, A. and Kumar, R. (1995) *Journal of the Chemical Society. Chemical Communications*, 349.
90 Tatsumi, T., Asano, K. and Yanagisawa, K. (1994) *Stud Surf Sci Catal*, **84**, 1861.
91 Tatsumi, T., Yanagisawa, K., Asano, K. et al. (1994) *Stud Surf Sci Catal*, **83**, 417.
92 Corma, A., Esteve, P. and Martinez, A. (1996) *Journal of Catalysis*, **161**, 11.
93 Bonino, F., Damin, A., Ricchiardi, G. et al. (2004) *The Journal of Physical Chemistry B*, **108**, 3575.
94 Prestipino, C., Bonino, F., Usseglio, S. et al. (2004) *ChemPhysChem*, **5**, 1799.
95 Bordiga, S., Bonino, F., Damin, A. and Lamberti, C. (2007) *Physical Chemistry – Chemical Physics*, **9**, 4854.
96 Bhaumik, A., Kumar, R. and Ratnasamy, P. (1994) *Stud Surf Sci Catal*, **84**, 1883.
97 Kumar, S.B., Mirajkar, S.P., Pais, G.C.G. et al. (1995) *Journal of Catalysis*, **156**, 163.
98 Fan, W., Wu, Peng and Tatsumi T. (2008) *Journal of Catalysis*, **256**, 62.
99 Bellussi, G., Carati, A., Clerici, M.G. et al. (1992) *Journal of Catalysis*, **133**, 220.
100 Baerlocher, Ch., McCusker, L.B. and Olson, D.H. (2007) *Atlas of Zeolite Framework Types*, (6th Revised Edition) Elsevier, Amsterdam, pp. 213 and 235.
101 Hari Prasad Rao, P.R., Kumar, R., Ramaswany, A.V. and Ratnasamy, P. (1992) *Journal of Catalysis*, **137**, 225.
102 Tatsumi, T., Hirasawa, Y. and Tsuchiya, J. (1996) *ACS Symposium Series*, **638**, 374.
103 Gallot, J.E., Fu, H., Kapoor, M.P. and Kaliaguine, S. (1996) *Journal of Catalysis*, **161**, 798.

104 Sen, T., Ramaswamy, V., Ganapathy, S. et al. (1996) *The Journal of Physical Chemistry*, **100**, 3809.

105 Dzwigaj, S., Peltre, M.J., Massiani, P. et al. (1998) *Chemical Communications*, 87.

106 Reddy, K.R., Ramaswany, A.V. and Ratnasamy, P. (1992) *Journal of the Chemical Society. Chemical Communications*, 1613.

107 Mal, N.K., Ramaswany, V., Ganapathy, S. and Ramaswamy, A.V. (1995) *Applied Catalysis A*, **125**, 233.

108 Corma, A., Nemeth, L.T., Renz, M. and Valencia, S. (2001) *Nature*, **412**, 423.

109 Jayachandran, B., Sasidhran, M., Sudalai, A. and Ravindranathan, T. (1995) *Journal of the Chemical Society. Chemical Communications*, 1341.

110 Raju, S.V.N., Upadhya, T.T., Ponrathanam, S. et al. (1996) *Journal of the Chemical Society. Chemical Communications*, 1969.

111 Lempers, H.E.B. and Sheldon, R.A. (1997) *Stud Surf Sci Catal*, **105**, 1061.

112 Kharitonov, A.S., Sheveleva, G.A., Panov, G.I. et al. (1993) *Applied Catalysis A*, **98**, 33.

5
Design of Well-Defined Active Sites on Crystalline Materials for Liquid-Phase Oxidations
Kiyotomi Kaneda and Takato Mitsudome

5.1
Introduction

Many transition metal complexes act as homogeneous catalysts, promoting oxidation of a wide range of organic compounds in the liquid phase via coordination of substrates to the metal center [1]. More sophisticated oxidation processes can be achieved by using heterogeneous catalysts that possess catalytically active species on a solid surface [2], which is advantageous in terms of reuse, circulation of resources and elimination of waste at source [3]. In particular, inorganic crystalline materials have received much attention as potential heterogeneous catalysts because they allow precise control of the creation of catalytically active metal species responsible for targeted oxidation – such as monomers, bimetals, multinuclear species and clusters – and their tolerance of high temperatures and acidity and/or basicity. The design of well-defined active metal sites on crystalline material not only opens up an avenue to a significant boost in catalytic performance, but also aids our understanding of heterogeneous catalysis on the molecular basis.

The recent development of inorganic crystalline-supported metal catalysts for various liquid-phase oxidation reactions such as alcohol oxidation, epoxidation, Baeyer–Villiger oxidation and oxidation via C–H activation using molecular oxygen (O_2) or hydrogen peroxide (H_2O_2) as an oxidant are reviewed in this chapter.

5.2
Oxidation of Alcohols

The selective oxidation of alcohols is widely recognized as one of the most fundamental transformations in both laboratory and industrial synthetic chemistry because the resulting carbonyl compounds serve as important and versatile intermediates for the synthesis of fine chemicals [4]. Many oxidizing reagents, including permanganate and dichromate, have traditionally been employed to achieve this transformation [5]. These stoichiometric oxidants, however, have serious drawbacks in that they are expensive and/or toxic and produce large amounts of waste. In the light of ever-increasing environmental concerns, much attention has been directed towards the development of promising catalytic protocols employing O_2 as a primary oxidant, which is readily available and produces water as a sole by-product (5.1) [6]:

$$R^1-CH(OH)-R^2 + 1/2\, O_2 \xrightarrow{\text{catalyst}} R^1-CO-R^2 + H_2O \quad (5.1)$$

Accordingly, the synthesis of highly active and selective heterogeneous catalysts has been explored in recent years. Among these, ruthenium-based catalysts have been found to show interesting catalytic activity, and various types of ruthenium-based heterogeneous catalyst have been developed [7].

5.2.1
Ru Catalyst

Recently, Kaneda and coworkers developed a new strategy for the design of high-performance heterogeneous catalysts utilizing hydroxyapatite as a macroligand for catalytically active centers [8]; an efficient hydroxyapatite-bound Ru catalyst (RuHAP) was developed for selective oxidation of various alcohols using O_2 [9].

Apatites and related compounds, most notably hydroxyapatite [HAP, $Ca_{10}(PO_4)_6(OH)_2$], are of considerable interest due to their potential as biomaterials, adsorbents and ion-exchange materials [10]. Their hexagonal apatite structure consists of Ca^{2+} sites surrounded by PO_4^{3-} tetrahedra; the OH^- ions occupy columns parallel to the hexagonal axis (Figure 5.1). The crystal structure features two nonequivalent Ca^{2+} sites: one set of Ca^{2+} ions is aligned in columns (site I, Ca_I), while the other set forms equilateral triangles centered on the screw axis (site II, Ca_{II}).

The use of hydroxyapatites as a catalyst support has the following advantages: (1) well-defined monomeric active species can be immobilized on the surface based on multiple functionalities, for example, cation exchange ability, adsorption capacity and nonstoichiometry; (2) their hydrophilic character allows smooth reactions under aqueous conditions; and (3) due to their robust structure, no leaching of metals occurs.

RuHAP was synthesized from a stoichiometric HAP, $Ca_{10}(PO_4)_6(OH)_2$, with $RuCl_3 \cdot nH_2O$. Analysis by powder X-ray diffraction (XRD), X-ray photoelectron spectroscopy (XPS), energy-dispersive X-ray (EDX), IR and Ru K-edge X-ray absorption fine structure (XAFS) showed that a monomeric Ru phosphate species is created on the HAP surface. Figure 5.2a shows a proposed surface structure of RuHAP.

Oxidation of various alcohols using the RuHAP catalyst at 80 °C under an O_2 atmosphere proceeded efficiently to give the corresponding carbonyl compounds without additives. The most important catalytic property of RuHAP is its applicability to a wide range of alcohols. For example, the non-activated alcohol 1-octanol was smoothly oxidized to afford 1-octanal without any formation of the corresponding

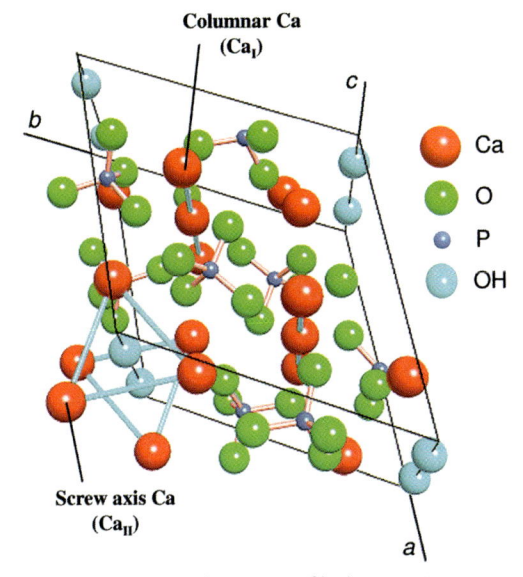

Figure 5.1 Hexagonal structure of hydroxyapatite.

carboxylic acid or ester. Moreover, this catalyst system is useful for the oxidation of heterocyclic alcohols containing nitrogen and sulfur atoms, such as 2-pyridine-methanol and 2-thiophenemethanol, giving the corresponding aldehydes in high yields (Scheme 5.1). Even when air is used in place of pure O_2, the above oxidation reactions proceeded smoothly.

The proposed catalytic cycle of this reaction is shown in Figure 5.3. Oxidation is initiated by ligand exchange between the alcohol and the Cl species of RuHAP to give a Ru-alcoholate, which undergoes β-hydride elimination to produce the corresponding carbonyl compound and a Ru-hydride species. Reaction of the hydride species with O_2 affords a Ru-hydroperoxide species, followed by ligand exchange to

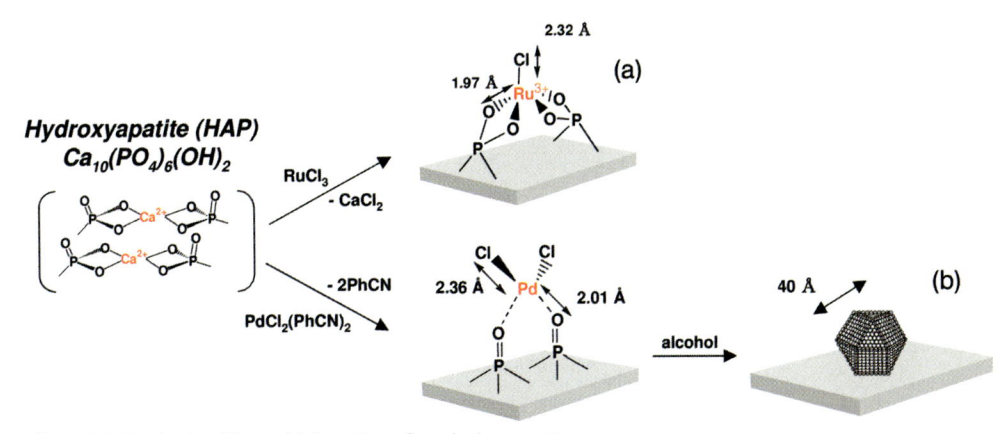

Figure 5.2 Synthesis of Ru and Pd catalysts from hydroxyapatite.

5 Design of Well-Defined Active Sites on Crystalline Materials for Liquid-Phase Oxidations

$$R\diagdown OH \xrightarrow[\text{toluene (5 mL), 60-80 °C, O}_2]{\text{RuHAP (0.2 g, Ru:17 mol \%)}} R\diagdown\!\!\!\overset{O}{\diagdown} H$$

2 mmol

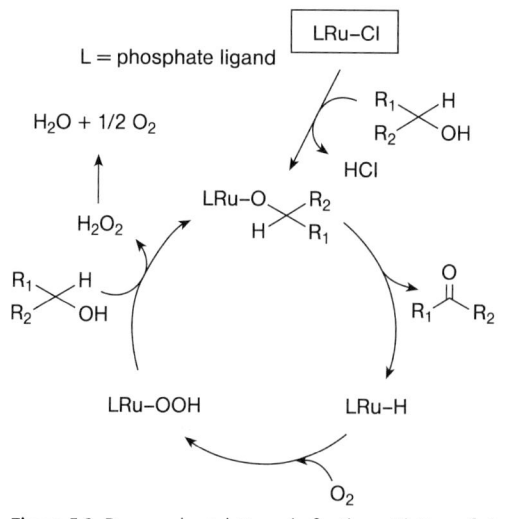

~~~~~~OH   (pyridyl)~~OH   (thienyl)~~OH

16 h   10 h   2 h
conversion 95 %   conversion 100 %   conversion 100 %
yield   94 %   yield   >99 %   yield   94 %

Scheme 5.1

regenerate the Ru-alcoholate species together with the formation of O_2 and H_2O from H_2O_2.

More recently, highly functionalized RuHAP materials with magnetic properties have been developed [11]: magnetic γ-Fe_2O_3 nanocrystallites dispersed in a hydroxy-apatite matrix (HAP-γ-Fe_2O_3) have been synthesized as a new catalyst support. The cation exchange ability of the external HAP surface enables equimolar substitution of Ru for Ca to form a catalytically active center (RuHAP-γ-Fe_2O_3). Characterization by several spectroscopic methods showed the formation of γ-Fe_2O_3 nanocrystallites with a mean diameter of 8.0 nm within the HAP matrix and a monomeric Ru species on the HAP surface.

RuHAP-γ-Fe_2O_3 exhibits superior catalytic activity in the oxidation of various alcohols to the corresponding carbonyl compounds using O_2 as a primary oxidant. The magnetic properties of RuHAP-γ-Fe_2O_3 provide a convenient route for separation of the catalyst from the reaction mixture by application of an external

Figure 5.3 Proposed catalytic cycle for the oxidation of alcohols using a RuHAP catalyst.

Figure 5.4 Separation of the RuHAP-γ-Fe$_2$O$_3$ catalyst from a reaction mixture by application of an external permanent magnet.

permanent magnet, and the spent catalyst can be recycled without appreciable loss of catalytic activity (Figure 5.4).

The RuHAP-γ-Fe$_2$O$_3$ catalyst was also found to be applicable to oxidation of sterically bulky alcohols: 3,5-dibenzyloxybenzyl alcohol and cholestanol were successfully converted into the corresponding carbonyl compounds with excellent yields, (5.2) and (5.3), respectively. In particular, the oxidation of 7-hexadecyn-1-ol proceeded quantitatively without any effect on the alkynyl group (5.4):

$$\text{HO}\diagup\!\!\diagdown\!\!\diagup\!\!\diagdown_{7}\!\!\equiv\!\!\diagdown\!\text{H} \quad\xrightarrow[\text{O}_2,\ \text{toluene},\ 90\ °\text{C},\ 10\ \text{h}]{\text{RuHAP-}\gamma\text{-Fe}_2\text{O}_3\ (5.0\ \text{mol\%})}\quad \text{OHC}\diagup\!\!\diagdown\!\!\diagup\!\!\diagdown_{7}\!\!\equiv\!\!\diagdown\!\text{H}$$
>99 % yield
(5.4)

An important advantage of this catalyst system over other systems is its smooth oxidation of alcohols even at room temperature (Table 5.1). Benzylic and secondary alcohols were efficiently oxidized to give the corresponding carbonyl compounds in high yields.

Another highly efficient ruthenium-based heterogeneous catalyst system for aerobic oxidation of alcohols has been reported by Mizuno and coworkers [12]. Highly dispersed ruthenium(III) hydroxide supported on alumina (Ru/Al$_2$O$_3$) was found to have a large substrate scope. Ru/Al$_2$O$_3$ has one of the highest catalytic activities among reported ruthenium catalyst systems. In the oxidation of 2-octanol and 1-phenylethanol under solvent-free conditions at 150 °C, turnover frequency (TOF) values of 300 and 340 h^{-1} and turnover numbers (TONs) of 950 and 980, respectively, were obtained (Scheme 5.2).

RuHAP [13] and Ru/Al$_2$O$_3$ [14] catalysts were also able to oxidize primary amines to nitriles in high yield using O$_2$ as an oxidant (Scheme 5.3).

Table 5.1 Oxidation of various alcohols catalyzed by RuHAP-γ-Fe$_2$O$_3$ at room temperature[a].

Entry	Substrate	Mol.%	Conversion (%)[b]	Yield (%)[b]
1	benzyl alcohol	1.25	>99	98
2	piperonyl alcohol	4	>99	>99
3	1-phenylethanol	1.4	94	89
4	diphenylmethanol	1	99	98
5	2-adamantanol	2	>99	99

[a] Alcohol (0.5–1 mmol), RuHAP-γ-Fe$_2$O$_3$ (1–2 mol.%), toluene (5 mL), O$_2$ flow, 24 h.
[b] Determined by GC using an internal standard technique.

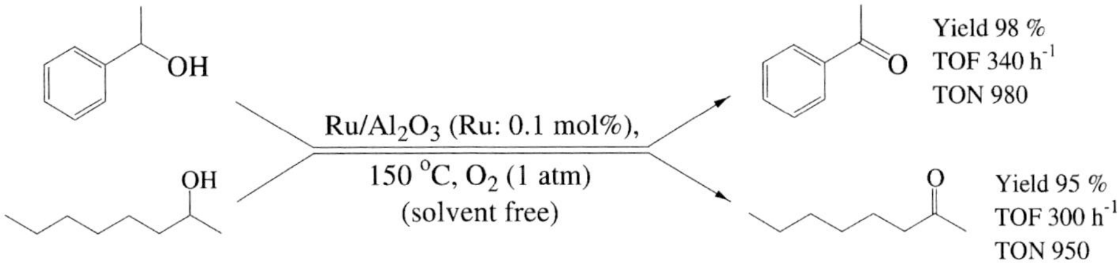

Scheme 5.2

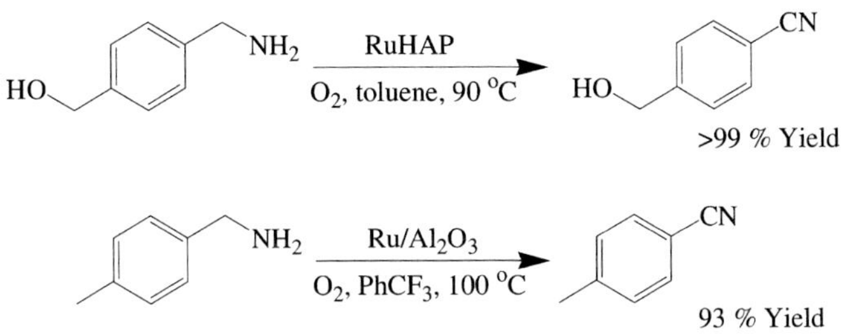

PhCH(OH)CH₃ → (Ru/Al₂O₃ (Ru: 0.1 mol%), 150 °C, O₂ (1 atm), solvent free) → acetophenone, Yield 98 %, TOF 340 h⁻¹, TON 980

2-octanol → 2-octanone, Yield 95 %, TOF 300 h⁻¹, TON 950

Scheme 5.3

4-(hydroxymethyl)benzylamine → (RuHAP, O₂, toluene, 90 °C) → 4-(hydroxymethyl)benzonitrile, >99 % Yield

4-methylbenzylamine → (Ru/Al₂O₃, O₂, PhCF₃, 100 °C) → 4-methylbenzonitrile, 93 % Yield

5.2.2 Pd Catalyst

A remarkable number of palladium-catalyzed aerobic oxidation reactions of alcohols have been reported to date [15]. Unfortunately, although some progress has been made with heterogeneous Pd catalysts, such as Pd on activated carbon [16], Pd on pumice [17], Pd-hydrotalcite [18], Pd on TiO_2 [19] and Pd/SBA-15 [20], most of these systems suffer from low catalytic activities and a limited substrate scope.

In an attempt to overcome these issues, a new type of palladium-grafted solid catalyst was developed by Kaneda and coworkers [21]. Treatment of a stoichiometric hydroxyapatite with $PdCl_2(PhCN)_2$ gives palladium-grafted hydroxyapatite (PdHAP). Analysis by several spectroscopic methods revealed that a monomeric $PdCl_2$ species is chemisorbed on the HAP surface and is readily transformed into Pd nanoclusters with a narrow size distribution in the presence of alcohol (Figure 5.2b). Nanocluster Pd^0 species effectively promote alcohol oxidation under atmospheric O_2 pressure, giving a remarkably high TONs of up to 236 000 with an excellent TOF of approximately 9800 h^{-1} for a 250-mmol-scale oxidation of 1-phenylethanol under solvent-free conditions (Scheme 5.4). The Pd catalyst does not require additives to complete the catalytic cycle. The diameter of the generated Pd nanoclusters can be controlled by the alcohol molecules used. Calculations on the palladium crystallites have shown that oxidation of alcohols occurs primarily on low-coordination sites within a regular arrangement of the Pd nanoclusters (Figure 5.5).

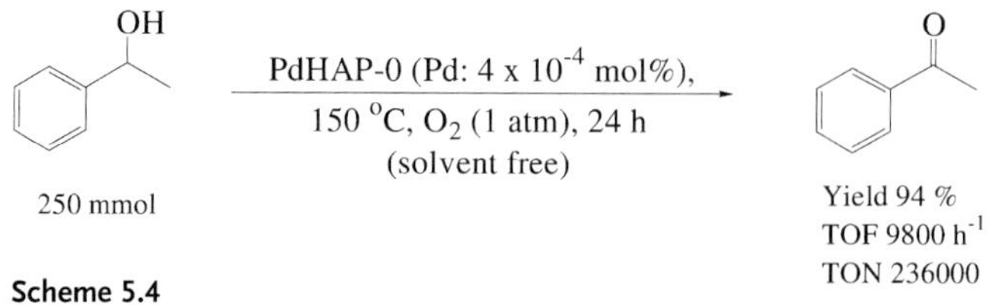
OH
PdHAP-0 (Pd: 4 x 10^{-4} mol%),
150 °C, O_2 (1 atm), 24 h
(solvent free)
O
250 mmol
Yield 94 %
TOF 9800 h^{-1}
TON 236000

Scheme 5.4

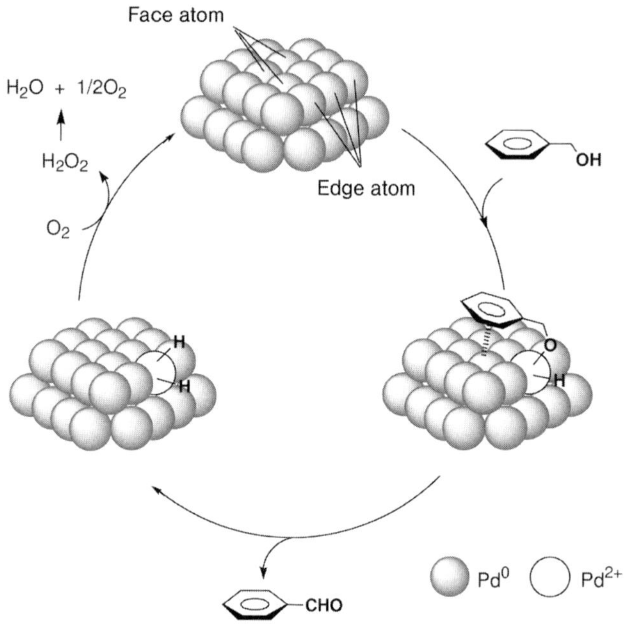
Face atom
H_2O + 1/2O_2
H_2O_2
O_2
Edge atom
OH
H
H
O
H
CHO
Pd^0
Pd^{2+}

Figure 5.5 Catalytic cycle for the oxidation of benzylalcohol using a Pd^0 nanocluster catalyst.

5.2.3
Au Catalyst

Current topics in this field include unexpected catalysis by gold nanoparticles immobilized with inorganic oxides [22], active carbon [23] or polymers [24]. Among these, the most efficient and powerful heterogeneous catalyst is probably gold nanoparticles combined with nanocrystalline cerium oxide ($Au \subset CeO_2$), introduced by Corma and coworkers [25]. The interaction between gold and ceria gives rise to a significant population of positively charged gold and Ce^{3+} species, which converts nanocrystalline cerium oxide from a stoichiometric oxidant into a catalytic system.

The $Au \subset CeO_2$ catalyst shows high performance in the selective oxidation of primary and secondary alcohols without requiring the use of solvents or high oxygen pressure (Table 5.2). The reaction has a high TOF of 12 500 h^{-1}, and the total TON reached 250 000 after three recycling experiments (Scheme 5.5). The proposed mechanism is shown in Figure 5.6. According to the authors, the alcohol reacts

Table 5.2 Aerobic alcohol oxidation catalyzed by AuCeO$_2$[a].

Entry	Substrate	Time (h)	Conversion (%)	Product	Selectivity (%)
1[b]	3-Octanol	3.5	97	3-Octanone	>99
2[b]	1-Phenylethanol	2.5	92	Acetophenone	97
3[b]	Cinnamyl alcohol	7	66	Cinnamaldehyde	73
4[c]	Vanillin alcohol	2	96	Vanillin	98
5[c]	Cinnamyl alcohol	3	>99	Cinnamic acid	98
6[d]	n-Hexanol	10	>99	Hexanoic acid	>99

[a] Conversion and selectivity were determined by GC using nonane and nitrobenzene as external standards.
[b] Substrate (4.85 mmol), Au⊂CeO$_2$ (0.5 mol.%), 80 °C, O$_2$ (1 atm, flow: 25 mL min^{-1}).
[c] Substrate (0.4 mmol), Au⊂CeO$_2$ (0.66 mol.%), H$_2$O (5 mL), Na$_2$CO$_3$ (0.3 g), 50 °C, O$_2$ (1 atm, flow; 25 mL min^{-1}).
[d] Substrate (1 mmol), Au⊂CeO$_2$ (0.25 mol.%), H$_2$O (5 mL), Na$_2$CO$_3$ (0.71 g), 100 °C, air (25 atm).

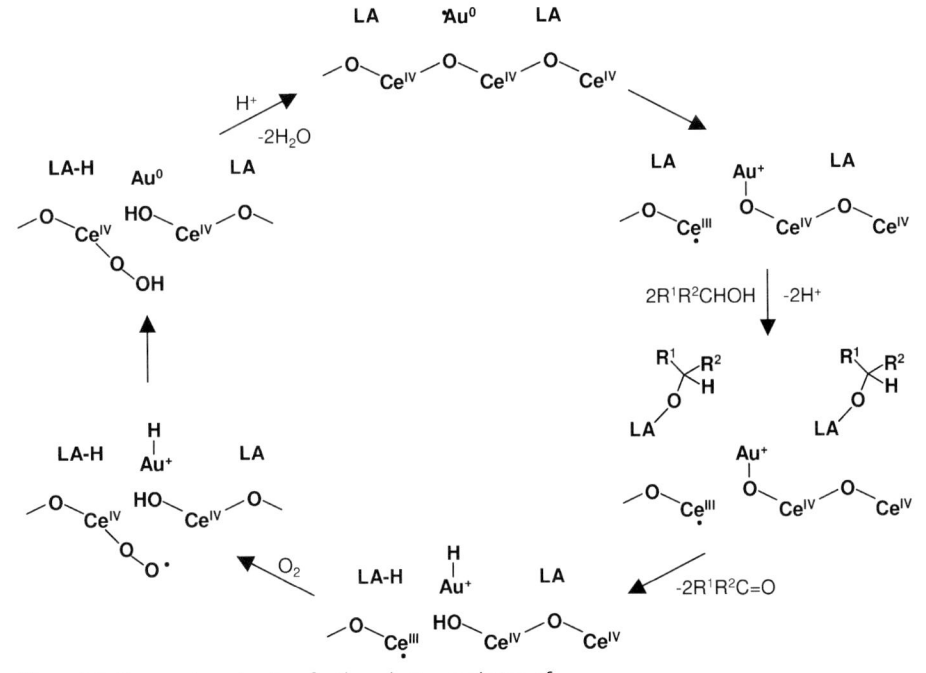

Figure 5.6 Proposed mechanism for the selective oxidation of primary and secondary alcohols using a Au⊂CeO$_2$ catalyst.

with the Lewis acid sites of ceria to give a metal alkoxide, which subsequently undergoes hydride transfer from C–H to Ce^{3+} and Au$^+$ to give a ketone, Ce–H and Au–H. Upon admission of oxygen into the system and its coordination to the oxygen-deficient sites of ceria, a cerium-coordinated superoxide (Ce–OO$^•$) species is formed. These superoxide species evolve into cerium hydroperoxide by hydrogen abstraction from Au–H, and are responsible for the formation of the initial Au$^+$ species.

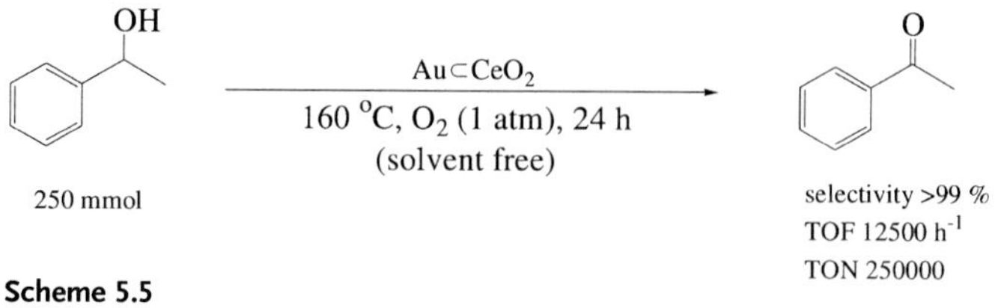

Scheme 5.5

PhCH(OH)CH₃ →[Au⊂CeO₂, 160 °C, O₂ (1 atm), 24 h, (solvent free)] PhC(O)CH₃

250 mmol

selectivity >99 %
TOF 12500 h^{-1}
TON 250000

5.2.4
Au-Pd Catalyst

One of the disadvantages associated with the systems described above is limited activity for selective oxidation of primary alcohols to the corresponding aldehydes. Hutchings and coworkers have developed a highly efficient heterogeneous catalyst, Au-Pd/TiO$_2$, for solvent-free oxidation of primary alcohols to aldehydes [26]. The addition of Au to Pd nanocrystals improved the selective production of aldehydes in the oxidation of primary alcohols. TEM combined X-ray photoelectron spectroscopy showed that the Au-Pd nanocrystals were made up of a Au-rich core with a Pd-rich shell, and indicated that Au electronically influences the catalytic properties of Pd. Although the PdHAP and Au ⊂ CeO$_2$ catalysts gave high TOFs of 9800 h^{-1} and 12 500 h^{-1}, respectively, for the oxidation of 1-phenylethanol, notably, Au-Pd/TiO$_2$ showed a much higher TOF, reaching almost 270 000 for the oxidation of 1-phenylethanol under solvent-free conditions at 160 °C under 1 atm of O$_2$ (Scheme 5.6).

All of the above heterogeneous catalysts are recyclable without any loss of activity or selectivity after removal from the reaction mixture by filtration.

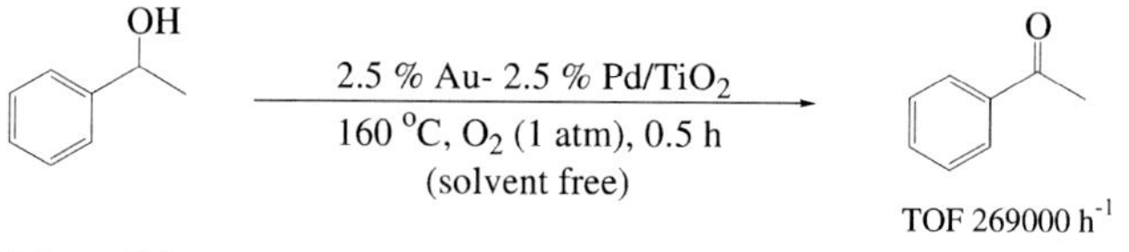

Scheme 5.6

PhCH(OH)CH₃ →[2.5 % Au- 2.5 % Pd/TiO₂, 160 °C, O₂ (1 atm), 0.5 h, (solvent free)] PhC(O)CH₃

TOF 269000 h^{-1}

5.3
Epoxidation of Olefins

Epoxidation is one of the most fundamental and important oxygenation reactions in organic synthesis. Various methods of epoxidation have been developed and exploited, and the search for new environmentally-friendly methods using H_2O_2 or O_2 as the sole oxidant has attracted much interest [27]. Catalyst systems for this reaction can have several disadvantages, such as difficulties with catalyst recovery from the reaction mixture and reuse. With the aim of overcoming these issues, the use of heterogeneous catalysts containing various transition metals such as Ti, Mn,

W, V and Co has been widely investigated [28]. This section will highlight some novel and unique approaches to epoxide synthesis using environmentally benign oxidants and heterogeneous metal catalysts such as solid-supported Ti, polyoxometalates and hydrotalcite.

5.3.1
Epoxidation with Hydrogen Peroxide

5.3.1.1 Titanium-Based Catalysts

Titanium catalysts have been widely investigated for the epoxidation of olefins using organic peroxides such as *tert*-butyl hydroperoxide (TBHP) and ethylbenzene hydroperoxide (EBHP) [29]. However, an efficient titanium catalyst system employing H_2O_2 has not been reported since the development of the heterogeneous titanium (IV)-silicate (TS-1) catalyst. TS-1 showed high activity and selectivity for epoxidation of small terminal olefins using H_2O_2 [30]. To extend the scope of substrates, several different titanium-containing silicate zeolites have been prepared. Among these, Al-free Ti-beta, which has larger pores than TS-1, was found to catalyze efficiently the epoxidation of larger olefins such as 1-hexene and methyl cyclohexene with high selectivity (>90%) [31].

5.3.1.2 Tungsten-Based Catalysts

Since Payne and Williams reported selective epoxidation using a catalytic amount of sodium tungstate with H_2O_2 as an oxidant in 1959 [32], many highly efficient homogeneous tungsten-based catalyst systems have been reported [33]. However, there are few examples of heterogenization of tungsten catalysts despite the significant advantages of heterogeneous catalysis from an environmental standpoint.

Hydrotalcite (HT, $Mg_6Al_{12}(OH)_{16}CO_3 \cdot nH_2O$) is a layered anionic clay consisting of a positively charged two-dimensional brucite layer with anionic species such as hydroxide and carbonate located in the interlayer [34]. Figure 5.7 shows a structural model of hydrotalcite. Transition metals of various types can easily be introduced into

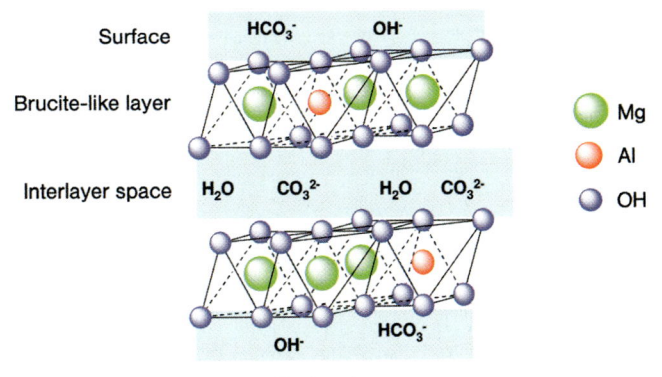

Figure 5.7 Structural model of hydrotalcite.

the brucite-like layer or interlayer space or onto the surface due to the following characteristics of hydrotalcite: (1) the cation-exchange ability of the brucite layer; (2) the anion-exchange ability of the interlayer; (3) the tunable basicity of the surface; and (4) its adsorption capacity [35].

Jacobs and coworkers succeeded in heterogenization of tungstate catalysts by using HT [36]. WO_4^{2-} ions, derived from Na_2WO_4, are intercalated into the interlayer of the HT by anion exchange reaction, and two types of tungsten-intercalated HT (hydrophilic WO_4^{2-}-HT catalyst and hydrophobic WO_4^{2-} containing p-toluenesulfonate HT catalyst) are synthesized. These heterogeneous catalysts show high selectivity for epoxidation of allylic and homoallylic alcohols and non-functionalized olefins. Especially, the hydrophobic HT catalyst shows superior selectivity towards epoxide compared to hydrophilic catalysts. For example, when a hydrophobic catalyst was used in place of a hydrophilic one, the selectivity toward 2,3-dimethyl-2-butene oxide and cyclohexene oxide increased from 39 and 65% to 79 and 91%, respectively.

A color change of the catalyst from white to yellow upon contact with H_2O_2 indicates the *in situ* formation of active peroxotungsten species at the HT surface. The same research group also synthesized phosphoramide-grafted MCM-41 and used it to immobilize tungsten derived from H_2WO_4 [37]. Figure 5.8 shows the possible active site of the catalyst, based on Raman and NMR spectroscopy. The catalytic activity of MCM-41-immobilized tungsten is lower than the hydrophobic WO_4^{2-}-HT catalyst described above, but the advantage of the latter is low tungsten leaching (less than 2%).

Mizuno and coworkers have synthesized the dinuclear peroxotungstate [{W(=O)$(O_2)_2(H_2O)$}$_2$(μ-O)]$^{2-}$ catalyst (W2). This homogeneous catalyst shows highly efficient utilization of H_2O_2 and high selectivity in the epoxidation of olefins [38]. This group also recently succeeded in heterogenizing W2 by using ionic liquid-modified SiO_2 [39]. Reaction of the ionic liquid **1** [1-octyl-3-(3-triethoxysililpropyl)-4,5-dihydroimidazolium hexafluorophosphate] with SiO_2 afforded a SiO_2-supported

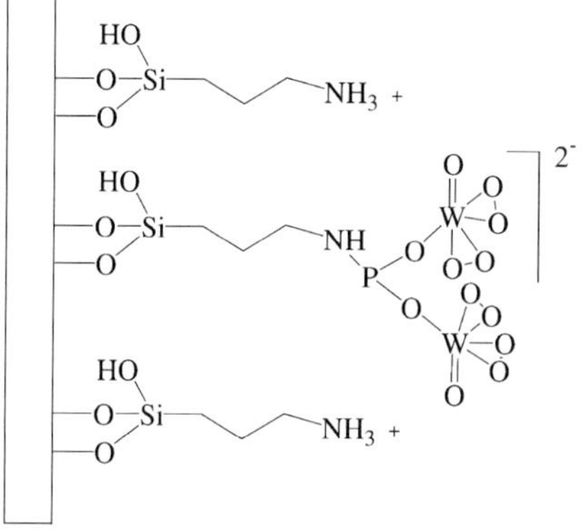

Figure 5.8 Possible active site of an MCM-41-immobilized peroxotungsten species.

Table 5.3 Epoxidation of olefins with H_2O_2 catalyzed by $W2/1\text{-}SiO_2$[a)].

Entry	Olefin	Time (h)	Yield of epoxide (%)
1	cis-2-Octene	3	>99[b)]
2	trans-2-Octene	7	73[c)]
3	cis-β-Methylstyrene	2	>99[b)]
4	Cyclohexene	3	80
5	2-Norbornene	2.5	80[d)]
6	Cycloheptene	2	>99
7	Cyclooctene	1	>99
8	Geraniol	4	95[e)]
9	4-Methyl-3-pentene-2-ol	4	90[f)]
10	(Z)-3-Methyl-3-pentene-2-ol	4	90[g)]

[a)] Reaction conditions: Olefin (1 mmol), catalyst (W2: 1 mol.% with respect to olefin), H_2O_2 (30% aq. solution, 0.2 mmol), CH_3CN (0.5 mL), 60 °C. Yield and selectivities were determined by GC or ^1H NMR with an internal standard technique. Selectivities to the corresponding epoxides were >95% in all cases. Yield (%) = epoxide (mmol) per initial H_2O_2 (mmol) × 100.
[b)] Only cis.
[c)] Only trans.
[d)] Only exo.
[e)] Only 2,3-epoxide.
[f)] Threo/erythro = 90 : 10.
[g)] Threo/erythro = 40 : 60.

ionic liquid (2-SiO_2). Treatment of 1-SiO_2 with an aqueous solution of the potassium salt of the dinuclear peroxotungstate W2 afforded W2/1-SiO_2 by anion exchange (see Chapter 6). The W2/1-SiO_2 catalyst is capable of heterogeneous epoxidation of a wide range of olefins while retaining the catalytic activity of its homogeneous analogue (Table 5.3).

This catalyst overcomes the following disadvantages of previous reported catalyst systems: (1) a certain amount of tungsten leaching; (2) low accessibility for organic substrates and H_2O_2 due to hydrophilicity and/or low surface area of the catalyst; (3) decomposition of H_2O_2 on the basic sites of the support; and/or (4) decrease in activity due to immobilization. The epoxidation reaction stopped immediately upon removal of the catalyst, and no tungsten species could be found in the filtrate. These results rule out the possibility that the reaction was being catalyzed by leached tungsten species; thus, the observed catalysis is truly heterogeneous. Furthermore, the catalyst was reusable without loss of catalytic performance.

5.3.1.3 Base Catalyst

Basic materials catalyze oxidation reactions by activation of either the substrate or the oxidant, giving rise to electrophilic attack on carbanions or nucleophilic attack by peroxoanions [40].

Kaneda and coworkers found that HT acted as a heterogeneous base catalyst for the epoxidation of various α,β-unsaturated ketones with 30% aqueous hydrogen peroxide as the oxidant [41]. Notably, oxidation systems using HT give epoxides exclusively,

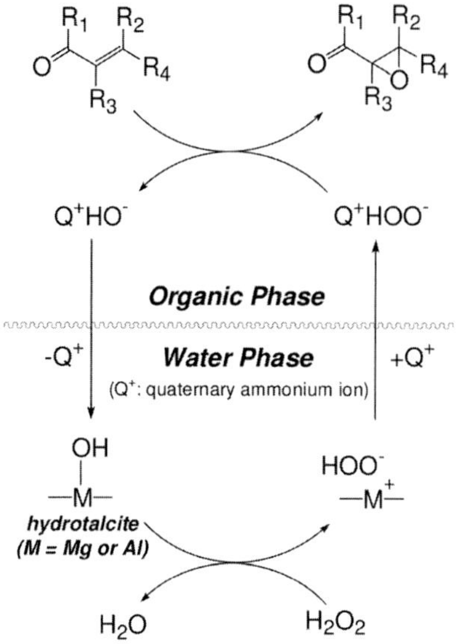

Scheme 5.7

[Scheme 5.7 depicts a catalytic cycle with an organic phase containing Q^+HO^- and Q^+HOO^- reacting with an α,β-unsaturated ketone (substituents R_1, R_2, R_3, R_4) to form an epoxyketone, and a water phase where hydrotalcite (−M−OH, M = Mg or Al) reacts with H_2O_2 to give −M−OOH and H_2O. Q^+ is a quaternary ammonium ion that shuttles between phases.]

without other oxidation products. The reactivity of the α,β-unsaturated ketones can be explained in terms of orbital energies of the LUMO, that is the electrophilicity of enones.

The reaction mixture in base-catalyzed epoxidation consists of three phases: an organic phase of enones, an aqueous phase of H_2O_2 and solid HT. Under these conditions, a HOO^- anion may move into the organic phase by forming an ion pair such as a Q^+HOO^- species to react with a lipophilic enone substrate (Scheme 5.7). The long alkyl chains of cationic surfactants bring about an increase in lipophilicity, which facilitates the transfer of lipophilic enones from the organic phase to the interface at the reaction zone.

Highly active base HT catalysts were prepared from MgO and Al_2O_3 [42], or calcined HT (MO = mixed oxide), for the epoxidation of various α,β-unsaturated ketones using H_2O_2 [41b]. The Mg-Al MO showed much higher catalytic activity for many α,β-unsaturated ketones than conventional HT. Notably, even with low concentrations of H_2O_2, this mixed oxide catalyst showed high activity for epoxidation without organic solvents. For example, the reaction of 2-cyclopenten-1-one using 1.5 equivalents of 5% H_2O_2 for 2 h gave the corresponding epoxyketone in 94% yield (Scheme 5.8).

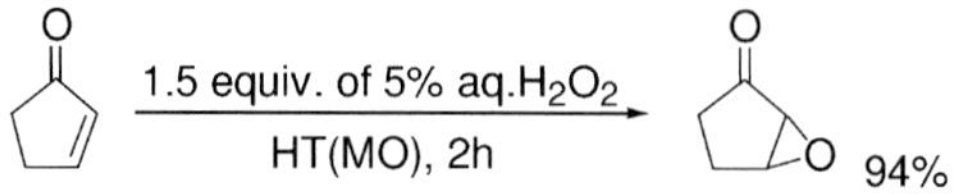

Scheme 5.8

[Scheme 5.8: 2-cyclopenten-1-one + 1.5 equiv. of 5% aq. H_2O_2, HT(MO), 2 h → 2,3-epoxycyclopentan-1-one, 94%.]

Scheme 5.9

In a previous epoxidation of α,β-unsaturated ketones, the reaction of H_2O_2 with base sites on the HT surface gave a HOO^- anion species that could react in the presence of nitriles to form peroxycarboximidic acid, with the oxygen being transferred to an olefin to produce the epoxide together with a carboxamide [43]. Kaneda's group found that HT catalyzed the oxidation of various common olefins using H_2O_2

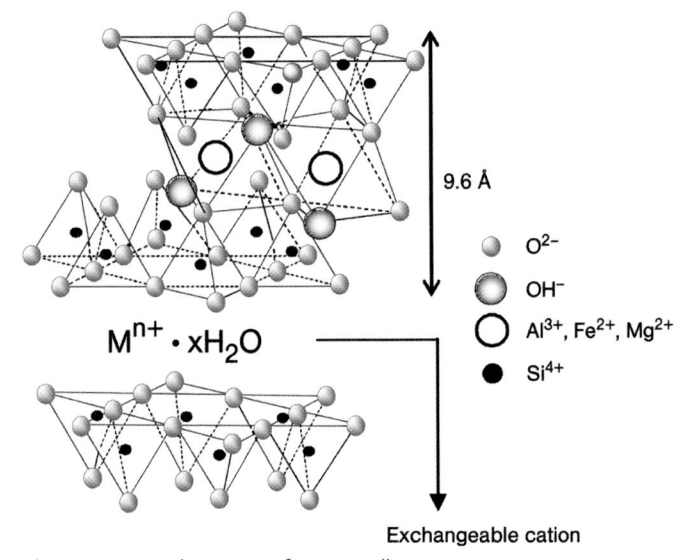

Figure 5.9 Layered structure of montmorillonite.

in the presence of nitriles to afford the corresponding epoxides in excellent yields [44]. Cis- and trans-olefins gave the corresponding epoxides stereospecifically, with retention of the double bond configuration.

Scheme 5.9 shows a possible mechanism for this epoxidation reaction. First, H_2O_2 reacts with a base site on the HT surface to form a HOO^- species, which attacks a nitrile to generate peroxycarboximidic acid as an active intermediate oxidant. The oxygen of the peroxycarboximidic acid is transferred to an olefin molecule. Interestingly, the resulting amide can be further employed for the epoxidation of olefins in the presence of a HT catalyst [44b].

5.3.2
Epoxidation with Molecular Oxygen

The search for new environmentally-friendly epoxidation methods using O_2 as a sole oxidant has attracted much interest. Although there has been some success with O_2 and homogeneous catalyst systems in the liquid phase without the use of reducing reagents, there have been few reports concerning heterogeneous epoxidation of olefins [45, 46].

Co^{2+}-exchanged faujasite zeolite is a unique heterogeneous catalyst for liquid-phase epoxidation using O_2 [45]. This catalyst is active only for styrene, and the conversion and yield of styrene oxide were 65 and 45%, respectively. The TON, based on Co ions, reached 12. The Co^{2+} ions, located in supercages, are thought to cause activation of O_2 for epoxidation.

More recently, vanadium-exchanged montmorillonite (V-mont) has been found to catalyze the epoxidation of several olefins using O_2 as a sole oxidant [47]. Montmorillonites are layered clay minerals, composed of octahedral alumina sheets sandwiched by two tetrahedral silica layers (Figure 5.9). The aluminum atoms in the center of the octahedral sheet are partly replaced by magnesium or iron atoms, resulting in a cation deficiency over the whole clay. To compensate for this, some cations, typically sodium ions, are present in the expansive interlamellar spaces between the two-dimensional silicate sheets. Various metal cations with high potential as catalytic active centers can be introduced into the narrow interlayers by simple ion-exchange methods [48]. V-mont was synthesized from Na-mont using aqueous VCl_3 solution, and XAFS analysis suggested that a highly dispersed monomeric dioxo V^{5+} species surrounded by four oxygen atoms was created on the montmorillonite.

Epoxidation of cyclooctene proceeded under 1 atm of O_2 in α,α,α-trifluorotoluene solvent. The yield of cyclooctene oxide reached 80%, with >99% selectivity after 72 h. The V-mont selectively catalyzed the epoxidation of several olefins, affording the corresponding epoxides as the major products. Upon completion of cyclooctene epoxidation, the V-mont was separated from the reaction mixture by simple filtration, and could be reused without any appreciable loss of its high catalytic activity and selectivity (Table 5.4). Thus, the development of highly efficient heterogeneous catalysts for the epoxidation of olefins using O_2 is still of great importance.

Table 5.4 Epoxidation of various olefins by V-mont in the presence of O_2 [a].

Entry	Olefin	Yield of epoxide (%)[b]	Selectivity (%)
1	Cyclooctene	80	>99
2[c]	Cyclooctene	79	98
3[d]	Cyclooctene	78	98
4	Cyclododecene	40	93
5	Cyclopentene	35	54[e]
6	2-Octene	21	80[f]
7	1-Decene	11	85[g]

[a] Substrate (3 mmol), V-mont (0.1 g, V: 0.019 mmol), α,α,α-trifluorotoluene (1 mL), 90 °C, 72 h, O_2 atmosphere.
[b] Determined GC analysis using an internal standard technique.
[c] Reuse 1.
[d] Reuse 2.
[e] 2-Cyclopentene-1-ol was formed.
[f] Small amount of octanal was formed.
[g] Small amount of decanal was formed.

5.4
Cis-Dihydroxylation

Cis-dihydroxylation of olefins is a powerful synthetic reaction that can be applied to pharmaceuticals, agrochemicals and fine chemicals [49]. Osmium tetroxide (OsO_4) has been widely used as an efficient catalyst for this reaction [50]; however, the OsO_4 catalyst has the disadvantages of high toxicity and cost and the possibility of metal-leakage in the products. Several groups have attempted to resolve these issues by heterogenization of the osmium catalyst [51] using polymers [52], organomolecule-modified silica [53] or hydrotalcite (HT) [54, 55]. Particularly efficient heterogeneous dihydroxylation catalyst systems using crystalline materials have been developed independently by Jacobs and Choudary.

Jacobs and coworkers succeeded in immobilizing the Os catalyst by using SiO_2-anchored tetrasubstituted olefin [53a]. Treatment of Os^{8+} with this material gives the corresponding osmate ester, and immobilized Os^{6+} can be smoothly reoxidized to the cis-dioxo Os^{8+} complex by N-methylmorpholine N-oxide (NMO) without release of the diol (Scheme 5.10). The Os-supported heterogeneous catalyst shows high activity for cis-dihydroxylation of olefins with NMO. The cis-dihydroxylation of twelve olefins substantiates the value of the catalyst (Table 5.5). The reusability of the heterogeneous catalyst and the possibility of using more environmentally benign oxidants are issues that remained to be resolved. Choudary and coworkers developed a recoverable and reusable Os catalyst immobilized on HT by ion-exchange of OsO_4^{2-} for asymmetric dihydroxylation of olefins using NMO as an oxidant [54]. The same research group also reported the development of a trifunctional hydrotalcite catalyst (LDH-PdOsW), containing $PdCl_4^{2-}$, OsO_4^{2-} and WO_4^{2-}, for a one-pot synthesis of chiral diols from olefins via Heck coupling, N-oxidation and asymmetric dihydroxylation reactions using a combined catalytic amount of

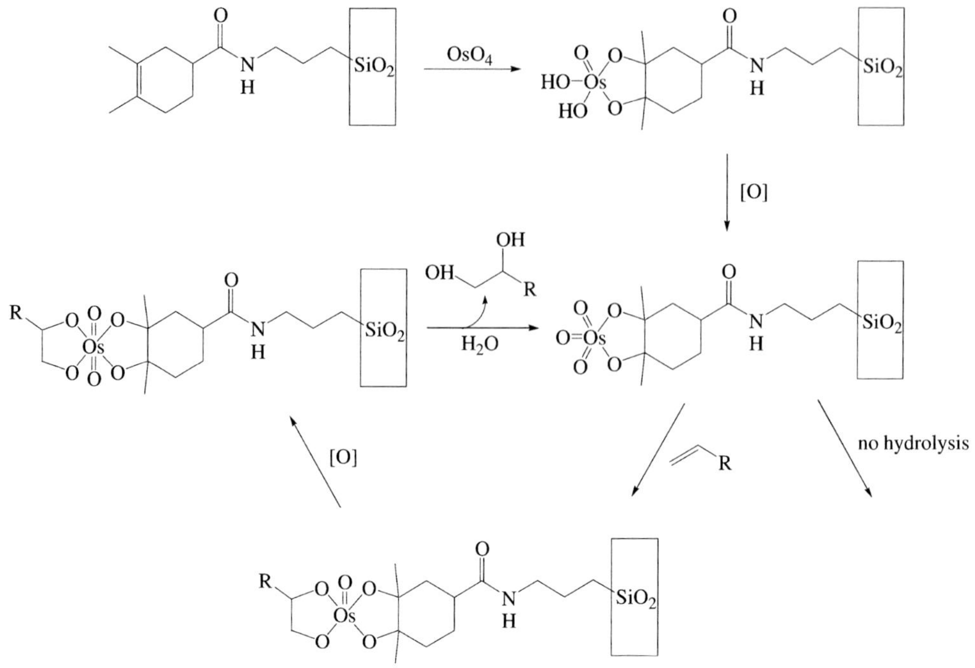

Scheme 5.10

Table 5.5 Cis-dihydroxylation of olefins with NMO[a].

Entry	Olefin	Time (h)	Conversion (%)	Selectivity (%)
1	1-Pentene	48	83	99
2	1-Hexene	48	99	98
3	1-Heptene	48	96	99
4	Cyclopentene	48	83	98
5	Cyclohexene	48	99	99
6	cis-2-Hexene	48	99	99
7	trans-2-Hexene	48	98	98
8	Styrene	48	99	96
9	Indene	48	72	99
10	2-Methyl-2-pentene	48	50	99
		150	99	99
11	Ethyl trans-cinnamate	24	65	99
12	Ethyl trans-crotonate	48	85	99

[a] Reaction conditions: 0.1 g heterogeneous catalyst (4×10^{-6} mol Os), olefin (1.6 mmol), NMO (1.6 mmol), solvent (3 mL), H_2O (200 mL), room temperature.

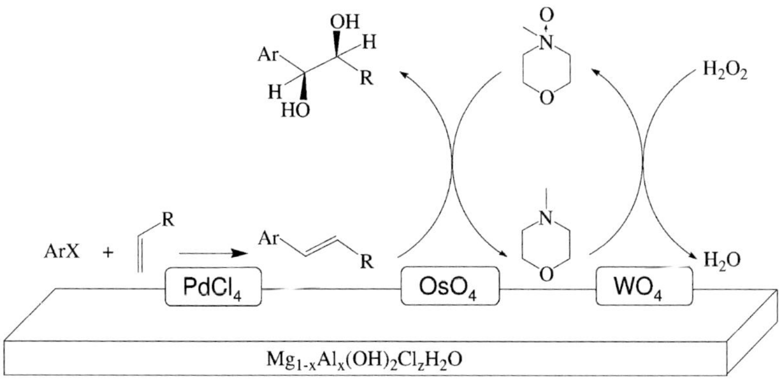

Scheme 5.11

N-methylmorpholine (NMM) and the cheaper oxidant H_2O_2 rather than stoichiometric amounts of NMO (Scheme 5.11) [55]. In this process, NMM is reoxidized into NMO by H_2O_2 together with tungsten catalyst. LDH-PdOsW asymmetrically catalyzed a one-pot synthesis of chiral diols from aryl halides and olefins (Figure 5.10).

5.5
Baeyer–Villiger Oxidation

The Baeyer–Villiger reaction has become one of the most well-known and widely applied reactions in organic syntheses [56]. Its success is largely due to its versatility [57]: (1) various carbonyl compounds can be oxidized: that is, ketones are converted into esters, cyclic ketones into lactones, benzaldehydes into phenols, and carboxylic acids and α-diketones into anhydrides; (2) numerous functional groups are tolerated; (3) the reaction is generally stereoselective: that is, the migrating group retains its configuration; (4) a wide range of oxidants may be used, with activity decreasing in the order $CF_3CO_3H > m\text{-CPBA} > C_6H_5CO_3H \gg H_2O_2 > t\text{-BuOOH}$ [58]. Although more than a century has passed since its discovery, the Baeyer–Villiger reaction is far from being at the end of its development.

Free-radical autoxidation of aldehydes with O_2 is facile and affords the corresponding peracids, which are used as oxidants for carbonyl compounds. The peracid can transfer an oxygen atom to a substrate such as an olefin or ketone, resulting in the formation of one equivalent of epoxide or ester and acid as a co-product in the absence of metal catalysts [59]. Kaneda and coworkers have developed several HT materials that are active for heterogeneous Baeyer–Villiger reactions with O_2/aldehyde [60]. Combination with Lewis acidic metals improved the reaction by allowing coordination of the peracid and the intermediate.

Supported Pt [61] and Ti [62] catalysts have been used in heterogeneous catalyst systems with H_2O_2 as an oxidant. However, these catalysts suffered from low TON values and low selectivity. A Sn-substituted beta zeolite was recently applied as a

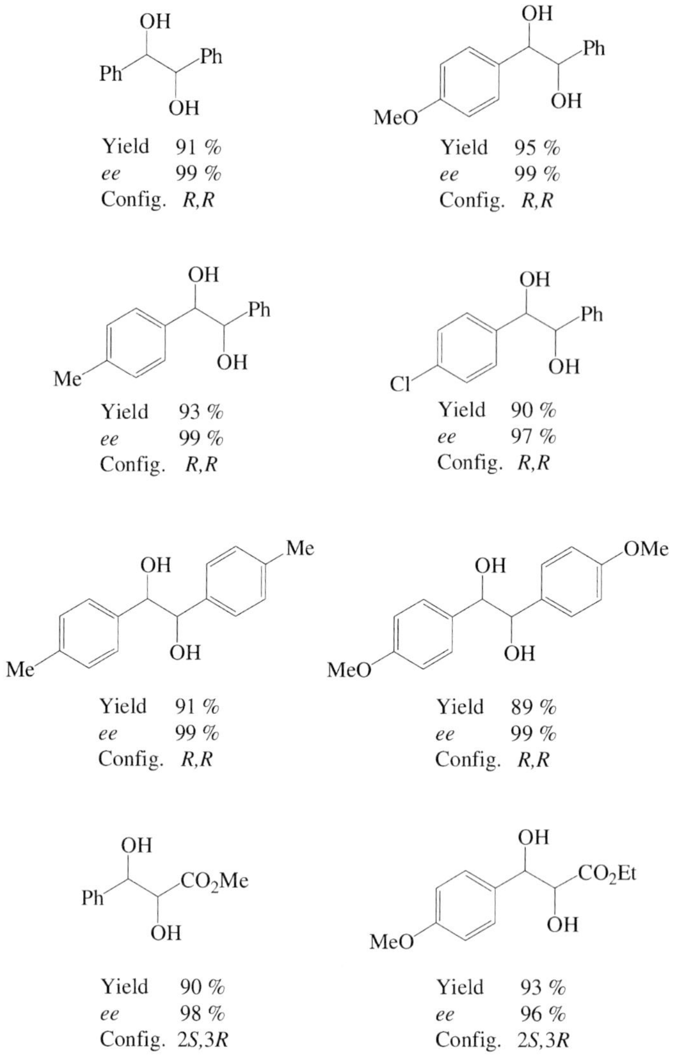

Figure 5.10 Chiral diols synthesized from aryl halides and olefins in one pot by asymmetric catalysis with LDH-PdOsW.

water-stable heterogeneous Lewis acid catalyst for Baeyer–Villiger reaction of saturated and unsaturated ketones to the desired lactones with >98% selectivity using H_2O_2 as an oxidant (Scheme 5.12) [63].

The activation of carbonyl compounds by Lewis acid sites in the tin species has been proposed to play an important role in the oxidation reaction (Figure 5.11). The pore size of beta zeolite (~6.5 Å × 7.5 Å) may pose some restrictions on the substrate size. For much bulkier substrates, a new tin-MCM-41 (pore size around 20 Å) catalyst has been developed by the same research group [64].

Scheme 5.12

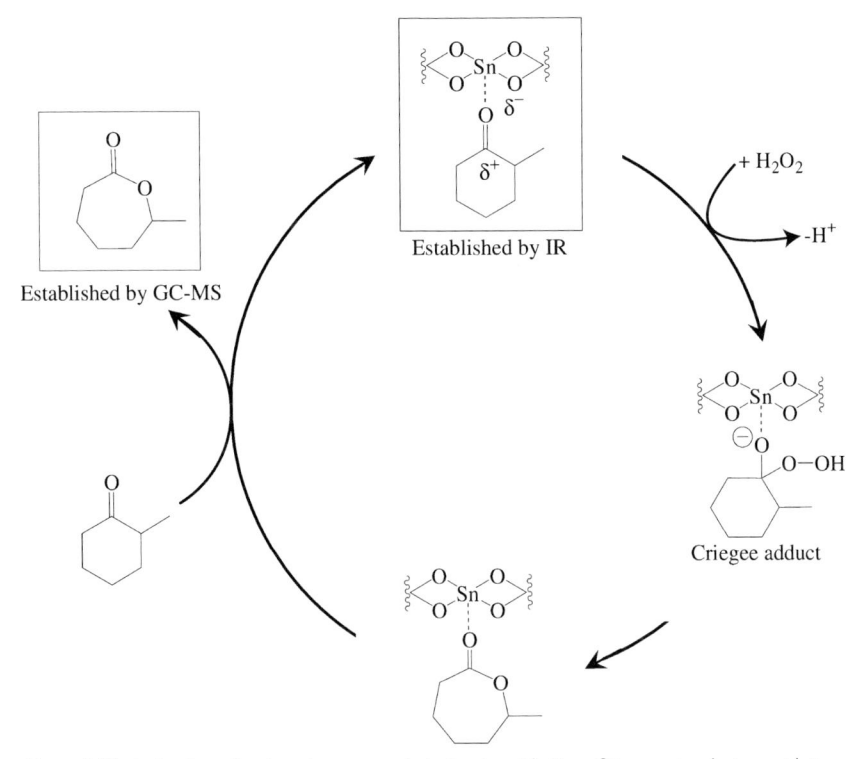

Reaction: R₁-C(O)-R₂ + Sn-zeolite beta, 35 wt% H$_2$O$_2$, methyl t-butyl ether or dioxane → R₁-C(O)-O-R₂

Products shown:
- Yield 94 %, Sel. >98 %
- Yield 52 %, Sel. >98 %
- Conv. >95 %, Sel. 100 %
- Conv. 68 %, Sel. 100 %

Figure 5.11 Activation of carbonyl compounds by Lewis acid sites of tin species during oxidation.

5.6
C–H Activation Using Molecular Oxygen

Direct C–H activation of saturated hydrocarbons by heterogeneous catalysts using O$_2$ as an oxidant under mild reaction conditions is of great importance in terms of the development of chemical processes employing abundant natural resources. For this reason, considerable efforts have been made to develop efficient heterogeneous oxidation systems using O_2 [65]. One of the few successful examples of selective oxidation of saturated hydrocarbons was reported by Thomas and coworkers, who found that aluminophosphate molecular sieves (AlPO) substituted with transition metal cations such as Co^{3+}, Mn^{3+} and Fe^{3+} acted as efficient heterogeneous catalysts for liquid-phase oxyfunctionalization of saturated hydrocarbons under an O_2 atmosphere at relatively low temperatures (~130 °C) [66]. These catalysts can oxidize alkanes in a regioselective and stereoselective manner, thereby favoring enhanced reactivity of the terminal methyl groups. Framework-isolated divalent metal cations, which are present initially in the sieve when synthesized, that are then converted into the trivalent state are vital in this unique catalytic system.

5.7
Conclusions

Crystalline materials-immobilized active metal species can offer significant benefits in achieving simple and clean liquid-phase oxidations because they have the following advantages: (1) robust ligands under the oxidation conditions; (2) milder reaction conditions than that in the gas-phase reactions, which enable high selectivity for the desired products; (3) tunable surface properties such as acidity and basicity for the targeted oxidations; (4) multifunctional catalysis by cooperative actions by several immobilized active species on solid surface; and (5) a simple work-up procedure and easy recovery of the catalysts, and the recyclability. Design of well-defined active species that are uniform in composition and distribution on crystalline materials can allow understanding of structure-activity relationships, which will bridge the gap between homogeneous and heterogeneous catalysis and contribute the development of new and high-performance catalysts.

References

1 Gates, B.C., Katzer, J.R. and Schuit, G.C. (1979) *Chemistry of Catalytic Process*, McGraw-Hill, New York.

2 (a) Ertl, G., Knözinger, H. and Weitkamp, J. (eds) (1997) *Handbook of Heterogeneous Catalysis*, VCH, Weinheim; (b) Arends, I.W.C.E. and Sheldon, R.A. (2001) *Applied Catalysis A – General*, **212**, 175; (c) Mallat, T. and Baiker, A. (2004) *Chemical Reviews*, **104**, 3037.

3 (a) Sheldon, R.A. and Kochi, J.K. (1994) *Metal-Catalyzed Oxidations of Organic Compounds*, Academic Press, New York; (b) Barton, D.H.R., Martell, A.E. and Sawyer, D.T. (1993) *The Activation of Dioxygen and Homogeneous Catalytic Oxidation*, Plenum Press, New York; (c) Hudlucky, M. (1990) *Oxidations in Organic Chemistry*, ACS Monograph, Washington, DC; (d) Bäckvall, J.-E. (2004) *Modern Oxidation Methods*, Wiley-VCH, Weinheim.

4 (a) Sheldon, R.A. and Kochi, J.K. (1981) *Metal-Catalyzed Oxidations of Organic Compounds*, Academic Press, New York; (b) Hill, C.L. (1988) Advances in Oxygenated Processes, Vol. 1 (ed. A.L. Baumstark), JAI Press, London, 1; (c) Hudlucky, M. (1990) *Oxidations in Organic Chemistry*, ACS Monograph Series,

American Chemical Society, Washington, DC; (d) Trost, B.M. (1991) *Comprehensive Organic Synthesis* I, Fleming Pergamon, Oxford; (e) Parshall, G.W. and Ittel, S.D. (1992) *Homogeneous Catalysis*, 2nd edn, John Wiley & Sons, New York.

5 (a) Cainelli, G. and Cardillo, G. (1984) *Chromium Oxidants in Organic Chemistry*, Springer, Berlin; (b) Lee, D.G. and Spitzer, U.A. (1970) *The Journal of Organic Chemistry*, **35**, 3589; (c) Menger, F.M. and Lee, C. (1981) *Tetrahedron Letters*, **22**, 1655.

6 (a) Trost, B.M. (1991) *Science*, **254**, 1471; (b) Anastas, P.T. and Warner, J.C. (1998) *Green Chemistry in Theory and Practice*, Oxford University Press, Oxford; (c) Sheldon, R.A., Wallau, M., Arends, I.W.C.E. and Schuhardt, U. (1998) *Accounts of Chemical Research*, **31**, 485; (d) Clark, J.H. (1999) *Green Chemistry*, **1**, 1; (e) Sheldon, R.A. (2000) *Green Chemistry*, **2**, G1; (f) Anastas, P.T., Bartlett, L.B., Kirchhoff, M.M. and Williamson, T.C. (2000) *Catalysis Today*, **55**, 11.

7 (a) Hinzen, B. and Ley, S.V. (1997) *Journal of the Chemical Society-Perkin Transactions 1*, 1707; (b) Vocanson, F., Guo, Y.P., Namy, J.L. and Kagan, H.B. (1998) *Synthetic Communications*, **28**, 2577; (c) Kaneda, K., Yamashita, T., Matsushita, T. and Ebitani, K. (1998) *The Journal of Organic Chemistry*, **67**, 1750; (d) Matsushita, T., Ebitani, K. and Kaneda, K. (1999) *Chemical Communications*, 265; (e) Bleloch, A., Johnson, B.F.G., Ley, S.V. et al. (1999) *Chemical Communications*, 1907; (f) Choi, E., Lee, C., Na, Y. and Chang, S. (2002) *Organic Letters*, **4**, 2369; (g) Ji, H.-B., Mizugaki, T., Ebitani, K. and Kaneda, K. (2002) *Tetrahedron Letters*, **43**, 7179; (h) Ji, H.-B., Ebitani, K., Mizugaki, T. and Kaneda, K. (2002) *Catalysis Communications*, **3**, 511; (i) Zhan, B.-Z., White, M.A., Sham, T.-K. et al. (2003) *Journal of the American Chemical Society*, **125**, 2195; (j) Yamaguchi, K. and Mizuno, N. (2003) *Chemistry – A European Journal*, **9**, 4353; (k) Ciriminna, R. and Pagliaro, M. (2003) *Chemistry – A European Journal*, **9**, 5067; (l) Ebitani, K., Ji, H.-B., Mizugaki, T. and Kaneda, K. (2004) *Journal of Molecular Catalysis A-Chemical*, **212**, 161; (m) Ebitani, K., Motokura, K., Mizugaki, T. and Kaneda, K. (2005) *Angewandte Chemie – International Edition*, **44**, 3423.

8 Kaneda, K., Ebitani, K., Mizugaki, T. and Mori, K. (2006) *Bulletin of the Chemical Society of Japan*, **79**, 981, and references cited therein.

9 Yamaguchi, K., Mori, K., Mizugaki, T., Ebitani, K. and Kaneda, K. (2000) *Journal of the American Chemical Society*, **122**, 7144.

10 Elliott, J.C. (1994) *Structure and Chemistry of the Apatites and Other Calcium Orthophosphates*, Elsevier, Amsterdam.

11 Mori, K., Kanai, S., Hara, T. et al. (2007) *Chemistry of Materials*, **19**, 1249.

12 Yamaguchi, K. and Mizuno, N. (2002) *Angewandte Chemie – International Edition*, **41**, 4538.

13 Mori, K., Yamaguchi, K., Mizugaki, T. et al. (2001) *Chemical Communications*, 461.

14 Yamaguchi, K. and Mizuno, N. (2003) *Angewandte Chemie – International Edition*, **42**, 1479.

15 (a) Blackburn, T.F. and Schwartz, J. (1977) *Journal of the Chemical Society-Chemical Communications*, 157; (b) Gómez-Bengoa, E., Noheda, P. and Echavarren, A.M. (1994) *Tetrahedron Letters*, **35**, 7097; (c) Aiet-Mohand, S., Henin, F. and Muzart, J. (1995) *Tetrahedron Letters*, **36**, 2473; (d) Kaneda, K., Fujii, M. and Morioka, K. (1996) *The Journal of Organic Chemistry*, **61**, 4502; (e) Kaneda, K., Fujie, Y. and Ebitani, K. (1997) *Tetrahedron Letters*, **38**, 9023; (f) Peterson, K.P. and Larock, R.C. (1998) *The Journal of Organic Chemistry*, **63**, 3185; (g) Nishimura, T., Onoue, T., Ohe, K. and Uemura, S. (1998) *Tetrahedron Letters*, **39**, 6011; (g) Nishimura, T., Onoue, T., Ohe, K. and Uemura, S. (1999) *The Journal of Organic Chemistry*, **64**, 6750; (i) ten Brink, G.-J., Arends, I.W.C.E. and Sheldon, R.A. (2000) *Science*, **287**, 1636; (j) Stahl, S.S., Thorman, J.L., Nelson, R.C. and Kozee, M.A. (2001) *Journal of the American Chemical Society*, **123**, 7188; (k) Steinhoff, M.A., Fix, S.R. and Stahl, S.S. (2002)

Journal of the American Chemical Society, **124**, 766; (l) ten Brink, G.-J., Arends, I.W.C.E. and Sheldon, R.A. (2002) *Advanced Synthesis and Catalysis*, **344**, 355; (m) Steinhoff, B.A. and Stahl, S.S. (2002) *Organic Letters*, **4**, 4179; (n) Schultz, M.J., Park, C.C. and Sigman, M.S. (2002) *Chemical Communications*, 3034; (o) Jensen, D.R., Schultz, M.J., Mueller, J.A. and Sigman, M.S. (2003) *Angewandte Chemie – International Edition*, **42**, 3810; (p) Paavola, S., Zetterberg, K., Privalov, T. et al. (2004) *Advanced Synthesis and Catalysis*, **346**, 237; (q) Iwasawa, T., Tokunaga, M., Obora, T. and Tsuji, Y. (2004) *Journal of the American Chemical Society*, **126**, 6554.

16 Mallat, T. and Baiker, A. (1994) *Catalysis Today*, **19**, 247.

17 Liotta, L.F., Venezia, A.M., Deganello, G. et al. (2001) *Catalysis Today*, **66**, 271.

18 (a) Nishimura, T., Kakiuchi, N., Inoue, M. and Uemura, S. (2000) *Chemical Communications*, 1245; (b) Kakiuchi, N., Maeda, Y., Nishimura, T. and Uemura, S. (2001) *The Journal of Organic Chemistry*, **66**, 6220.

19 (a) Ebitani, K., Fujie, Y. and Kaneda, K. (1907) *Langmuir*, **1999**, (b) Choi, K.-M., Akita, T., Mizugaki, T. et al. (2003) *New Journal of Chemistry*, **27**, 324.

20 Karimi, B., Abedi, S., Clark, J.H. and Budarin, V. (2006) *Angewandte Chemie – International Edition*, **45**, 4776.

21 Mori, K., Hara, T., Mizugaki, T. et al. (2004) *Journal of the American Chemical Society*, **126**, 10657.

22 (a) Prati, L. and Rossi, M. (1998) *Journal of Catalysis*, **176**, 552; (b) Milone, C., Ingoglia, R., Pistone, A. et al. (2003) *Catalysis Letters*, **87**, 201; (c) Biella, S., Prati, L. and Rossi, M. (2003) *Inorganica Chimica Acta*, **349**, 253.

23 (a) Carretin, S., McMorn, P., Johnston, P. et al. (2002) *Chemical Communications*, 696; (b) Carretin, S., McMorn, P., Johnston, P. et al. (2004) *Topics in Catalysis*, **27**, 131; (c) Biella, S., Prati, L. and Rossi, M. (2002) *Journal of Catalysis*, **206**, 242.

24 (a) Tsunoyama, H., Sakurai, H., Negishi, Y. and Tsukuda, T. (2005) *Journal of the American Chemical Society*, **127**, 9374; (b) Miyamura, H., Matsubara, R., Miyazaki, Y. and Kobayashi, S. (2007) *Angewandte Chemie – International Edition*, **46**, 4151; (c) Kanaoka, S., Yagi, N., Fukuyama, Y. et al. (2007) *Journal of the American Chemical Society*, **129**, 12060.

25 Abad, A., Concepción, C., Corma, A. and García, H. (2005) *Angewandte Chemie – International Edition*, **44**, 4066.

26 Enache, D.I., Edwards, J.K., Landon, P. et al. (2006) *Science*, **311**, 362.

27 (a) Bauer, K., Garbe, D. and Surburg, H. (1997) *Common Fragrance and Flavor Materials*, Wiley-VCH, Weinheim; (b) (1998) *Ullmann's Encyclopedia of Industrial Chemistry*, 6th edn, Wiley-VCH, Weinheim; (c) Lane, B.S. and Burgess, K. (2003) *Chemical Reviews*, **103**, 2457; (d) Grigoropoulou, G., Clark, J.H. and Elings, J.A. (2003) *Green Chemistry*, **5**, 1.

28 For a review, see De Vos, D.E., Sels, B.F. and Jacobs, P.A. (2003) *Advanced Synthesis and Catalysis*, **345**, 457.

29 (a) Sheldon, R.A. (1981) *Aspects of Homogeneous Catalysts*, **4**, R. Ugo Reidel, Dordrecht 3; (b) Katsuki, T. (1999) *Comprehensive Asymmetric Catalysis II*, (eds E.N., Jacobsen, A., Pfaltz and H., Yamamoto), Springer, Heidelberg, 621.

30 Notari, B. (1993) *Catalysis Today*, **18**, 163.

31 van der Waal, J.C., Rigutto, M.S. and van Bekkum, H. (1998) *Appl Catal A*, **167**, 331.

32 Payne, G.B. and Williams, P.H. (1959) *The Journal of Organic Chemistry*, **24**, 54.

33 (a) Venturello, C., D'Aloiso, R., Bart, J.C. and Ricci, M. (1985) *J Mol Catal*, **32**, 107; (b) Ishii, Y., Yamawaki, K., Yoshida, T. et al. (1987) *The Journal of Organic Chemistry*, **52**, 1868; (c) Venturello, C. and D'Aloiso, R. (1988) *The Journal of Organic Chemistry*, **53**, 1553; (d) Ishii, Y., Yamawaki, K., Ura, T. et al. (1988) *The Journal of Organic Chemistry*, **53**, 3587; (e) Aubry, C., Chottard, G., Platzer, N. et al. (1991) *Inorganic Chemistry*, **30**, 4409; (f) Dengel, C., Griffith, W.P., and Parkin, B.C. (1993) *Journal of The Chemical Society – Dalton Transactions*,

2683; (g) Salles, L., Aubry, C., Thouenot, R. et al. (1994) *Inorganic Chemistry*, **33**, 871; (h) Duncan, D.C., Chambers, C., Hecht, E. and Hill, C.L. (1995) *Journal of the American Chemical Society*, **117**, 681; (i) Sato, K., Aoki, M., Ogawa, M. et al. (1996) *The Journal of Organic Chemistry*, **61**, 8310; (j) Sato, K., Aoki, M. and Ogawa, M. (1997) *Bulletin of the Chemical Society of Japan*, **70**, 905.

34 (a) Miyata, S. (1980) *Clays and Clay Minerals*, **28**, 50; (b) Cavani, F., Trifiró, F. and Vaccari, A. (1991) *Catalysis Today*, **11**, 173; (c) Sels, B.F., De Vos, D.E. and Jacobs, P.A. (2001) *Catalysis Reviews: Science and Engineering*, **43**, 443.

35 (a) Nakatsuka, T., Kawasaki, H., Yamashita, S. and Kohjiya, S. (1979) *Bulletin of the Chemical Society of Japan*, **52**, 2449; (b) Reichle, W.T. (1985) *Journal of Catalysis*, **94**, 547; (c) Suzuki, E. and Ono, Y. (1988) *Bulletin of the Chemical Society of Japan*, **61**, 1008; (d) Corma, A., Fornes, V., Martin-Aranda, R.M. and Rey, F. (1992) *Journal of Catalysis*, **134**, 58; (e) Tatsumi, T., Yamamoto, K., Tajima, H. and Tominaga, H. (1992) *Chemistry Letters*, 815; (f) Sels, B., De Vos, D.E., Buntinx, M. et al. (1999) *Nature*, **400**, 855; (g) Yamaguchi, K., Ebitani, K., Yoshida, T. et al. (1999) *Journal of the American Chemical Society*, **121**, 4526; (h) Nishimura, T., Kakiuchi, N., Inoue, M. and Uemura, S. (2000) *Chemical Communications*, 1245; (i) Choudary, B.M., Choudary, N., Madhi, S. and Kantam, M. (2001) *Angewandte Chemie – International Edition*, **40**, 4620; (j) Motokura, K., Nishimura, D., Mori, K. et al. (2004) *Journal of the American Chemical Society*, **126**, 5662; (k) Motokura, K., Mizugaki, T., Ebitani, K. and Kaneda, K. (2004) *Tetrahedron Letters*, **45**, 6029; (l) Motokura, K., Fujita, N., Mori, K. et al. (2005) *Tetrahedron Letters*, **46**, 5507.

36 Sels, B.F., De Vos, D.E. and Jacobs, P.A. (2001) *Journal of the American Chemical Society*, **123**, 8350.

37 Hoegaerts, D., Sels, B.F., de Vos, D.E. et al. (2000) *Catalysis Today*, **60**, 209.

38 (a) Kamata, K., Yamaguchi, K. and Mizuno, N. (2004) *Chemistry – A European Journal*, **10**, 4728; (b) Kamata, K., Yamaguchi, K., Hikichi, S. and Mizuno, N. (2003) *Advanced Synthesis and Catalysis*, **345**, 1193.

39 Yamaguchi, K., Yoshida, C., Uchida, S. and Mizuno, N. (2005) *Journal of the American Chemical Society*, **127**, 530.

40 (a) Pines, H. and Stalick, W.M. (1977) *Base-Catalyzed Reactions of Hydrocarbons and Related Compounds*, Academic Press, New York; (b) Fraile, J.M., Garcia, J.I. and Mayoral, J.A. (2000) *Catalysis Today*, **57**, 3.

41 (a) Yamaguchi, K., Mori, K., Mizugaki, T. et al. (2000) *The Journal of Organic Chemistry*, **65**, 6897; (b) Honma, T., Nakajo, M., Mizugaki, T. et al. (2002) *Tetrahedron Letters*, **43**, 6229.

42 Rajamathi, M., Nataraja, G.D., Ananthamurthy, S. and Kamath, P.V. (2000) *Journal of Materials Chemistry*, **10**, 2754.

43 Payne, G.B., Deming, P.H. and Williams, P.H. (1961) *The Journal of Organic Chemistry*, **26**, 651.

44 (a) Ueno, S., Yamaguchi, K., Yoshida, K. et al. (1998) *Chemical Communications*, 295; (b) Yamaguchi, K., Ebitani, K. and Kaneda, K. (1999) *The Journal of Organic Chemistry*, **64**, 2966; (c) Yamaguchi, K., Mizugaki, T., Ebitani, K. and Kaneda, K. (1999) *New Journal of Chemistry*, **23**, 799.

45 Tang, Q., Wang, Y., Liang, J. et al. (2004) *Chemical Communications*, 440.

46 (a) Groves, J.T. and Quinn, R. (1985) *Journal of the American Chemical Society*, **107**, 5790; (b) Khan, M.M.T. and Rao, A.P. (1987) *J Mol Catal*, **39**, 331; (c) Goldstein, A.S., Beer, R.H. and Drago, R.S. (1994) *Journal of the American Chemical Society*, **116**, 2424; (d) Nishiyama, Y. Nakagawa, Y. and Mizuno, N. (2001) *Angewandte Chemie – International Edition*, **40**, 3639; (e) Yin, C.-X. and Finke, R.G. (2005) *Inorganic Chemistry*, **44**, 4175.

47 Mitsudome, T., Mori, K., Mizugaki, T. et al. (2005) *Chemistry Letters*, **34**, 1626.

48 (a) Ryland, L.B., Tamele, M.W. and Wilson, J.N. (1960) *Catalysis*, Vol. 70, P.H. Emmett Reinhold, New York (b) Pinnavaia, T.J. (1983) *Science*, **220**, 365; (c) Laszlo, P. (1986) *Accounts of Chemical Research*, **19**, 121; (d) Laszlo, P. (1987) *Science*, **235**, 1473; (e) Izumi, Y. and Onaka, M. (1992) *Advances in Catalysis*, **38**, 245; (f) Clark, J.H. and Macquarrie, D.J. (1996) *Chemical Society Reviews*, **25**, 303; (g) Ebitani, K., Ide, M., Mitsudome, T. *et al.* (2002) *Chemical Communications*, 690; (h) Kawabata, T., Mizugaki, T., Ebitani, K., and Kaneda, K. (2003) *Tetrahedron Letters*, **44**, 9205; (i) Kawabata, T., Mizugaki, T., Ebitani, K. and Kaneda, K. (2003) *Journal of the American Chemical Society*, **125**, 10486; (j) Kawabata, T., Kato, M., Mizugaki, T. *et al.* (2005) *Chemistry – A European Journal*, **11**, 288; (k) Shanbhag, G.V. and Halligudi, S.B. (2005) *Journal of Molecular Catalysis A – Chemical*, **236**, 139; (l) Motokura, K., Nakagiri, N., Mori, K. *et al.* (2006) *Organic Letters*, **8**, 4617; (m) Shanbhag, G.V., Kumbar, S.M., Joseph, T. and Halligudi, S.B. (2006) *Tetrahedron Letters*, **47**, 141.

49 Beller, M. and Bolm, C. (1998) *Transition Metals for Organic Synthesis*, Wiley-VCH, Weinheim.

50 Kolb, H.C., Van Nieuwenhze, M.S. and Sharpless, K.B. (1994) *Chemical Reviews*, **94**, 2483.

51 (a) Sundermeier, U., Döbler, C. and Beller, M. (2004) *Modern Oxidation Methods*, J.-E., Bäckvall, Wiley-VCH, Weinheim, 12–17; (b) Kobayashi, S. and Sugiura, M. (2006) *Advanced Synthesis and Catalysis*, **348**, 1496.

52 (a) Nagayama, S., Endo, M. and Kobayashi, S. (1998) *The Journal of Organic Chemistry*, **63**, 6094; (b) Kobayashi, S., Endo, M. and Nagayama, S. (1999) *Journal of the American Chemical Society*, **121**, 11229; (c) Kobayashi, S., Ishida, T. and Akiyama, R. (2001) *Organic Letters*, **3**, 2649; (d) Ishida, T., Akiyama, R. and Kobayashi, S. (2003) *Advanced Synthesis and Catalysis*, **345**, 576; (e) Ishida, T., Akiyama, R. and Kobayashi, S. (2005) *Advanced Synthesis and Catalysis*, **347**, 1189.

53 (a) Severeyns, A., De Vos, D.E., Fiermans, L. *et al.* (2001) *Angewandte Chemie – International Edition*, **40**, 586; (b) Huang, K., Liu, H.-W., Dou, X. *et al.* (2003) *Polymers for Advanced Technologies*, **14**, 364.

54 (a) Choudary, B.M., Chowdari, N.S., Kantam, M.L. and Raghavan, K.V. (2001) *Journal of the American Chemical Society*, **123**, 9220; (b) Choudary, B.M., Chowdari, N.S., Jyothi, K. and Kantam, M.L. (2002) *Journal of the American Chemical Society*, **124**, 5341; (c) Choudary, B.M., Jyothi, K., Madhi, S. and Kantam, M.L. (2003) *Advanced Synthesis and Catalysis*, **345**, 1190.

55 (a) Choudary, B.M., Chowdari, N.S., Madhi, S. and Kantam, M.L. (2001) *Angewandte Chemie – International Edition*, **40**, 4619; (b) Choudary, B.M., Chowdari, N.S., Madhi, S. and Kantam, M.L. (2003) *The Journal of Organic Chemistry*, **68**, 1736.

56 Baeyer, A. and Villiger, V. (1899) *Berichte der Deutschen Chemischen Gesellschaft*, **32**, 3625.

57 ten Brink, G.-J., Arends, I.W.C.E. and Sheldon, R.A. (2004) *Chemical Reviews*, **104**, 4105.

58 Krow, G.R. (1993) *Organic Reactions*, **43**, 251.

59 (a) Kaneda, K., Haruna, S., Imanaka, T. *et al.* (1992) *Tetrahedron Letters*, **33**, 6827; (b) Kaneda, K., Ueno, S., Imanaka, T. *et al.* (1994) *The Journal of Organic Chemistry*, **59**, 2915.

60 (a) Kaneda, K., Ueno, S., and Imanaka, T. (1994) *Journal of the Chemical Society. Chemical Communications*, 797; (b) Kaneda, K., Ueno, S. and Imanaka, T. (1995) *Journal of Molecular Catalysis A – Chemical*, **102**, 135; (c) Kaneda, K. and Ueno, S. (1996) Heterogeneous Hydrocarbon Oxidation *ACS Symposium Series*, 300; (d) Kaneda, K. and Yamashita, T. (1996) *Tetrahedron Letters*, **37**, 4555; (e) Ueno, S., Ebitani, K., Ookubo, A. and Kaneda, K. (1997) *Applied Surface Science*, **121/122**, 366.

61 Palazzi, C., Pinna, F. and Strukul, G. (2000) *J Mol Catal A*, **151**, 245.
62 Bhaumik, A., Kumar, P. and Kumar, R. (1996) *Catalysis Letters*, **40**, 47.
63 (a) Corma, A., Nemeth, L.T., Renz, M. and Valencia, S. (2001) *Nature*, **412**, 423; (b) Corma, A., Domine, M.E., Nemeth, L. and Valencia, S. (2002) *Journal of the American Chemical Society*, **124**, 3194; (c) Bare, S.R., Kelly, S.D., Sinkler, W. *et al.* (2005) *Journal of the American Chemical Society*, **127**, 12924.
64 Corma, A., Navarro, M.T., Nemeth, L. and Renz, M. (2001) *Chemical Communications*, 2190.
65 Sheldon, R.A. (2000) *Green Chemistry*, **2**, G1.
66 (a) Thomas, J.M. (1985) *Nature*, **314**, 670; (b) Ito, T. and Lunsford, J.N. (1985) *Nature*, **314**, 721; (c) Maschmeyer, T., Oldoyd, R.D., Sanker, G. *et al.* (1997) *Angewandte Chemie – International Edition*, **36**, 1639; (d) Thomas, J.M., Raja, R., Sanker, G. and Bell, R.G. (1999) *Nature*, **398**, 227; (e) Raja, R., Sanker, G. and Thomas, J.M. (1999) *Journal of the American Chemical Society*, **121**, 11926; (f) Thomas, J.M., Raja, R., Sanker, G. and Bell, B.G. (2001) *Accounts of Chemical Research*, **34**, 191.

6
Liquid-Phase Oxidations with Hydrogen Peroxide and Molecular Oxygen Catalyzed by Polyoxometalate-Based Compounds

Noritaka Mizuno, Keigo Kamata, Sayaka Uchida, and Kazuya Yamaguchi

6.1
Introduction

Chemical industries have contributed to worldwide economic development over the past century and chemical products make an enormous contribution to the quality of our lives. However, the manufacturing processes of chemicals have also led to vast amounts of wastes. Especially for fine chemicals production, antiquated methodologies, for example, stoichiometric oxidations with permanganate or chromium(VI) reagents and stoichiometric reductions with metals or metal hydrides, are still widely used. Today, the reduction (or elimination) of these wastes is a central issue [1–4]. To minimize wastes in chemical manufacturing, the catalytic method is a reliable solution, which replaces synthetic processes of low atom efficiency using (hazardous) stoichiometric reagents. For example, the atom efficiencies of the stoichiometric production of propylene oxide (the chlorohydrin process) and catalytic epoxidation with H_2O_2 are compared in Figure 6.1. In contrast with the stoichiometric process, the catalytic process offers economical and environmental advantages due to the high atom efficiency and no output of chloride laden wastes.

To date, much attention has been paid to the development of catalytic oxidations – the choice of the oxidant determines the practicability and efficiency of the systems. Numerous oxidants have been investigated extensively for catalytic liquid-phase oxidation processes. H_2O_2 and O_2 (or air) are regarded as "green oxidants" because of their high contents of active oxygen species, high atom efficiencies and co-production of only water [5–8]. The ideal system for greener, cleaner oxidations is the use of H_2O_2 or O_2 together with recyclable catalysts in nontoxic solvents. Therefore, the current goal of basic research and industry is the development of effective metal catalysts that can activate H_2O_2 or O_2 under ambient conditions and transfer the active oxygen species to various substrates with high chemo-, regio-, diastereo- and stereoselectivity.

New analytical methods, high-level theoretical studies and various mechanistic studies are producing important information on the factors related to the catalytic activity, selectivity and stability. In particular, valuable information on the correlations

---

Modern Heterogeneous Oxidation Catalysis: Design, Reactions and Characterization
Edited by Noritaka Mizuno
Copyright © 2009 WILEY-VCH Verlag GmbH & Co. KGaA, Weinheim
ISBN: 978-3-527-31859-9

Stoichiometric:

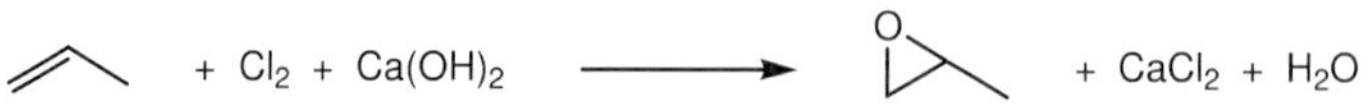

$$\text{CH}_2\text{=CHCH}_3 + \text{Cl}_2 + \text{Ca(OH)}_2 \longrightarrow \text{epoxide} + \text{CaCl}_2 + \text{H}_2\text{O}$$

> Atom efficiency
> = (molecular weight of the desired product)/(total molecular weights of all substances) × 100
> = (58.1)/(42.1+70.9+74.1) × 100 = **31.1%**

Catalytic:

$$\text{CH}_2\text{=CHCH}_3 + \text{H}_2\text{O}_2 \xrightarrow{\text{catalyst}} \text{epoxide} + \text{H}_2\text{O}$$

> Atom efficiency
> = (58.1)/(42.1+34.0) × 100 = **76.3%**

Figure 6.1 Atom efficiencies of stoichiometric and catalytic epoxidation.

between active sites and catalytic performance can lead to a new generation of catalysts. For the oxidation reactions, mono-, di- and polynuclear sites of various catalysts such as well-defined molecular catalysts and biomimetic organometallic catalysts relating to the heme enzyme of cytochrome P-450 and the non-heme enzyme of methane monooxygenase can effectively activate oxidants and/or substrates, resulting in specific activity and selectivity [5–17]. However, degradation of organic ligands invariably occurs under oxidative conditions and their catalytic activity and lifetime are limited. Therefore, "inorganic catalysts" with structurally well-defined active sites at the atomic or molecular level can contribute to the development of efficient, green and long-lived oxidation processes.

Polyoxometalates (POMs) are a large family of anionic metal–oxygen clusters of early transition metals that have stimulated much current research activity in broad fields such as catalysis, materials and medicine because their chemical properties, such as redox potentials, acidities and solubilities in various media, can be finely tuned by choosing constituent elements and counter-cations [18–34]. An additional attractive aspect of POMs is their stability under thermal and oxidative conditions. Therefore, POMs have especially received much attention in the area of oxidation catalysis [26–34]. Various types of POMs can act as effective catalysts for H_2O_2- and O_2-based green oxidations. This chapter summarizes comprehensively recent progress in the development of homogeneously as well as heterogeneously catalyzed selective H_2O_2- and O_2-based liquid-phase oxidations by POMs and related compounds.

6.2
Molecular Design of Polyoxometalates for H_2O_2- and O_2-Based Oxidations

The structures and properties of POMs are briefly mentioned here. The details have been summarized in excellent books and review articles [18–25].

Isopolyoxometalates and heteropolyoxometalates are formulated as $[M_mO_y]^{p-}$ and $[X_xM_mO_y]^{q-}$ ($x<m$), respectively, where M is the addenda atom and X is the heteroatom. The most common addenda atoms are the d^0-early-transition-metal cations such as W^{6+}, Mo^{6+} and V^{5+}. The heteroatoms can be p-, d- or f-block elements such as P^{5+}, As^{5+}, Si^{4+}, Ge^{4+} and B^{3+}. Among a wide variety of heteropolyoxometalates, the Keggin structures are the most stable and more easily available. Keggin anions, typically represented by the formula $[X^{n+}M_{12}O_{40}]^{(8-n)-}$, contain one central heteroatom and twelve addenda atoms (four M_3O_{13} triads). Each of the M_3O_{13} triads can be rotated by 60° on its threefold axis, leading to geometrical isomers. Rotation of one, two, three and all four M_3O_{13} triads of the most common α-Keggin isomer produces the β, γ, δ and ε isomers, respectively. Another typical heteropolyoxometalate is the Wells-Dawson type POM. The Wells-Dawson structure, typically represented by the formula $[X_2^{n+}M_{18}O_{62}]^{(16-2n)-}$, consists of two trivacant A-type lacunary Keggin species, which are generated by the loss of corner-shared group of MO_6 octahedra, linked directly across the lacunae.

The lacunary POMs can be obtained by the removal of one or more addenda atoms from fully occupied POMs. The most stable lacunary compounds are obtained with Si^{4+} as the heteroatom. While the dodecatungstosilicates are stable in acidic solution, hydrolytic cleavages of W—O bonds occur and well-defined lacunary POMs with eleven, ten and nine tungsten atoms are produced when the pH increases.

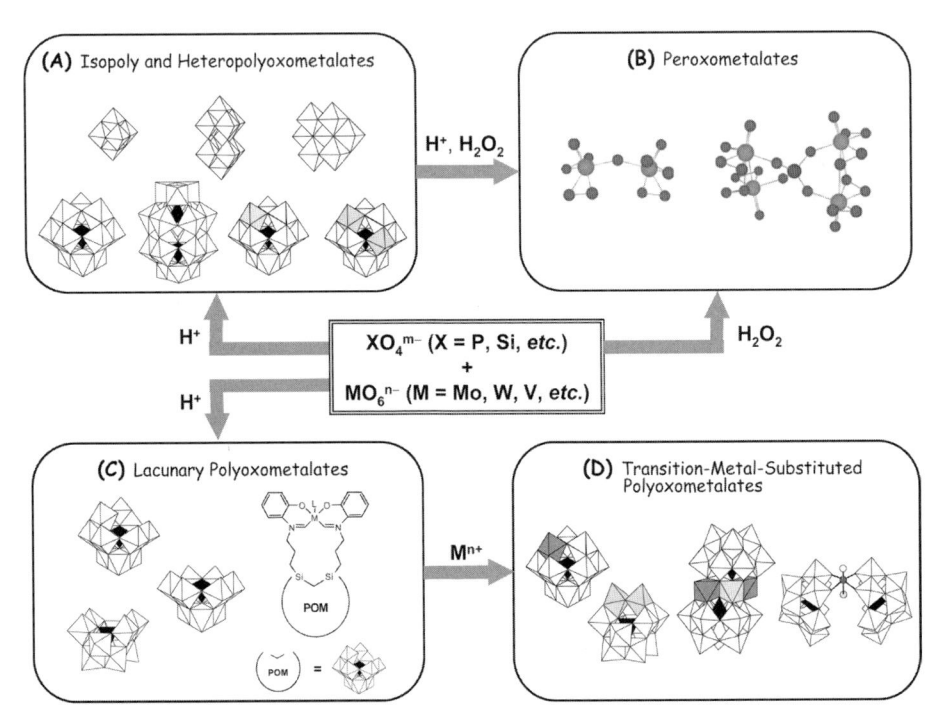

Figure 6.2 Classification of POM-based compounds.

Lacunary species can act as "structural motifs" with numerous metal cations, leading to mono-, di- and trinuclear transition-metal-substituted POMs according to the number of vacant sites. The metal incorporation involves the reaction of aquated first-row and occasionally second-row d-block metal ions, $M(H_2O)_6^{z+}$, with a lacunary POM derived from the Keggin, Wells-Dawson and other POM families. In addition, lacunary species assemble into large POM structures, either directly or with incorporation of metal ion linkers.

POM-based catalysts for H_2O_2- and O_2-based oxidations can be classified into four groups according to the structures of POMs (Figure 6.2): (A) isopoly and heteropolyoxometalates, (B) peroxometalates, (C) lacunary POMs and (D) transition-metal-substituted POMs. This section describes liquid-phase homogeneous oxidations by functionalized POM-based compounds with H_2O_2 and O_2 according to the above classification.

6.2.1
Isopoly- and Heteropolyoxometalates

The tungsten-catalyzed oxidation systems with H_2O_2 have attracted much attention because of their high reactivities compared with molybdenum analogues and inherent poor activity for non-productive decomposition of H_2O_2. Although an effective H_2O_2-based epoxidation of terminal alkenes catalyzed by $H_3[PW_{12}O_{40}]$ in the presence of cetylpyridinium chloride (CPC) as a phase transfer agent has been reported (6.1) [35], other POMs such as $H_3[PMo_{12}O_{40}]$, $H_4[SiW_{12}O_{40}]$ and $H_3[PMo_6W_6O_{40}]$ were much less active:

$$\text{alkene} \xrightarrow[\text{CHCl}_3,\ 333\ \text{K},\ 5\ \text{h}]{H_3[PW_{12}O_{40}]/CPC,\ H_2O_2} \text{epoxide} \quad 80\%$$

(6.1)

The above system could further be applied to the epoxidation of allylic alcohols [35, 36], monoterpenes [37] and α,β-unsaturated carboxylic acids [38], the oxidation of alcohols [35, 36], amines [39] and alkynes [40], and the oxidative transformation of diols [35, 41–44]. The subsequent spectroscopic and kinetic studies by many researchers show that $[PO_4\{WO(O_2)_2\}_4]^{3-}$ is a true catalytically active species formed by the reaction of $H_3[PW_{12}O_{40}]$ with excess H_2O_2 (Figure 6.3) [45–48].

Isopolyoxometalates and heteropolyoxometalates such as $[W_6O_{19}]^{2-}$, $[W_{10}O_{32}]^{4-}$ and $[PW_{12}O_{40}]^{3-}$ could act as efficient, stable, selective photooxidation catalysts

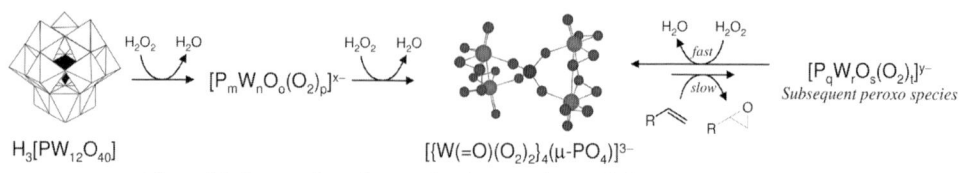

Figure 6.3 Proposed mechanism for the epoxidation of alkenes with H_2O_2 catalyzed by $H_3[PW_{12}O_{40}]$ [45–48].

under UV irradiation [49–51]. The products generated in the photocatalytic oxidations depended on the kinds of POMs, especially on their ground-state redox potentials.

A wide variety of hydrocarbons, including alkanes, alcohols, amines and arenes, could be oxidized with O_2 in the presence of mixed-addenda POMs, $[PV_nMo_{12-n}O_{40}]^{(3+n)-}$ (6.2 and 6.3) [52–67]:

$$\text{2,6-dimethylphenol} \xrightarrow[\text{n-hexanol, 333 K, 4 h}]{H_5[PV_2Mo_{10}O_{40}],\ O_2} \text{3,3',5,5'-tetramethyldiphenoquinone (80%)} \quad (6.2)$$

$$\text{9,10-dihydroanthracene} \xrightarrow[\text{ClCH}_2\text{CH}_2\text{Cl, 343 K, 16 h}]{H_5[PV_2Mo_{10}O_{40}]/\text{tetraglyme},\ O_2} \text{anthracene (98%)} \quad (6.3)$$

In these cases, the oxidation includes one (or more) electron transfer from substrates to $[PV_nMo_{12-n}O_{40}]^{(3+n)-}$ to afford the corresponding oxidized products with the formation of the reduced form of $[PV_nMo_{12-n}O_{40}]^{(3+n)-}$. The reduced POMs can be reoxidized by O_2. The $[PV_nMo_{12-n}O_{40}]^{(3+n)-}/O_2$/hydrocarbon redox system can be useful for the aerobic oxidations of a wide variety of hydrocarbon substrates since the redox potential of $[PMo_{12-n}V_nO_{40}]^{(3+n)-}$ (about 0.7 V vs. SHE) is lower than that of O_2 and is higher than those of common hydrocarbons.

The oxidation of CO efficiently proceeded at ambient temperature in an aqueous medium with a combined catalyst of $[PMo_{12}O_{40}]^{3-}$ and supported metal nanoparticles (Au, Pt, Pd and Ir) [68]. Especially, the catalysts containing Au nanoparticles selectively catalyzed the oxidation of CO in H_2-rich gas mixtures using a reversible redox property of the POM without consuming significant amounts of H_2, in contrast with the conventional processes. Thus, this system has advantages in the production and purification of H_2 for proton-exchange membrane fuel-cell applications. Thermodynamically controlled self-assembly of an equilibrated ensemble of POMs with $[AlVW_{11}O_{40}]^{6-}$ as the main component could act as a catalyst for the selective delignification (oxidative degradation of lignin derivatives) of lignocellulose fibers [69]. Equilibration reactions typical of POMs kept the pH of the system near 7 during the catalysis, which avoided acid or base degradation of cellulose.

6.2.2
Peroxometalates

Various peroxo species have been isolated and characterized crystallographically. Peroxotungstates containing phosphorous or arsenic ligands are generally much more active than $[W_2O_3(O_2)_4]^{2-}$ for the catalytic epoxidation of terminal alkenes [70–75], while the sulfate species, $[SO_4\{WO(O_2)_2\}_2]^{2-}$, was the most active for stoichiometric epoxidation of (R)-(+)-limonene among $[XO_4\{WO(O_2)_2\}_2]^{2-}$ (X = HAs, HP, and S) peroxotungstates [75]. The catalytic system consisting of

$[PO_4\{WO(O_2)_2\}_4]^{3-}$ and $Pd(OAc)_2$ in methanol was also effective for epoxidation of propylene with O_2 instead of H_2O_2 (6.4) [76]:

$$\text{CH}_2=\text{CHCH}_3 \xrightarrow[\text{CH}_3\text{OH, 373 K, 6 h}]{\text{THA}_3[\text{PO}_4\{\text{WO}(\text{O}_2)_2\}_4],\ \text{Pd}(\text{OAc})_2,\ \text{O}_2} \text{propylene oxide} \quad (6.4)$$
$$35\%$$

The tetra-n-hexylammonium (THA) salt of $[PO_4\{WO(O_2)_2\}_4]^{3-}$ was isolated and characterized crystallographically by Venturello and coworkers [77]. The anion consisted of the PO_4^{3-} and two $[W_2O_2(O_2)_4]$ species. As mentioned above, $[PO_4\{WO(O_2)_2\}_4]^{3-}$ has been postulated to be a catalytically active species for the $H_3[PW_{12}O_{40}]/H_2O_2$ system developed by Ishii and coworkers [35–44] because $[PO_4\{WO(O_2)_2\}_4]^{3-}$ exhibited a very similar catalytic activity and selectivity to those of the Ishii system [77, 78].

The isolated $K_2[\{WO(O_2)_2(H_2O)_2\}(\mu\text{-O})]$ efficiently catalyzed the chemo-, regio- and diastereoselective and stereospecific epoxidation of various allylic alcohols with only 1 equiv of H_2O_2 in water (6.5) [79, 80]:

$$\text{HO-CH}_2\text{CH=CH}_2 \xrightarrow[\text{water, 305 K, 10 h}]{K_2[\{WO(O_2)_2(H_2O)\}_2(\mu\text{-O})]\cdot 2H_2O,\ H_2O_2} \text{HO-CH}_2\text{-epoxide} \quad (6.5)$$
$$95\%$$

The reactivity of the dinuclear peroxotungstate with H_2O_2 in organic solvents was quite different from that in water [81]. The reaction of the tetra-n-butylammonium (TBA) salt of dinuclear peroxotungstate $[\{WO(O_2)_2\}_2(\mu\text{-O})]^{2-}$ with H_2O_2 in acetonitrile gave the novel $\mu\text{-}\eta^1\text{:}\eta^1$-peroxo-bridging dinuclear tungsten species $[\{WO(O_2)_2\}_2(\mu\text{-}O_2)]^{2-}$. Only $[\{WO(O_2)_2\}_2(\mu\text{-}O_2)]^{2-}$ was active for the epoxidation of cyclic, internal and terminal alkenes, whereas $[\{WO(O_2)_2\}_2(\mu\text{-O})]^{2-}$ was almost inactive.

6.2.3
Lacunary Polyoxometalates

The oxo ligands at the vacant sites of a Keggin-type silicodecatungstate $[\gamma\text{-SiW}_{10}O_{36}]^{8-}$ are basic enough to react with H^+. A novel tetraprotonated divacant silicodecatungstate has been synthesized and the single-crystal X-ray structural analysis revealed that two of the four oxo ligands located at the vacant sites were selectively protonated to afford two aquo ligands ($[\gamma\text{-SiW}_{10}O_{34}(H_2O)_2]^{4-}$). The TBA salt of $[\gamma\text{-SiW}_{10}O_{34}(H_2O)_2]^{4-}$ catalyzed oxygen-transfer reactions of various substrates, including alkenes, allylic alcohols and sulfides with H_2O_2 (6.6) [82–84]:

$$\text{cyclohexene} \xrightarrow[\text{CH}_3\text{CN, 305 K, 4 h}]{\text{TBA}_4[\gamma\text{-SiW}_{10}O_{34}(H_2O)_2],\ H_2O_2} \text{cyclohexene oxide} \quad (6.6)$$
$$83\%$$

The reaction of $[\gamma\text{-SiW}_{10}O_{34}(H_2O)_2]^{4-}$ with H_2O_2 led to the generation of the diperoxo species $[\gamma\text{-SiW}_{10}O_{32}(O_2)_2]^{4-}$. The two oxo groups ($O^{2-}$) were replaced by two peroxo groups (O_2^{2-}) on the divacant lacunary site with retention of the γ-Keggin framework.

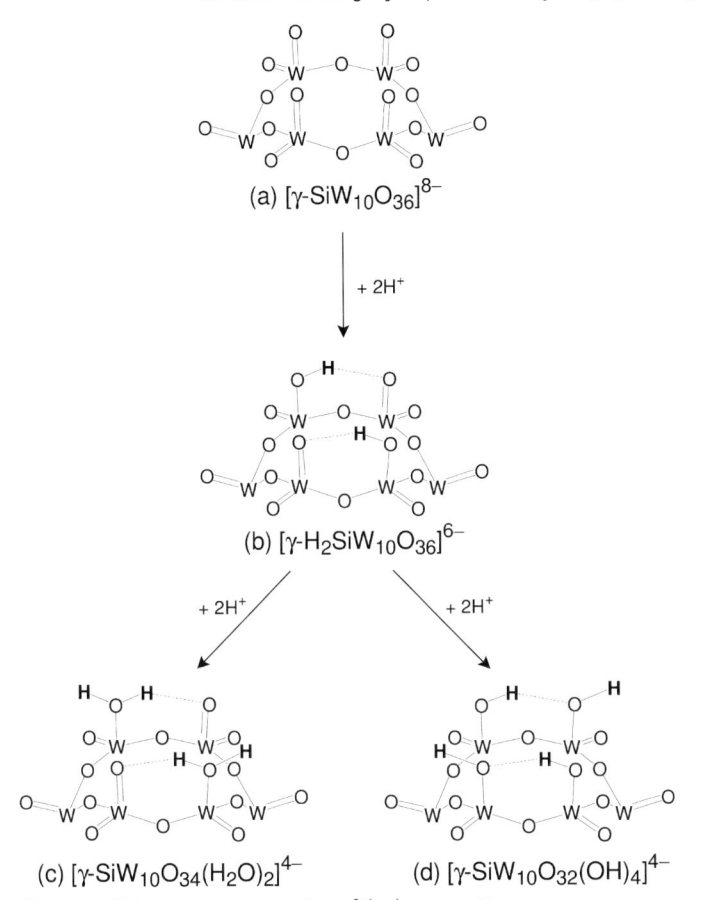

6.2 Molecular Design of Polyoxometalates for H_2O_2- and O_2-Based Oxidations | 191

Figure 6.4 Schematic representation of the lacunary sites calculated on (a) $[\gamma\text{-SiW}_{10}O_{36}]^{8-}$, (b) $[\gamma\text{-H}_2\text{SiW}_{10}O_{36}]^{6-}$, (c) $[\gamma\text{-SiW}_{10}O_{34}(H_2O)_2]^{4-}$, and (d) $[\gamma\text{-SiW}_{10}O_{32}(OH)_4]^{4-}$ [82–87].

Subsequently, experimental and theoretical studies on the structure of $[\gamma\text{-SiW}_{10}O_{34}(H_2O)_2]^{4-}$ and the epoxidation mechanism have reported by several research groups (Figure 6.4) [85–87]. While it was suggested that its formula was $[\gamma\text{-SiW}_{10}O_{34}(H_2O)_2]^{4-}$ with two aqua ligands, the structural assignment of the four protons on the lacunary sites remains unclear. Protons on the vacant sites are key features of the catalytic epoxidation mechanism, which may affect the formation and reactivity of the tungsten peroxide species. Density functional theory (DFT) calculations on the structure of tetraprotonated form of $[\gamma\text{-H}_4\text{SiW}_{10}O_{36}]^{4-}$ were presented by Musaev and coworkers [85, 86]. Based on DFT calculations at the B3LYP/Lanl2dz + d(Si) level, the structure with four hydroxo ligands, that is, $[\gamma\text{-SiW}_{10}O_{32}(OH)_4]^{4-}$, was calculated to be more stable than that with two aqua and two oxo(terminal) ligands, $[\gamma\text{-SiW}_{10}O_{34}(H_2O)_2]^{4-}$. In contrast, Bonchio and coworkers have insisted that the tetraprotonated form should be formulated as $[\gamma\text{-SiW}_{10}O_{34}(H_2O)_2]^{4-}$ [87]. The titration of $[\gamma\text{-SiW}_{10}O_{34}(H_2O)_2]^{4-}$ by TBA hydroxide (TBAOH) indicated that only two out of four acidic protons on the POM surface played a major role in promoting oxygen transfer. Upon addition of 2 equiv of TBAOH relative to $[\gamma\text{-SiW}_{10}O_{34}(H_2O)_2]^{4-}$, the ^{183}W NMR spectrum showed a C_{2v} structure, which necessarily implies fast exchange between the remaining protons. Using relativistic DFT calculations employing the ZORA-BP86/TZP level theory, including solvent effects with the COSMO method, inspection of the molecular electrostatic potentials for the calculated structure of $[\gamma\text{-H}_2\text{SiW}_{10}O_{36}]^{6-}$ showed that the mono-protonated lacunary oxygens still retained a significant electron density that did not prevent a regioselective double protonation of each site to form $[\gamma\text{-SiW}_{10}O_{34}(H_2O)_2]^{4-}$. The energies of $[\gamma\text{-SiW}_{10}O_{34}(H_2O)_2]^{4-}$ and $[\gamma\text{-SiW}_{10}O_{32}(OH)_4]^{4-}$ were so close as to be very sensitive to the method/basis set combination adopted. However, the optimized geometry of $[\gamma\text{-SiW}_{10}O_{34}(H_2O)_2]^{4-}$ fitted better with the X-ray structure [82].

Inorganic–organic hybrid POMs have attracted much attention because the electron and steric properties of the POMs significantly affect the catalytic properties of a metal center [88–91]. One hybrid catalyst consisted of a metallo-salen complex attached covalently to a lacunary POM $[\text{SiW}_{11}O_{39}]^{8-}$ via two propyl spacers. In contrast with an usual Mn(III) salen species, the metallosalen complex modified with the POM afforded a more oxidized Mn(IV) salen–POM or Mn(III) salen cation radical–POM compound [89].

6.2.4
Transition-Metal-Substituted Polyoxometalates

Transition-metal-substituted POMs are oxidatively and hydrolytically stable compared with organometallic complexes, and their active sites can be controlled. These advantages have been applied to the development of biomimetic catalysis relating to enzyme analogues. To date, various kinds of transition-metal-substituted POMs have been synthesized and certain kinds of them can efficiently activate O_2 and H_2O_2, as described below.

The ruthenium-substituted sandwich-type POM $[\text{WZnRu}_2(OH)(H_2O)(XW_9O_{34})_2]^{11-}$ ($X = Zn^{2+}$ or Co^{2+}) catalyzed the selective hydroxylation of adamantane with O_2 as an oxidant [92–94]. Finke and Weiner have reported that vanadium- and iron-containing polyanion-based precatalysts show high catalytic activity for catechol dioxygenation with O_2 [95]. For example, 3,5-di-*tert*-butylcatechol was oxidized in the presence of $[n\text{-Bu}_4N]_5[(CH_3CN)_x\text{Fe·SiW}_9V_3O_{40}]$ with an extremely high turnover number (100 000) – a value far superior to those for man-made and natural enzymes. Recent reinvestigation concluded that the vanadyl semiquinone catecholate dimer complex $[\text{VO}(DBSQ)(DTBC)]_2$ (DBSQ = 3,5-di-*tert*-butylsemiquinone anion and DTBC = 3,5-di-*tert*-butylcatecholate anion) formed from V-leaching from the precatalyst is the catalytically active species [96].

The γ-isomer Keggin-type silicodecatungstate $[\gamma\text{-SiW}_{10}O_{36}]^{8-}$ has been used as a "structural motif" for di-metal-substituted POMs with bis-μ-oxo(or hydroxo)-dimetal cores [97–102]. The di-iron-substituted silicotungstate $[\gamma\text{-SiW}_{10}\{\text{Fe}(OH_2)\}_2O_{38}]^{6-}$ was synthesized by the reaction of $[\gamma\text{-SiW}_{10}O_{36}]^{8-}$ with $Fe(NO_3)_3$ in acidic aqueous solution [103]. This POM catalyzed the selective oxidation of alkenes and alkanes with high efficiency of H_2O_2 utilization [104] and the aerobic epoxidation of simple alkenes [105]. With the di-iron-substituted γ-isomer Keggin silicodecatungstate trimer, the aerobic oxidation of 2-mercaptoethanol to disulfide proceeded in water at ambient temperature [106]. The di-vanadium-substituted silicotungstate $[SiW_{10}O_{38}V_2(\mu\text{-}OH)_2]^{4-}$ with a {VO-(μ-OH)$_2$-VO} core could catalyze the epoxidation of various alkenes using H_2O_2 with a high epoxide yield and a high efficiency of H_2O_2 utilization under very mild reaction conditions [107]. Notably, the $[SiW_{10}O_{38}V_2(\mu\text{-}OH)_2]^{4-}$-catalyzed system required only 1 equiv of H_2O_2 and showed unique stereospecificity, diastereoselectivity and regioselectivity that are quite different from those of conventional catalysts and polyoxometalates (6.7) [107]:

$$\text{1-butene} \xrightarrow[\text{CH}_3\text{CN}/t\text{BuOH, 293 K, 24 h}]{\text{TBA}_4[\gamma\text{-SiW}_{10}\text{O}_{36}\text{V}_2(\mu\text{-OH})_2], \text{H}_2\text{O}_2} \text{1,2-epoxybutane} \quad 91\%$$

(6.7)

The di-copper-substituted γ-Keggin silicotungstate $[\gamma\text{-}H_2SiW_{10}O_{36}Cu_2(\mu\text{-}1,1\text{-}N_3)_2]^{4-}$ could act as an effective homogeneous catalyst for the oxidative homo-coupling of various kinds of alkynes, including aromatic, aliphatic and heteroatom-containing ones (6.8) [108]:

$$\text{PhC≡CH} \xrightarrow[\text{PhCN, 373 K, 3 h}]{\text{TBA}_4[\gamma\text{-}H_2\text{SiW}_{10}\text{O}_{36}\text{Cu}_2(\mu\text{-}1,1\text{-N}_3)_2], \text{O}_2} \text{PhC≡C-C≡CPh} \quad 91\%$$

(6.8)

Although terminal oxo complexes of the late-transition-metal elements have been proposed as possible intermediates for oxidations catalyzed by these elements, late-transition-metal-oxo complexes were scarcely known. Hill and co-workers reported the synthesis and characterization of Pt^{4+}-, Pd^{4+}- and Au^{3+}-oxo complexes, $[M(O)(OH_2)\{WO(OH_2)\}_n(PW_9O_{34})_2]^{m-}$ (M = Pt, Pd and Au, n = 0–2), stabilized by electron-accepting polyoxotungstate ligands [109–111]. The stoichiometric reaction of the Au-oxo complex $[Au(O)(OH_2)\{WO(OH_2)\}_2(PW_9O_{34})_2]^{9-}$ with triphenylphosphine led to the formation of triphenylphosphine oxide.

6.3
Heterogenization of Polyoxometalates

To date, many soluble transition metal-based homogeneous catalysts including POMs have been developed for H_2O_2- and O_2-based green oxidations. They are usually dissolved in reaction solutions, making all catalytic sites accessible to substrates, and show high catalytic activity and selectivity. Despite these advantages, homogeneous catalysts have a share of only about 20% in industrial processes because catalyst/product(s) separation, that is, problem of product contamination, and reuse of expensive catalysts are very difficult [112]. Therefore, the development of easily recoverable and recyclable heterogeneous catalysts has been a subject of a particular research interest for fine chemicals syntheses.

While many efficient green oxidations based on POMs have been developed, as mentioned above, most of them are homogeneous and share common drawbacks of catalyst/product separation and reuse of catalysts. The practical application of POM-based oxidations will need the development of easily recoverable and recyclable catalysts. Recently, the development of heterogeneous oxidation catalysts based on POMs and the related compounds has been attempted and the strategies used can approximately be classified into two categories (Figure 6.5): (A) "solidification" of POMs (formation of insoluble solid ionic materials with appropriate counter-cations) and (B) "immobilization" of POMs through adsorption, covalent linkage and ion-exchange. This section summarizes comprehensively the developments of heterogeneous POM catalysts for H_2O_2- and O_2-based green oxidations according to the above classification. Table 6.1 summarizes typical examples for the oxidations with heterogeneous POM catalysts.

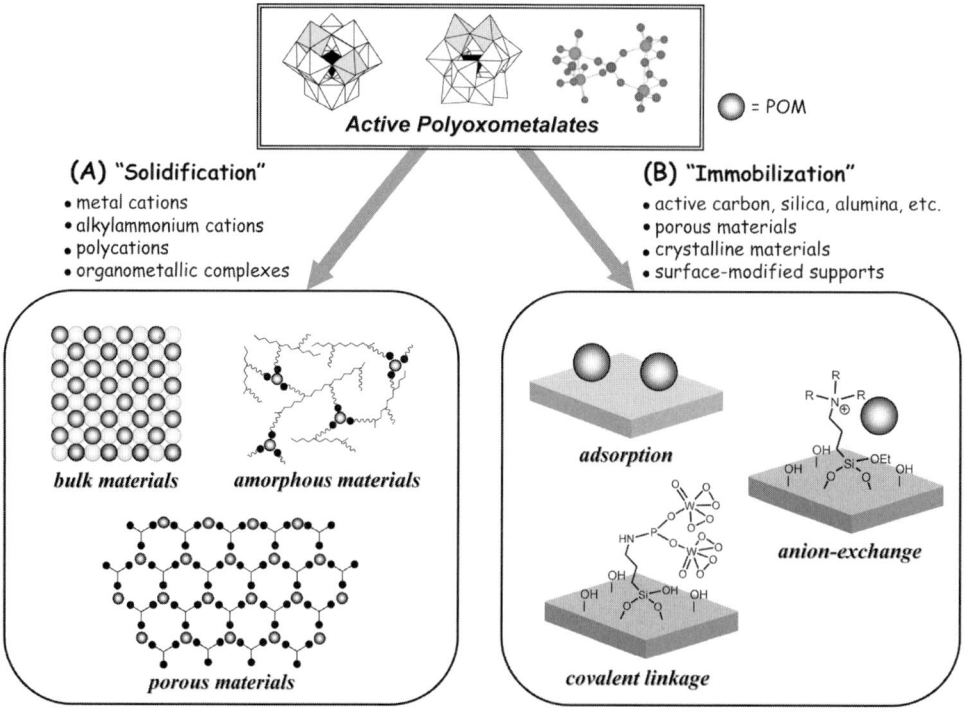

Figure 6.5 Schematic representation of the strategies employed for heterogenization of POM-based compounds.

Table 6.1 Selected examples of heterogeneously catalyzed H_2O_2- and O_2-based oxidations by POMs.

Category[a]	Oxidant	Catalyst	Reaction (% yield)	Solvent	Temperature (K)	Ref.
A	O_2	$Ag_5[PV_2Mo_{10}O_{40}]$	(30%)	2,2,2-Trifluoroethanol	RT[b]	[113]
A	O_2	$Fe_xH_{5-x}[PV_2Mo_{10}O_{40}]$	(92%)	CH_3CN	353	[114]
A	O_2	$[Mo_{12}O_{39}(OH)_{10}H_2\{Cu(H_2O)_3\}_4]$	(39%)	Toluene	348	[115]
A	O_2	$TBA_4H[SiW_{11}O_{39}Ru(H_2O)]$ (TBA = tetra-n-butylammonium)	(99%)	AcOiBu	373	[116]
A	H_2O_2	$CP_3[PO_4(WO_3)_4]$ (CP = cetylpyridinium)	(88%)	1,2-Dichloroethane	308	[117]
B	H_2O_2	BTC-[ZnWZn$_2$(H$_2$O)$_2$(ZnW$_9$O$_{34}$)$_2$]$^{12-}$ (BTC = tripodal cation)	(98%)	CH_3CN	294	[122]
B	H_2O_2	NIPAM-[PW$_{12}$O$_{40}$]$^{3-}$ (NIPAM = N-isopropylacrylamide-based polymer)	(>99%)	Solvent-free	RT[b]	[124]
B	H_2O_2	NIPAM-[PW$_{12}$O$_{40}$]$^{3-}$	(70%)	Solvent-free	RT[b]	[125]

Table 6.1 (Continued)

Category[a]	Oxidant	Catalyst	Reaction (% yield)	Solvent	Temperature (K)	Ref.
B	H_2O_2	PEI-[ZnWZn$_2$(H$_2$O)$_2$(ZnW$_9$O$_{34}$)$_2$]$^{12-}$ (PEi = ethyleneimine-based polymer)	Ph–S–Ph → Ph–S(O$_2$)–Ph (72%)	Water	295	[129]
B	H_2O_2	PEI-[ZnWZn$_2$(H$_2$O)$_2$(ZnW$_9$O$_{34}$)$_2$]$^{12-}$	Ph–CH=CH$_2$ → PhCHO (96%)	Water	295	[129]
C	H_2O_2	[Ni(tacn)$_2$]$_2$[SiW$_{10}$O$_{38}$V$_2$(OH)$_2$] (tacn = 1,4,7-triazacyclononane)	cyclooctene → cyclooctene oxide (28%)	CH$_3$CN/tBuOH	283	[133]
D	O_2	Na$_5$[PV$_2$Mo$_{10}$O$_{40}$]/C	PhCH$_2$OH → PhCHO (97%)	Toluene	373	[135]
D	O_2	Na$_5$[PV$_2$Mo$_{10}$O$_{40}$]/C	PhCH$_2$NH$_2$ → PhCH=N–Ph (>99%)	Toluene	373	[135]
D	O_2	(NH$_4$)$_5$H$_6$[PV$_8$Mo$_4$O$_{40}$]/C	2,6-dimethylphenol → 2,6-dimethyl-1,4-benzoquinone (68%)	AcOH/water	333	[136]
D	O_2	(NH$_4$)$_5$H$_6$[PV$_8$Mo$_4$O$_{40}$]/C	isophorone → ketoisophorone (75%)	Toluene	373	[137]
D	O_2	(NH$_4$)$_5$H$_6$[PV$_8$Mo$_4$O$_{40}$]/C	isochroman → isochromanone (71%)	Toluene	373	[137]

		Substrate	Product	Yield	Solvent	T (K)	Ref.
D	O_2	$H_5[PV_2Mo_{10}O_{40}]$/fabrics	CH_3CHO	CH_3COOH	(50–52%)	1,2-Dichloroethane	RT[b]	[138]
D	O_2	$Cs_6H_2[P_2W_{17}O_{61}Co(H_2O)]/SiO_2$ Dichloromethane	butanal	butanoic acid	(76%)			[141]
D	O_2	$H_{14}Na[P_5W_{30}O_{110}]/SiO_2$	cyclohexanol	cyclohexanone	(99%)	AcOH	Reflux	[142]
D	O_2/IBA (IBA = isobutyraldehyde)	$H_5[PV_2Mo_{10}O_{40}]$/MCM-41	cyclooctene	cyclooctene oxide	(63%)	1,2-Dichloroethane	298	[143]
D	H_2O_2	$Cs_{2.5}H_{0.5}[PW_{12}O_{40}]$/K-10 (K-10 = montmorillonite)	mandelate	phenylglyoxylate	(52%)	CH_3CN	323	[144]
D	O_2	$K_8[\{Fe(OH_2)_2\}_3(P-W_9O_{34})_2]/(Si/AlO_2)$	THT	sulfoxide	(28%)	CH_3CN	348	[146]
D	O_2	$[Pt(Mebipym)Cl_2]$ $H_4[PV_2Mo_{10}O_{40}]/SiO_2$ (Mebipym = N-methyl-2,2′-pyrimidine)	CH_4	$CH_3OH + HCHO + CH_3CHO$	(TON = 33)	Water	323	[149]
D	Urea–H_2O_2	H_2WO_4/FAP (FAP = fluoroapatite)	allylic alcohol	epoxy alcohol	(99%)	Solvent-free	RT[b]	[155]

Table 6.1 (Continued)

Category[a]	Oxidant	Catalyst	Reaction (% yield)	Solvent	Temperature (K)	Ref.
D	Urea–H$_2$O$_2$	TBA$_3$[PMo$_{12}$O$_{40}$]/FAP	ArSMe → ArS(O$_2$)Me (99%)	Solvent-free	298	[157]
D	O$_2$	Ru-H$_5$[PV$_2$Mo$_{10}$O$_{40}$]/α-Al$_2$O$_3$	cyclohexene → cyclohexene oxide (90%)	Trifluorotoluene	403	[159]
E	O$_2$/IBA	H$_5$[PV$_2$Mo$_{10}$O$_{40}$]/PE-SiO$_2$ (PE = polyether)	cyclooctene → cyclooctene oxide (40%)	1,2-Dichloroethane	296	[161]
E	H$_2$O$_2$	MCM-NHP[OWO(O$_2$)$_2$]$_2$	geraniol → 2,3-epoxygeraniol (95%)	CH$_3$CN	323	[162]
E	O$_2$	Pd/FSM-NHP[OWO(O$_2$)$_2$]$_2$	1-butene → 1,2-epoxybutane (28%)	CH$_3$OH	373	[165]
F	H$_2$O$_2$	[{WO(O$_2$)$_2$(H$_2$O)}$_2$O]$^{2-}$/IM-SiO$_2$ (IM = N-octylimidazolium)	allylbenzene → 2-benzyloxirane (>99%)	CH$_3$CN	333	[169]
F	H$_2$O$_2$	[SiW$_{10}$O$_{38}$V$_2$(OH)$_2$]$^{4-}$/IM-SiO$_2$	limonene → limonene epoxide (78%)	CH$_3$CN/tBuOH	293	[170]
F	H$_2$O$_2$	[SiW$_{10}$O$_{38}$V$_2$(OH)$_2$]$^{4-}$/IM-SiO$_2$	propyl allyl sulfide → sulfoxide (89%)	CH$_3$CN/tBuOH	293	[170]

6.3 Heterogenization of Polyoxometalates

	Oxidant	Catalyst	Substrate → Product	Yield	Solvent	T (K)	Ref.
F	O_2	$[PO_4\{WO(O_2)_2\}_4]^{3-}$ / ODMA-SiO$_2$ [ODMA = (octyldimethyl)ammonium]	cyclooctene → cyclooctene oxide	(80%)	Solvent-free	Reflux	[172]
F	O_2	$[Fe_4(H_2O)_{10}(SeW_9O_{32})]^{4-}$ / SBA-15	C$_{14}$ alkane → ketones	(9%)	Solvent-free	423	[173]
F	O_2/IBA	$[PVMo_{11}O_{40}]^{4-}$ / N^+Me_3-SiO$_2$	1-octanol → octanal	(10%)	CH_3CN	358	[174]
F	O_2	$[PW_{11}CoO_{39}]^{5-}$ / N^+H_3-SiO$_2$	isobutyraldehyde → isobutyric acid	(89%)	CH_3CN	293	[175]
F	H_2O_2	ZnAl-LDH-$[H_2W_{12}O_{40}]^{6-}$	cyclooctene → cyclooctene oxide	(17%)	C_2H_5OH	333	[180]
F	H_2O_2 + NH_4Br	NiAl-LDH-$[WO_4]^{2-}$	bisphenol S → tetrabromo bisphenol S	(TOF = 48 h^{-1})	Water/ CH_3OH/THF	298	[182]
F	H_2O_2 + NH_4Br	NiAl-LDH-$[WO_4]^{2-}$	2,6-dimethylphenol → 4-methoxy-2,6-dimethyl-cyclohexa-2,5-dienone	(91%)	Water/CH_3OH/THF	333	[184]

a) Category A: solidification by metal and alkylammonium cations, B: solidification by polycations, C: solidification by cationic organometallic complexes, D: immobilization by wet impregnation, E: immobilization by solvent-anchoring and covalent linkage, F: immobilization by anion-exchange.
b) RT = room temperature.

6.3.1
Solidification of Polyoxometalates with Appropriate Cations

6.3.1.1 Metal and Alkylammonium Cations

POMs form by a self-assembly process, typically in an acidic aqueous solution and can be isolated as powder or crystals with counter-cations. Appropriate selection of counter-cations can control the solubility of POMs in various reaction media. For homogeneous system, alkylammonium cations, generally TBA, are selected as counter-cations of POM anions for the dissolution in organic solvents such as acetonitrile, DMF, DMSO and 1,2-dichloroethane. POMs with metal counter-cations such as Na^+, K^+, Rb^+, Cs^+ and Ag^+ are not soluble in common organic solvents.

A mixed-addenda POM with Ag^+ as a counter-cation, $Ag_5[PV_2Mo_{10}O_{40}]$, has been synthesized by a simple metathesis precipitation between $Na_5[PV_2Mo_{10}O_{40}]$ and $AgNO_3$ in water [113]. It was confirmed by thermogravimetric analysis that $Ag_5[PV_2Mo_{10}O_{40}]$ was stable up to 1103 K and was more stable than the precursor, $Na_5[PV_2Mo_{10}O_{40}]$. $Ag_5[PV_2Mo_{10}O_{40}]$ could act as an efficient heterogeneous catalyst for selective aerobic oxidation of 2-chloroethyl ethyl sulfide (mustard simulant) to the corresponding harmless sulfoxide. In this system, 2,2,2-trifluoroethanol was chosen as solvent because the substrate was soluble but $Ag_5[PV_2Mo_{10}O_{40}]$ was completely insoluble in it. A similar mixed-addenda POM with Fe^{3+} as a counter-cation showed high catalytic activity for the aerobic oxidation of various kinds of alcohols [114].

Neumann and coworkers have synthesized novel ε-Keggin POMs $[Mo_{12}O_{39}(\mu\text{-OH})_{10}H_2\{X(H_2O)_3\}_4]$ ($X = Co^{2+}$, Mn^{2+} and Cu^{2+}) [115]. These POMs are insoluble in any solvents and could be used as heterogeneous catalysts for the aerobic oxidation of aldehydes to carboxylic acids in toluene. Aliphatic linear aldehydes showed >99% selectivities. Aldehydes with various functional groups such as myrtenal and 5-norbornene-2-carboxylic acid afforded significant amounts of formate esters and allylic oxidation products. Interestingly, with TBA salts of the same transition-metal-substituted α-Keggin POMs (soluble, thus homogeneous) both the heterogeneous and homogeneous systems showed similar catalytic activity and selectivity.

The selection of reaction media (solvents) is also important in constructing the heterogeneous oxidation systems. The TBA salt of mono-ruthenium-substituted POM $TBA_4H[SiW_{11}O_{39}Ru(H_2O)]$ was insoluble in ester solvents such as isobutyl acetate and tert-butyl acetate and thus could act as a heterogeneous catalyst for the aerobic oxidation of alkanes and alcohols in these solvents [116]. After several catalytic turnovers of substrates (up to 100 000 TON per surface exposed POM for adamantane oxidation), the catalyst was removed by filtration. No further conversion was observed after the removal and no POM species was detected in the filtrate. When the oxygenation was repeated with the recovered catalyst by simple filtration under the same conditions, the reaction proceeded at almost the same rate and selectivity as those for the first run. Therefore, any POM species that leached into the reaction solution is not an active homogeneous catalyst and the observed catalysis is truly heterogeneous.

Zuwei and coworkers have developed the concept of a "reaction-controlled phase-transfer catalyst" [117]. They used $CP_3[PO_4(WO_3)_4]$ (CP = cetylpyridinium) as a catalyst for the H_2O_2-based epoxidation. Although $HDP_3[PO_4(WO_3)_4]$ (HDP = hexadecylpyridinium) itself was not soluble in the reaction medium (1,2-dichloroethane) it could form a soluble active species, $HDP_3[PO_4\{WO_2(O_2)\}_4]$, by the reaction with H_2O_2 that could act as a homogeneous catalyst to readily and selectively produce the corresponding epoxides. When H_2O_2 was completely consumed, active species returned to the original form and became insoluble again, and could then be easily recovered and reused. When coupled with the 2-ethylanthraquinone(EAQ)/2-ethylanthrahydroquinone(EAHQ) redox process for H_2O_2 production, propylene oxide was obtained in 85% yield (based on EAHQ) when using O_2 as a terminal oxidant [117]. A similar behavior, that is, a self-precipitation of catalysts at the end of the reaction, is observed among ionic complexes with weakly coordinating counter ions, where the charges are delocalized over large molecular fragments and the crystal packing forces are weakened [118].

6.3.1.2 Polycations

POMs have diverse self-assembly properties and can act as building units for organic–inorganic hybrid supramolecular complexes. The suitable design and selection of organic polycation components, for example, multipodal cations, cross- and non-crosslinking copolymers, and dendrimers, can control the structure and formation of the n-dimensional organic–inorganic hybrid networks in self-assembly processes. Such an approach has been applied to prepare micro- or mesoporous materials for catalysts (see below) and POM-containing thin films for electrochromic devices [119–121], for example. These hybrid materials sometimes show specific selectivities and properties for oxidation catalysis.

Neumann and coworkers have synthesized novel building units of tripodal polyammonium cations such as benzene-1,3,5-tricarboxylic acid tris(2-trimethylammonium ethyl) ester (BTC) and benzene-1,3,5-[tri(phenyl-4-carboxylic acid)] tris(2-trimethylammonium ethyl) ester (BTPC) [122]. Three-dimensional, perforated, coral-shaped, mesoporous materials were obtained almost immediately simply by mixing DMSO solutions of a sandwich-type POM, $TBA_7Na_5[ZnWZn_2(H_2O)_2(ZnW_9O_{34})_2]$, and 4 equiv of methylsulfate salts of BTC or BTPC (Figure 6.6) [122]. These hybrid materials had moderate BET surface areas (51 and 27 $m^2 g^{-1}$, respectively) with average pore sizes of 36 ± 6 Å. The materials showed high catalytic activity for the epoxidation of allylic alcohols and secondary alcohols with H_2O_2. While the oxidation of secondary allylic alcohols by these heterogeneous catalysts exhibited similar diastereoselectivities to those of the corresponding homogeneous analogue (methyltrioctylammonium salt of $[ZnWZn_2(H_2O)_2(ZnW_9O_{34})_2]^{12-}$ dissolved in 1,2-dichloroethane) [123], the chemoselectivities towards epoxidation were somewhat lower than those of homogeneous catalysts (possibly because of the steric hindrance of the template intermediates [123]). The catalyst recycling experiments and characterization of recovered catalysts and filtrate showed that the above catalysis is truly heterogeneous. However, it remains unclear as to whether the reaction occurs in the mesopore or only on the surface of the catalysts.

Figure 6.6 Tripodal polyammonium cations BTC and BTPC [122].

A networked supramolecular POM-based hybrid catalyst has been synthesized by Ikegami and coworkers via the self-assembly of $[PW_{12}O_{40}]^{3-}$ and non-crosslinked copolymer based on N-isopropylacrylamide (NIPAM) and ammonium cations (Figure 6.7) [124, 125]. This heterogeneous NIPAM-POM hybrid catalyst showed very high catalytic activity for the epoxidation of allylic alcohols with H_2O_2 (e.g., up to 35 000 TON for phytol). In addition, the oxygenation of amines and sulfide could be

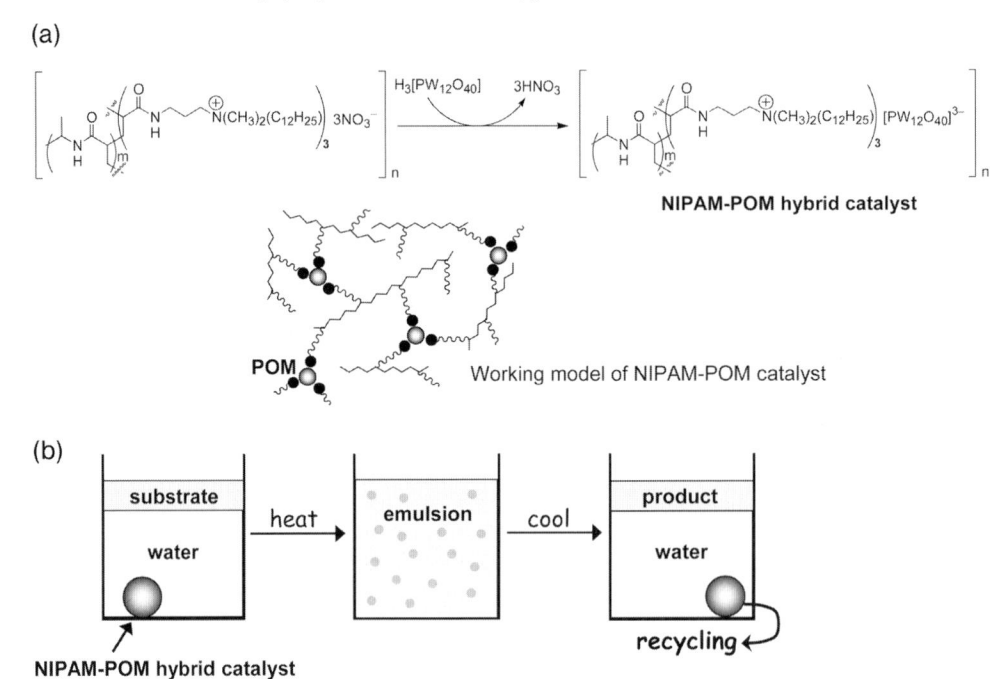

Figure 6.7 (a) NIPAM-POM hybrid catalyst [124, 125] and (b) simplified representation of a thermomorphic system with a NIPAM-POM hybrid catalyst [128].

realized in the presence of this catalyst. The activity of the recovered catalyst was somewhat reduced, possibly because of the partial pulverization of polymer and/or degradation of the POM during catalysis. NIPAM-based polymers intrinsically exhibit temperature-responsive behavior [126, 127] and the thermomorphic system based on a similar NIPAM-POM hybrid catalyst has been developed by Ikegami and coworkers [128]. In the oxidation of alcohols with H_2O_2 in water, the thermoregulated formation of a stable emulsion was detected at high temperature (363 K), which exhibited high catalytic activity. After the reaction was completed, the reaction mixture was cooled to room temperature, and then the catalyst was precipitated off (Figure 6.7).

After Ikegami's first report in 2001 [124], this strategy was extended and applied to the development of heterogeneous oxidation catalysts [129, 130]. For example, an alkylated polyethyleneimine-POM hybrid heterogeneous catalyst has been prepared from branched polyethyleneimine, a crosslinking reagent of n-octylamine-epichlorohydrin and a sandwich-type POM $[ZnWZn_2(H_2O)_2(ZnW_9O_{34})_2]^{12-}$ [130]. This assembly was active for the H_2O_2-based oxidation of 2-alkanols in water. Interestingly, more lipophilic substrates were preferentially oxidized even though the higher molecular weight homologues are much less soluble in the aqueous phase that contains the POM catalyst (distinctive "lipophiloselectivity") (6.9):

$$\begin{array}{c} \text{OH} \\ \bigwedge\!\!{}_{C_4H_9} \\ + \\ \text{OH} \\ \bigwedge\!\!{}_{C_{12}H_{25}} \end{array} \quad \xrightarrow[\text{water, 333 K, 19 h}]{\text{catalyst, } H_2O_2} \quad \begin{array}{c} \text{O} \\ \bigwedge\!\!{}_{C_4H_9} \\ + \\ \text{O} \\ \bigwedge\!\!{}_{C_{12}H_{25}} \end{array} \quad (6.9)$$

catalyst	TON C_{14}	C_6	lipophiloselect.
$Na[ZnWZn_2(H_2O)_2(ZnW_9O_{34})_2]^{12-}$	3	125	0.02
$PEI-[ZnWZn_2(H_2O)_2(ZnW_9O_{34})_2]^{12-}$	495	410	1.2

Thus, it is concluded that the reaction likely proceeds at the hydrophobic domain (inside cavity of the catalytic assembly).

6.3.1.3 Cationic Organometallic Complexes

The combination of POMs with the small-sized and highly charged metal cations, for example, first-row transition metal cation, results in the formation of solid that is soluble in water and polar organic solvents because of the large solvation enthalpy of the ionic components [27, 131]. The complexation of transition metal cations with the organic ligands can control the charge, size, shape and hydrophilicity/hydrophobicity of the resulting organometallic cations [132]. The combination of such organometallic cations with POMs may lead to the formation of insoluble fine particles because of the strong ionic interaction between the ionic components and the hydrophobicity of the organic moieties, which would facilitate the nucleation and prevent aggregation of the particles.

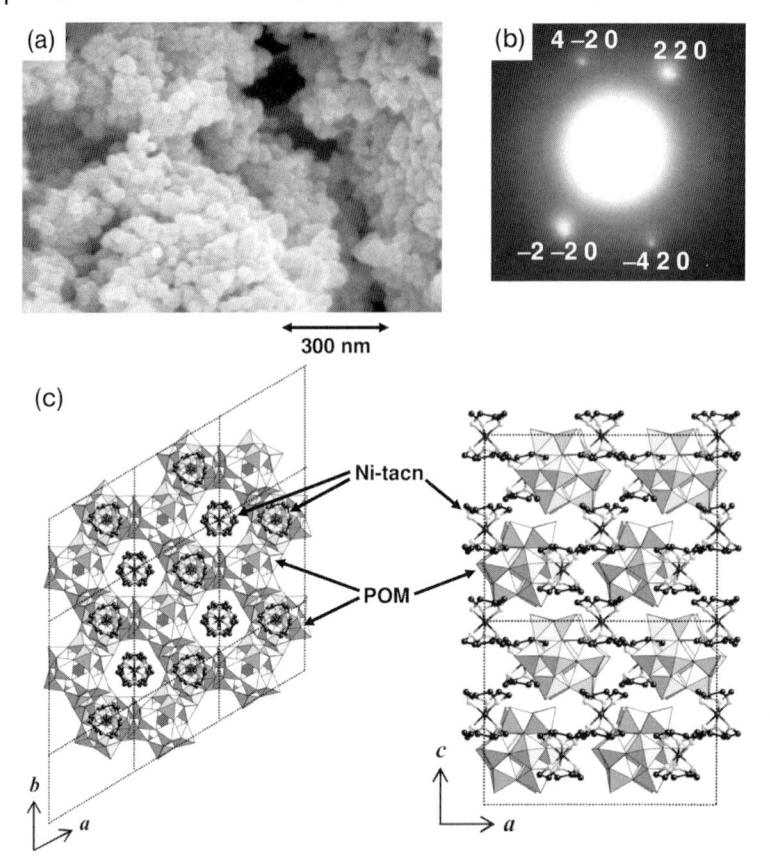

Figure 6.8 (a) SEM image, (b) electron diffractogram and (c) crystal packing structure of [Ni(tacn)$_2$]$_2$[α-SiW$_{12}$O$_{40}$] [133].

The reaction of H$_4$[SiW$_{12}$O$_{40}$] with Ni^{II}-tacn (tacn = 1,4,7-triazacyclononane) afforded monodispersed fine particles of the complex [Ni(tacn)$_2$]$_2$[α-SiW$_{10}$O$_{40}$] with an average particle size of about 60 nm (Figure 6.8). The electron diffractogram of a single particle exhibits discrete spots, showing that the particle is crystalline. The spot was reasonably assignable to the crystal packing structure shown in Figure 6.8c. Also, the combination of catalytically active [γ-SiW$_{10}$O$_{38}$V$_2$(μ-OH)$_2$]$^{4-}$ with Ni^{II}-tacn yielded crystalline monodispersed fine particles of [Ni(tacn)$_2$]$_2$[γ-SiW$_{10}$O$_{38}$V$_2$(μ-OH)$_2$] with an average particle size of about 100 nm [133]. An investigation of the catalytic activity of [Ni(tacn)$_2$]$_2$[γ-SiW$_{10}$O$_{38}$V$_2$(μ-OH)$_2$] for the epoxidation of 1-octene and cyclooctene found selectivities to 1,2-epoxyoctane and cyclooctene oxide of >99%. For the competitive epoxidation of *cis*- and *trans*-2-octenes, the initial rates for the epoxidation of *cis*- and *trans*-2-octenes were 0.16 and <0.001 mM h^{-1}, respectively. For *trans*-1,4-hexadiene, the more accessible terminal C=C moiety was oxygenated in preference to the electron-rich inner C=C bond. These stereo- and regioselectivities were consistent with those for the epoxidation by the TBA salt of [γ-SiW$_{10}$O$_{38}$V$_2$(μ-OH)$_2$]$^{4-}$ in a homogeneous liquid phase [107].

The combination of [α-SiW$_{12}$O$_{40}$]$^{4-}$ with [Cr$_3$O(OOCH)$_6$(H$_2$O)$_3$]$^+$ yielded ionic crystals of K$_3$[Cr$_3$O(OOCH)$_6$(H$_2$O)$_3$][α-SiW$_{12}$O$_{40}$] with molecular-sized space in the crystal lattice [134]. The large size and low charge of the ionic components, reducing the ionic interaction, were crucial to the formation of the molecular sized space. The ionic crystal sorbed only methanol from a mixture of methanol/ethanol/n-propanol. The shape-selective sorption property was reflected in the oxidation of a mixture of alcohols with hydrogen peroxide: only methanol was oxidized to formaldehyde whereas no oxidation of ethanol and n-propanol proceeded [134].

6.3.2
Immobilization of Polyoxometalate-Based Compounds

6.3.2.1 Wet Impregnation

Most frequently used procedures for the immobilization of POM-based compounds are classic wet impregnation of POMs onto inert solid supports. Active carbon, SiO$_2$ and Al$_2$O$_3$ are representative supports [27]. Adsorption of POMs onto crystalline materials such as zeolites, mesoporous materials and apatites is of considerable interests due to their high surface areas, morphologies and unique pore systems. In this way, mixed-addenda POMs, for example, [PV$_2$Mo$_{10}$O$_{40}$]$^{5-}$ and [PV$_8$Mo$_4$O$_{40}$]$^{11-}$, have been studied most frequently and immobilized on various kinds of supports such as active carbon [135–138], textile fibers [139, 140], SiO$_2$ [141, 142], MCM-41 [143] and clays [144]. These supported mixed-addenda POM catalysts were active for various kinds of oxidations such as benzylic and allylic alcohols to the corresponding carbonyl compounds [135, 136], benzylic amines to the corresponding N-alkylimines [135,136], aldehydes to carboxylic acids [138], sulfides to sul-foxides (or sulfones) [139, 140] and alkenes to epoxides with aldehydes as sac-rificial reagents [142]. Alkyl-substituted phenols [137] and alkanes [142] were also oxidized to afford the corresponding oxygenated products. Isophorene was selectively oxidized to 3-formyl-5,5-dimethyl-2-cyclohexen-1-one in the presence of (NH$_4$)$_5$H$_6$[PV$_8$Mo$_4$O$_{40}$] supported on active carbon [137]. This regioselectivty was quite different from that observed with the corresponding homogeneous analogue that produces the 1,4-diketone as a major product (6.10):

$$\text{isophorone} \xrightarrow[\text{toluene, 373 K, 20 h}]{\text{catalyst, O}_2} \mathbf{a} + \mathbf{b}$$

(6.10)

catalyst	conv./%	select./%	
		a	b
(NH$_4$)$_5$H$_6$[PV$_8$Mo$_4$O$_{40}$]	10	50	<1
(NH$_4$)$_5$H$_6$[PV$_8$Mo$_4$O$_{40}$]/C	74	12	65

This difference was explained by the pore size effect of the active carbon [137]. Non-polar solvents such as toluene are generally choice of solvents for the above-mentioned systems, as they can prevent measurable leaching of POM species during the oxidations.

Originally, it was believed that the roles of the supports are simply to increase the surface areas of POM species (the surface areas of unsupported POMs are generally very low and below $10\,m^2\,g^{-1}$) and to improve their stabilities. For the certain kinds of aerobic oxidations, however, the catalytic activity of POMs supported on active carbon was much superior to those on other supports with high surface areas such as SiO_2 and Al_2O_3, suggesting that active carbon is not an inert matrix but plays a significant role in the catalytic cycle. A subsequent study by Neumann and coworkers suggests the formation of "quinone(s)" on the surface of active carbon through the presence of POMs and O_2 (during the catalyst preparation or oxidation catalysis) [145]. For the aerobic alcohol oxidation, the quinone species formed on the surface of active carbon might oxidize alcohols to the corresponding carbonyl compounds and the reduced quinine (hydroquinone) is then reoxidized by POMs and O_2 [145].

Reaction of the sandwich-type POM $[(Fe(OH_2)_2)_3(A\text{-}\alpha\text{-}PW_9O_{34})_2]^{9-}$ with a colloidal suspension of silica/alumina nanoparticles ($(Si/AlO_2)Cl$) resulted in the production of a novel supported POM catalyst [146–148]. In this case, about 58 POM molecules per cationic silica/alumina nanoparticle were electrostatically stabilized on the surface. The aerobic oxidation of 2-chloroethyl ethyl sulfide (mustard simulant) to the corresponding harmless sulfoxide proceeded efficiently in the presence of the heterogeneous catalyst and the catalytic activity of the heterogeneous catalyst was much higher than that of the parent POM. In addition, this catalytic activity was much enhanced when binary cupric triflate and nitrate $[Cu(OTf)_2/Cu(NO_3)_2 = 1.5]$ were also present [148].

The aerobic oxidation of methane in water catalyzed by $[Pt(Mebipym)Cl_2]$ $[PV_2Mo_{10}O_{40}]^{5-}$ (Mebipym = N-methy-2,2′-bipyrimidine) complex supported on SiO_2 was reported [149]. The conjugation of $[PV_2Mo_{10}O_{40}]^{5-}$ to a known Pt^{2+}-bipyrimidine complex by electrostatic interaction could facilitate the oxidation of the Pt^{2+} intermediate to a Pt^{4+} intermediate by O_2, resulting in the catalytic aerobic oxidation of methane to methanol in water and then surprisingly further to acetaldehyde via a carbon–carbon coupling reaction.

Calcium apatite, $Ca_{10}(PO_4)_6X_2$ ($X=OH^-$ or F^-), which forms the mineral component of bone and teeth, is handled as a solid that is harmless to the environment, and has strong affinity to various organic substances [150]. Various kinds of transition metal cations can be readily accommodated into the apatite framework based on their large cation exchange ability [150]; such cations have a high potentiality to function as the catalytic active center. Thus, apatite has generated great interest as a stable and recyclable support for the preparation of heterogeneous catalysts employed in various functional group transformations [151–154]. Various kinds of POMs, including $(CP)_{10}[H_2W_{12}O_{42}]$, $(NH_4)_3[PMo_{12}O_{40}]$ and $H_{3+n}[PV_nMo_{12}O_{40}]$, could be supported on apatite [155–158]. Oxidation of alkenes and sulfides with urea–H_2O_2 (UHP) [155–157] and aerobic oxidative dehydrogenation of α-terpinene [158] efficiently proceeded with this solid catalyst under solvent-free conditions. The recovered solid catalysts could be reused several times with the maintenance of their catalytic activity.

Neumann and coworkers have developed a novel method to prepare size-controlled metal nanoparticles [159]. The method is based on the coupling of the two reactions as shown in (6.11) and (6.12):

$$H_5[PV^{5+}_2Mo_{10}O_{40}] + Zn \longrightarrow ZnH_5[PV^{4+}_2Mo_{10}O_{40}] \quad (6.11)$$

$$ZnH_5[PV^{5+}_2Mo_{10}O_{40}] + M^{n+} \longrightarrow M^0\text{-}ZnH_3[PV^{5+}_2Mo_{10}O_{40}]$$
(M-POM, M = Ru or Ag)
$$(6.12)$$

First, two-electron reduction of $H_5[PV^V_2Mo_{10}O_{40}]$ with zinc metal affords $ZnH_5[PV^{IV}_2Mo_{10}O_{40}]$ followed by re-oxidation with the desired metal cations such as Ag^+ and Ru^{3+} to give the metal nanoparticles stabilized by $H_5[PV_2Mo_{10}O_{40}]$ (M-POM, M = Ag and Ru). Average particle sizes are about 5 nm. The M-POMs were supported by wet impregnation and were homogeneously dispersed on α-Al_2O_3. These supported metal catalysts showed high catalytic performance for the aerobic epoxidation of cyclohexene and 1-methylcyclohexene. Although these substrates are known to be highly sensitive to allylic oxidation, no allylic oxidation products were detected in this oxidation system. This is likely because of the inhibition of allylic oxidation by metal nanoparticles.

6.3.2.2 Solvent-Anchoring and Covalent Linkage

It is well known that acid form POMs can interact with ethers to form crown ether-type complexes (etherates). The acid form POMs, for example, $H_5[PV_2Mo_{10}O_{40}]$, in combination with polyethylene glycol as a solvent (PEG, PEG-200 or PEG-400) can construct an attractive catalyst separation/recovery system [160]. Since the POM-PEG phase in intrinsically immiscible with common organic phases, the desired product(s) can be easily recovered, and the retained POM-PEG phase can be recycled [160]. Beyond the simple use of PEG as a solvent, PEG-anchored supported POM catalysts have also been developed [161]. Polyglycols such as hydrophilic PEG and hydrophobic polypropylene glycol (PPG) were covalently attached on the surface of SiO_2, acting as a solvent-like phase and complexing agent for POMs (Figure 6.9). Aerobic epoxidation of cyclooctene in combination with isobutyraldehyde as a sacrificial reagent proceeded efficiently with a $H_5[PV_2Mo_{10}O_{40}]$ based heterogeneous catalyst [161], for example. The hydrophilic–hydrophobic balance is very important for control of activity.

Various kinds of supports with phosphorylated spacers were synthesized to form covalent bonds with peroxotungstate species that act as effective catalysts for the H_2O_2-based epoxidation [162–165]. The peroxo compound $[HPO_4\{W(O)(O_2)_2\}_2]^{2-}$ was synthesized on the surface of mesoporous HMS by reacting HMS-$CH_2CH_2CH_2NH(PO_3H_2)$ with $[\{WO(O_2)_2(H_2O)_2\}(\mu\text{-}O)]^{2-}$, and then palladium ions were exchanged into the channels of HMS [165]. In the presence of the catalyst in methanol, aerobic oxidation of propylene efficiently proceeded to give propylene oxide in 83% selectivity (at 34% conversion). The actual epoxidation catalyst is a peroxo tungstate species, and H_2O_2 formed by the reaction of methanol and O_2 over

6.3.2.3 Anion Exchange

POMs can be immobilized onto anion-exchange resins and surface-modified metal oxides with quaternary ammonium cation- or amino-functional groups via anion-exchange. Jacobs and coworkers tethered Venturello's catalyst $[PO_4(WO(O_2)_2)_4]^{3-}$ on a commercially available nitrate-form resin with alkylammonium cations and have carried out the epoxidation of allylic alcohols and terpenes with this supported catalyst [166, 167]. The regio- and diastereoselectivity of the parent homogeneous catalysts were preserved in the supported catalyst. For bulky alkenes, the reactivity of the POM catalyst was superior to that of Ti-β zeolite with a large pore size. The catalytic activity of the recycled catalyst was maintained completely after several cycles.

In certain cases, pulverization and degradation of the organic polymer supports are frequently observed for the polymer supported POM catalysts. To overcome this shortcoming, organic–inorganic hybrid supports have been developed. By using silane coupling reagents such as trialkoxy- and trichloroorganosilanes, various kinds of organic–inorganic hybrid supports can be synthesized by grafting organic moieties (anion-exchange parts) onto silanol-containing surfaces (post-modification procedure) or by direct synthesis into the final composite materials (sol–gel synthesis) [168]. The surface properties, for example, hydrophobicity and hydrophilicity, can be controlled by changing the organic moieties. The grafted organic moieties are intrinsically attached through covalent bonds, and thus the hybrid materials are more stable than polymer-based resins.

One such organic–inorganic hybrid support has been synthesized by covalently anchoring a N-octyldihydroimidazolium cation fragment onto SiO_2 (denoted as

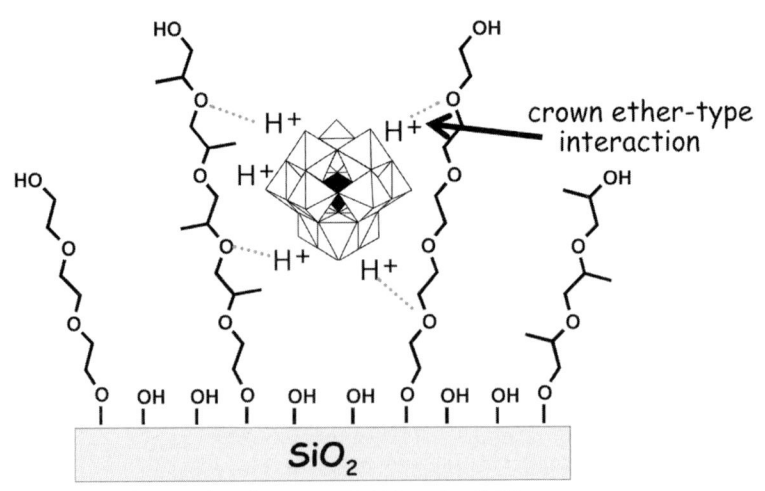

Figure 6.9 Solvent-anchored supported POM catalyst [161].

IM-SiO$_2$) [169–171]. Figure 6.10 shows the procedure for the synthesis of IM-SiO$_2$. Characterization of modified support IM-SiO$_2$ by ^{13}C, ^{29}Si and ^{31}P solid-state NMR, IR spectroscopy and elemental analysis showed that the structure of the dihydroimidazolium skeleton was preserved on the surface of SiO$_2$. The modified support IM-SiO$_2$ could act as a good anion exchanger, and catalytically active peroxo- and polyoxometalate anions such as [{WO(O$_2$)$_2$(H$_2$O)}$_2$(μ-O)]$^{2-}$ and [SiW$_{10}$O$_{38}$V$_2$(μ-OH)$_2$]$^{4-}$ could be immobilized onto the support with the stoichiometric anion exchange (Figure 6.10). The structure of polyanions was preserved after the anion exchange. These supported catalysts showed high catalytic performance for the above-mentioned oxidations without loss of the intrinsic catalytic nature, for example, reaction rate, chemo-, regio- and diastereoselectivities, and efficiencies of H$_2$O$_2$ utilization, of the corresponding homogeneous analogues (6.13) [107, 169, 170]:

$$\text{cyclohexene} \xrightarrow[\text{CH}_3\text{CN}/t\text{BuOH, 293 K, 24 h}]{\text{catalyst, H}_2\text{O}_2} \text{cyclohexene oxide} \quad (6.13)$$

		select./%	
catalyst	yield/%	syn	anti
TBA$_4$[SiW$_{10}$O$_{38}$V$_2$(OH)$_2$]	91	5	95
[SiW$_{10}$O$_{38}$V$_2$(OH)$_2$]$^{4-}$/IM-SiO$_2$	82	4	96

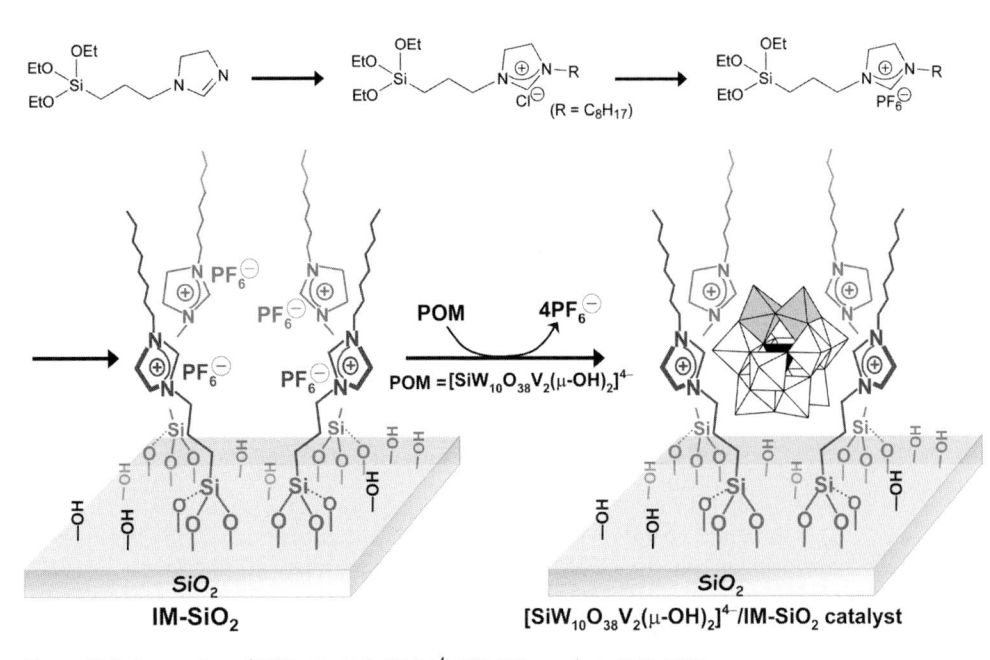

Figure 6.10 Preparation of [SiW$_{10}$O$_{38}$V$_2$(μ-OH)$_2$]$^{4-}$/IM-SiO$_2$ catalysts [169–171].

The oxidation was immediately stopped by the removal of the solid catalysts, and POM species could hardly be found in the filtrate after removal of the catalysts. These results could rule out any contribution to the observed catalysis from POM species that leached into the reaction solution, and show that the observed catalysis was truly heterogeneous in nature. In addition, the catalysts could be reused several times without an appreciable loss of their high catalytic performance.

For tethering POMs, quaternary ammonium cation-functionalized hybrid supports have been most frequently used. Neumann and Miller grafted POMs such as $[PO_4(WO(O_2)_2)_4]^{3-}$ and $[ZnWMn_2(H_2O)_2(ZnW_9O_{34})_2]^{12-}$ on modified SiO_2 particles functionalized with various quaternary ammonium cation moieties, which were prepared by copolymerization of tetraethyl orthosilicate and trialkoxy-organosilanes using the sol–gel technique, for application to heterogeneous epoxidation using H_2O_2 [172]. The catalytic activities were greatly influenced by the type of quaternary ammonium cation moieties introduced into the modified SiO_2. Octyldimethyl (benzyl)ammonium cations showed the best results. Many catalytic oxidations of alkanes [173], alkenes [162], alcohols [174], aldehydes [175] and aromatic compounds [176] with analogous catalysts have also been reported.

All-inorganic anion-exchangers, for example, layered double hydroxides (LDHs), are attractive from the standpoint of their stabilities. LDHs have the general formula $[M^{2+}{}_nM^{3+}{}_m(OH)_{2(n+m)}]^{m+}A^{x-}{}_{m/x}\cdot yH_2O$, where M^{2+} and M^{3+} are divalent and trivalent metal cations, respectively [177]. The A^{x-} anion required to compensate the net positive charge of brucite-like layers are located in the interlayer space [177]. LDHs have anion-exchange ability and various kinds of anions, including POMs, can be intercalated. For example, $Zn_2Al(OH)_6NO_3 4H_2O$ (ZnAl-LDH) underwent facile and complete intercalation with POMs such as $[SiW_{11}O_{39}]^{8-}$, $[SiV_3W_9O_{40}]^{7-}$, $[H_2W_{12}O_{40}]^{6-}$ and $[W_4Nb_2O_{19}]^{4-}$ by an anion exchange reaction [178–180]. These ZnAl-LDH-POM catalyzed the epoxidation of cyclohexene with H_2O_2 [178–180]. Tatsumi and coworkers applied MgAl-LDH intercalated with POMs such as $[Mo_7O_{24}]^{6-}$ or $[H_2W_{12}O_{42}]^{10-}$ to the shape selective epoxidation of 2-hexene and cyclohexene with H_2O_2 [181]. Monomeric oxoanions such as WO_4^{2-} and MoO_4^{2-} have also been intercalated into the interlayer of LDHs to give species that can act as efficient heterogeneous catalysts for H_2O_2-based oxidations [182–184].

6.4
Conclusion

Various kinds of POMs and related compounds, including isopolyoxometalates, heteropolyoxometalates, peroxometalates, lacunary POMs and transition-metal-substituted POMs, can homogeneously catalyze liquid-phase oxidations with environmentally-benign H_2O_2 or O_2 as an oxidant. Intrinsically, H_2O_2-based oxidations catalyzed by tungsten-based compounds are very effective. Mixed-addenda POMs, $[PV_nMo_{12-n}O_{40}]^{(3+n)-}$, are very useful and the most frequently studied for O_2-based oxidations because their redox potentials are higher than that of common hydrocarbons. A strictly designed active site structure with lacunary POMs as "structural motifs" can efficiently activate oxidants and can readily transfer the active oxygen species to hydrocarbon substrates. In certain cases, perfect efficiency of H_2O_2 utilization (>99%) and aerobic selective epoxidation can be realized with well-designed POM catalysts.

To date, many methodologies for the design of "heterogeneous POM-based catalysts," in which the observed catalyses are truly heterogeneous in nature, have been developed as summarized in this chapter. In most cases, catalysts can easily be recovered by simple filtration or centrifugation and the recovered catalysts can be recycled without appreciable loss of catalytic performance. Another strategy for catalyst recovery and recycling is the use of liquid–liquid biphasic systems, where the catalyst and product phases can be separated and then the catalyst phase can be reused. Several liquid–liquid POM-based biphasic systems, with POM catalysts in aqueous [79, 80, 185, 186], fluorous [187], ionic liquid phases [188–190] and polyethylene glycol [160], have been reported recently.

Although efficient recovery and recycling of catalysts have been achieved in most cases, the catalytic activities and selectivities of the parent homogeneous POMs are somewhat, or much, lowered by the heterogenization. At present, only a few successful examples have been developed in which the catalytic activities and selectivities of heterogenized catalysts are comparable to (or even higher than) those of the corresponding homogeneous analogues [113, 169, 172, 182]. Our own impression is that this area (the development of POMs-based heterogeneous oxidation catalysts) has not progressed greatly for some years. Thus, future targets will need strategies to overcome such problems so as to open up a new avenue for the use of POM-based catalysts for many practical oxidations.

References

1 Anastas, P.T. and Warner, J.C. (1998) *Green Chemistry: Theory and Practice*, Oxford University Press, Oxford.
2 Sheldon, R.A. (2000) *Pure and Applied Chemistry*, **72**, 1233.
3 Clark, J.H. (1999) *Green Chemistry*, **1**, 1.
4 Special issues on "Green Chemistry": *Accounts of Chemical Research*, **35**, 685–816 (2002).
5 Sheldon, R.A. and Kochi, J.K. (1981) *Metal Catalyzed Oxidations of Organic Compounds*, Academic Press, New York.
6 Hill, C.L. (1988) *Advances in Oxygenated Processes*, Vol. 1 (ed. A.L. Baumstark), JAI Press, London, pp. 1–30.
7 Hudlucky, M. (1990) *Oxidations in Organic Chemistry*, ACS Monograph Series, American Chemical Society, Washington, DC.
8 Bäckvall, J.-E. (2004) *Modern Oxidation Methods*, Wiley-VCH, Weinheim.
9 Romao, C.C., Kuehn, F.E. and Herrmann, W.A. (1997) *Chemical Reviews*, **97**, 3197.
10 Thomas, J.M. and Raja, R. (2001) *Chemical Communications*, 675.
11 Ishii, Y., Sakaguchi, S. and Iwahama, T. (2001) *Advanced Synthesis and Catalysis*, **343**, 393.
12 de Vos, D.E., Sels, B.F. and Jacobs, P.A. (2001) *Advances in Catalysis*, **46**, 1.
13 Corma, A. and Garcia, H. (2002) *Chemical Reviews*, **102**, 3837.
14 Chen, K., Costas, M. and Que, L. Jr (2002) *Journal of the Chemical Society – Dalton Transactions*, 672.
15 Brégeault, J.-M. (2003) *Journal of the Chemical Society – Dalton Transactions*, 3289.

16 De Bruin, B., Budzelaar, P.H.M. and Gal, A.W. (2004) *Angewandte Chemie – International Edition*, **43**, 4142.
17 Sibbons, K.F., Shastri, K. and Watkinson, M. (2006) *Journal of the Chemical Society – Dalton Transactions*, 645.
18 Thematic issue on "Polyoxometalates" (1998) *Chemical Reviews*, **98** (1), 1–389.
19 Pope, M.T. (1983) *Heteropoly and Isopoly Oxometalates*, Springer-Verlag, Berlin.
20 Pope, M.T. and Müller, A. (eds) (2001) *Polyoxometalate Chemistry From Topology via Self-Assembly to Applications*, Kluwer Academic Publishers, Dordrecht.
21 Yamase, T. and Pope, M.T. (eds) (2002) *Polyoxometalate Chemistry for Nano-Composite Design*, Kluwer Academic/Plenum Publishers, New York.
22 Pope, M.T. (2004) *Comprehensive Coordination Chemistry II*, Vol. 4 (eds J.A. McCleverty and T.J. Meyer), Elsevier Pergamon, Amsterdam, p. 635.
23 Mialane, P., Dolbecq, A. and Sécheresse, F. (2006) *Chemical Communications*, 3477.
24 Long, D.-L., Burkholder, E. and Cronin, L. (2007) *Chemical Society Reviews*, **36**, 105.
25 Uchida, S. and Mizuno, N. (2007) *Coordination Chemistry Reviews*, **251**, 2537.
26 Hill, C.L., Chrisina, C. and Prosser-McCartha, M. (1995) *Coordination Chemistry Reviews*, **143**, 407.
27 Okuhara, T., Mizuno, N. and Misono, M. (1996) *Advances in Catalysis*, Vol. 42, **41**, 113.
28 Neumann, R. (1998) *Progress in Inorganic Chemistry*, **47**, 317.
29 Mizuno, N. and Misono, M. (1998) *Chemical Reviews*, **98**, 199.
30 Kozhevnikov, I.V. (1998) *Chemical Reviews*, **98**, 171.
31 Kozhevnikov, I.V. (2002) *Catalysis by Polyoxometalates*, John Wiley & Sons, Chichester, England.
32 Mizuno, N., Yamaguchi, K. and Kamata, K. (2005) *Coordination Chemistry Reviews*, 249. 1944.
33 Hill, C.L. (2004) *Comprehensive Coordination Chemistry II*, Vol. 4 (eds J.A. McCleverty and T.J. Meyer), Elsevier, Amsterdam, p. 679.
34 Mizuno, N., Kamata, K. and Yamaguchi, K. (2006) *Surface and Nanomolecular Catalysis* (ed. R. Richards), Taylor and Francis Group, New York, p. 463.
35 Ishii, Y., Yamawaki, K., Ura, T. et al. (1988) *The Journal of Organic Chemistry*, **53**, 3587.
36 Ishii, Y., Yamawaki, K., Yoshida, T. et al. (1987) *The Journal of Organic Chemistry*, **52**. 1868.
37 Sakaguchi, S., Nishiyama, Y. and Ishii, Y. (1996) *The Journal of Organic Chemistry*, **61**, 5307.
38 Oguchi, T., Sakata, Y., Takeuchi, N. et al. (1989) *Chemistry Letters*, 2053.
39 Sakaue, S., Sakata, Y., Nishiyama, Y. and Ishii, Y. (1992) *Chemistry Letters*, 289.
40 Ishii, Y. and Sakata, Y. (1990) *The Journal of Organic Chemistry*, **55**, 5545.
41 Ishii, Y., Yoshida, T., Yamawaki, K. and Ogawa, M. (1988) *The Journal of Organic Chemistry*, **53**, 5549.
42 Sakata, Y. and Ishii, Y. (1991) *The Journal of Organic Chemistry*, **56**, 6233.
43 Sakata, Y., Katayama, Y. and Ishii, Y. (1992) *Chemistry Letters*, 671.
44 Iwahara, T., Sakaguchi, S., Nishiyama, Y. and Ishii, Y. (1995) *Tetrahedron Letters*, **36**, 1523.
45 Aubry, C., Chottard, G., Platzer, N. et al. (1991) *Inorganic Chemistry*, **30**, 4409.
46 Dengel, A.C., Griffith, W.P. and Parkin, B.C. (1993) *Journal of the Chemical Society – Dalton Transactions*, 2683.
47 Salles, L., Aubry, C., Thouvenot, R. et al. (1994) *Inorganic Chemistry*, **33**, 871.
48 Duncan, D.C., Chambers, R.C., Hecht, E. and Hill, C.L. (1995) *Journal of the American Chemical Society*, **117**, 681.
49 Chambers, R.C. and Hill, C.L. (1989) *Inorganic Chemistry*, **28**, 2509.
50 Renneke, R.F., Pasquali, M. and Hill, C.L. (1990) *Journal of the American Chemical Society*, **112**, 6585.
51 Attanasio, D., Suber, L. and Thorslund, K. (1991) *Inorganic Chemistry*, **30**, 590.
52 Neumann, R. and Khenkin, A.M. (2006) *Chemical Communications*, 2529.
53 Matveev, K.I. (1977) *Kinetika i Kataliz*, **18**, 862.

54 Khenkin, A.M., Rosenberger, A. and Neumann, R. (1999) *Journal of Catalysis*, **182**, 82.
55 Neumann, R. and Lissel, M. (1989) *The Journal of Organic Chemistry*, **54**, 4607.
56 Atlamsani, A., Brégeault, J.-M. and Ziyad, M. (1993) *The Journal of Organic Chemistry*, **58**, 5663.
57 Hamamoto, M., Nakayama, K., Nishiyama, Y. and Ishii, Y. (1993) *The Journal of Organic Chemistry*, **58**, 6421.
58 Lissel, M., in de Wal, H.J. and Neumann, R. (1992) *Tetrahedron Letters*, **33**, 1795.
59 Neumann, R., Khenkin, A.M. and Vigdergauz, I. (2000) *Chemistry – A European Journal*, **6**, 875.
60 Monflier, E., Blouet, E., Barbaux, Y. and Mortreux, A. (1994) *Angewandte Chemie – International Edition*, **33**, 2100.
61 Passoni, L.C., Cruz, A.T., Buffon, R. and Shuchardt, U. (1997) *J Mol Catal A*, **120**, 117.
62 Yokota, T., Tani, M., Sakaguchi, S. and Ishii, Y. (2003) *Journal of the American Chemical Society*, **125**, 1476.
63 Yokota, T., Sakaguchi, S. and Ishii, Y. (2002) *Advanced Synthesis and Catalysis*, **344**, 849.
64 Khenkin, A.M., Weiner, L., Wang, Y. and Neumann, R. (2001) *Journal of the American Chemical Society*, **123**, 8531.
65 Khenkin, A.M., Weiner, L. and Neumann, R. (2005) *Journal of the American Chemical Society*, **127**, 9988.
66 Kozhevnikov, I.V. and Matveev, K.I. (1983) *Applied Catalysis*, **5**, 135.
67 Grate, J.H. (1996) *J Mol Catal A*, **106**, 57.
68 Kim, W.B., Voitl, T., Rodriguez-Rivera, G.J. et al. (2005) *Angewandte Chemie – International Edition*, **44**, 778.
69 Weinstock, I.A., Barbuzzi, E.M.G., Wemple, M.W. et al. (2001) *Nature*, **414**, 191.
70 Gresley, N.M., Griffith, B.C., Parkin, B.C. et al. (1996) *Journal of the Chemical Society – Dalton Transactions*, 2039.
71 Brégeault, J.-M., Vennat, M., Salles, L. et al. (2006) *Journal of Molecular Catalysis A – Chemical*, **250**, 177.
72 Piquemal, J.-Y., Salles, L., Chottard, G. et al. (2006) *European Journal of Inorganic Chemistry*, 939.
73 Bailey, A.J., Griffith, W.P. and Parkin, B.C. (1995) *Journal of the Chemical Society – Dalton Transactions*, 1833.
74 Salles, L., Piquemal, J.-Y., Thouvenot, R., Minot, C. and Brégeault, J.-M. (1997) *Journal of Molecular Catalysis A – Chemical*, **117**, 375.
75 Salles, L., Robert, F., Semmer, V. et al. (1996) *Bulletin de la Societe Chimique de France*, **133**, 319.
76 Liu, Y., Murata, K. and Inaba, M. (2004) *Chemical Communications*, 582.
77 Venturello, C., D'Aloisio, R., Bart, J.C.J. and Ricci, M. (1985) *Journal of Molecular Catalysis*, **32**, 107.
78 Venturello, C., Alneri, E. and Ricci, M. (1983) *The Journal of Organic Chemistry*, **48**, 3831.
79 Kamata, K., Yamaguchi, K., Hikichi, S. and Mizuno, N. (2003) *Advanced Synthesis and Catalysis*, **345**, 1193.
80 Kamata, K., Yamaguchi, K. and Mizuno, N. (2004) *Chemistry – A European Journal*, **10**, 4728.
81 Kamata, K., Kuzuya, S., Uehara, K. et al. (2007) *Inorganic Chemistry*, **46**, 3768.
82 Kamata, K., Yonehara, K., Sumida, Y. et al. (2003) *Science*, **300**, 964.
83 Kamata, K., Nakagawa, Y., Yamaguchi, K. and Mizuno, N. (2004) *Journal of Catalysis*, **224**, 224.
84 Kamata, K., Kotani, M., Yamaguchi, K. et al. (2007) *Chemistry – A European Journal*, **13**, 639.
85 Musaev, D.G., Morokuma, K., Geletii, Y.V. and Hill, C.L. (2004) *Inorganic Chemistry*, **43**, 7702.
86 Prabhakar, R., Morokuma, K., Hill, C.L. and Musaev, D.G. (2006) *Inorganic Chemistry*, **45**, 5703.
87 Sartorel, A., Carraro, M., Bagno, A. et al. (2007) *Angewandte Chemie – International Edition*, **46**, 3255.
88 Carraro, M., Sandei, L., Sartorel, A. et al. (2006) *Organic Letters*, **8**, 3671.
89 Bar-Nahum, I., Cohen, H. and Neumann, R. (2003) *Inorganic Chemistry*, **42**, 3677.

90 Bar-Nahum, I. and Neumann, R. (2003) *Chemical Communications*, 2690.
91 Bar-Nahum, I., Narasimhulu, K.V., Weiner, L. and Neumann, R. (2005) *Inorganic Chemistry*, **44**, 4900.
92 Neumann, R. and Dahan, M. (1997) *Nature*, **388**, 353.
93 Neumann, R. and Dahan, M. (1998) *Journal of the American Chemical Society*, **120**, 11969.
94 Yin, C.-X. and Finke, R.G. (2005) *Inorganic Chemistry*, **44**, 4175.
95 Weiner, H. and Finke, R.G. (1999) *Journal of the American Chemical Society*, **121**, 9831.
96 Yin, C.-X. and Finke, R.G. (2005) *Journal of the American Chemical Society*, **127**, 9003.
97 Zhang, X., O'Connor, C.J., Jameson, G.B. and Pope, M.T. (1996) *Inorganic Chemistry*, **35**, 30.
98 Canny, J., Thowenot, R., Tézé, A. et al. (1991) *Inorganic Chemistry*, **30**, 976.
99 Botar, B., Geletii, Y.V., Kögerler, P. et al. (2006) *Journal of the American Chemical Society*, **128**, 11268.
100 Nakagawa, Y., Uehara, K. and Mizuno, N. (2005) *Inorganic Chemistry*, **44**, 14.
101 Goto, Y., Kamata, K., Yamaguchi, K. et al. (2006) *Inorganic Chemistry*, **45**, 2347.
102 Geletii, Y.V., Botar, B., Kögerler, P. et al. (2008) *Angewandte Chemie – International Edition*, **47**, 3896.
103 Nozaki, C., Kiyoto, I., Minai, Y. et al. (1999) *Inorganic Chemistry*, **38**, 5724.
104 Mizuno, N., Nozaki, C., Kiyoto, I. and Misono, M. (1998) *Journal of the American Chemical Society*, **120**, 9267.
105 Nishiyama, Y., Nakagawa, Y. and Mizuno, N. (2001) *Angewandte Chemie – International Edition*, **40**, 3639.
106 Botar, B., Geletii, Y.V., Kögerler, P. et al. (2006) *Journal of the American Chemical Society*, **128**, 11268.
107 Nakagawa, Y., Kamata, K., Kotani, M. et al. (2005) *Angewandte Chemie – International Edition*, **44**, 5136.
108 Kamata, K., Yamaguchi, S., Kotani, M. et al. (2008) *Angewandte Chemie – International Edition*, **47**, 2407.
109 Anderson, T.M., Neiwert, W.A., Kirk, M.L. et al. (2004) *Science*, **306**, 2074.
110 Anderson, T.M., Cao, R., Slonkina, E. et al. (2005) *Journal of the American Chemical Society*, **127**, 11948.
111 Cao, R., Anderson, T.M., Piccoli, P.M.B. et al. (2007) *Journal of the American Chemical Society*, **129**, 11118.
112 Hagen, J. (1999) *Industrial Catalysis: A Practical Approach*, Wiley-VCH, Weinheim.
113 Phule, J.T., Neiwert, W.A., Hardcastle, K.I. et al. (2001) *Journal of the American Chemical Society*, **123**, 12101.
114 Nagaraju, P., Pasha, N., Prasad, P.S.S. and Lingaiah, N. (2007) *Green Chemistry*, **9**, 1126.
115 Sloboda-Rozner, D., Neimann, K. and Neumann, R. (2007) *Journal of Molecular Catalysis A – Chemical*, **262**, 109.
116 Yamaguchi, K. and Mizuno, N. (2002) *New Journal of Chemistry*, **26**, 972.
117 Zuwei, X., Ning, Z., Yu, S. and Kunlan, Li. (2001) *Science*, **292**, 1139.
118 Dloumaev, V.K. and Bullock, R.M. (2003) *Nature*, **424**, 530.
119 Clemente-León, M., Mingotaud, C., Agricole, B. et al. (1997) *Angewandte Chemie – International Edition*, **36**, 1114.
120 Clemente-León, M., Coronado, E., Delhaes, P. et al. (2001) *Advanced Materials*, **13**, 574.
121 Kurth, D.G., Volkmer, D., Kuttorf, M. and Müller, A. (2000) *Chemistry of Materials*, **12**, 2829.
122 Vasylyev, M.V. and Neumann, R. (2004) *Journal of the American Chemical Society*, **126**, 884.
123 Adam, W., Alsters, P.L., Neumann, R. et al. (2003) *The Journal of Organic Chemistry*, **68**, 1721.
124 Yamada, Y.M.A., Ichinohe, M., Takahashi, H. and Ikegami, S. (2001) *Organic Letters*, **3**, 1837.
125 Yamada, Y.M.A., Tabata, H., Ichinohe, M. et al. (2004) *Tetrahedron*, **60**, 4087.
126 Schild, H.G. (1992) *Progress in Polymer Science*, **17**, 163.
127 Bergbreiter, D.E. (2002) *Chemical Reviews*, **102**, 3345.

128 Hamamoto, H., Suzuki, Y., Yamada, et al. (2005) *Angewandte Chemie – International Edition*, **44**, 4536.

129 Haimov, A., Cohen, H. and Neumann, R. (2004) *Journal of the American Chemical Society*, **126**, 11762.

130 Haimov, A. and Neumann, R. (2008) *Journal of the American Chemical Society*, **128**, 15697.

131 MacMonagle, J.B. and Moffat, J.B. (1984) *Journal of Colloid and Interface Science*, **101**, 479.

132 Chaudhuri, P. and Wieghardt, K. (1987) *Progress in Inorganic Chemistry, Vol 47*, **35**, 329.

133 Uchida, S., Hikichi, S., Akatsuka, T. et al. (2007) *Chemistry of Materials*, **19**, 4694.

134 Uchida, S. and Mizuno, N. (2003) *Chemistry – A European Journal*, **9**, 5850.

135 Neumann, R. and Levin, M. (1991) *The Journal of Organic Chemistry*, **56**, 5707.

136 Fujibayashi, S., Nakayama, K., Hamamoto, et al. (1996) *Journal of Molecular Catalysis A*, **110**, 105.

137 Hanyu, A., Sakurai, Y., Fujibayashi, S. et al. (1997) *Tetrahedron Letters*, **38**, 5659.

138 Xu, L., Boring, E. and Hill, C.L. (2000) *Journal of Catalysis*, **195**, 394.

139 Gall, R.D., Hill, C.L. and Walker, J.E. (1996) *Journal of Catalysis*, **159**, 473.

140 Gall, R.D., Hill, C.L. and Walker, J.E. (1996) *Chemistry of Materials*, **8**, 2523.

141 Kharat, A.N., Pendleton, P., Badalyan, A. et al. (2001) *Journal of Molecular Catalysis A – Chemical*, **175**, 277.

142 Heravi, M.M., Zadsirjan, V., Bakhtiari, K. et al. (2007) *Catalysis Communications*, **8**, 315.

143 Khenkin, A.M., Neumann, R., Sorokin, A.B. and Tuel, A. (1999) *Catalysis Letters*, **63**, 189.

144 Yadav, G.D. and Bhagat, R.D. (2004) *Organic Process Research & Development*, **8**, 879.

145 Neumann, R., Khenkin, A.M. and Vigdergauz, I. (2000) *Chemistry – A European Journal*, **6**, 875.

146 Okun, N.M., Anderson, T.M. and Hill, C.L. (2003) *Journal of the American Chemical Society*, **125**, 3194.

147 Okun, N.M., Ritorto, M.D., Anderson, T.M. et al. (2004) *Chemistry of Materials*, **16**, 2551.

148 Okun, N.M., Anderson, T.M. and Hill, C.L. (2003) *Journal of Molecular Catalysis A – Chemical*, **197**, 283.

149 Bar-Nahum, I., Khenkin, A.M. and Neumann, R. (2004) *Journal of the American Chemical Society*, **126**, 10236.

150 Elliott, C.J. (1994) *Structure and Chemistry of the Apatites and Other Calcium Orthophosphates*, Elsevier, Amsterdam.

151 Yamaguchi, K., Mori, K., Mizugaki, T. et al. (2000) *Journal of the American Chemical Society*, **122**, 7144.

152 Mori, K., Yamaguchi, K., Hara, T. et al. (2002) *Journal of the American Chemical Society*, **124**, 11572.

153 Mori, K., Hara, T., Mizugaki, T. et al. (2003) *Journal of the American Chemical Society*, **125**, 11460.

154 Mori, K., Hara, T., Mizugaki, T. et al. (2004) *Journal of the American Chemical Society*, **126**, 10657.

155 Ichihara, J. (2001) *Tetrahedron Letters*, **42**, 695.

156 Ichihara, J., Yamaguchi, S., Nomoto, T. et al. (2002) *Tetrahedron Letters*, **43**, 8231.

157 Sasaki, Y., Ushimaru, K., Iteya, K. et al. (2004) *Tetrahedron Letters*, **45**, 9513.

158 Iteya, K., Ichihara, J., Sasaki, Y. and Itoh, S. (2007) *Catalysis Today*, **111**, 349.

159 Maayan, G. and Neumann, R. (2005) *Chemical Communications*, 4595.

160 Haimov, A. and Neumann, R. (2002) *Chemical Communications*, 876.

161 Neumann, R. and Cohen, M. (1997) *Angewandte Chemie – International Edition*, **36**. 1738.

162 Hoegaerts, D., Sels, B.F., de Vos, D.E. et al. (2000) *Catalysis Today*, **60**, 209.

163 Gelbard, G., Breton, F., Quenard, M. and Sherrington, D.C. (2000) *Journal of Molecular Catalysis A – Chemical*, **153**, 7.

164 Duprey, E., Maquet, J., Man, P.P. et al. (1995) *Applied Catalysis A*, **128**, 89.
165 Liu, Y., Murata, K. and Inaba, M. (2004) *Green Chemistry*, **6**, 510.
166 Villa, A.L., Sels, B.F., de Vos, D.E. and Jacobs, P.A. (1999) *The Journal of Organic Chemistry*, **64**, 7267.
167 Sels, B.F., Villa, A.L., Hoegaerts, D. et al. (2000) *Topics in Catalysis*, **13**, 223.
168 Wight, A.P. and Davis, M.E. (2002) *Chemical Reviews*, **102**, 3589.
169 Yamaguchi, K., Yoshida, C., Uchida, S. and Mizuno, N. (2005) *Journal of the American Chemical Society*, **127**, 530.
170 Kasai, J., Nakagawa, Y., Uchida, S. et al. (2006) *Chemistry – A European Journal*, **12**, 4176.
171 Yamaguchi, K., Imago, T., Ogasawara, Y. et al. (2006) *Advanced Synthesis and Catalysis*, **348**, 1516.
172 Neumann, R. and Miller, H. (1995) *Journal of the Chemical Society, Chemical Communications*, 2277.
173 Chen, L., Zhu, K., Bi, L.-H. et al. (2007) *Inorganic Chemistry*, **46**, 8457.
174 Nozaki, C.K., Tanabe, A., Negishi, S. et al. (2005) *Chemistry Letters*, **34**, 238.
175 Kholdeeva, O.A., Vanina, M.P., Timofeeva, M.N. et al. (2004) *Journal of Catalysis*, **226**, 363.
176 Bordoloi, A., Lefebvre, F. and Halligudi, S.B. (2007) *Journal of Catalysis*, **247**, 166.
177 Cavani, F., Trifiró, F. and Vaccari, A. (1991) *Catalysis Today*, **11**, 173.
178 Narita, E., Kaviratna, P. and Pinnavaia, T.J. (1991) *Chemistry Letters*, 805.
179 Kwon, T., Tsigdinos, G.A. and Pinnavaia, T.J. (1988) *Journal of the American Chemical Society*, **110**, 3653.
180 Carriazo, D., Lima, S., Martín, C. et al. (2007) *Journal of Physics and Chemistry of Solids*, **68**, 1872.
181 Tatsumi, T., Yamamoto, K., Tajima, H. and Tominaga, H. (1992) *Chemistry Letters*, 815.
182 Sels, B.F., de Vos, D.E., Buntinx, M. et al. (1999) *Nature*, **400**, 855.
183 Sels, B.F., de Vos, D.E. and Jacobs, P.A. (2001) *Journal of the American Chemical Society*, **123**, 8350.
184 Sels, B.F., de Vos, D.E. and Jacobs, P.A. (2005) *Angewandte Chemie – International Edition*, **44**, 310.
185 Sloboda-Rozner, D., Alster, P.L. and Neumann, R. (2003) *Journal of the American Chemical Society*, **125**, 5280.
186 Sloboda-Rozner, D., White, P., Alster, P.L. and Neumann, R. (2004) *Advanced Synthesis and Catalysis*, **346**, 339.
187 Maayan, G., Fish, R.H. and Neumann, R. (2003) *Organic Letters*, **5**, 3547.
188 Liu, L., Chen, C., Hu, X. et al. (2008) *New Journal of Chemistry*, **32**, 283.
189 Bhilare, S.V., Deorukhkar, A.R., Darvatkar, N.B. et al. (2007) *Journal of Molecular Catalysis A – Chemical*, **270**, 123.
190 Rasalkar, M.S., Bhilare, S.V., Deorukhkar, A.R. et al. (2007) *Canadian Journal of Chemistry*, **85**, 77.

7
Nitrous Oxide as an Oxygen Donor in Oxidation Chemistry and Catalysis

Gennady I. Panov, Konstantin A. Dubkov, and Alexander S. Kharitonov

7.1
Introduction

Dioxygen is an ideal oxidant for oxidation reactions: it is ecologically benign, cheap and with unlimited and renewable supply. Oxygen is the most available element on the Earth. Among more than 100 elements composing our planet the amount of oxygen alone represents nearly 54 atomic% [1]. One may say that we live in the world of oxygen, and therefore it is not surprising that oxygen reactions play such an outstanding role in this world, especially in nature, where they are controlled by enzymes – unique catalysts perfected by nature over millions of years.

Although man-made catalysts are not as perfect as enzymes, oxidation reactions play an important role in our technical activity, too. They are widely practiced in industry and are thoroughly studied in academic and industrial laboratories. During recent decades, major achievements have been attained in this field, resulting in the development of a significant number of new selective oxidation processes using O_2 [2–7]. But if we compare this number with the great number of desired and possible oxidation reactions, which we fail to perform, it will be evident that our achievements in this difficult field are quite modest, regardless of many efforts.

Oxidation by dioxygen has a fundamental difficulty. The molecule has a diatomic structure, while in most cases only one atom is needed for selective oxidation of organic compounds. Even in the case of more complex reactions, the stoichiometry of which requires several (and sometimes many) oxygen atoms, the oxidation process on a catalyst surface is likely to proceed step by step, involving consecutively one oxygen atom after another.

The creation of selective catalysts for such complex reactions seems to be an especially difficult problem. Nevertheless, surprisingly, selective catalysts have been developed for complex reactions, which can be exemplified by the oxidation and ammoxidation of propylene, oxidation of butene and even butane to maleic anhydride (which requires seven oxygen atoms). Such reactions are usually performed over V and Mo oxide systems [4, 6, 8–10]. High selectivity of these systems is presumably provided by a special structure of the catalyst surface that allows control

Modern Heterogeneous Oxidation Catalysis: Design, Reactions and Characterization
Edited by Noritaka Mizuno
Copyright © 2009 WILEY-VCH Verlag GmbH & Co. KGaA, Weinheim
ISBN: 978-3-527-31859-9

of the number of oxygen atoms supplied by the active site (or sites) for the oxidation of one molecule of the reactant.

However, attempts to develop similar selective catalysts failed in the case of reactions that require one oxygen atom, like the oxidation of methane, ethane and other alkanes to alcohols, aromatic compounds to phenols, alkenes to epoxides, and many others. These mechanistically simple reactions assume one difficult condition: the presence of active sites that upon obtaining two atoms from gas-phase O_2 can transfer only one of them to the molecule to be oxidized, reserving the second atom for the next catalytic cycle with another molecule. This problem remains a hard challenge for chemical catalysis.

A new possibility for solving this problem is suggested by monatomic oxygen donors. High valence metal oxides as well as various organic peroxides have long been in use as stoichiometric oxidants in organic synthesis. However, these oxidants have significant drawbacks: beside their high cost they give a large amount of wastes. This is incompatible with the modern strategy of green chemistry, which aims, finally, to develop zero-waste technologies [11, 12]. This implies the use of ecologically benign oxygen donors. Among them, the most attractive are hydrogen peroxide (H_2O_2) and nitrous oxide (N_2O). Upon donating the oxygen, they turn into H_2O and N_2, which are natural components of the environment.

This chapter is devoted to nitrous oxide. Extensive studies in the field carried out during the two last decades have uncovered the great potential of this remarkable oxygen donor. Beside catalytic reactions, we shall discuss also a very interesting type of thermal oxidation by N_2O to draw the attention of the catalytic community to this new field of oxidation chemistry that is not covered yet by catalysis.

7.2
Molecular Structure and Physical Properties of Nitrous Oxide

N_2O is a linear molecule, with an electron structure that can be described by the following three resonance forms [13]:

$$:N\equiv \overset{+}{N}-\overset{..}{\underset{..}{O}}:^{-} \longleftrightarrow ^{-}\overset{..}{\underset{..}{N}}=\overset{+}{N}=\overset{..}{\underset{..}{O}} \longleftrightarrow :\overset{+}{N}=N-\overset{..}{\underset{..}{O}}:^{-} \qquad (7.1)$$
$$A B C$$

The N-oxide form A provides the main contribution to the structure. Table 7.1 presents the effective atomic charges in form A. An opposite distribution of charges in the N_2O resonance forms explains the relatively low dipole moment of the molecule (1.161 dB). Bond lengths in N_2O are $N-N = 1.128$ Å and $N-O = 1.184$ Å, which correspond to bond orders of 2.73 and 1.61, respectively [13].

Form C is a 1,3-dipole. Owing to this feature N_2O can perform a very specific type of oxidation involving its cycloaddition to alkenes, which will be discussed later.

Nitrous oxide was discovered by J. Priestly at the end of the eighteenth century. In 1800, aged 21, H. Davy published a 580-page book entitled *Nitrous Oxide, or De-Phlogisticated Nitrous Air, and its Respiration*, in which he described his experiments

Table 7.1 Effective charges of nitrogen and oxygen atoms in N_2O [from A in (7.1)].

Calculation method	q_N	$q_{N(O)}$	q_O	Reference
CNDO/2	−0.14	0.47	−0.33	[14]
4-31G	−0.07	0.62	−0.55	[15]
MP3	−0.11	0.44	−0.33	[16]

with the physiological effects of N_2O [17]. Since N_2O inhalation left people in a joyful mood, he called it "laughing gas." Thanks to a light narcotic effect, nitrous oxide is widely used as anesthetic, especially in maternity homes.

Thermodynamically, N_2O is an endothermic molecule ($\Delta H^\circ_{f,298} = +82.1$ kJ mol^{-1}), which, however, is quite stable kinetically. Its thermal decomposition to the elements:

$$2\, N_2O \rightarrow 2\, N_2 + O_2 \tag{7.2}$$

becomes noticeable above 873 K.

Nitrous oxide is non-explosive itself, but can form flammable mixtures with organic compounds. This should be taken into account for providing a safe working environment when using N_2O oxidant. Table 7.2 gives the flammability limits of some compounds in mixtures with N_2O. These data are quite scarce and reported

Table 7.2 Flammability limits of some organic compounds in the mixture with N_2O at an initial pressure 0.1 MPa.

Compound	Initial temperature (K)	Concentration limits (vol.%)		Reference
		Lower	Upper	
Methane	293	3.9	40	[18]
	293	—	46.5	[19]
	473	—	49	[19]
Propane	293	2.1	24.8	[18]
Butane	293	1.8	21.0	[18]
		0.17	21.5	[20]
Propylene	293	1.8	26.8	[18]
Cyclopropane	293	1.6	30.3	[21]
Diethyl ether	293	1.5	24.2	[21]
Cyclohexene	323	—	19.6	[19]
	373	—	18.6	
Benzene	373	1.6	26.0	[19]
	423	1.25	26.5	[19]
	473	0.75	23	[19]
Xylene	293	0.19	13	[20]
Phenol	493	1.75	29.5	[19]

Table 7.3 Some physicochemical properties of N_2O.

Melting point	182.14 K
Boiling point	184.52 K
Critical temperature	309.43 K
Critical pressure	7.17 MPa
Critical density	0.452 g cm^{-3}
Density as liquid	1.226 g cm^{-3} (at 184.5 K)
Enthalpy of formation, $\Delta H_{f,298}^\circ$	82.1 kJ mol^{-1}
Free energy of formation, $\Delta G_{f,298}^\circ$	103.1 kJ mol^{-1}
Entropy, S°	220.1 J/K^{-1} mol^{-1}

mainly in publications that are not easily accessible. The limits with N_2O are somewhat narrower than in the mixtures with O_2, but broader than in the mixtures with air.

Table 7.3 gives some physicochemical parameters of N_2O [17].

Having a high critical temperature, nitrous oxide can exist as a liquid at room temperature (5.01 MPa at 294 K). Commonly, it is stored and transported in the liquefied state under pressure in steel cylinders or refrigerated tanks.

7.3
Catalytic Oxidation by Nitrous Oxide in the Gas Phase

7.3.1
Oxidation of Lower Alkanes Over Oxide Catalysts

Nitrous oxide first attracted considerable attention in the 1980s in relation to the search for ways for the oxidative conversion of methane into valuable chemical products. Although practical application of N_2O for solving this problem was economically not feasible, such studies may be useful for better understanding the activation mechanism of methane, thus improving the possibility of getting better results with O_2.

Numerous works on the oxidation of methane to methanol and/or formaldehyde as well as on the oxidative dimerization of methane were reviewed by many authors [22–27]. First, high selectivity of methane oxidation by N_2O was reported by Lunsford et al. [28–30]. Over a supported Mo oxide [30], the total selectivity to methanol and formaldehyde at low methane conversions attained 100%, although this rapidly dropped as the conversion increased (Table 7.4). High selectivity for this reaction was obtained also with supported vanadium oxide [31].

When N_2O is replaced by O_2, the selectivity sharply decreases. This is the conclusion of all researchers in the field [31–34]. As for the reactivity of these oxidants, opinions differ. Thus, Zhen et al. [31] and Barbaux et al. [34] observed a higher methane conversion in the presence of O_2, while Chen and Wilcox [33], conversely, saw greater conversion in the presence of N_2O.

Table 7.4 Methane oxidation by N_2O over 1.7% MoO_3/SiO_2[a)].

Temperature (K)	CH_4 conversion (%)	Selectivity (%)			
		CH_3OH	HCHO	CO	CO_2
823	1.6	20.5	79.5	—	—
833	1.9	19.9	80.1	—	—
843	2.9	13.8	64.3	19.1	2.8
853	4.0	10.0	58.8	27.7	3.4
867	6.0	7.8	49.5	38.1	4.6

[a)] Reaction conditions: catalyst weight = 1 g; GHSV = 4400 L kg-h^{-1}; feed composition: 10% CH_4, 37% N_2O, 34% H_2O, He balance.

Lunsford and coauthors suggested the following reaction mechanism [30], which is now generally accepted:

$$Mo^V + N_2O \rightarrow Mo^{VI}O^- + N_2 \quad (7.3)$$

$$Mo^{VI}O^- + CH_4 \rightarrow Mo^{VI}OH^- + {}^\bullet CH_3 \quad (7.4)$$

$$Mo^{VI}O^{2-} + {}^\bullet CH_3 \rightarrow Mo^V OCH_3^- \quad (7.5)$$

$$Mo^V OCH_3^- + H_2O \rightarrow Mo^V OH^- + CH_3OH \quad (7.6)$$

$$Mo^{VI}O^{2-} + Mo^V OCH_3^- \rightarrow Mo^V OH^- + HCOH + Mo^{IV} \quad (7.7)$$

$$Mo^{VI}OH^- + Mo^V OH^- \rightarrow Mo^V + Mo^{VI}O^{2-} + H_2O \quad (7.8)$$

The reaction is initiated by the formation of anion radical O^-, which then activates a methane molecule by the abstraction of a hydrogen atom. The resulting methyl radical reacts with the surface oxygen of oxide to form a methoxy complex. This complex, reacting with water, gives methanol, or upon reacting with another atom of surface oxygen gives formaldehyde. The desorption of H_2O closes the catalytic cycle.

There are some arguments for the involvement of O^- radicals in the selective oxidations of methane. The O^- is registered by ESR upon adsorption of N_2O on partially reduced surfaces of supported Mo and V oxides [28, 35]. The radical exhibits a very high reactivity. At room temperature, it readily interacts with methane to yield methanol as the main product, which can be registered by the IR *in situ* [36] or desorbed from the catalyst surface by heating [29, 30].

Unlike N_2O, adsorption of O_2 leads not only to the O^- species, but also to O_2^- species. Many authors consider the latter radical as a reason for decreased selectivity in the presence of O_2.

Beside V_2O_5 and MoO_3, many other catalytic systems were tested in selective oxidation of methane by N_2O, including various metal oxides [37, 38], zeolites [39], phosphates [40, 41] and polyoxometalates [42].

Beside methanol and formaldehyde, the oxidation of methane may be directed to another route, leading to the formation of its condensation products, for example, ethane, ethylene and benzene. This route may provide an alternative way for the chemical use of natural sources of methane. Here, various catalysts were also tested using both O_2 and N_2O as the oxidants [22]. The general picture observed by most authors was similar to that with methane oxidation to oxygenates. The conversion of methane was always higher with O_2 than with N_2O. However, the selectivity to the coupling products showed an opposite trend.

An interesting effect of mechanical activation of supported V and Mo oxides on the oxidation of methane with N_2O was reported by Firsova et al. [43].

Beside methane, nitrous oxide was repeatedly used for selective oxidation of ethane [40, 44–47]. Along with the oxygenation (mainly to acetaldehyde), oxidative dehydrogenation proceeds also in this case, yielding ethylene. Similar to the case of methane, Mo oxide supported on SiO_2 proved to be among the most effective systems. At 3% ethane conversion, total selectivity to acetaldehyde and ethylene attains 85% [44]. A similar performance was obtained with polyoxometalates containing V, Mo or W [45]. With virtually all the catalytic systems studied, a common regularity was observed: when nitrous oxide is replaced with dioxygen, a considerable drop in the selectivity occurs due to the intense process of deep oxidation.

To summarize, N_2O studies have contributed significantly to a better understanding of the activation mechanism of lower alkanes. This facilitates progress in the oxidation by O_2, which is being made step by step [49–51].

7.3.2
Oxidation Over Zeolites

7.3.2.1 Oxidation by Dioxygen

Zeolites are aluminosilicate materials having a very specific feature, that is, an intracrystalline micropore system of a molecular size. When considering the application of zeolites in oxidation catalysis, one should take into account that oxidation reactions are usually catalyzed by transition metals. The oxidation activity is provided by a more or less easy electron transfer that results in a reversible change of the oxidation state of the metal at its interaction with dioxygen or the substance to be oxidized. Zeolites with their aluminosilicate composition cannot function as redox catalysts, but their application as supports has always been of great interest. The regular structure of zeolites, high adsorption potential, availability of the micropores as well as of Brønsted and Lewis acid sites may provide various states of a supported component, especially favoring the formation of small clusters and isolated metal ions. Such a variety is hardly probable with other systems and seems to promise particular prospects for oxidation catalysis.

The results of many studies carried out with zeolites, mostly in the 1970s and reviewed in ref. [52], did not meet these expectations. Zeolites cannot compete with conventional catalytic systems both in the selective and complete oxidations. The main reason is that the transition metals introduced into a zeolite matrix lose their ability to activate dioxygen, as was evidenced by isotopic O_2 exchange [53, 54]. The decrease of atomic catalytic activity may be several orders of magnitude if compared to bulk oxides. To rationalize this result, one should take into account that O_2 activation proceeds by its dissociative adsorption on the catalyst surface, requiring two neighboring sites and the transfer of four or, at least, two electrons from the surface to the oxygen molecule. This can hardly occur in the case of small clusters and is even more unlikely in the case of isolated metal ions.

But the situation is dramatically changed if monooxygen donors are used instead of dioxygen. The two best known examples are titanosilicalites TS-1, which proved to be excellent catalysts for liquid-phase oxidation by H_2O_2, and FeZSM-5 zeolites, which are efficient catalysts for gas-phase oxidation by N_2O. Numerous works with H_2O_2 are well known and discussed in several reviews [55, 56]. We shall consider the oxidation by N_2O.

7.3.2.2 Oxidation of Benzene to Phenol by N_2O

The introduction of hydroxyl groups into an aromatic nucleus is one of the most difficult problems in the synthesis of fine chemicals and intermediates. Such hydroxylation is typically performed via multistage technologies, often using aggressive reagents. It can be exemplified by the simplest reaction of this type, that is, the oxidation of benzene to phenol, which is one of the most prominent bulk chemicals with a world production about 8 MMT per year. Various phenol processes used historically in industry have been reviewed in ref. [57]. Currently, phenol is mainly produced by the cumene process, which consists of three stages. Although the process is a well-established technology, it has important disadvantages: poor ecology, an explosive intermediate (cumene hydroperoxide) and acetone as by-product, the market for which is smaller than that of phenol.

An ideal solution would be the direct oxidation of benzene to phenol by dioxygen:

$$C_6H_6 + \tfrac{1}{2}O_2 \rightarrow C_6H_5OH \tag{7.9}$$

However, many attempts to accomplish this reaction have failed. The interaction with O_2 leads to destruction of the aromatic nucleus and low phenol selectivity.

Oxidation over vanadia catalyst In 1983, Iwamoto and co-authors [58] were the first to use N_2O for the oxidation of benzene to phenol:

$$C_6H_6 + N_2O \rightarrow C_6H_5OH + N_2 \tag{7.10}$$

This reaction over vanadia catalyst showed much better selectivity than before. At 823 K the selectivity exceeded 70%. This result was recognized as a promising lead to a long awaited direct process. However, a pilot test did not meet expectation since the selectivity proved to be too low for commercialization of the process.

Selectivity is among the most important parameters for the partial oxidation, especially in the case of non-O_2 oxidants. Usually, for O_2 oxidations, the selectivity is considered only with respect to the organic reagent. But this approach is not acceptable with monooxygen donors, the cost of which may be of the same order as that of the organic reagent. In this case, selectivity based on the oxidant becomes an equally important parameter for estimating process efficiency. Thus, in the case of benzene oxidation, one molecule of nitrous oxide is needed to convert it into phenol (7.10), but 15 molecules are needed for its complete oxidation:

$$C_6H_6 + 15N_2O \rightarrow 6CO_2 + 3H_2O + 15N_2 \qquad (7.11)$$

Therefore, at 70% benzene selectivity, the selectivity of N_2O is only 12–13%, which makes the process unfeasible.

Oxidation over zeolites The results of Iwamoto et al. on the oxidation of benzene to phenol [58] stimulated further efforts in the search for new and more efficient catalytic systems. As a result, in 1988, ZSM-5 zeolites were shown to be the best catalysts for this reaction [59–61]. Over zeolites, the reaction proceeded at much lower temperature and, which was even more important, with a very high selectivity, approaching 100%. Further studies involving many other groups [62–80] contributed much to the improvement of ZSM-5 catalysts. Some other type zeolites and $FePO_4$ were shown to be also active [69, 70, 73, 80, 81], although their efficiency was inferior to that of ZSM-5.

Based on FeZSM-5 zeolites, a new one-step phenol process (the AlphOx) has been developed jointly by Solutia Inc. and the Boreskov Institute of Catalysis [82]. The process was successfully tested with a pilot plant constructed at Solutia facilities in Pensacola (Florida). The process runs in an adiabatic reactor with the parameters shown in Table 7.5. It provides a 97–98% yield of phenol, with 100% N_2O conversion per pass and a recycle of benzene. A 1% yield of dihydroxybenzenes (DHB) is also obtained. This valuable by-product is mainly hydroquinone. Periodically, the catalyst is subjected to regeneration by burning-off coke deposits. Its lifetime is 1.5 years. More details on the process are given elsewhere [82, 83].

7.3.2.3 Nature of Zeolite Activity, α-Sites

Zeolite activity, being a quite untypical phenomenon for oxidation catalysis, gave rise to several fundamental questions. The origin of the activity caused most discussion.

Table 7.5 Performance of the AlphOx pilot plant adiabatic reactor.

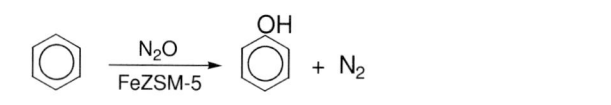

$$C_6H_6 \xrightarrow[\text{FeZSM-5}]{N_2O} C_6H_5OH + N_2$$

Reaction parameter	Value
Temperature (K)	673–723
Contact time (s)	1–2
Degree of conversion (mol.%):	
benzene to phenol	97–98
benzene to CO_x	0.2–0.3
benzene to DHBs	1
N_2O to phenol	85
Phenol productivity [kg (kg cat h)$^{-1}$]	0.4

Two different approaches were considered. One of them associated the activity with Brønsted and Lewis acid sites. However, experimental attempts to establish some correlation with the acidity have failed. The other approach related to the conventional redox catalysis, which supposedly may be provided by an admixed transition metal. After many efforts, this intriguing admixture has been identified [84–86]. It proved to be iron, which is present always in zeolites, at least at the level of a few hundred ppm, being introduced at the preparation step with the starting reagents. In the micropore space of the zeolite matrix, iron can form special very active complexes (called α-sites) that catalyze the reaction. Later, several groups [87–90] conducted most conclusive studies on this subject, including the preparation of virtually Fe-free zeolites, which proved to be quite difficult. Presently, the catalytic role of iron is generally recognized, although the detailed structure and composition of α-sites are not perfectly clear and are the subject of many theoretical [91–95] and experimental [81, 96–106] studies.

To provide effective performance in the oxidation of benzene to phenol, FeZSM-5 should have a rather small iron content, all of which is ideally in the form of α-sites. Iron is usually introduced at the zeolite synthesis step or in a post-synthetic way. To transform the iron into the active state, the zeolite should be activated by high-temperature treatment (870–1270 K) in steam, vacuum or inert atmosphere. Mössbauer spectroscopy has been used successfully to study the state of iron [81, 107–110]. Figure 7.1 shows the Mössbauer spectra of FeZSM-5 (0.31 wt% Fe) obtained after different degrees of activation, with a progressive increase of the α-sites concentration [108]. One can see that activation causes dramatic changes in the iron state, leading to its progressive reduction into a special bivalent state that makes up the α-sites, $Fe^{III} \rightarrow Fe^{II}_\alpha$. In this state, the iron is not oxidized by O_2 (even at 900–1000 K), but is readily oxidized by N_2O at 420–520 K, providing a trivalent state with stoichiometric deposition of anion radical species O^- bound to the iron [108, 111, 112]:

$$(Fe^{II})_\alpha + N_2O \rightarrow (Fe^{III} - O^-)_\alpha + N_2 \tag{7.12}$$

The O^- radicals on FeZSM-5 are called the α-oxygen, denoted further as O_α. Reaction (7.12) proceeds with an activation energy of 42 kJ mol^{-1} and is described by the following kinetic equation:

$$W_{N_2O} = C_\alpha \cdot k_0 e^{-42000/RT} \cdot P_{N_2O} \tag{7.13}$$

Desorption of α-oxygen into the gas phase starts at above 520 K. The concentration of α-oxygen (and hence the concentration of α-sites, C_α) can be reliably measured based on the amount of N_2 evolved by (7.12), the isotopic exchange of $^{18}O_2$ with O_α, or its reaction with CO at room temperature [113].

Figure 7.2 shows a supposed mechanism of α-site formation [83, 111], starting from an initial binuclear iron complex (A). The existence of such complexes was documented by many authors [81, 109, 110, 114–116]. Upon oxygen desorption at elevated temperature (step 1), complex A may transform into the reduced complex B. If the oxygen desorption is reversible, α-site formation does not occur and, upon cooling, the system returns to its initial state, restoring complex A. Such reversibility

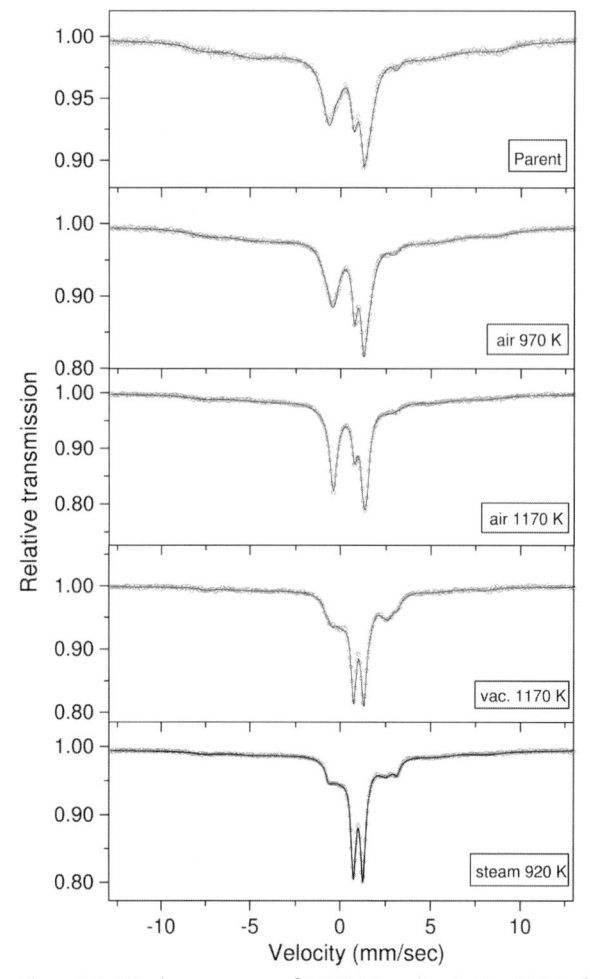

Figure 7.1 Mössbauer spectra of FeZSM-5 zeolite (0.31 wt% Fe) after various types of activation.

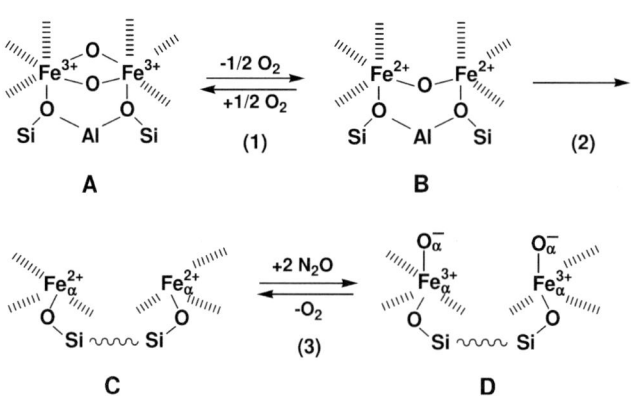

Figure 7.2 Supposed mechanism of α-site formation.

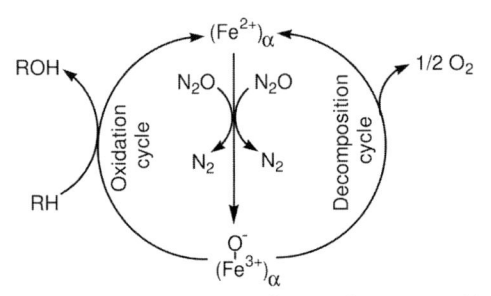

Figure 7.3 Redox transition of iron in the α-site enables catalytic cycles involving N_2O.

of the iron redox transformation was noted in several works with FeZSM-5 zeolites, especially with those having a high concentration of iron. For α-site formation to occur, the reduced Fe^{II} atoms should stabilize by step 2 into a new very stable state Fe^{II}_α (complex C), in which they lose the ability to react with O_2, thus making the step irreversible. The step of Fe^{II}_α stabilization is strongly facilitated by water vapor [68, 74, 77, 108]. The stabilization may occur due to a strong binding of Fe^{II}_α entities by silicate fragments of the zeolite matrix.

The reduced iron atoms of complex C, being inert to dioxygen, are readily oxidized by nitrous oxide into complex D to give adsorbed species of α-oxygen, O_α. As Figure 7.3 shows, the reversible redox transition $Fe^{II}_\alpha \leftrightarrow Fe^{III}_\alpha$ provides the catalytic activity of FeZSM-5: both the oxidation cycle due to the oxygen transfer from N_2O to a substrate and the decomposition cycle of N_2O into N_2 and O_2 due to recombination of α-oxygen into the gas phase. The decomposition is an environmentally important process, and FeZSM-5 zeolites are considered to be the best catalysts for this reaction (see review [117] and references therein).

7.3.2.4 N_2O specificity, α-Oxygen and its Stoichiometric Reactions

A specificity of N_2O oxidant compared to O_2 is one of the most interesting points arising from benzene oxidation over FeZSM-5 zeolites. The specificity is clearly seen from the results presented in Table 7.6 [118]. With nitrous oxide, benzene conversion is 27% at 623 K, whereas with dioxygen it is only 0.3% at 773 K. Moreover, the reaction route changes totally: N_2O leads to selective formation of phenol, while O_2 leads only to the products of complete oxidation.

It is reasonable to assume that such a strong specificity relates to the ability of N_2O to generate α-oxygen. Many studies were devoted to this remarkable species, including reactivity tests, IR-spectroscopy, isotope labeling and step-response experiments [95, 97–105, 119–122]. Calorimetric measurements showed that the bond energy of α-oxygen to the surface is about 250 kJ mol^{-1} [123]. This is a very low value compared to other types of oxygen bonds. It explains why O_α can be formed from the endothermic molecule N_2O ($\Delta H^\circ_{f,298} = 82.1$ kJ mol^{-1}), but cannot be formed from O_2, which has a bond energy of 494 kJ mol^{-1}.

Recently, interesting effects of NO and water on the formation and properties of α-oxygen were discovered. Even small amounts of NO facilitated desorption of O_α from the surface, strongly increasing the rate of catalytic decomposition of N_2O to the

Table 7.6 Effect of oxidant on the oxidation of benzene to phenol over FeZSM-5 and Fe_2O_3[a].

Catalyst	N_2O oxidant			O_2 oxidant		
	Reaction temperature (K)	Benzene conversion (%)	Phenol selectivity (%)	Reaction temperature (K)	Benzene conversion (%)	Phenol selectivity (%)
FeZSM-5 (0.055 wt% Fe)	623	27.0	98.0	773	0.3	0.0
Fe_2O_3	623	5.5	0.0	623	24.5	0.0

[a] Reaction conditions: 5% benzene, 20% N_2O (O_2), balance helium; contact time 2 s.

elements [117, 124–129]. Conversely, water vapor exhibits an opposite effect, causing deactivation of α-sites and an increase in the activation energy of N_2O decomposition [98, 101, 125, 130–132]. This poorly understood effect of water, observed sometimes at its admixed amount, may be a reason for some contradictory data in the literature concerning the formation and, especially, the reactivity of α-oxygen.

Like O^- radicals on V and Mo oxides discussed above, O_α exhibits a very high reactivity. At room temperature, it readily oxidizes various organic molecules, including methane. This allows one to conduct single turnover reactions on the catalyst surface, providing in particular the synthesis of phenol according to the following scheme [119]:

$$N_2O + (\ )_\alpha \xrightarrow{470-520\ K} (O^-)_\alpha + N_2 \quad (7.14)$$

$$C_6H_6 + (O^-)_\alpha \xrightarrow{295\ K} (C_6H_5OH)_\alpha \quad (7.15)$$

$$(C_6H_5OH)_\alpha \xrightarrow[\text{extraction}]{295\ K} C_6H_5OH + (\ )_\alpha \quad (7.16)$$

This scheme includes α-oxygen loading (7.14), its interaction with benzene at room temperature (7.15) and product extraction from the catalyst surface (7.16). A nearly theoretical yield of phenol was obtained, with no other products detected. These results proved clearly the α-oxygen participation, which was confirmed additionally by isotopic experiments using $^{18}O_\alpha$ [119, 133].

Similar stoichiometric reactions can be conducted with other organic substrates. Beside mechanistic importance, such reactions are a convenient way for estimating the potential of α-oxygen oxidation. For that, various organic substrates were tested for their room temperature interaction with α-oxygen to identify the primary oxidation products extracted from the surface. Substrates included alkanes, cycloalkanes, alkenes and aromatics [121, 122]. Analysis of products showed that in all cases selective formation of hydroxylated compounds took place.

Reactions (7.14)–(7.16) represent the main mechanistic steps not only of the stoichiometric, but also of the steady state catalytic oxidation of benzene to phenol. The involvement of α-oxygen in the catalytic oxidation is most convincingly evidenced by a linear dependence of the reaction rate on the concentration of α-sites [134, 135].

7.3.2.5 Hydroxylation of Alkanes and Benzene Derivatives

Selective oxidation of alkanes and benzene derivatives to alcohols and phenols, respectively, are among the most difficult reactions in oxidation catalysis. Therefore, the stoichiometric hydroxylation of alkenes and aromatics performed by α-oxygen at room temperature has aroused great interest as a potential way for developing new steady state catalytic processes for the preparation of these valuable products, similar to the hydroxylation of benzene to phenol.

The catalytic hydroxylation of methane [39, 95], ethane [136, 137], propane [138–141], cyclohexane [68] and cyclododecane [142] has been tested using the Fe-containing zeolites. In most cases the selectivity was low or even zero. With ethane and propane, instead of their hydroxylation to alcohols, oxidehydrogenation (ODH) takes place, yielding ethylene and propylene, respectively, in yields of up to 25–30%. According to Bulanek *et al.* [140] the ODH of propane proceeds via intermediate formation of propanol, which further is subjected to dehydration.

The hydroxylation of benzene derivatives proceeds much more selectively. Table 7.7 presents the main published results. With some substrates like chloroben-

Table 7.7 Hydroxylation of benzene derivatives by N_2O over Fe-containing zeolites.

Substrate	Temperature (K)	Substrate conversion (%)	Selectivity to hydroxylated products (%)	Reference
Toluene	623	24	27	[143]
	593	36.6	41.5	[144]
	723	25	40	[145]
	623	22	19	[146]
	623	3	72	[147]
Isopropylbenzene	623	35	20	[146]
para-Xylene	623	30	~1	[146]
Chlorobenzene	603	6.7	100	[59]
	623	23	58	[143]
	623	11	74	[146]
Fluorobenzene	623	9.2	90.4	[60]
	673	45	94	[63]
	623	19	68	[143]
Phenol	623	11.5	92	[118]
	748	8	97	[148]
	623	7	95	[146]
Benzaldehyde	573	12	0	[143]
	623	17	~1	[146]
Nitrobenzene	623	5	0	[143]
Anisole	623	53	~1	[143]
Aniline	773	13	29	[143]
Difluorobenzenes	670	25–30	80–85	[149]
1,3,5-Trimethylbenzene	623	0	—	[68]
Biphenyl	623	6	70	[143]
	673	12	95	[150]
Naphthalene	623	1	43	[143]

zene, fluorobenzene, phenol and biphenyl the selectivity is over 90–95%. However, in many cases the selectivity appeared to be much lower. Similar to the case with alkanes, the reason is a difficult desorption of the primary hydroxylated products into the gas phase. For that one needs to elevate the temperature, which may lead to degradation of the products as well as to catalyst deactivation due to coking. Consequently, further studies and improvements are needed in this promising field.

With alkylaromatics, Costine et al. [146] observed a competition effect between the aromatic and aliphatic attack of the N_2O that was influenced by the steric and electronic factors. A similar effect has been observed by Rodkin et al. [121] for the stoichiometric hydroxylation of alkylaromatics by α-oxygen at room temperature.

7.3.2.6 Other Types of Oxidation Reactions

Recently, some other oxidation reactions using N_2O oxidant such as the ammoxidation and epoxidation were successfully conducted. The ammoxidation of propane proceeds with rather high selectivity over FeZSM-5 zeolite [151]. Remarkably, the reaction most effectively proceeds in the presence of a N_2O–O_2 mixture.

The gas-phase epoxidation of propylene with N_2O seems to be a particularly interesting reaction. This reaction was first demonstrated by Duma and Hönicke using a sodium-modified Fe/SiO_2 [152]. Later, modification by rubidium was found to be more effective, providing up to 90% selectivity to propylene oxide [153]. High catalytic performance of Fe/SBA-15 modified by KCl was discovered by X. Wang et al. [154, 155]. This catalyst provided 65–72% selectivity at 3–5% propylene conversion. Costine et al. [147] showed that epoxidation of 1-butene is also possible over this catalyst. The above results may be a promising lead for developing new processes for the production of these valuable epoxides.

The problem of catalyst deactivation should be noted for all these cases. The epoxidation selectivity is usually calculated without taking into account the consumption of the reagents for coke formation. As Thömmes et al. showed recently with a Fe/SiO_2 catalyst [156], coking may significantly reduce the selectivity of the reaction.

Although the epoxidation by nitrous oxide proceeds over non-zeolite catalysts, they also include iron as an active element. One may think that in all these cases a special oxygen species generated by N_2O plays an important role, similar to the α-oxygen on FeZSM-5.

Among other N_2O reactions, one may mention the ODH of ethylbenzene to styrene over metal-modified mesoporous silica systems [157]. Fe-modified catalysts showed the best performance, providing 90% selectivity at 30% conversion.

7.4
Catalytic Oxidation by N_2O in the Liquid Phase

Technologically, gas-phase processes are more preferable than their liquid-phase counterparts. However, the gas-phase performance usually requires high temperatures and is feasible only with comparatively low-boiling and thermally stable reactants. Therefore, most organic syntheses are conducted in the liquid phase, with the oxidants represented by such active oxygen donors as H_2O_2, alkyl peroxides, peracids and high-valence metal oxides. The successful results of nitrous oxide in gas-phase oxidations has stimulated efforts to use it in various liquid-phase reactions, with transition metal complexes as catalysts [158–166]. Potentially the application of N_2O in the liquid phase clearly offers economic and environmental advantages. Both nitrous oxide itself and the product of its reaction, N_2, are easily removed from the reaction zone and do not leave impurities that are typical of all other oxygen donors.

Studies in this field are just beginning, and the number of publications hardly exceeds a dozen. The most interesting results were obtained by the research groups of Yamada [160–162], Neumann [163, 164] and Kozhevnikov [165, 166]. Using various type catalysts (Ru porphyrene complexes, polyoxometalates, supported metals), the authors conducted selective oxidations of various types. These include epoxidation of alkenes, oxidation of alcohols, oxidation of alkylaromatics, oxidation and aromatization of dihydroanthracenes, and some other reactions. The experiments were typically conducted at 373–423 K under 1.0 MPa pressure of nitrous oxide.

All the authors concluded that N_2O provides a very high selectivity, which is higher than generally achieved with H_2O_2 or O_2. In some cases, a virtually quantitative yield of the target product was obtained. This can be exemplified by the N_2O epoxidation of cholesteryl benzoate [162]:

$$\text{BzO-cholesteryl} \xrightarrow[\text{Ru(por)(O)}_2]{N_2O} \text{BzO-cholesteryl-}\beta\text{-epoxide} \tag{7.17}$$

The reaction was conducted with 0.20 mmol of cholesteryl benzoate and 5.0 mol % Ru(TMP)(O)$_2$ catalyst in fluorobenzene solvent. At the reaction temperature of 413 K and under 1.0 MPa N_2O pressure, the catalyst provided a 99% yield with >99% selectivity.

At the same time, N_2O proved to be a poor ligand and quite an inert oxidant. It resulted in a low reaction rate and needs a high reaction temperature. The lack of catalysts that could provide an effective activation of N_2O is the main factor limiting a widespread use of nitrous oxide in liquid-phase oxidations. The development of such catalysts is an important target in this field.

7.5
Non-Catalytic Oxidations by N_2O

7.5.1
Liquid-Phase Oxidation of Alkenes

The first extensive attempt to use nitrous oxide as a selective oxidant in the liquid phase was made in the early 1950s by ICI researchers Bridson-Jones et al. [167]. The gas-phase thermal combustion of hydrocarbons in an N_2O atmosphere had long been known. The idea of the authors was to try to decrease the reaction temperature so as to increase the probability of partial oxidation products. To keep the reaction rate at a measurable level, the temperature decrease was compensated by a strongly increased pressure. Typical reaction conditions were: temperature 573 K, pressure 50 MPa and no catalyst.

The authors [167] tested a significant number of organic substrates. The oxidation of alkenes leading to aldehydes and ketones was the most interesting discovery. However, the reaction selectivity was quite low, not exceeding 65% in best cases. Because of very the harsh reaction conditions, which are difficult to provide in laboratory practice, these modest results did not stimulate further studies and then virtually dropped out of the researchers' sight.

In 2002, this type of N_2O oxidation was re-discovered by Panov et al. [168]. Being guided by quite a different idea, the authors [168] used milder conditions and obtained much better selectivity, which in many cases exceeded 90%. Such a high selectivity was shown to relate to a non-radical type reaction mechanism as well as to a remarkable feature of the oxidant. N_2O reacts solely with alkene C=C bonds and is inert towards all other bonds. Therefore, reaction products having no double bonds are not subjected to overoxidation. Only non-oxidation side processes may be a reason for decreasing selectivity.

The ability of nitrous oxide to form a 1,3-dipole (Section 7.2) seems to be of critical importance for the reaction with alkenes. The oxygen transfer proceeds via the 1,3-dipolar cycloaddition mechanism, assuming intermediate formation of a 1,2,3-oxadiazoline complex, the decomposition of which leads to a carbonyl compound:

$$\text{\textbackslash C=C/} + \overset{+}{N}=N-O^- \longrightarrow \left[\begin{array}{c} N \overset{N}{\diagup} O \\ C-C \end{array} \right] \longrightarrow -\overset{O}{\overset{\|}{C}}-C\diagdown + N_2 \quad (7.18)$$

This mechanism, first suggested by Bridson-Jones et al. [167], explains all experimental results and recently was strongly supported by quantum chemical calculations [169–171].

Since the carbonyl groups are formed by oxidation, this reaction type was called "carboxidation" [83, 172]. In particular cases, depending on the main carbonyl product (ketone or aldehyde), one may call the reaction more specifically, that is, "ketonization" or "aldehydization."

Below we give an overview of carboxidation results published for alkenes of various types [168, 173–175].

7.5.1.1 Linear Alkenes

Carboxidation of linear alkenes is presented in Table 7.8 [173]. The location of the C=C bond and configuration of intermediate oxadiazoline complex are two important parameters with these substrates. For terminal alkenes, the complex may have configuration I and II (Figure 7.4). Complex I has the oxygen bound to the first carbon atom, and its decomposition leads to an aldehyde. Complex II has the oxygen bound to the second carbon atom, and its decomposition leads to a ketone. However, decomposition of the latter complex may occur also in a different way, involving

Table 7.8 Carboxidation of linear alkenes[a].

Entry	Alkene	Conversion (%)	Contribution of cleavage route (%)	Product (composition in mol.%)
Terminal alkenes				
1	Ethylene	27	7	Acetaldehyde (91) Cyclopropane (4) Cycloheptatriene (3)
2	Propylene	26	29	Propanal (23) Acetone (31) Acetaldehyde (22) Methylcyclopropane (4) Cycloheptatriene (15)
3	1-Butene	38	38	Butanal (13) Methyl ethyl ketone (34) Propanal (29) Ethylcyclopropane (13) Cycloheptatriene (9)
4	1-Hexene	35	38	Hexanal (15) 2-Hexanone (29) Pentanal (27) Butylcyclopropane (14) Cycloheptatriene (12)
5	1-Octene	30	39	Octanal (13) 2-Octanone (28) Heptanal (26) Hexylcyclopropane (9) Cycloheptatriene (18)
Internal alkenes				
6	2-Butene	23	8	Methyl ethyl ketone (85), i-butanal (3), acetaldehyde (7), trimethylcyclopropane (3)
7	2-Pentene	22	8	2-Pentanone (41) 3-Pentanone (47) Acetaldehyde (5) Propanal (3) Cyclopropanes (1)

[a] Reaction conditions: Parr reactor 100 mL capacity; initial amount of alkene and N_2O: 0.08 mol and 0.12 mol; benzene solvent 50 mL; 493 K; 12 h.

cleavage of the original double bond. It leads to an aldehyde with fewer carbon atoms and an equivalent amount of methylene, which is a very reactive carbene species. Upon interaction with alkenes the methylene forms cyclopropane, and upon interaction with benzene solvent it forms cycloheptatriene. These two by-products are always observed in the resulting reaction mixture of terminal alkenes.

Based on the product distribution, one may calculate the contribution of cleavage route to the total rate of carboxidation [173]. For the terminal alkenes, the cleavage

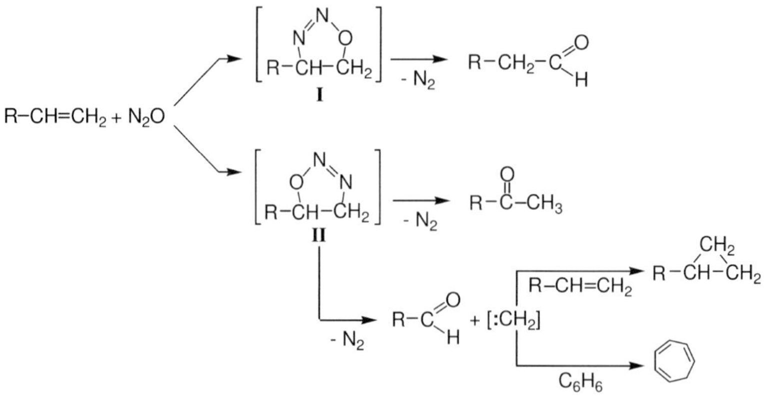

Figure 7.4 Carboxidation scheme of terminal alkenes.

contribution increases sharply on going from ethylene (7%) to propylene (29%) and 1-butene (38%), with virtually no further change (Table 7.8).

The carboxidation of internal 2-butene and 2-pentene proceeds with a much smaller cleavage (8%), yielding the ketones as major products. The reason for this difference in the behavior of terminal and internal alkenes is unclear. One may relate it to dissipation of the energy evolved at the oxidative attack to the double bond. The efficiency of this process may be higher for the more remote position of the double bond from the "end" of the molecule. However, the cleavage contribution for the ethylene is as low (7%) as that for the internal alkenes, which undermines the idea.

7.5.1.2 Cyclic Alkenes

Carboxidation of this type alkenes proceeds with a minor cleavage of C=C bonds, leading to selective formation of the corresponding cyclic ketones (Table 7.9). With unsubstituted cycloalkenes (Table 7.9, entries 1–4), the reaction provides a single ketone, which in all cases forms with high selectivity (94–99%). One may note a slight tendency towards an increasing contribution of side reactions in the cyclopentene to cyclododecene sequence, which is mainly due to ring opening leading to the formation of aldehydes. Minor products of aldol condensation are also detected in the resulting reaction mixture.

The presence of a substituent at a distant position from the double bond provides for the formation of isomeric ketones, in most cases keeping the high total selectivity. For example, 4-methyl-1-cyclohexene (entry 5) and 4-(3-cyclohexen-1-yl)pyridine (entry 6), having substituents in position 3, give the two expected isomeric ketones with total selectivities 95% and 97%, respectively. In the case of (6-methyl-3-cyclohexen-1-yl)methanol (entry 7), the selectivity is lower, which seems to be caused mainly by dehydration of the alcohol.

However, the oxidation of 1-methyl-1-cyclohexene, having a substituent in position 1 (entry 8), proceeds much less selectively, providing only 44% of a cyclic ketone in the reaction products. The reaction is accompanied by a significant cleavage of double bonds leading to 1-cyclopentyl-ethanone and 6-hepten-1-one. An aldehyde ($C_7H_{12}O$) is also found in the reaction products (~2–3%).

Table 7.9 Carboxidation of cyclic alkenes[a].

Entry	Substrate	T (K)	Time (h)	Conversion (%)	Product selectivity (mol.%)
1	cyclopentene	473	20	67	cyclopentanone, 99
2	cyclohexene	523	5	27	cyclohexanone, 97
3	cyclooctene	493	12	56	cyclooctanone, 95
4	cyclododecene	523	3	22	cyclododecanone, 94
5	methylcyclohexene	523	5	10	methylcyclohexanones, 41 + 54

Table 7.9 (Continued)

Entry	Substrate	T (K)	Time (h)	Conversion (%)	Product selectivity (mol.%)
6	4-(cyclohex-3-en-1-yl)pyridine	523	5	15	3-(pyridin-4-yl)cyclohexanone (45); 4-(pyridin-4-yl)cyclohexanone (52)
7	(6-methylcyclohex-3-en-1-yl)methanol	523	5	16	4-(hydroxymethyl)-3-methylcyclohexanone (HO-, 45); 3-(hydroxymethyl)-4-methylcyclohexanone (HO-, 75)
8	1-methylcyclohex-1-ene	523	12	33	2-methylcyclohexanone (44); 1-cyclopentylethanone (34); hept-6-en-2-one (5)

[a] Reaction conditions: Parr reactor 100 mL capacity; alkene amount 25 mL, $P^0_{N_2O} = 2.5$ MPa.

Cyclic ketones are suitable starting material for the preparation of corresponding oximes and dicarbonic acids, which are key intermediates for the production of various types of nylons. Recently, BASF announced [176] the development of a new process for cyclododecanone production via the ketonization of cyclododecene (entry 4, Table 7.9), for which commercialization is expected in 2009.

7.5.1.3 Cyclodienes

These compounds contain two double bonds, which can consecutively react with N_2O, yielding accordingly unsaturated monoketones and diketones. Figure 7.5 shows results for two cyclodienes having the isolated double bonds [175]. The distribution between mono- and diketones is an interesting feature of the reaction. For the carboxidation of 1,4-cyclohexadiene, 3- and 2-cyclohexen-1-ones are the main products, comprising 90% of the total amount of ketones. The concentration of diketones comprises only 3%, although at the given conversion ($X = 37\%$) its statistical value should be 9%. This low amount of diketones may indicate that the introduction of the first carbonyl group deactivates the remaining double bond and makes it more resistant to further oxidation.

Carboxidation of 1,5-cyclooctadiene also leads to the formation of mono- and diketones. However, in this case there is no deactivation effect of the C=O group, so that the diketone fraction (13%) is close to the statistical value. This is probably explained by a more distant location of the double bonds from each other in the latter diene.

In addition to the cyclodienes discussed above, the carboxidation of 1,3-cyclohexadiene having conjugated double bonds was also tested. In this case, the reaction is strongly complicated by the Diels–Alder side reaction. The main part of the diene is consumed by the dimerization process, and only 25–30% is involved in the oxidation, yielding cyclic ketones.

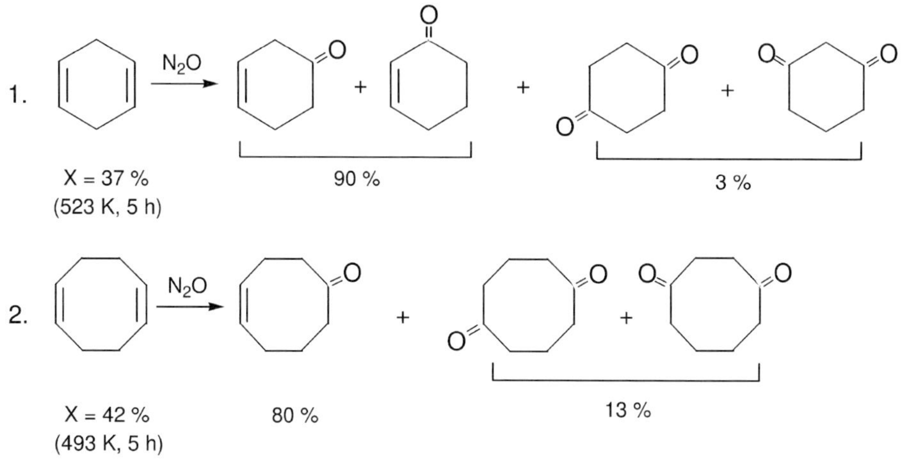

Figure 7.5 Carboxidation of cyclodienes.

Table 7.10 Carboxidation of bicyclic alkenes[a].

Entry	Substrate	T (K)	Time (h)	Conversion (%)	Product selectivity (mol.%)		
1	norbornylene 0.053 mol in 45 ml cyclohexane	453	4	42	28	29	
2	1,2-dihydronaphthalene 0.039 mol in 20 ml benzene	523	5	46	65	28	
3	indene 0.043 mol in 20 ml benzene	523	5	35	63	14	10

[a] Reaction conditions: Parr reactor 100 mL capacity; $P^0_{N_2O} = 1.0$ MPa.

7.5.1.4 Bicyclic Alkenes

Table 7.10 presents results for this alkene type. Carboxidation of norbornylene (entry 1) leads to the formation of both a ketone and an aldehyde in approximately equal amounts, indicating that a significant cleavage of the double bonds occurs. There is also a set of other reaction products as a consequence of the bond cleavage.

The oxidation of 1,2-dihydronaphthalene (entry 2) proceeds with a minor cleavage and yields α- and β-tetralones with a total selectivity of 93%. The greater part of the oxygen adds to the α-position.

In the oxidation of indene, where the double bond is located in the five-membered ring (entry 3), the preference of oxygen for the α-position is even more pronounced, with the ratio of α- to β-indanone being 4.5 : 1. Presumably, 2-ethenylbenzaldehyde (10%) forms as a result of double bond cleavage.

7.5.1.5 Heterocyclic Alkenes

To test the effect of a heteroatom, Starokon et al. [175] studied three five-membered heterocycles with a similar molecular structure, containing the oxygen (2,5-dihydrofuran), nitrogen (3-pyrrole) and sulfur (butadiene sulfone). Only the oxygen-containing cycle was carboxidized selectively, while the others showed a strong tendency towards side reactions resulting in a set of unidentified products.

The comparison of some oxygen-containing cycles presented in Table 7.11 allows one to reveal a significant effect of the double bond location with respect to the heteroatom. Results with 2,3-dihydrofuran (entry 1) and 3,4-dihydro-2H-pyran (entry 3) show that the double bonds having the nearest location to the oxygen exhibit strong

Table 7.11 Carboxidation of heterocyclic alkenes.

Entry	Substrate	$P^0_{N_2O}$ (MPa)	T (K)	Time (h)	Conversion (%)	Product selectivity (mol.%)
1	(2,3-dihydrofuran) 0.13 mol in 60 ml cyclohexane	1.0	473	5	34	allyl formate, 71
2	(2,5-dihydrofuran) 0.13 mol in 50 ml benzene	1.0	493	12	16	dihydrofuran-3-one, 94
3	(3,4-dihydro-2H-pyran) 0.11 mol in 50 ml benzene	2.5	513	5	20	3-butenyl formate, 48; δ-valerolactone, 31
4	(4,7-dihydro-1,3-dioxepin) 0.10 mol in 15 ml benzene	2.5	493	5	16	dioxepanone, 94

disposition toward cleavage. Indeed, the oxidation of 2,3-dihydrofuran proceeds exclusively via the cleavage route. It leads to allyl formate as the main product, with no sign of an expected butyrolactone. The oxidation of 3,4-dihydro-2H-pyran also proceeds primarily via the cleavage route, yielding 3-butenyl formate (48%). But in this case, a significant contribution of a non-cleavage route is also observed, providing δ-valerolactone (31%).

In contrast, the carboxidation of 2,5-dihydrofuran (entry 2) and 4,7-dihydro-1,3-dioxepin (entry 4), having a more distant location of the double bonds, proceeds with minor cleavage, leading in both cases to formation of the corresponding ketones with 94% selectivity.

In conclusion, we can say that the liquid-phase carboxidation of alkanes can be applied to various substrates, including linear, cyclic, heterocyclic alkenes and their derivatives, yielding the corresponding ketones and aldehydes with selectivities in many cases of >90%.

Recently, this approach was applied to more complex compounds, that is, fatty acids methyl esters and triacylglycerols, resulting in up to 99% selectivity of the ketonization products [177].

With some alkenes, carboxidation can be successfully performed also in the gas phase [172, 178].

7.5.2
Carboxidation of Polymers

The macromolecules of many polymers include C=C bonds and therefore can be considered as alkene analogs. This gives rise to the idea of applying the carboxidation approach to polymers. It was expected to open up a new way for their chemical modification so as to improve the adhesion and other physicochemical properties.

7.5.2.1 Carboxidation of Polyethylene

First, this idea was tested [179] with a linear polyethylene (PE). The material had a low molecular weight ($M_n = 960$, $M_w/M_n = 1.7$) and relatively high concentration of terminal C=C bonds (12 bonds per 1000 carbon atoms).

Figure 7.6 shows IR spectra of the parent sample and two PE samples after carboxidation at 503 and 523 K in a toluene solvent. It is seen that the reaction with nitrous oxide decreases the intensity of absorption bands at 909 and 990.6 cm^{-1} assigned to deformation vibrations of C=C bonds in vinyl groups. At 523 K, these bands virtually disappear from the spectrum (spectrum 3). Simultaneously, there appears a new intense band at 1723 cm^{-1}, which corresponds to stretching vibrations of C=O groups in the aldehydes and ketones (spectra 2 and 3). Thus, carboxidation results in a quantitative replacement of –CH=CH$_2$ groups with C=O groups, providing a 1.5 wt% oxygen amount in the resulting material. Using the NMR and IR data, the main transformations of polyethylene can be described by a scheme similar to the carboxidation scheme of terminal alkenes shown in Figure 7.4.

As in the case of alkenes, the carboxidation of PE may also include the cleavage of C=C bond. However, the decrease of PE macromolecule by one carbon atom does not

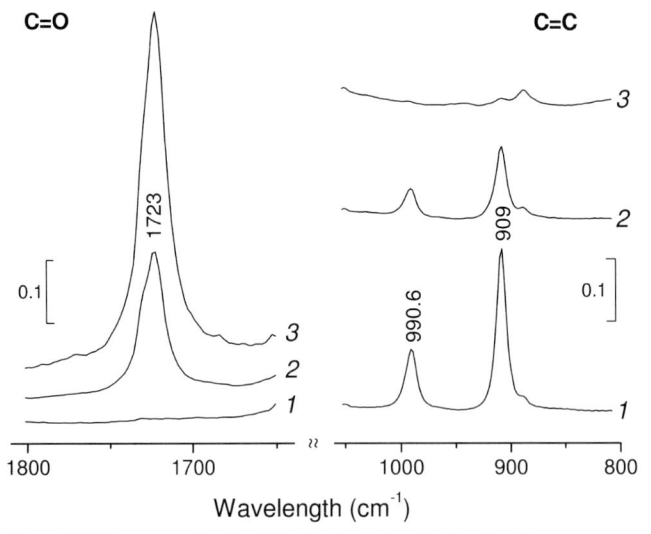

Figure 7.6 IR spectra of carboxidized polyethylene in the vibration region of C=O and C=C bonds: (1) parent sample; (2) after carboxidation at 503 K; (3) after carboxidation at 523 K.

lead to a noticeable decrease of its molecular weight measured by the high-temperature gel-permeation chromatography (GPC).

7.5.2.2 Carboxidation of Polybutadiene Rubber

Unlike PE, many polymers have a high concentration of the internal C=C bonds. The carboxidation of such materials led to quite unexpected results. This can be illustrated by the carboxidation of cis-1,4-polybutadiene rubber (poly-BD) studied in detail by Dubkov et al. [180]. The concentration of C=C bonds in this rubber is 250 bonds per 1000 carbon atoms. Possible oxygen content in the poly-BD can reach up to ~23 wt%, if all C=C bonds are transformed into the C=O groups.

Similar to internal alkenes, carboxidation of the rubber involves intermediate formation of an oxadiazoline cycle (Figure 7.7). Decomposition of the cycle without cleavage of the C=C bond (route 1) is accompanied by the formation of a ketone and does not lead to change in the molecular weight. Decomposition with cleavage (route 2) leads to fragmentation of the macromolecule with the formation of two smaller fragments: a linear aldehyde R_2-CH_2-CHO and a carbene R_1-CH_2-CH:, which further isomerizes into the terminal alkene $R_1-CH=CH_2$.

An estimation based on NMR data showed that the route 1 comprises 95% and route 2 only 5% of the total carboxidation rate. These data are close to the carboxidation results of 2-butene, for which the non-cleavage route was found to be 92% and cleavage route 8% [173]. However, in distinction to the individual alkene, where the cleavage route only slightly decreases the selectivity for ketone, in the case of poly-BD, as we shall see below, this route may have dramatic consequences even at a 5% contribution.

Figure 7.8 shows the GPC curves characterizing the molecular weight distribution (MWD) of the carboxidized poly-BD samples. With the growing degree of carboxidation, the MWD peak progressively moves to a lower MW region. Indeed, the carboxidation leads to a systematic decrease of both the number-average molecular weight (M_n) and the weight-average molecular weight (M_w) (Table 7.12). For a sample with an oxygen content 9.2 wt%, which corresponds to the maximum conversion of double bonds ($X_{C=C} = 34\%$), this decrease is about 50-fold. Regardless of that, the M_w/M_n ratio remains nearly unchanged, indicating quite a narrow MW distribution for all samples. Table 7.12 also gives an indication of the average number of cleavages (N_{cleav}) of an initial rubber macromolecule. As the oxygen content increases, the number of cleavages also increases, reaching 51 for the most oxygenated sample,

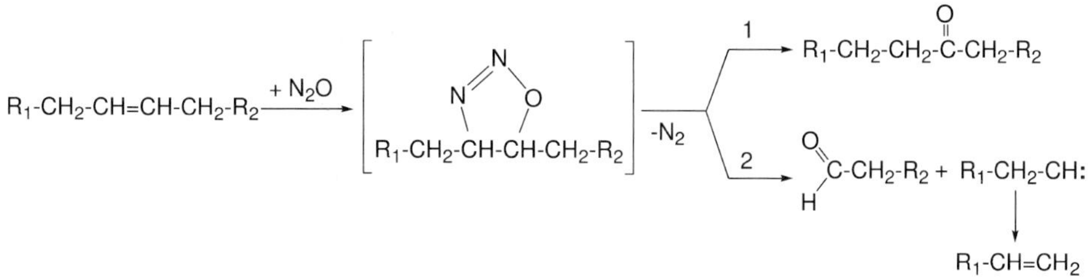

Figure 7.7 Carboxidation of polybutadiene rubber.

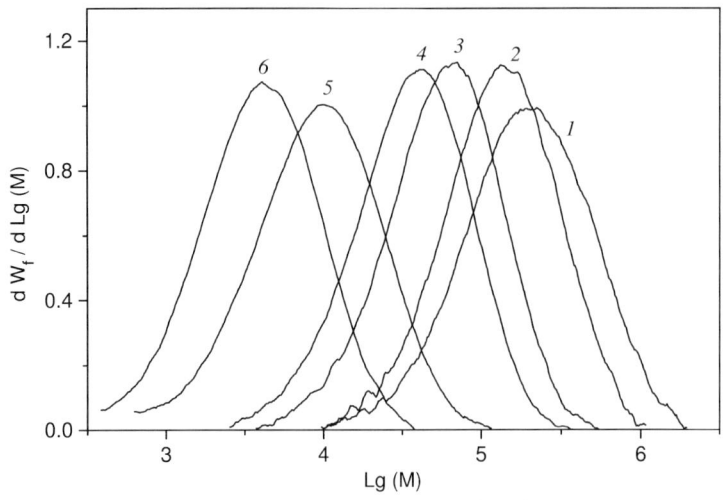

Figure 7.8 MWD curves of carboxidized poly-BD samples. The numbers of the curves correspond to the sample numbers in Table 7.12.

no. 6. Accordingly, the resulting fragments become progressively smaller, with the number of monomeric units decreasing from the initial 2370 to 45.

The carboxidation degree exhibits a dramatic effect on the consistency of the resulting samples. At the introduction of small amounts of oxygen (0.2–0.8 wt%), the rubber retained its consistency, but became sticky. At an oxygen content of 1.6 wt%, the material became fluid and lost the ability to retain its shape. At oxygen contents of 5.0 wt% or more, the samples turned into a bright viscous liquid.

For the most oxygenated sample, the carboxidation reaction may be presented by (7.19), having huge stoichiometric coefficients compared to conventional chemical reactions:

$$[-\text{CH}=\text{CH}-]_n + 810\, N_2O \longrightarrow 52\, [\text{oligomer}] + 810\, N_2$$

$$n = 2370 \qquad\qquad n = 45\ (16\ C{=}O + 29\ C{=}C)$$

(7.19)

Reaction (7.19) shows that one polymeric macromolecule, consisting on average of 2370 monomeric units (all containing C=C bonds), interacts with 810 molecules of N_2O. This leads to fragmentation of the macromolecule into 52 oligomeric fragments. Each fragment consists of 45 monomeric units, of which 16 units contain C=O groups and 29 units contain C=C bonds. Since the major part of the carbonyls is presented by ketone groups, the resulting product was called an "unsaturated polyketone." This is a new type of material that differs from solid saturated polyketones prepared by the copolymerization of carbon monoxide with ethylene [181–183]. The latter material was studied intensively in the 1980–1990s. However, the method of its production proved to be commercially inefficient [184].

Table 7.12 Characteristics of carboxidized poly-BD samples.

Sample no. (see Fig. 7.8)	Carboxidation condition ($P^0_{N_2O}$ = 2.5 MPa)	Oxygen content (wt%)	XC=C (%)	$10^{-3} M_n$	$10^{-3} M_w$	M_w/M_n	N_{cleav}	Number of monomeric units in a fragment	Consistency of sample
1	None (parent)	0.0	0	128	288	2.2	0.0	2370	Rubber
2	433 K, 12 h	0.2	0.7	92	178	1.9	0.4	1700	Sticky rubber
3	473 K, 5 h	0.8	2.6	38	76	2.0	2.4	698	Sticky rubber
4	473 K, 10 h	1.6	5.5	23	48	2.1	4.7	419	Fluid rubber
5	503 K, 6 h	5.0	18	6.3	14	2.2	20	111	Viscous liquid
6	503 K, 12 h	9.2	34	2.7	5.4	2.0	51	45	Viscous liquid

Carboxidation of the poly-BD and other rubbers by nitrous oxide opens a simple and effective way for preparing polyketones with a regulated oxygen content and molecular weight. The presence of C=O and C=C groups provides additional opportunities for application and modification of the material.

7.6
Economic Aspects of N_2O as Oxidant

Nowadays, the main practical application of nitrous oxide is related to medicine. It is also used also as a foaming agent in the food industry, for the preparation of lead and sodium azide propellants as well as gas generators in car air-bags. In addition, N_2O is used also to enhance the power of racing cars. The total consumption of nitrous oxide for all these applications is rather small, and its annual production does not exceed several tens of thousands tons.

The remarkable oxidation properties of nitrous oxide discussed above open a new field of its application. The capacity of this field will depend on the economic aspects, that is, on the availability and cost of N_2O. There are two sources of nitrous oxide: the recovery from off-gases and deliberate preparation.

7.6.1
Recovery of N_2O From Off-Gases

N_2O emission at chemical plants and the methods of its abatement have been considered in several reviews [185–187]. The major emission is related to the preparation of nitric acid and its use in oxidation processes, like those involved in the production of adipic acid, caprolactam, glyoxal, acrylonitrile, and so forth. Of them, the biggest emission is the off-gases of adipic acid: about 1 MMT N_2O per year with a concentration of 30–40%. Recovery and purification of N_2O from these off-gases for use in the oxidation of benzene to phenol are described by Uriarte [188]. Some companies use these off-gases to obtain medical-grade nitrous oxide.

When discussing N_2O emissions, one should keep in mind a detrimental effect of nitrous oxide on the environment. Since its discovery at the turn of the eighteenth century and up to the 1970s, nitrous oxide was considered as a quite harmless substance. However, in the last 2–3 decades, upon discovering its greenhouse and ozone depleting effects, waste N_2O has became a matter of significant concern because of its possible contribution to the global changes of climate. N_2O concentration in the atmosphere is very low compared to the main greenhouse gases (CO_2 and CH_4) and makes up only 310 ppbv. However, due to its long lifetime in the atmosphere (about 150 years), N_2O has a 310 and 21 times greater molecular warming potential than CO_2 and CH_4, respectively [187]. In this relation, the use of waste N_2O as an oxidant is especially attractive. Instead of spending money for its destruction, N_2O can be used as a valuable chemical feedstock, thus solving both economic and ecological problems.

7.6.2
Deliberate Preparation of N_2O

The classic way of N_2O preparation is the thermal decomposition of ammonium nitrate:

$$NH_4NO_3 \rightarrow N_2O + H_2O \tag{7.20}$$

The reaction proceeds at 493–523 K and is used by most producers.
Another way is direct catalytic oxidation of ammonia with dioxygen:

$$2NH_3 + 2O_2 \rightarrow N_2O + 3H_2O \tag{7.21}$$

This process is practiced by Mitsui Toatsu Chemical with a fixed bed reactor to make 450 tons of medical grade nitrous oxide [188]. Economically it is the most viable route. However, reaction (7.21) is strongly exothermic, which causes the difficult problem of heat removal from the catalyst bed. Therefore, a fixed bed reactor allows only a small scale production of N_2O. More recently, a new version of the technology has been developed [189, 190]. The oxidation of ammonia is conducted over a supported manganese-bismuth oxide catalyst designed for operation in a fluidized bed reactor. Owing to a good heat and mass transfer, this type of reactor makes possible a high production capacity of the unit. This new technology has been successfully tested with a pilot reactor [189]. In the temperature range 613–643 K, the unit provides virtually complete conversion of ammonia with 86% selectivity for nitrous oxide.

The economic estimation based on the pilot unit results showed that ammonia contributes 70% to the prime cost of the N_2O. Using this result and the ammonia price 0.37 \$ kg^{-1} [191], one can evaluate the N_2O price to be 0.53 \$ kg^{-1}. Certainly, this cost far exceeds that of dioxygen. Therefore, for reactions producing inexpensive products, like the oxidation of methane to methanol, the application of N_2O cannot be economically sound. However, this modest cost opens great N_2O prospects for the preparation of more expensive chemical products. For instance, the theoretical expenditure for N_2O in the oxidation of benzene to phenol is 17%, and in the oxidation of phenol to hydroquinone is 4%, of the cost of the target product. The commercial viability of such processes will depend primarily on their technological advantages rather than the cost of nitrous oxide.

Note that the price of hydrogen peroxide is 0.56 \$ kg^{-1} for a 35% solution, which equates to $\sim$1.6 \$ kg^{-1} for the neat H_2O_2. Therefore, the active oxygen in N_2O prepared by direct ammonia oxidation is several times cheaper than the active oxygen in H_2O_2.

Apart from a low cost, nitrous oxide has some additional advantages over other oxygen donors:

- wide range of possible reaction conditions;
- significant solubility in organic substrates;
- possibility to recycle;

- generally better safety;
- transportation of neat N_2O in the liquefied form.

Thus, from economic and technical points of view, nitrous oxide has excellent prospects for use in oxidation chemistry and catalysis.

7.7
Conclusion

Early references to catalytic reactions of nitrous oxide can be found in a book by Marek and Hahn published over 70 years ago [192]. Since that time, for many years interest in this molecule was rather occasional. But starting from the 1980s N_2O has been a subject of ever-increasing attention of researchers because of two factors. The first is an ecological concern and a necessity of abating N_2O emissions. Numerous works in this field are discussed in several reviews [185–187] and are not considered in this chapter.

The second factor is the unique feature of N_2O as an oxygen donor – first demonstrated with the gas-phase oxidation of lower alkanes over metal oxides. Later, the high catalytic efficiency of the FeZSM-5 zeolites was discovered, suggesting an opportunity for developing new processes of gas-phase oxidation with N_2O, especially the hydroxylation of benzene and other aromatics to the corresponding phenols.

Nitrous oxide offers a tempting possibility for the epoxidation of propylene and butylene in the gas phase. Encouraging results were obtained with Fe-modified SiO_2 and some zeolite-like catalysts. No other oxidant allows selective performance of these delicate reactions.

As well as in the gas phase, nitrous oxide gives promising results in liquid-phase oxidations. With many reactions using both homogeneous and heterogeneous catalysts, N_2O provides better selectivity than H_2O_2 or O_2. However, it proved to be quite an inert molecule that allows only rather small reaction rates. The development of effective catalysts able to activate N_2O at low temperature may provide a breakthrough in this field.

Of special interest is a new type reaction discovered with N_2O: direct oxidation of alkenes to carbonyl compounds, called carboxidation. Beside various individual alkenes, carboxidation can be applied effectively to unsaturated polymers, opening up a way for the preparation of new materials. Reactions of this type should receive special attention from the catalytic community, since currently they are conducted in a thermally. This oxidation area is waiting for the beneficial arrival of catalysis to provide better control of the activity and selectivity.

Summarizing, we may conclude that intensive studies conducted during the two last decades have uncovered a remarkable oxidation chemistry of nitrous oxide. This chemistry is far from being properly understood. Further studies may lead to new discoveries that are important not only for fundamental knowledge, but also for the development of new chemical technologies, which is strongly supported by the economic advantages of N_2O as oxidant.

References

1. Lide, D.R. (2004–2005) *Handbook of Chemistry and Physics, Section 14*, CRC Press.
2. Misono, M. Nojiri, N. (1990) *Applied Catalysis*, **64**, 1–30. *ibid*, (1993) **93**, 103–122.
3. Armor, J. (1991) *Applied Catalysis A – General*, **78**, 141–173. (2001) *Applied Catalysis A – General*, **222**, 407–426.
4. Centi, G., Cavani, F. and Trifiro, F. (2001) *Selective Oxidation by Heterogeneous Catalysis*, Kluwer Academic/Plenum Publishers.
5. Hodnett, B.K. (2000) *Heterogeneous Catalytic Oxidation*, John Wiley & Sons, New York.
6. Krylov, O.V. (2004) *Heterogeneous Catalysis*. Akademkniga (in Russian).
7. Bartholomew, C.H. and Farrauto, R.J. (2006) *Fundamentals of Industrial Catalytic Processes*, Wiley Interscience.
8. Grasselli, R.K. (1997) *Handbook of Heterogeneous Catalysis*, Vol. 5 (eds G. Ertl, H. Knösinger and J. Weitkamp), Wiley-VCH, Weinheim, pp. 2302–2326.
9. Boreskov, G.K. (2003) *Heterogeneous Catalysis*, Nova Science Publishers, New-York.
10. Belansky, A. and Haber, J. (1991) *Oxygen in Catalysis*, Marcel Dekker, New-York.
11. Sheldon, R.A. (1994) *Chemtech, March*, pp. 38–47; Sheldon, R.A. (2000) *Pure and Applied Chemistry*, **72**, 1233–1246.
12. Mallat, T. and Baiker, A. (2000) (Guest Editors of Special issue), Catalytic oxidation for the synthesis of specialty and fine chemicals. *Catalysis Today*, **57**, 1–166.
13. Leont'ev, A.V., Fomicheva, O.A., Proskurnina, M.V. and Zefirov, N.S. (2001) *Russian Chemical Reviews*, **70**, 91–104.
14. Houk, K.N., Sims, J., Duke, R.E. Jr *et al.* (1973) *Journal of the American Chemical Society*, **95**, 7287–7301.
15. Olah, G.A., Herges, R., Laali, K. and Segal, G.A. (1986) *Journal of the American Chemical Society*, **108**, 2054–2057.
16. Chaban, G.M., Klimenko, N.M. and Charkin, O.P. (1992) *Russian Chemical Bulletin*, **41**, 99–106.
17. Eger, E.I. II, (1985) *Nitrous Oxide N_2O*, Elsevier, Amsterdam.
18. Panetier, G. and Sicard, A. (1955) 5th Symposium on Combustion, pp. 620–628.
19. Zamaschikov, V.V., Bunev, V.A., Dubkov, K.A. *et al.* (2004) *Combustion and Plasmochemistry*, **2**, 13–21 (in Russian).
20. Brandt, B.B., Matov, L.A., Rozlovsky, A.I. and Khailov, V.S. (1960) *Chemical Industry*, **5**, 67–73 (in Russian).
21. Coward, H.F. and Jons, G.W. (1952) Limits of Flammability of Gases and Vapors, U.S. Bureau of Mines, Bull. 503, Washington.
22. Krylov, O.V. (1993) *Catalysis Today*, **18**, 209–302.
23. Hall, T.J., Hargreaves, J.S.J., Hutchings, G.J. *et al.* (1995) *Fuel Processing Technology*, **42**, 151–158.
24. Parmaliana, A. and Arena, F. (1997) *Journal of Catalysis*, **167**, 57–65.
25. Lunsford, J.H. (2000) *Catalysis Today*, **63**, 165–174.
26. Otsuka, K. and Wang, Y. (2001) *Applied Catalysis A – General*, **222**, 145–161.
27. Reitzmann, A., Häfele, M. and Emig, G. (1996) *Trends in Chemical Engineering*, **3**, 63–75.
28. Aika, K. and Lunsford, J.H. (1977) *The Journal of Physical Chemistry*, **81**, 1393–1398.
29. Liu, R..-S., Iwamoto, M. and Lunsford, J.H. (1982) *Journal of the Chemical Society. Chemical Communications* (1), 78–79.
30. Liu, H.-F., Liu, R.-S., Liew, K.Y. *et al.* (1984) *Journal of the American Chemical Society*, **106**, 4117–4121.
31. Zhen, K.J., Khan, M.M., Mak, C.H. *et al.* (1985) *Journal of Catalysis*, **94**, 501–507.
32. Kennedy, M., Sextone, A., Kartheuser, B. *et al.* (1992) *Catalysis Today*, **13**, 447–454.

33 Chen, S.Y. Willcox, D. (1993) *Industrial & Engineering Chemistry Research*, **32**, 584; (1993) *ibid*, **33**, 832–839.

34 Barbaux, Y., Elamrani, A.R., Payen, E. *et al.* (1988) *Applied Catalysis A – General*, **44**, 117–132.

35 Lipatkina, N.I., Shvets, V.A. and Kazansky, V.B. (1978) *Kinet Katal*, **19**, (4), 979–984.

36 Goto, A. and Aika, K. (1998) *Bulletin of the Chemical Society of Japan*, **71**, 95–98.

37 Otsuka, K. and Nakajima, T. (1986) *Inorganica Chimica Acta*, **120** (2), 27–38.

38 Solymosi, F., Tombacz, I. and Kutsan, G. (1985) *Journal of the Chemical Society. Chemical Communications* (20), 1455–1456.

39 Anderson, J.P. and Tsai, P. (1987) *Journal of the Chemical Society. Chemical Communications* (19), 1435–1436.

40 Wang, Y. and Otsuka, K. (1994) *Chemistry Letters* 1893–1896; Wang, Y. and Otsuka, K. (1997) *Journal of Catalysis*, **171**, 106–114.

41 Wang, X., Wang, Y., Tang, Q. *et al.* (2003) *Journal of Catalysis*, **217**, 457–467.

42 Kasztelan, S. and Moffat, J.B. (1989) *Journal of Catalysis*, **116**, 82–94.

43 Firsova, A.A., Vorobieva, G.A., Bobyshev, A.A. *et al.* (1991) *Kinet Katal*, **32**, 395–403.

44 Mendelovici, L. and Lunsford, J.H. (1985) *Journal of Catalysis*, **94** (1), 37–50.

45 Erdöhelyi, A., Mate, F. and Solymosi, F. (1992) *Journal of Catalysis*, **135**, 563–575.

46 Hong, S.S. and Moffat, J.B. (1994) *Applied Catalysis A – General*, **109**, 117–134.

47 Aika, K., Isobe, M., Kido, K. *et al.* (1987) *Journal of the Chemical Society – Faraday Transactions*, **83**, (10), 3139–3148.

48 Kustrowski, P., Segura, Y., Chmielarz, L. *et al.* (2006) *Catalysis Today*, **114**, 307–313.

49 Groothaert, M.H., Smeets, P.J., Sels, B.F. *et al* (2005) *Journal of the American Chemical Society*, **127**, 1394–1395. Smeets, P.J. Groothaert, M.H. Schoonheydt, R.A. (2005) *Catalysis Today*, **110**, 303–309.

50 Launay, H., Loridant, S., Pigamo, A. *et al.* (2007) *Journal of Catalysis*, **246**, 390–398.

51 Chempath, S. and Bell, A.T. (2007) *Journal of Catalysis*, **247**, 119–126.

52 Ione, K.G. (1982) *Polyfunctional Catalysis on Zeolites*, Nauka, Novosibirsk (in Russian).

53 Panov, G.I., Sobolev, V.I. and Kharitonov, A.S. (1990) *Journal of Molecular Catalysis*, **61**, 85–97.

54 Sobolev, V.I., Panov, G.I., Kharitonov, A.S. *et al.* (1993) *Journal of Catalysis*, **139**, 435–443.

55 Notari, B. (1996) *Advances in Catalysis*, **41**, 253–334.

56 Perego, C., Carati, A., Ingallina, P. *et al.* (2001) *Applied Catalysis*, **221**, 63–72.

57 Panov, G.I. (2000) *Cattech*, **4**, 18–32.

58 Iwamoto, M., Hirata, J., Matsukami, K. and Kagawa, S. (1983) *The Journal of Physical Chemistry*, **87**, 903–905.

59 Suzuki, E., Nakashiro, K. and Ono, Y. (1988) *Chemistry Letters*, 953–956.

60 Gubelmann, M. and Tirel, Ph. (1988) France Patent 2 630 735, Filed May.

61 Kharitonov, A.S., Aleksandrova, T.N., Vostrikova, L.A. *et al.* (1988) USSR Patent 1 805 127, Filed June.

62 Burch, R. and Howitt, C. (1992) *Applied Catalysis A – General*, **86** (2), 139–146.

63 Bogdan, V.I., Kustov, L.M., Batizat, D.B. *et al.* (1995) *Studies in Surface Science and Catalysis*, **94**, 635–647.

64 Häfele, N., Reitzmann, A., Roppelt, D. and Emig, G. (1997) *Applied Catalysis A – General*, **150**, 153–164.

65 Centi, G., Perathoner, S., Pino, F. *et al.* (2005) *Catalysis Today*, **110**, 211–220.

66 Vereshchagin, S.N., Kirik, N.P., Shishkina, N.N. and Anshits, A.G. (1998) *Catalysis Letters*, **56**, 145–148.

67 Reitzmann, A., Klemm, E. and Emig, G. (2002) *Chemical Engineering Journal*, **90**, 149–164.

68 Ribera, A., Arends, I.W.C.E., de Vries, S. *et al.* (2000) *Journal of Catalysis*, **195**, 287–297.

69 Shevade, S.S. and Rao, B.S. (2000) *Catalysis Letters*, **66**, 99–103.

70 Perathoner, S., Pino, F., Genti, G. *et al.* (2003) *Topics in Catalysis*, **23**, 125–136.

71 Wichterlova, B., Sobalik, Z. and Dedecek, J. (2003) *Applied Catalysis B – Environmental*, **41**, 97–114.

72 Vereschagin, S.N., Kirik, N.P., Shishkina, N.N. *et al.* (2000) *Catalysis Today*, **61**, 129–136.

73 Ren, T., Yan, L., Zhang, X. and Suo, J. (2003) *Applied Catalysis A – General*, **244**, 11–17.

74 Hensen, E.J.M., Zhu, Q., Hendrix, M.M.R.M. *et al.* (2004) *Journal of Catalysis*, **221**, 560–574.

75 Zhu, Q., van Teeffelen, R.M., van Santen, R.A. and Hensen, E.J.M. (2004) *Journal of Catalysis*, **221**, 575–583.

76 Waclaw, A., Nowinska, K. and Schwieger, W. (2004) *Applied Catalysis A – General*, **270**, 151–156.

77 Pillai, K.S., Jia, J. and Sachtler, W.M.H. (2004) *Applied Catalysis A – General*, **264**, 133–139.

78 Reshetnikov, S.I., Ilyin, S.B., Ivanov, A.A. and Kharitonov, A.S. (2004) *Reaction Kinetics and Catalysis Letters*, **83**, 157–164.

79 Ivanov, A.A., Chernyavsky, V.S., Gross, M.J. *et al.* (2003) *Applied Catalysis A – General*, **249**, 327–343.

80 Yuranov, I., Bulushev, D.A., Renken, A. and Kiwi-Minsker, L. (2006) *Applied Catalysis A – General*, **319**, 128–137.

81 Mauvezin, M., Delahay, G., Coq, B. *et al.* (2001) *The Journal of Physical Chemistry B*, **105**, 928–935.

82 Uriarte, A.K., Rodkin, M.A., Gross, M.J. *et al.* (1997) *Studies in Surface Science and Catalysis*, **110**, 857–864.

83 Parmon, V.N., Panov, G.I., Uriarte, A.K. and Noskov, A.S. (2005) *Catalysis Today*, **100**, 115–131.

84 Panov, G.I., Sheveleva, G.A., Kharitonov, A.S. *et al.* (1992) *Applied Catalysis A – General*, **82**, 31–36.

85 Sobolev, V.I., Panov, G.I., Kharitonov, A.S. *et al.* (1993) *Journal of Catalysis*, **139**, 435–443.

86 Kharitonov, A.S., Sheveleva, G.A., Panov, G.I. *et al.* (1993) *Applied Catalysis A – General*, **98**, 33–43.

87 Pirutko, L.V., Chernyavsky, V.S., Uriarte, A.K. and Panov, G.I. (2002) *Applied Catalysis. A – General*, **227**, 143–157.

88 Kubanek, P., Wichterlova, B. and Sobalik, Z. (2002) *Journal of Catalysis*, **211**, 109–118.

89 Meloni, D., Monaci, R., Solinas, V. *et al.* (2003) *Journal of Catalysis*, **214**, 169–178.

90 Perez-Ramirez, J., Kapteijn, F., Groen, J.C. *et al.* (2003) *Journal of Catalysis*. **214**, 33–45.

91 Kachurovskaya, N.A., Zhidomirov, G.M., Hensen, E.J.M. and van Santen, R.A. (2003) *Catalysis Letters*, **86**, 25–34.

92 Yoshizawa, K., Shiota, Y., Yumura, T. and Yamabe, T. (2000) *The Journal of Physical Chemistry. B*, **104**, 734–740.

93 Ryder, J.A. Chakraborty, A.K. Bell, A. (2003) *Journal of Catalysis*, **220**, 84–91. (2003) *The Journal of Physical Chemistry*, **106**, 7059–7064.

94 Kachurovskaya, N.A., Zhidomirov, G.M. and van Santen, R.A. (2004) *The Journal of Physical Chemistry. B*, **108**, 5944–5950.

95 Wood, B.R., Reimer, J.A., Bell, A. *et al.* (2004) *Journal of Catalysis*, **224**, 148–155.

96 Berlier, G., Zecchina, A., Spoto, G. *et al.* (2003) *Journal of Catalysis*, **215**, 264–270.

97 Kiwi-Minsker, L., Bulushev, D.A. and Renken, A. (2003) *Journal of Catalysis*, **219**, 273–285.

98 Kiwi-Minsker, L., Bulushev, D.A. and Renken, A. (2004) *Catalysis Today*, **91–92**, 165–170.

99 Ates, A. and Reitzmann, A. (2005) *Journal of Catalysis*, **235**, 164–174.

100 Pirngruber, G.D. (2003) *Journal of Catalysis*, **219**, 456–463; Pirngruber, G.D. Luechinger, M. Roy, P.K. *et al.* (2004) *Journal of Catalysis*, **224**, 429–440.

101 Roy, P.K. and Pirngruber, G.D. (2004) *Journal of Catalysis*, **227**, 164–174.

102 Kondratenko, E.V. and Perez-Ramirez, J. (2006) *Applied Catalysis B – Environmental*, **64**, 35–41.

103 Sun, K., Zhang, H., Xia, H. *et al.* (2004) *Chemical Communication*, 2480–2481.

104 Sun, K., Xia, H. Hensen, E. *et al.* (2006) *Journal of Catalysis*, **238**, 186–195.

105 Komeoka, S., Nobukova, T., Tanaka, S.-I. *et al.* (2003) *Physical Chemistry Chemical Physics*, **5**, 3328–3333.

106 Volodin, A.M., Zhidomirov, G.M., Dubkov, K.A. et al. (2005) *Catalysis Today*, **110**, 247–254.

107 Taboada, J., Hensen, E.J.M., Arends, I.W.C.E. et al. (2005) *Catalysis Today*, **110**, 221–227.

108 Dubkov, K.A., Ovanesyan, N.S., Shteinman, A.A. et al. (2002) *Journal of Catalysis*, **207**, 341–352.

109 Fejes, P., Lazar, K., Marsi, I. et al. (2003) *Applied Catalysis A – General*, **252**, 75–90.

110 Taboada, J., Overweg, A., Kooyman, P.J. et al. (2005) *Journal of Catalysis*, **231**, 56–66.

111 Starokon, E.V., Dubkov, K.A., Pirutko, L.V. and Panov, G.I. (2003) *Topics in Catalysis*, **73**, 137–143.

112 Panov, G.I., Dubkov, K.A. and Starokon, E.V. (2006) *Catalysis Today*, **117**, 148–155.

113 Panov, G.I., Uriarte, A.K., Rodkin, M.A. and Sobolev, V.I. (1998) *Catalysis Today*, **41**, 365–385.

114 Marturano, P., Drozdova, L., Kogelbauer, A. and Prins, R. (2000) *Journal of Catalysis*, **192**, 236–247.

115 Battiston, A.A., Bitter, J.H., Heijboer, W.M. et al. (2003) *Journal of Catalysis*, **215**, 279–293.

116 Berlier, G., Spoto, G., Bordiga, S. et al. (2002) *Journal of Catalysis*, **208**, 64–82.

117 Perez-Ramirez, J., Kapteijn, F., Schöffel, F. and Moulijn, J.A. (2003) *Applied Catalysis B – Environmental*, **44**, 117–151.

118 Panov, G.I., Kharitonov, A.S. and Sobolev, V.I. (1993) *Applied Catalysis A – General*, **98**, 1–20.

119 Panov, G.I., Sobolev, V.I., Dubkov, K.A. and Kharitonov, A.S. (1996) *Studies in Surface Science and Catalysis*, **101**, 493–502.

120 Panov, G.I., Sobolev, V.I., Dubkov, K.A. et al. (1997) *Reaction Kinetics and Catalysis Letters*, **61** (2), 251–258.

121 Rodkin, M.A., Sobolev, V.I., Dubkov, K.A. et al. (2000) *Studies in Surface Science and Catalysis*, **130**, 875–880.

122 Knops-Gerrits, P.P. Smith, W.J. (2000) *Studies in Surface Science and Catalysis*, **130**, 3531–3537. Knops-Gerrits, P.P. Goddard, W.A. III (2001) *Journal of Molecular Catalysis*, **166**, 135–145.

123 Sobolev, V.I., Kovalenko, O.N., Kharitonov, A.S. et al. (1991) *Mendeleev Commun*, **1**, 29–30.

124 Osch, T. and Turek, T. (1999) *Chemical Engineering Science*, **54**, 4513–4523.

125 El-Malki, E.M., van Santen, R.A. and Sachtler, W.M.H. (2000) *Journal of Catalysis*, **196**, 212–223.

126 Sang, Ch., Kim, B.H. and Lund, C.R.F. (2005) *The Journal of Physical Chemistry*, **109**, 2295–2301.

127 Grubert, G., Hudson, M.J., Joyner, R.W. and Stockenhuber, M. (2000) *Journal of Catalysis*, **196**, 126–133.

128 Bulushev, D.A. and Kiwi-Minsker, L. (2006) *The Journal of Physical Chemistry B*, **110**, 10691–10700.

129 Novakova, J. and Sobalik, Z. (2005) *Catalysis Letters*, **105**, 169–177.

130 Heyden, A., Peters, B., Bell, A.T. and Keil, F.J. (2005) *The Journal of Physical Chemistry. B*, **109**, 1857–1873.

131 Hansen, N., Heyden, A., Bell, A.T. and Keil, F.J. (2007) *Journal of Catalysis*, **248**, 213–225.

132 Bulushev, D.A., Prechtl, P.M., Renken, A. and Kiwi-Minsker, L. (2007) *Industrial & Engineering Chemistry Research*, **46**, 4178–4185.

133 Dubkov, K.A., Sobolev, V.I., Talzi, E.P. et al. (1997) *Journal of Molecular Catalysis A – Chemical*, **123**, 155–161.

134 Yuranov, I., Bulushev, D., Renken, A. and Kiwi-Minsker, L. (2004) *Journal of Catalysis*, **227**, 138–147.

135 Chernyavsky, V.S., Pirutko, L.V., Uriarte, A.K. et al. (2007) *Journal of Catalysis*, **245**, 466–469.

136 Vereschagin, S.N., Baikalova, L.I. and Anshits, A.G. (1988) *Izvastiya AN SSS, Seria Khimiches-kaya*, **8**, 1718–1723 (in Russian).

137 Held, A., Kowalska, J. and Nowinska, K. (2006) *Applied Catalysis B – Environmental*, **64**, 201–208.

138 Bulanek, R., Wichterlova, B., Novoveska, K. and Kreibich, V. (2004) *Applied Catalysis A – General*, **264**, 13–22.

139 Kondratenko, E.V. and Perez-Ramirez, J. (2004) *Applied Catalysis A – General*, **267**, 181–189.

140 Bulanek, R., Adam, J., Novoveska, K. *et al.* (2005) *Studies in Surface Science and Catalysis*, **158**, 1977–1984.

141 Perez-Ramirez, J. and Gallardo-Llamos, A. (2005) *Applied Catalysis A – General*, **279**, 117–123.

142 Stockenhuber, M., Joyner, R.W., Dixon, J.M. *et al.* (2001) *Microporous and Mesoporous Materials*, **44–45**, 367–375.

143 Motz, J.L., Heinichen, H. and Hölderich, W.F. (1998) *Journal of Molecular Catalysis A – Chemical*, **136** 175–184.

144 Imre, B., Halas, J., Frey, K. *et al.* (2001) *Reaction Kinetics and Catalysis Letters*, **74**, 377–383.

145 Vogel, B., Schneider, C. and Klemm, E. (2002) *Catalysis Letters*, **79**, 107–114.

146 Costine, A., O'Sullivan, T. and Hodnett, B.K. (2005) *Catalysis Today*, **99**, 199–208.

147 Costine, A., O'Sullivan, T. and Hodnett, B.K. (2006) *Catalysis Today*, **112**, 103–106.

148 Ivanov, D.P., Sobolev, V.I., Pirutko, L.V. and Panov, G.I. (2002) *Advanced Synthesis and Catalysis*, **344**, 986–995.

149 Kustov, L.M., Tarasov, A.L., Bogdan, V.I. *et al.* (2000) *Catalysis Today*, **61**, 123–128.

150 Ivanov, D.P., Pirutko, L.V. and Sobolev, V.I. (2004) *Petroleum Chemistry*, **44**, 322–327.

151 Perez-Ramirez, J., Blangenois, N. and Ruiz, P. (2005) *Catalysis Letters*, **104**, 163–167.

152 Duma, V. and Hönicke, D. (2002) *Journal of Catalysis*, **191**, 93–104.

153 Moens, B., De Winne, H., Corthals, S. *et al.* (2007) *Journal of Catalysis*, **247**, 86–100.

154 Wang, X., Zhang, Q., Guo, Q. *et al.* (2004) *Chemical Communication*, 1396–1397.

155 Wang, X., Zhang, Q., Yang, S. and Wang, Y. (2005) *The Journal of Physical Chemistry B*, **109**, 23500–23508.

156 Thömmes, T., Zürcher, S., Wix, A. *et al.* (2007) *Applied Catalysis A – General*, **318**, 160–169.

157 Kustrowski, P. Chmielars, L. Dziembaj, R. *et al.* (2005) *The Journal of Physical Chemistry*, **109**, 330–336 and 9808–9815.

158 Ohtani, B., Takamiyk, S., Hirai, Y. *et al.* (1992) *Journal of the Chemical Society – Perkin Transactions*, **2**, 175–179.

159 Groves, J.T. and Roman, J.S. (1995) *Journal of the American Chemical Society*, **117**, 5594–5595.

160 Yamada, T., Hashimoto, K., Kitaichi, Y. and Suzuki, K. (2001) *Chemistry Letters*, 922–923.

161 Hashimoto, K., Tanaka, H., Ikeno, T. and Yamada, T. (2002) *Chemistry Letters*, 582–583.

162 Tanaka, H., Hashimoto, K., Suzuki, K. *et al.* (2004) *Bulletin of the Chemical Society of Japan*, **77**, 1905–1914.

163 Ben-Daniel, R., Weiner, L. and Neumann, R. (2002) *Journal of the American Chemical Society*, **124**, 8788–8789.

164 Ben-Daniel, R. and Neumann, R. (2003) *Angewandte Chemie – International Edition*, **42**, 92–95.

165 Stuchinskaya, T.L. and Kozhevnikov, I. (2003) *Catalysis Communications*, **4**, 609–614.

166 Stuchinskaya, T.L., Muzavir, M., Kozhevnikova, E.F. and Kozhevnikov, I.V. (2005) *Journal of Catalysis*, **231**, 41–47.

167 Bridson-Jones, F.S. Buckley, G.D. Cross, L.H. and Driver, A.P. (1951) *Journal of the Chemical Society*, 2999–3008; Bridson-Jones and F.S. Buckley, G.D. (1951) *ibid*, 3009–3016.

168 Panov, G.I., Dubkov, K.A., Starokon, E.V. and Parmon, V.N. (2002) *Reaction Kinetics and Catalysis Letters*, **76**, 401–406.

169 Avdeev, V.I., Ruzankin, S.Ph. and Zhidomirov, G.M. (2003) *Chemical Communication*, 42–43.

170 Avdeev, V.I., Ruzankin, S.F. and Zhidomirov, G.M. (2005) *Kinetics and Catalysis*, **46**, 177–188.

171 Hermans, I., Moens, B., Peeters, J., Jacobs, P. and Sels, B. (2007) *Physical Chemistry Chemical Physics*, **9**, 4269–4274.

172 Starokon, E.V., Dubkov, K.A., Parmon, V.N. and Panov, G.I. (2005) *Reaction Kinetics and Catalysis Letters*, **84**, 383–388.

173 Semikolenov, S.V., Dubkov, K.A., Starokon, E.V., Babushkin, D.E. and Panov, G.I. (2005) *Russian Chemical Bulletin*, **54**, 948–956.

174 Dubkov, K.A., Panov, G.I., Starokon, E.V. and Parmon, V.N. (2002) *Reaction Kinetics and Catalysis Letters*, **77**, 197–205.

175 Starokon, E.V., Dubkov, K.A., Babushkin, D.E. et al. (2004) *Advanced Synthesis and Catalysis*, **346**, 268–274.

176 BASF News Release (2006) April 03, press.kontact@basf.com.

177 Hermans, I., Jansen, K., Moens, B. et al. (2007) *Advanced Synthesis and Catalysis*, **349**, 1604–1608.

178 Starokon, E.V., Shubnikov, K.S., Dubkov, K.A. et al. (2007) *Kinetics and Catalysis*, **48**, 376–380.

179 Semikolenov, S.V., Dubkov, K.A., Echevskaya, L.G. et al. (2004) *Polymer Science, Series B*, **46**, 308–311.

180 Dubkov, K.A., Semikolenov, S.V., Babushkin, D.E. et al. (2006) *Journal of Polymer Science Part A Polymer Chemistry*, **44**, 2510–2520.

181 Sen, A. (1986) *Advances in Polymer Science*, **73**, 125–144.

182 Drent, E. and Budzelaar, P.H.M. (1996) *Chemical Reviews*, **96**, 663–682.

183 Belov, G.P. and Novikova, E.V. (2004) *Russian Chemical Reviews*, **73**, 267–291.

184 Consiglio, G. (2003) in *Late Transition Metal Polymerization Catalysis* (eds B. Rieger, L.S. Baugh, S. Kacker and S. Striegler), Wiley-VCH, Weinheim, pp. 279–305.

185 Kapteijn, F., Rodriguez-Mirasol, J. and Moulijn, J.A. (1996) *Applied Catalysis B – Environmental*, **9**, 25–64.

186 Centi, G., Parathoner, S. and Vazzana, F. (2000) *Chemtech*, pp. 38–47.

187 Perez-Ramirez, J., Kapteijn, F., Schöffel, K. and Moulijn, J.A. (2003) *Applied Catalysis B – Environmental*, **44**, 117–151.

188 Uriarte, A.K. (2000) *Studies in Surface Science and Catalysis*, **130**, 743–748.

189 Kashkin, V.N., Lakhmostov, V.S., Zolotarskii, I.A. et al. (2003) *Chemical Engineering Journal*, **91**, 215–218.

190 Noskov, A.S., Zolotarskij, I.A., Pokrovskaya, S.A. et al. (2003) *Chemical Engineering Journal*, **91**, 235–242.

191 http://www.icispricing.com/il.shared/Samples/SubPage149.asp.

192 Marek, L.F. and Hahn, D.A. (1932) *The Catalytic Oxidation of Organic Compounds in the Vapour Phase*, The Chemical Catalog Company Inc., New York.

8
Direct Synthesis of Hydrogen Peroxide: Recent Advances

Gabriele Centi, Siglinda Perathoner, and Salvatore Abate

8.1
Introduction

8.1.1
Industrial Production

Hydrogen peroxide (H_2O_2), a weakly acidic colorless liquid, is the simplest peroxide. It was discovered about two centuries ago, but only began to be used industrially from the mid-nineteenth century [1]. Initially, it was produced by electrolysis of ammonium bisulfate, but nearly all commercial production is now based on catalytic hydrogenation followed by auto-oxidation of a suitable organic molecule, predominantly alkylated anthraquinone. This process is the dominant one for large-scale productions. Alternative process technologies are based on the electrochemical reduction of oxygen in a dilute sodium hydroxide solution (Huror-Dow process), which generates dilute H_2O_2 solutions (around 4%) suitable for onsite pulp bleaching, and alcohol oxidation (isopropyl alcohol by Shell Chemicals and methylbenzyl alcohol by Lyondel–ARCO). The alcohol oxidation process, however, has not been in use from about thirty years – instead the methyl anthraquinone route is preferred, which was first commercialized by Du Pont in 1953 [1–3]. The original alcohol oxidation process was non-catalytic, but recent advances in using catalysts in this process has improved significantly the performances and thus this route is seeing renewed interest, particularly in the case of methylbenzyl alcohol oxidation [4].

The alkylated anthraquinone process accounts for over 95% of the world production of H_2O_2, mainly because the it operates under mild conditions and direct contact of O_2 and H_2 is avoided. In this process, 2-alkylanthraquinone (the alkyl group is typically an ethyl, *tert*-butyl or amyl group) is dissolved in a mixture of a non-polar solvent (C_9–C_{11} alkylbenzene) and a polar solvent [Trioctyl phosphate (TOP), or tetrabutyl urea (TBU) or diisobutyl carbinol (DIBC)] and then hydrogenated over a precious metal (Pd or Ni) catalyst in a three-phase reactor (trickle bed or slurry bubble column) under mild reaction conditions (<5 bar, <80 °C) to generate 2-alkylanthrahydroquinone [1–3, 5]. The latter is then auto-oxidized with air in a

Modern Heterogeneous Oxidation Catalysis: Design, Reactions and Characterization
Edited by Noritaka Mizuno
Copyright © 2009 WILEY-VCH Verlag GmbH & Co. KGaA, Weinheim
ISBN: 978-3-527-31859-9

two-phase reactor (packed column or bubble column) to produce H_2O_2 and regenerate 2-alkylanthraquinone. The main drawbacks of this process are the following [5, 6]:

- Significant impact on the environment and generation of wastes (mainly associated with the less than 100% selectivity in the reduction and oxidation steps – formation of hydroxyanthrones, anthrones, anthracenes, and epoxide, some solvent stripping by air used in the oxidation step, and in the crude H_2O_2 stream).

- Low efficiency (limited solubility of alkylanthraquinone in the solvent – typically 120–150 g per liter of working solution; low H_2O_2 concentration in the working solution – typically <1.5 wt%; and limited hydrogenation conversion – typically <70% to minimize side reactions) and high energy consumption. For a 50% H_2O_2 solution the volume of working solution is about 40–50× higher or a high recirculation rate of the working solution has to be used; in addition, the purification of crude H_2O_2 is very energy intensive.

- Process complexity and risks of explosion.

- Presence of mass-transport limitations in the hydrogenation and oxidation reactors.

Owing to the high degree of complexity, this process is suited especially for large-scale productions, but now an increasing number of processes (from chemical to environmental applications) need a small onsite production to avoid the risks of transport of concentrated H_2O_2 and the presence of stabilizers in H_2O_2 solutions that are necessary for the transport. As commented below, the use of H_2O_2 as a clean oxidant is expanding worldwide. Furthermore, the cost of H_2 and O_2 reactants with respect to the total cost of hydrogen peroxide is about 30–35% (depending mainly on the H_2 cost). There are thus several incentives to consider a process of direct reaction of H_2 and O_2 to form H_2O_2. This process could significantly reduce the impact on the environment, the energy consumption and the process cost with respect to the alkylanthraquinone route. In addition, it is suited for small–medium scale dedicated productions of H_2O_2.

8.1.2
Outlook for H_2O_2 Production

H_2O_2 production is rising and global consumption is expected to be 2.2 million metric tons per year by 2009 [7] and over 4 m.t. y^{-1} by 2012. Figure 8.1 shows the projected demand of H_2O_2 for the various sectors of application of hydrogen peroxide, which will be still driven by use in pulp bleaching. However, demand in new applications, particularly propene oxide (PO), is rising and is expected to reach 150 000 m.t. by 2009 [7].

Different growth rates are present in the different regions. In North America H_2O_2 demand growth was 1% in 2006–2007, due to closure of several paper mills. However, in South America a strong pulp and paper market is raising demand and an expansion of H_2O_2 plant capacity is under way. In Europe, demand growth reached 3% for 2007,

**Global Consumption H$_2$O$_2$:
2009 projected total 2.2 million m.t.**

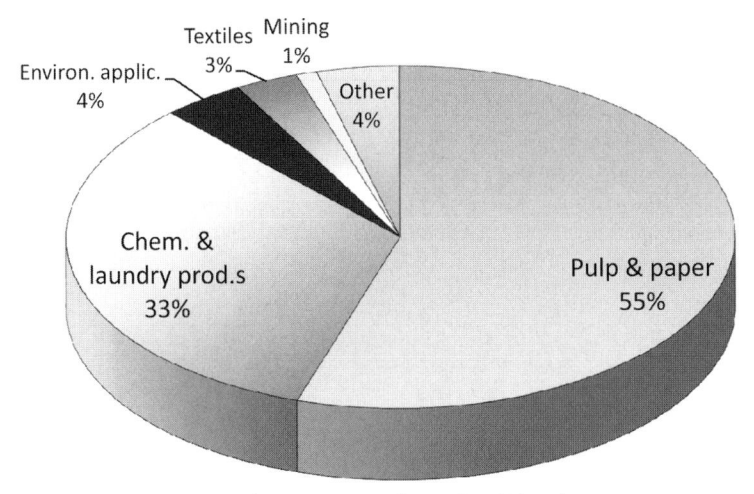

- Textiles 3%
- Mining 1%
- Environ. applic. 4%
- Other 4%
- Chem. & laundry prod.s 33%
- Pulp & paper 55%

Figure 8.1 Global projected consumption of H$_2$O$_2$ by 2009 and the main areas of use [7]. Source: SRI Consulting (Menlo Park, CA).

while in 2006 the rate was 4–5%. However, the outlook is positive and several plant expansions are under way in the region. Solvay is building a 230 000 m.t. y^{-1} plant in Antwerp (Belgium) to provide feedstock to a BASF-Dow Chemical joint venture for PO. The Asian market is rapidly expanding, particularly in China.

Table 8.1 overviews the H$_2$O$_2$ capacity of the main producers. The market is dominated by two companies (Solvay and Degussa) with about 30% and 20% of the share of the main producers. Two other companies (FMC and Arkema) have a share above 10%, while all the other are present with a lower market percentage. Note, however, that global production capacity is higher because only the main producers appear in Table 8.1.

8.1.3
Uses of Hydrogen Peroxide

H$_2$O$_2$ is a strong, nonpolluting oxidizing agent, and most of its uses and the uses of its derivatives depend on this property. Applications for hydrogen peroxide fall into the following broad categories: (1) pulp/paper bleaching, (2) water/waste and effluent treatment, (3) chemical synthesis, (4) textile bleaching, (5) mining/metallurgy, (6) electronics (semiconductors), (7) propulsion (satellite and rockets), (8) desulfurization and (9) food and miscellaneous. Hydrogen peroxide is available in three common product strengths (e.g., as 35, 50 and 70 wt% aqueous solutions). While transportation costs of contained hydrogen peroxide decrease with increasing concentration (lower volume for the same oxygen content), the cost of handling

Table 8.1 H_2O_2 capacity of the main producers. (Adapted from [7]).

Company	US	Canada	South America	Europe	Asia/Pacific	Total	(%)
			(thousand m.t. per year)				
Solvay	186		120	370		676	27.7
Degussa	87	93	70	206		456	18.7
Arkema	68	93		145		306	12.5
FMC	110	51		130		291	11.9
Kemira		41		145		186	7.6
Eka Chemicals				175		175	7.2
Mitsubishi Gas Chem.					126	126	5.2
EkO Peroxide	64					64	2.6
DC Chemical					60	60	2.5
Degussa-Headwaters					50	50	2.0
Hansol					40	40	1.6
Ecros				12		12	0.5
Total	515	278	190	1183	276	2442	100.0

this hazard (severity of tissue burns and ease of ignition of combustibles) increase with increasing concentration. Technical grade hydrogen peroxide is stabilized with low levels of tin-based stabilizers and phosphates. Organic-based stabilizers are used in hydrogen peroxide grades destined for use in organic synthesis reactions, due to possible interference by inorganic stabilizers, in some types of metal treatment and for electronic uses.

Hydrogen peroxide is an environmentally safe oxidizing agent in the pulp bleaching process, because it does not generate chlorine derivatives in the pulp suspension and therefore is displacing chlorine, chlorine dioxide and other reagents in several pulp bleaching operations. Hydrogen peroxide is used in the treatment of a wide variety of industrial wastes and waste waters, providing a versatile treatment chemical that can solve many waste treatment problems, such as the removal of cyanide, thiocyanate and nitrite, chlorine or hypochlorite, or organic compounds. In addition, H_2O_2 is also used as a source of hydroxyl radicals in more complex Advanced Oxidation Processes [8]. *In situ* biorestoration of contaminated soil with H_2O_2 offers the advantage that soil can be treated in place, saving removal/replacement costs and minimizing worker exposure.

Hydrogen peroxide has numerous applications in the chemical process industry. High purity ferric sulfate, hydrazine, perborates and percarbonates are typical targets of inorganic synthesis with hydrogen peroxide. Applications in organic synthesis include epoxidation (preparation of propene oxide) and hydroxylation (manufacture of plasticizers and stabilizers in the polymer industry, production of aromatic diphenols), oxohalogenation (flame retardants) and initiation of emulsion polymerization (MEK peroxide, benzoyl peroxide, lauryl peroxide). A growth in H_2O_2 use will also derive from the larger use of the Sumitomo's route to caprolactam. Good prospects also exist in terms of the introduction of a new process of direct synthesis of phenol from benzene using H_2O_2.

Further details on the different uses of H_2O_2 can be found in refs. [1–3], while more specific aspects of the uses of H_2O_2 for catalytic clean industrial oxidation processes in are given in refs. [6, 9, 10].

8.2
Direct Synthesis of H_2O_2 from an Industrial Perspective

8.2.1
Status of Development and Perspectives of Industrial Production

Direct synthesis of hydrogen peroxide from its elements is thermodynamically feasible and it is a known reaction from the beginning of the last century [11]. However, intense research activity in many companies on the direct synthesis of H_2O_2 only began in 1980. Some 100 patents were issued in the 1980s and 1990s from companies such as Air Products, BASF, Dow Chemicals, Du Pont, ENI, Eka Nobel, FMC, Solvay Interox, Mitsubishi Gas Chemical and Shell. Table 8.2 summarizes some selected comparative results.

After the explosion of a pilot-scale reactor at DuPont, however, research on the direct oxidation of H_2 to H_2O_2 declined. In fact, H_2/O_2 mixtures are explosive over a wide range of concentrations [17]. When the concentration of H_2 in air and oxygen is in the range 4–75 and 4–94 mol% respectively, at 1.0 atm pressure, the resulting gas mixture is flammable/explosive, if ignited. The flammability/explosive range is further widened with increasing pressure. A low concentration of H_2, however, results in a decrease of productivity and of the maximum concentration of H_2O_2 that can be reached. In the last decade, due both to new catalysts and reactor solutions, industrial interest and also academic research has begun again. A novel reactor configuration was patented by Advanced Peroxide Technology and later by Princeton Advanced Technology [18]. The reactor design centers on dispersing tiny bubbles of oxygen and hydrogen in a controlled manner while surrounded by enough liquid to suppress any runaway reaction within the bubbles themselves. As pointed out in Table 8.2, the Du Pont patent operates clearly within the explosion region. Similar reaction conditions are present in other patents issued during the 1980–1990s. Conversely, during the last decade, most patents report results outside the explosion region (patents of ENI, HTI and Degussa in Table 8.2), while some companies (BASF in Table 8.2) operate still inside the explosion region, but with a particular reactor design (using a catalyst monolith constructed from wovens, particularly weavable metal wires) and diluting H_2 along the catalytic bed (Figure 8.2) [14]. In this upright reactor, three layers of a woven catalyst monolith are inserted, with intermediate feed of H_2, while O_2 is fed only from the bottom where a mixing jet disperser creates the fine bubbles and turbulent mixing necessary to quench the radicals that may start radical-chain reactions leading to explosion. Nevertheless, operations inside the explosion limits are potentially risky, particularly for large reactors.

Table 8.2 Comparison of selected examples, taken from patents, of performance in direct H_2O_2 synthesis.

Company	Catalyst	H_2O_2 (wt%)	Selectivity (%)	Reaction conditions
DuPont [12]	Pd-Pt (Pt/Pd + Pt = 0.08) colloidal on alumina	19.6	69	136 bar, 5–8 °C, 18% H_2 in O_2, aqueous acid solution (0.1 N HCl)
ENI [13]	1% Pd-0.1% Pt on carbon	7.3	74	100 bar, 8 °C (autoclave, after 600 h), 3.6% H_2, 11% O_2 in inert 95:5 methanol: H_2O solution (+additives)
BASF [14]	Pd on monolith	7.0	84	144 bar, 10% H_2 in O_2, methanol (+additives)
HTI [15]	Pd(-Pt) on carbon black (140 m² g⁻¹)[a]	9.1 [276 g (g-Pd.h)$^{-1}$]	99	~120 bar, ~35 °C, (autoclave, after 600 h), 3% H_2 in air, solvent and additives not indicated
Degussa [16]	2.5% Pd-Au (95:5) on a-Al_2O_3	5.1 [13.8 g (g-Pd h)$^{-1}$]	72	50 bar, 25 °C (trickle bed), 3% H_2, 20% O_2, methanol (+additives)

[a] It is claimed that using a precursor solution containing an ionic polymer (Na acrylate) it is possible to deposit Pd nanocrystals exposing preferentially the (110) and (220) faces that contain the active sites.

An elegant solution is to use microchanneled reactors [19], where the high surface to volume ratio allows an effective quenching of radical-chain reactions, thus allowing safe operation with H_2 concentrations higher than 5%. This solution is suited for small-scale applications, while the cost is too high for larger scale reactors. There are also problems related to plugging of microchannels in long-term operations. Figure 8.3 shows an example of the microchannel reactor developed by Velocys (a spin-off from Battelle, US), and one of the proposed configurations for this microchanneled reactor in the patent of the same company for a process of direct H_2O_2 synthesis [20].

FMC, one of the main producers of hydrogen peroxide (Table 8.1), is also active in this area. Working with the Stevens Institute of Technology, FMC is developing a direct microchannel route to hydrogen peroxide from hydrogen and oxygen for distributed hydrogen peroxide production. The US DoE has awarded $1.6 million in

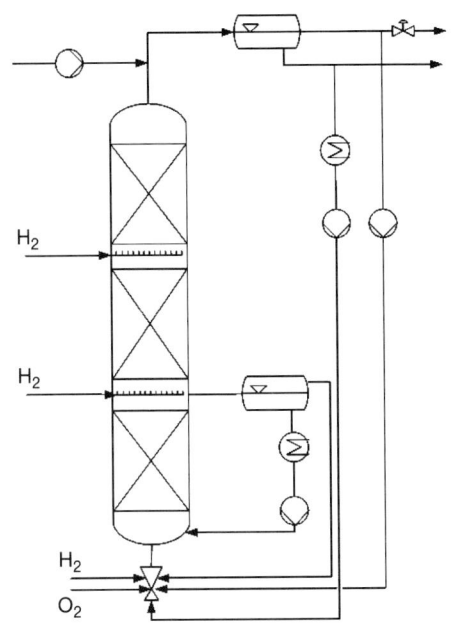

Figure 8.2 Scheme of the reactor proposed by BASF in the patent US 6,375,920 [14] for the direct synthesis of H_2O_2.

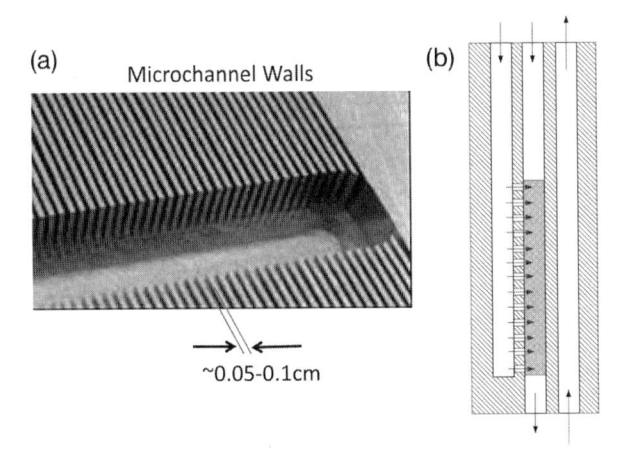

Figure 8.3 Scheme of the microchannel reactor developed by Velocys (a) and one of the proposed configurations for this microchanneled reactor in the patent of the same company for a process of direct H_2O_2 synthesis (b) [20].

funding to the Stevens' Center for MicroChemical Systems to design and demonstrate a microchannel reactor system for on-site production of hydrogen peroxide (H_2O_2) by controlled reaction between H_2 and O_2 [21]. Figure 8.4 shows the conceptual process flow diagram for the microchannel reactor system. Considerable potential exists for

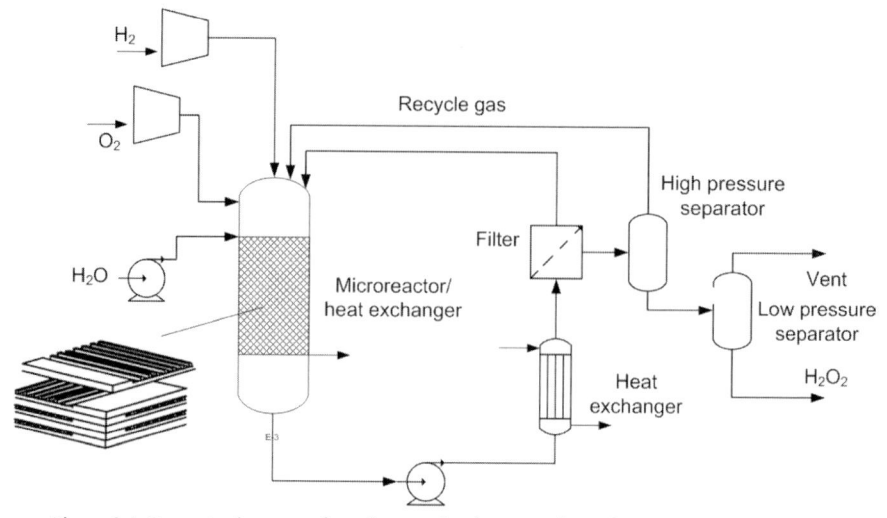

Figure 8.4 Conceptual process flow diagram for the microchannel reactor system [21].

on-demand production of toxic and explosive chemicals that may be possible with reactor system miniaturization [22]. FMC has been able to produce more than 2 wt% H_2O_2 in a single channel reactor and is undergoing process optimization of the pilot unit to achieve 5 wt% H_2O_2 concentrations. The 5 wt% H_2O_2 meets the requirement for the production of propene oxide (PO), which is a high commodity chemical used for various applications.

Degussa has performed a pilot demonstration (8 ton y^{-1} scale) of microreactor technology in the case of the gas-phase epoxidation of propene using hydrogen peroxide. Degussa's results show very stable and safe reactor operations, despite the fact that the reactants are well within the explosive regime. The combination of direct H_2O_2 synthesis in a microchanneled reactor for propene epoxidation to propene oxide (PO) is an expected extension in the near future, particularly for onsite small-scale production. However, the actual cost to productivity ratio is still too high, but at the same time a specific interest to accelerate the introduction of cleaner synthesis processes based on H_2O_2 is expected.

Degussa thus created a joint venture with Headwaters Inc. in 2004 to develop and commercialize a direct synthesis process for hydrogen peroxide (DSHP) [23]. The initial phase of activity by Degussa–Headwaters was the construction and operation of a DSHP pilot plant. The pilot plant has successfully operated since the beginning of 2005, leading to the next step in commercial development, construction of a DSHP demonstration plant. Figure 8.5 compares the conventional H_2O_2 process with the direct synthesis of H_2O_2, and also shows a photograph of the commercial demonstration plant in Hanau Wolfgang (Germany) [23, 24]. The process simplification and intensification possible with the direct synthesis process can be clearly seen.

Together with the engineering company Uhde (Germany), Degussa (now Evonik Industries) has also developed an innovative process for producing propene oxide (PO) from H_2O_2, the so-called HPPO process. The Korean company SKC, Seoul, has

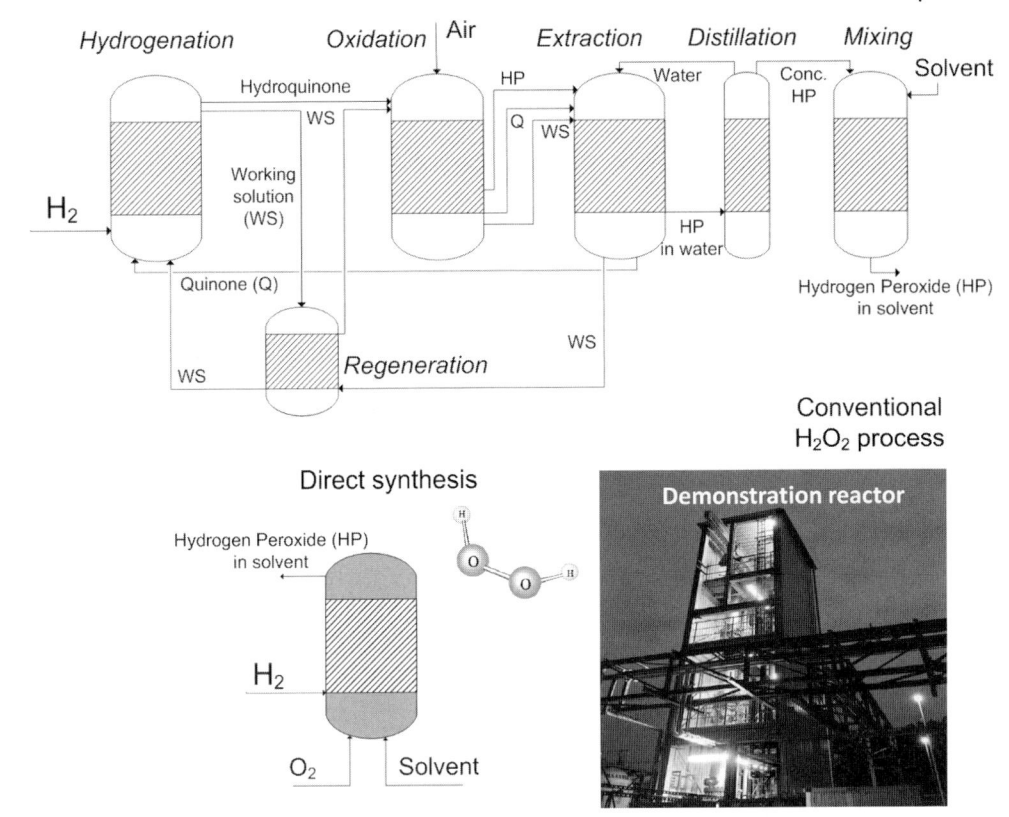

Figure 8.5 Comparison of conventional versus direct process of H_2O_2 synthesis. (Adapted from [22, 24]).

acquired a license for this process to bring a HPPO facility on stream at its site in the Korean town of Ulsan at the beginning of 2008 with an annual capacity of 100 000 tonnes. Degussa–Headwaters Korea will exclusively supply the new PO facility in Ulsan with hydrogen peroxide initially from the conventional alkylanthraquinone technology but in the future with the new DSHP technology. DOW-BASF is also planning to start, from 2010, operations in Thailand of a 390 kton y^{-1} PO plant based on propene oxidation with H_2O_2. A 230 kton y^{-1} PO plant with the same technology should start operations from 2009. Evonik Industries and Sibur have also created a joint venture to produce PO in Russia using H_2O_2. There are other large-scale chemical processes, such as the synthesis of caprolactam that have now shifted to the use of H_2O_2 as clean selective oxidant to reduce the environmental impact of the process. Sumitomo has already started production of caprolactam with this process, although again using H_2O_2 produced by the conventional route. ENI and other companies have also obtained, at a micropilot plant scale, interesting results in direct phenol synthesis from benzene and H_2O_2. Therefore, the prospects for expansion of the market of H_2O_2 are promising and driven mainly by (1) commercialization of new processes for production of large-scale chemicals (propene oxide, caprolactam and phenol) as well as of fine/specialty chemicals (diphenols, epoxide, lactam, etc.), (2) expansion of use for clean pulp/paper production as well as in the pretreatment of biomass and (3) an increase in utilization for wastewater and soil treatment/remediation.

Even if direct synthesis of H_2O_2 is still at pilot/demonstration level, it may be expected that from 2010 the first commercial processes will begin. Degussa–Headwaters, after completing in 2007 the first experimentation phase of the demonstration plant for direct H_2O_2 synthesis, has announced the design/construction of a 200 000 tons y^{-1} plant in 2008; the plant is expected to be completed in 2010.

Headwaters Technology Innovation (HTI) has received the Presidential Green Chemistry Award 2007 for their "Direct Synthesis of Hydrogen Peroxide by Selective Nanocatalyst Technology." The US EPAs Green Chemistry Program promotes the research, development and implementation of innovative chemical technologies that accomplish pollution prevention in a scientifically sound and cost-effective manner. HTI has achieved these goals by developing a robust nanocatalyst technology that enables the synthesis of H_2O_2 directly from hydrogen and oxygen. The technology, called NxCat, is based on a palladium-platinum catalyst that increases productivity and especially selectivity, cutting both energy use and costs (Table 8.2). This catalyst is the core of the new DSHP process. Although details are limited, the key aspect is the use of a special preparation method (based on the use of an ionic polymer) that leads to stable medium–small supported Pd particles (around 4 nm, particles with lower diameter are indicated as unselective) doped with Pt (Pd/Pt = 50) and especially with a preferential exposition of the (110) crystal face to which the very high selectivity and productivity is attributed. Table 8.2 shows, in fact, that the patented performances of this catalyst are superior both in terms of productivity and especially selectivity, even if in terms of composition and metal particle size the catalyst features are not different from those reported in patents by other companies or in the open literature. Figure 8.6A shows a model reported in a patent [15] of HTI to explain the superior selectivity of the (110) surface of Pd. Figure 8.6B shows one of the few TEM images of the NxCat catalyst [24], which, however, do support unequivocally the claim of the preferential exposition of the (110) face of Pd. The question of the higher performances of NxCat catalyst is thus still open.

8.2.2
Recent Patents on the Direct Synthesis of H_2O_2

Most recent patents focus on the catalytic direct synthesis of H_2O_2, but a few deal also on the electrocatalytic synthesis of hydrogen peroxide. As an example of the possible performances following this route, the recent patent issued by Electrosynthesis Co. may be cited [25], which reports a maximum concentration of H_2O_2 of about 7%, with a current efficiency of about 62%. A 1N H_2SO_4 electrolyte should be used and thus recovery of H_2O_2 is complex and costly. In addition, the costs of the overall process appears to be significantly higher than for the catalytic process. Other patents deal also with the cathodic reduction of oxygen at gas-diffusion electrodes to produce H_2O_2 [26]. They could be used for onsite small-scale production, particularly for

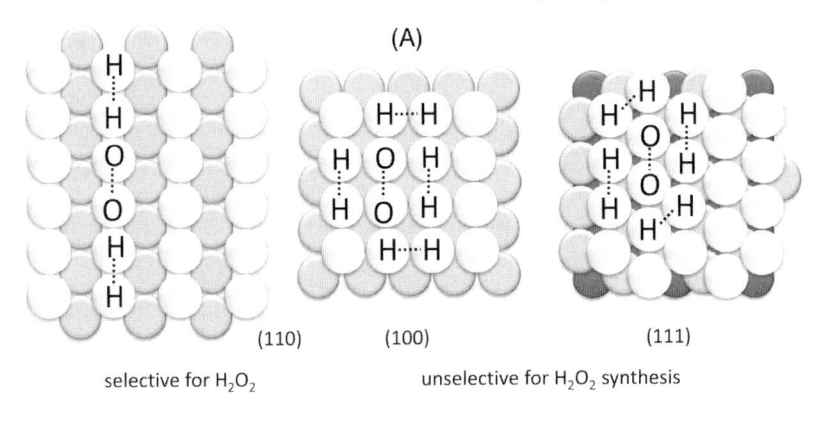

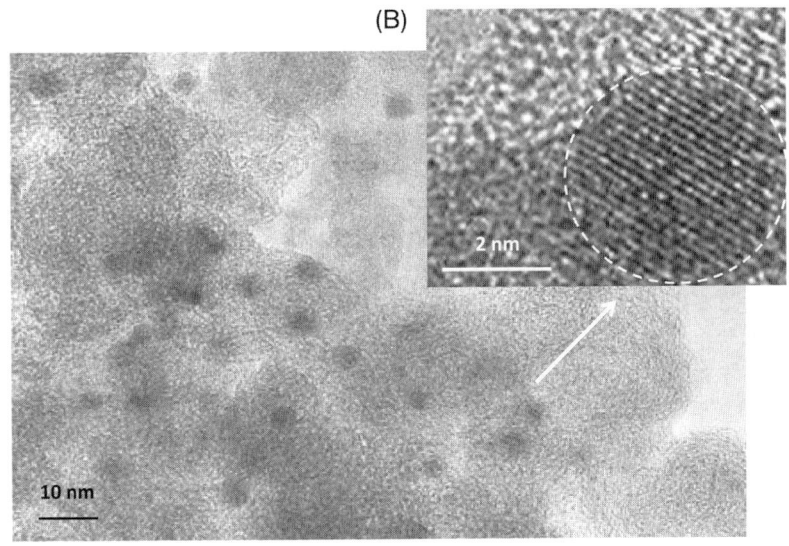

Figure 8.6 (A) Model reported in patent [15] to explain the superior selectivity of the (1 1 0) surface of Pd; (B) TEM images of the NxCat catalyst [24].

wastewater treatment (electro-Fenton) [27], but in general the catalytic route appears to be preferable and reliable in terms of costs and scale-up. Other methods such as H_2/O_2 non-thermal plasmas (produced at dielectric barrier discharge) have also been shown to be effective in producing H_2O_2. Guo et al. [28] reported a ca. 32% yield of H_2O_2 with a 56% selectivity via the gas-phase reaction of H_2/O_2 in non-equilibrium plasma. Also in this case, the applicability to industrial production has to be demonstrated.

In the direct catalytic synthesis of H_2O_2 the main findings of the earlier patents can be summarized as follows:

- Pd is the active element, but doping with another noble metal (Pt, typically in a 1 to 8–10 ratio with respect to Pd; Au was also reported) improves the selectivity; the typical loading of the noble metal over the support (alumina, carbon) is around 2–5 wt%. For example, Brill [29] disclosed the method of producing H_2O_2 by reacting H_2 and O_2 at greater than atmospheric pressure (40–150 bar) in an acidic aqueous solution in the presence of a supported noble metal catalyst. Gosser and Schwartz [12] disclosed that doping with Pt improves significantly the performances with a strong maximum in the amount of formed H_2O_2.

- The use of mixtures of H_2 and O_2 in concentrations within the explosivity range and with high O_2 concentrations improves both the productivity and selectivity, but requires special and complex reactors to operate safely. For example, Izumi et al. [30] disclosed an optimal ratio of O_2 to H_2 within the range from 5 to 20, for example, in large excess with respect to the stoichiometric value. Huckins [31] described a continuous process that operates within the explosivity range, but where the inner design allows the separation of tiny bubbles to avoid explosion. However, the reactor operations are costly. Notably, it is possible to use a "trickle bed" reactor with a feed consisting of about 5% H_2 and about 60% O_2, which is outside (above) the explosion limits [32]. However, this is possible only using aqueous solutions that may be used for pulp/paper bleaching. In organic syntheses (e.g., propene oxide – PO) a solvent such as methanol is necessary. In fact, a TS-1 catalyst is much less selective in propene oxidation to PO using H_2O_2 in water with respect to methanol solution. Recovery of H_2O_2 from the aqueous solution is costly and risky. Therefore, the direct synthesis of H_2O_2 in methanol is necessary to reduce the costs. However, when using methanol as solvent during the synthesis of H_2O_2 the use of high concentrations of oxygen is not allowed, owing to an increase in the explosivity limits (methanol/O_2/H_2 mixture).

- The use of small amounts of bromide ions (typically 5–30 ppm) and an acidic medium (addition of H_2SO_4, H_3PO_4, while HCl should be avoided) improves the performances, particularly limiting the consecutive decomposition of H_2O_2. However, the addition of acids and halides causes the reaction medium to become strongly corrosive towards metallic materials, particularly stainless steel. Hastelloy or glass-lined reactors could thus be necessary.

- The use of organic solvents instead of water increases the productivity/selectivity and in some cases also the catalyst lifetime. For example, Dalton and Skinner [33] disclosed an improvement in H_2O_2 synthesis from H_2/O_2 in an acidic medium containing an anhydrous oxygenated solvent (methanol, acetone) and/or a nitrogenous organic compound (acetonitrile). They also suggested the use of inorganic (sodium meta- and pyro-phosphate) or organic (formaldehyde) stabilizers.

Several patents were issued on this topic. For example, Rueter [34] in his patent cited about 120 US patents and about 40 foreign patents in the period 1987–2004. The actual number of patents is higher, but already these numbers indicate that many aspects of catalyst and process technology are covered. Patents in the last five years (2003–2008) were thus focused either on the improvement of catalyst preparation and/or improvement of reaction/reactor operations, particularly with strong attention on the safety of operations. However, often in the cited patents operations are still inside explosivity limits. As a general trend, the H_2 concentration is kept low, except for patents issued from research institutions (Council of Scientific and Industrial Research – CSIR, India; see Table 8.3 below) which use H_2 concentrations inside the explosion region. The oxygen concentration is typically maintained below the lower explosion limit for operations with methanol as solvent (the preferred organic solvent) or above the upper explosion limit for operations in acid solutions (typically water/H_2SO_4 or water/H_3PO_4).

The use of methanol (typically CH_3OH/H_2O solutions in a 95:5 ratio) allows us to improve of one order of magnitude the productivity. Zhou et al. [35, 36] (formerly Hydrocarbon Techn. Inc. – HTI, and then Headwaters Nanokinetix Inc.) reported a productivity of about 900 g-H_2O_2 g-Pd^{-1} h^{-1} in methanol and the presence of NaBr, and about 400 g-H_2O_2 g-Pd h^{-1} in the same solvent without NaBr. They related the productivity to a parameter indicated as the solvent selection parameter (SSP), which is defined as:

$$\text{SSP} = \sum_i (w_i \cdot S_i) \tag{8.1}$$

In Equation (8.1) SSP is the sum for each i component of the products of the weight fraction of solvent component (w) per the solubility of H_2 in the pure component at standard conditions (S). The SSP is thus related to the concentration of dissolved H_2 in solution. SSP is 1.6 for pure methanol and 0.14 for pure water. The productivity to H_2O_2 was found to depend linearly on this parameter. Increasing the alcohol's carbon chain increases the H_2 solubility. For pure isopropanol the SSP is 2.7. The effect of promoters such as NaBr on H_2 solubility is unclear, but at equivalent SSP the catalyst productivity is nearly half, as indicated above.

Few patents report the catalyst productivity. In addition, different reaction conditions and catalysts characteristics (e.g., loading of metal) do not allow a reliable comparison between data. However, data in the patent by Zhou et al. [35, 36] may be compared with those reported by Haas et al. [16] and Paparatto et al. [13, 37] in terms of feed composition, pressure and solvent. They appear to be effectively nearly one order of magnitude higher in terms of productivity [16] or time necessary in autoclave operations to reach a similar H_2O_2 concentration [13, 37]. Selectivity is also apparently higher. Therefore, it appears that, effectively, the method of preparation patented by Headwaters Nanokinetix Inc. (former HTI) leads to more active and selective catalysts. Notably, the patent by Fierro et al. [38], assigned to Repsol Q., also reports results that, in terms of time necessary to reach a similar H_2O_2 concentration in a methanol/water mixture, are comparable to those of Headwaters Nanokinetix. The method of catalyst preparation adopted in the Repsol patent is similar to that of Headwaters Nanokinetix, for example, a resin containing an acid group is used to support the metal. Paparatto et al. [37] of the ENI group also claimed that, by using polyolefin additives during the preparation, the catalyst performances could be significantly enhanced. Therefore, the interaction of Pd with organic linkers during the preparation is a critical parameter that emerges from patent analysis.

Zhou et al. [36] suggested the formation of an intermediate catalytic complex such as that reported in Figure 8.7. The idea is that functional groups in the polymer (or molecular backbone) anchor the metal and, during the heat treatment to remove

Table 8.3 Comparison of selected recent patents on the direct catalytic synthesis of H_2O_2.

Assignee and Ref.	Main feature	Catalyst	H_2O_2 concentration (wt%)	H_2O_2 selectivity (%)	Feed (H_2/O_2) (%)	P (bar)/T (°C)	Solvent	Notes
CSIR [39]	Incorporation of halogen promoters in the catalyst	Pd(5%) on alumina	0.72	41	50/50	1/30	Aq. ($+ Br^-$, H_3PO_4)	
Degussa [40]	Avoid corrosion limiting contact time	0.25% wt Pd-Au (95:5) on α-Al_2O_3	5.2	74	3/20	50/25	Methanol ($+ Br^-$, H_3PO_4)	2 h, 14.1 g H_2O_2 per g Pd.h
Repsol [38]	Non-corrosive sol., use metal supported on an acid halogen-free resin	1.5% Pd on Lewatit K2641	5.8	77	2/48	100/40	CH_3OH : H_2O (96:4) + 12 ppm HBr	2 h
ENI [37]	Sequence of deposition Pd and Pt on sulfonated carbon	Pd(1%) + Pt(0.1%) on carbon	6.6	76	3.6/10	100/6	CH_3OH : H_2O (95:5) + 9 ppm HBr + 300 ppm H_2SO_4	50 h, deposition: first Pd then Pt, (4.5% H_2O_2 conc. and 86% sel. inverting deposition order)

8.2 Direct Synthesis of H_2O_2 from an Industrial Perspective

Ref.	Approach	Catalyst				Solvent/additives	Notes
Headwaters [41]	Controlled surface coordination catal., water-soluble organic additives	Pd-Pt/C (from resin)	1.25	56	3.3/20	51/35	Aq. + 1% H_2SO_4 + 5 ppm NaBr + 2% CH_3OH	Using a 50% H_2–50% O_2 feed 2.3% conc. H_2O_2 & 60% select.
Headwaters [34]	Staged or sequential feed of H_2	0.75% Pd on carbon	10	90	See note	28/45	Methanol, 1% H_2SO_4, 5 ppm NaBr	O_2:H_2 ratio about 2 per stage, 17% O_2 excess on overall basis
ENI [42]	Use of polyolefin additives to improve performances	Pd(1%)-Pt(0.03%)/ carbon funct. with SO_3H (0.35% S) + 3% polySTY	5.9	80	3.6/11	60/25	CH_3OH : H_2O (95:5) + 6 ppm HBr + 200 ppm H_2SO_4	50 h
BASF [43]	Electroless deposition of Pd & Pt on steatite	Pd-Pt on steatite	10.2	72	3/97	50/40	Aq. sol H_3PO_4 (+HBr)	
Arkema [44]	Aq. solution + surfactants	0.7% Pd–0.03% Pt on silica (500 m^2 g^{-1})	10.7	92	4/96	63/60	Aq. sol 3.4% H_3PO_4, 90 ppm NaBr, 5 ppm Br_2 + 5 mg surfactant	$C_6F_{13}C_2H_4S$-O_3H as surfactant

Table 8.3 (Continued)

Assignee and Ref.	Main feature	Catalyst	H_2O_2 concentration (wt%)	H_2O_2 selectivity (%)	Feed (H_2/O_2) (%)	P (bar)/T (°C)	Solvent	Notes
Degussa [16]	Non-explosive mixture, trickle-bed reactor, Pd-Au catalyst	Pd-Au(95:5) on a-Al_2O_3; 2.5% and 0.25% wt Pd for H_2O and CH_3OH as solvent, respectively	5.1	72	3/20	50/25	Methanol + 2 ppm NaBr + 100 ppm H_2SO_4	10 h, 67% H_2 conv. 13.8 gH_2O_2 per gPd.h in methanol [1.6 g (g h)$^{-1}$ in water]
Headwaters [36]	Controlled coordination of catal., solvent selection parameter (SSP)	1% Pd + 0.02% Pt on carbon black, controlled heat treatment	2.5	91	3/97	97/35	CH_3OH + 1% H_2SO_4 + 5 ppm NaBr	150 h, 40% H_2 conv., dependence product. from SSP
Polimeri Eur. [45]	Prod. conc. H_2O_2 (>15%) with stage evaporation/distillation	Pd-Pt/carbon	6.8–7.2	72–75	3.6/13	130/4	CH_3OH:H_2O (97:3) + 4 ppm HBr + 200 ppm H_2SO_4	110 h, conc. H_2O_2 60% after two stage distillation

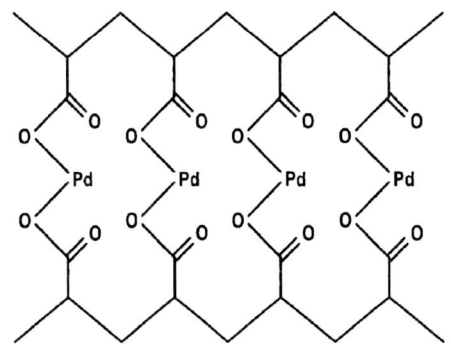

Figure 8.7 Model of the intermediate catalytic complex reported in patent [36] to explain the superior performances of their catalysts [possible precursor after heat treatment of the preferential formation of the selective (1 1 0) face in Pd].

the organic, lead to a controlled growth of Pd particles with straight-chained surface arrangement, such as that indicated as selective in Figure 8.6a. This treatment should be achieved in a careful controlled way and, in fact, the final performance of the catalyst is drastically influenced by this step. Although the concept is not proved, and some inconsistencies also exist in the chemical structure of this intermediate catalytic complex, the general idea that the organic template could determine the nature of the surface of the final Pd particles is probably correct. This aspect has been not investigated in the literature, but as commented below, there are results that are in line with this interpretation.

H_2 conversion was also typically lower than total, ranging from 30 to 70%. A H_2 recycle is thus necessary. Staged (sequential) addition of H_2 to maintain a more uniform $O_2:H_2$ ratio in the reactor and avoid excess O_2 has also been shown to improve performances. Batch-type autoclave or continuous fixed bed (trickle-bed) or stirred reactors have been used. Operations were typically under pressure in the 50–100 bar range, again with the exception of the cited CSIR patents. The reaction temperature ranged from 4 to 605 °C. Upon decreasing the temperature, H_2 solubility increases, but the catalyst specific activity decreases. The productivity should thus pass through a maximum; nevertheless this depends from case to case. Table 8.3 summarizes selected results from recent patents.

Attention has been also paid to the use of non-corrosive solutions. Although bromide ions are typically present, their concentration is kept low at about 4–6 ppm, except in a few cases. NaBr or HBr are used as additives. The use of acids (H_2SO_4, H_3PO_4 principally) also typically promotes performance. In some cases, relatively high acid concentrations are used, but this could create problems of corrosion on one side and purification of produced H_2O_2 on the other side. Relatively low acid concentrations (100–200 ppm) have been demonstrated to be enough for good selectivities, especially using an acidic support. For example, functionalization of carbon with sulfonic groups promotes performance. The presence of sulfur in the carbon also promotes dispersion of the noble metal.

The use of alternative solvents, in particular supercritical CO_2, has also been patented [46]. H_2, O_2 and a CO_2-philic catalyst are solubilized in compressed carbon dioxide to form hydrogen peroxide and use *in situ* the generated H_2O_2 to oxidize organic molecules (in particular, propene to PO). However, reported results are not particularly promising, with a maximum selectivity in H_2O_2 formation of about 40% and a PO yield of about 20% [47]. More aspects on the synthesis of H_2O_2 in compressed CO_2 by direct catalytic oxidation or via the anthraquinone route and the use of generated H_2O_2 for *in situ* green selective oxidation reactions have been discussed by Hancu *et al.* [48].

8.3
Fundamental Studies

Unlike industrial aspects, fundamental studies on the catalytic direct H_2O_2 synthesis have been reviewed recently [3, 48, 49]. We will thus not report a systematic analysis of the state-of-the-art but instead focus the discussion only on some aspects that are necessary for a better understanding of the outlook in this field or which are not clarified sufficiently in the cited reviews. Moreover, we will not discuss here general aspects related to the use of H_2O_2 for sustainable (green) chemistry (integrated or *in situ* production) but will only give some useful selected references [50–60]. The catalytic direct H_2O_2 synthesis offers the opportunity not only to decrease the environmental impact of chemical (e.g., propene oxide, caprolactam, phenol, adipic acid products) and non-chemical (e.g., pulp/paper production) productions, but also to decrease process complexity, energy consumption and risk of operations, as discussed in the previous section.

In addition, being suited also for small-scale integrated productions it offers the practical possibility for development of efficient on-site production of chemicals. Down-sizing chemical processes and development of modular-type integrated chemical processes is one of the objectives of sustainable (green) chemistry [61] and has been inserted into the priorities of the chemical industry by the European Technology Platform of Sustainable Chemistry [62]. In fact, this concept allows us to (i) minimize transport and storage, (ii) avoid large plants and concentration of several processes in a single site (e.g., the environmental impact will exceed its capacity for self-depuration), (iii) save energy and (iv) improve safety. But, in addition, it is an essential component to accelerate the introduction of new sustainable chemical processes, which are often limited by the investments necessary for the development and construction of large-scale processes. A modular design reduces the investment and especially makes possible a more flexible and faster scale-up to different production requirements [63]. It is also more flexible in production – a relevant aspect when market forecasts are difficult.

Notably, advances in the direct, safe and continuous synthesis of neutral H_2O_2 solutions with concentrations of up to 8 wt% from O_2 and water have been reported recently [64] using an improved electrolysis method involving a solid-polymer electrolyte (SPE). The formation and accumulation of neutral H_2O_2 were strongly enhanced by exposing the cathode to a stream of O_2. Although our attention here is on the catalytic direct synthesis of H_2O_2 but several notable advances have been reported on the electrochemical or plasma direct synthesis of H_2O_2 [65–70]. Quite interesting results have been obtained. For example, aqueous H_2O_2 solutions are formed with >93% current efficiency by electrochemical redox reaction of O_2 with H_2 in a fuel cell system using a porous cathode [70]. Using carbon black-type additives for the cathode and a Nafion 117 membrane, concentrations of H_2O_2 up to 7 wt.% are possible, with a rate of formation of $8.3\,L\,h^{-1}\,m^{-2}$ [70]. Moreover, phenol co-generation with electricity by using *in situ* generated H_2O_2 in a H_2–O_2 PEMFC (polymer electrolyte fuel cell) reactor is reported [69]. Therefore, this electrochemical approach could be an interesting opportunity to use H_2 side streams to produce chemicals and electrical energy.

Non-thermal plasma (generated, for example, by pulsed underwater corona discharges) generates H_2O_2 and $^\bullet OH$ radicals and thus represents a new emerging advanced oxidation process (AOP) for water disinfection [67]. Therefore, interesting fields of possible applications of these alternative routes of direct H_2O_2 synthesis are emerging, but actually they represent niche areas and the main industrial application area will be the catalytic direct synthesis of H_2O_2, as noted in the previous section.

8.3.1
Intrinsically Safe Operations and Microreactors

The analysis of patents reported in the previous section evidences the importance given, particularly in most recent years, to the issue of safety in the direct catalytic synthesis of H_2O_2. Even though some patents reported operations inside the explosion region, in most cases attention has been paid to operate below or above the lower and upper limit, respectively. This problem is even more critical for operations using organic solvents and higher pressure, which entail wider explosion limits. Even operations out of this range show safety issues, because in the mixing region or when a dead zone forms in the reactor, local explosion conditions may be present.

The use of catalytic membranes offers the potential of intrinsically safe operations, because a physical separation between H_2 and O_2 in the gas phase is established. The concept is shown in Figure 8.8. The direct synthesis of H_2O_2 is carried out in a catalytic membrane contactor working on the principle of a catalytic diffuser. The latter represents an asymmetric tubular ceramic membrane (e.g., made from alumina) into which a catalytically active material (Pd, Pd/Au, Pd/Pt) is incorporated in highly dispersed form only in the fine-porous surface layer. The reactants are supplied separately on the opposite sides of the membrane, that is, gaseous oxygen on the coarse-porous support side and hydrogen dissolved in a solvent wetting the membrane on the surface-layer side. By applying a controlled overpressure on the gas side, the gas/liquid contact plane is established inside the membrane close to the fine-porous layer with the active catalyst, where the reaction takes place.

This approach offers several advantages over a conventional suspended catalyst: It guarantees safe operation of the process due to the separated supply of both gaseous

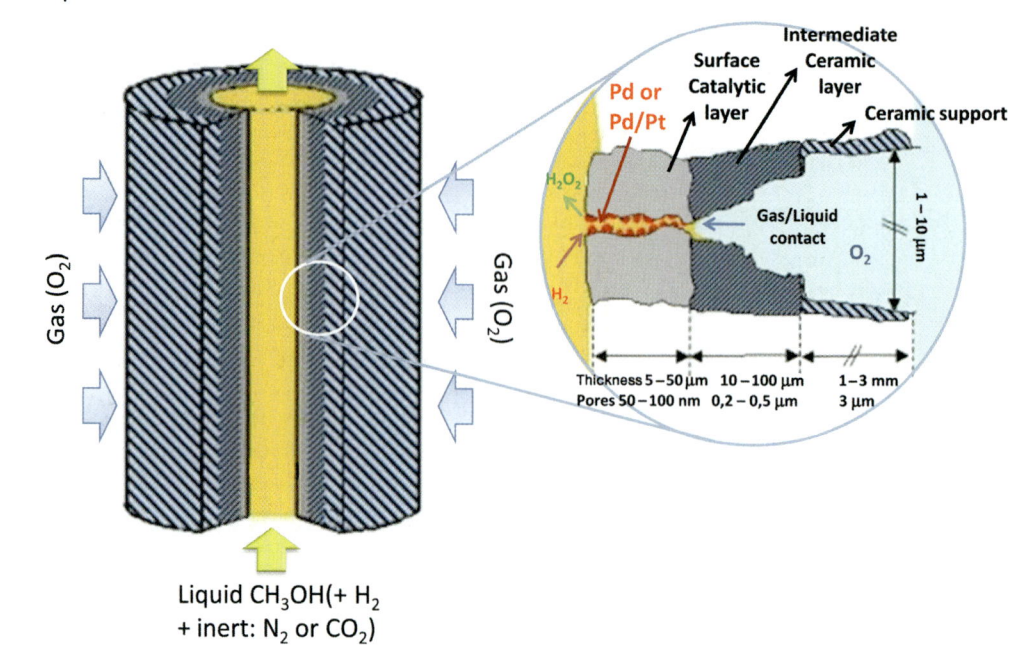

Figure 8.8 Conceptual model of operation of a catalytic diffuser for intrinsically safe direct synthesis of H_2O_2. (Adapted from [71, 72]).

reactants as well as reduced mass transfer limitations by maintaining a thin catalytic zone. At the same time it allows an efficient contact of the gaseous reactants on the surface of the wetted solid catalyst. It can be scaled up rather easily by employing multichannel tubes or capillary bundles, and it is well suited for small-scale, on-site production of H_2O_2.

Figure 8.9 shows the results obtained using this approach, when O_2 is supplied from the dry part of the membrane, that is, from the support side, while diluted H_2 is introduced to the liquid phase [71]. Liquid pressure in these tests was 64 bar, while the pressure differential through the membrane was kept at 3 bar. Liquid phase (methanol) was saturated with a $H_2/N_2 = 9:91$ feed and operations were carried out at room temperature. Total Pd loading on the membrane was about 8 mg. The concentration profile within the catalytic zone is shown in the figure inset and evidences that the molar ratio of O_2 versus H_2 remains above the critical value of 2 everywhere in the reaction zone. In fact, operations in the inverse mode, e.g., saturating the solution with O_2 and feeding H_2 through the membrane, give rise to worse performances (particularly in terms of selectivity, which under similar reaction conditions decreases from above 80% to about 30%), evidencing the critical role of the oxygen to H_2 concentration ratio at the catalyst surface more than in solution, which is essentially governed by the relative solubility. This result suggests that O_2 vs. H_2 coverage on the Pd surface is the critical factor determining selectivity. As H_2O_2 synthesis is a very fast reaction, this surface coverage depends on the relative concentration of O_2 and H_2 in

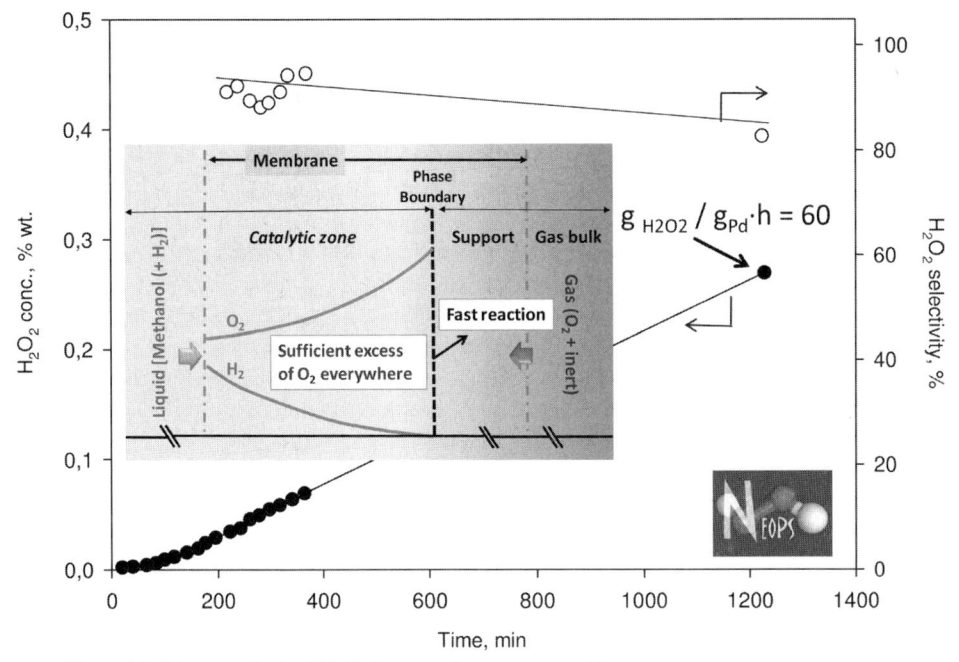

Figure 8.9 Direct synthesis of H_2O_2 in a membrane reactor with supply of pure O_2 through the membrane and partial saturation of the liquid with H_2. Inset: illustration of concentration profiles of H_2 and O_2 within the reaction zone. (Adapted from [71]).

solution (which is a function of the solvent and O_2 and H_2 partial pressures in the gas phase) and on the kinetic diffusivity of H_2 and O_2 across the boundary liquid/solid catalyst interface, as well as the catalyst characteristics. This issue of boundary transport is usually not considered, but simulations indicate that this phenomenon can be an important factor in determining the catalyst performances.

Catalytic reactivity results (H_2O_2 productivity and selectivity) summarized in Figure 8.9 are comparable, taking into account the differences in the reaction conditions, with patented results reported in Table 8.3. Although, for the direct synthesis of H_2O_2, the costs of membrane operations are still higher than those of conventional catalytic reactors, the possibility of safer operations is an incentive, particularly for smaller scale applications.

Various studies and some patents have been published on the use of membrane catalysts for the direct synthesis of H_2O_2 [73–81]. The redox treatment of the membrane influences the properties both in the synthesis and decomposition of H_2O_2. Formation of a hydrophobic layer improves the selectivity, because it limits the consecutive decomposition of hydrogen peroxide, limiting the chemisorption of H_2 and re-adsorption of H_2O_2 [73]. Either polymeric or ceramic-type membranes could be used, but the latter are preferable to allow more robust operations. The mono- or bi-metallic Pd-based active component could be deposited either in the form of dispersed particles (e.g., by precipitation-deposition) or of a thin film (e.g., by electroless plating deposition – EPD). Note that the largest part of patents and literature data refer to dispersed Pd-based particles, even if the use of EPD to prepare catalysts for direct H_2O_2 synthesis has been also patented by BASF [43]. The results obtained on membrane reactors point out that good performances could be obtained adopting both operation methods, but the sensitivity to reaction conditions and the effect of an eventual second metal in alloy with Pd [77–80] are different. A problem with the EPD method could be the formation of β-PdH, when oxygen is in defect, which could cause detachment of the Pd-based thin film [81]. The use of an Pd-Ag interlayer between the ceramic support and the Pd-film reduces this effect [81b]. Formation of β-PdH should be avoided anyway because it catalyzes the hydrogenolysis of H_2O_2 to water, thereby reducing selectivity [79]. These supported Pd-based thin-film catalysts, however, offer the possibility of gas-phase operation for the *in situ* generation of H_2O_2 and simultaneous use to convert benzene into phenol [82]. Direct hydroxylation of aromatic compounds was effectively achieved by using a Pd membrane reactor in which the Pd membrane is thin enough (about 1 μm) to allow permeation of hydrogen below 500 K. In this reactor, the active oxygen species are formed on the surface of Pd via the reaction between oxygen and permeated hydrogen from opposite sides of the membrane. Hydroxylation occurs on the surface of Pd via reaction of the aromatic compound and active oxygen. At a reaction temperature of 423 K, benzene conversion of 15% with a phenol selectivity of 95% was obtained. An increase in reaction temperature, however, caused simultaneous hydrogenation. Figure 8.10 shows the concept.

Reproducibly of results, however, was questioned, because the total oxidation reaction was underestimated [84]. Possibly, these discrepancies derive from the observation of structural changes of the Pd-based membrane during direct hydroxylation of benzene to phenol [85]. The surface state of Pd during the reaction could be divided into two major regions: the oxidized region near the gas

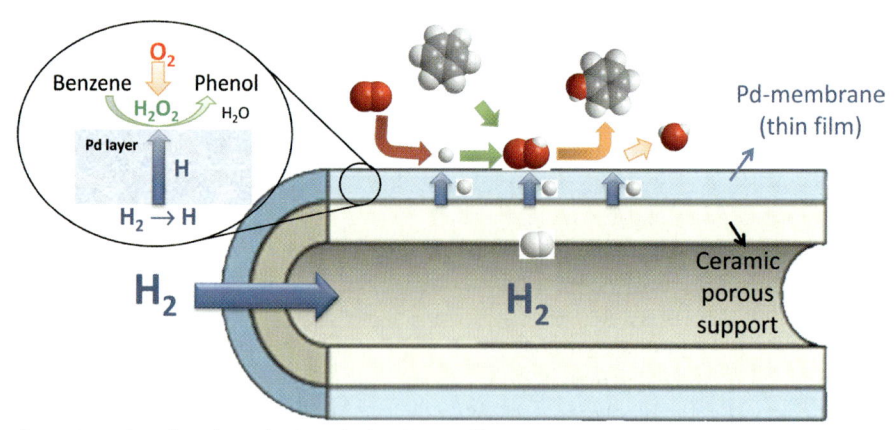

Figure 8.10 Phenol synthesis by direct hydroxylation of benzene using a hydrogen-permeable membrane. (Adapted from [83]).

entrance of the reactor, which favors complete oxidation, and the reduced region near the gas exit, which favors hydrogenation. These two surface states, during reaction, depended on the concentration of oxygen and hydrogen both inside and outside the membrane. After reaction or after both reductive or oxidative pretreatments, significant morphological changes of the membrane were not detected by SEM. Therefore, it may be possible that the poor reproducibility of the results is associated with different effective surface situations of the Pd-thin membrane, but probably more research is needed to clarify these aspects and verify the results. Nevertheless, this example shows how, using novel approaches that integrate *in situ* generation and use of H_2O_2 in a microreactor, it is possible to head towards novel sustainable processes on a small scale. This specific example evidences that phenol can be made on a desktop rather than in a massive chemical plant. The method can be used in the manufacture of pharmaceuticals and many other organic chemicals as well.

A few fundamental studies have also been reported on the topic of hydrogen peroxide synthesis by direct combination of H_2 and O_2 in a microreactor. A recent contribution of Voloshin *et al.* [86] from the Stevens Institute of Technology shows the first results on the role of reaction conditions. A maximum concentration of 1.3 wt% H_2O_2 was achieved. Further studies are necessary, but this interesting direction warrants further investigation.

8.3.2
Nature of the Catalyst and Reaction Network

The nature of the catalyst is a key question to improve further the performances in terms of both H_2O_2 productivity and selectivity. As shown in Section 8.2.2, most patents report a selectivity of around 70–80%, apart from the higher selectivity given by Headwaters Nanokinetix Inc. (formerly HTI), which, however, was not reproduced by other companies or in fundamental studies. Also in terms of catalyst productivity (g-H_2O_2 g-Pd^{-1} h^{-1}) the reported performances are definitively superior. The claim is for a specific surface structure of Pd, where the (1 1 0) crystal plane is preferentially exposed (Figure 8.6a), on which chemisorbed oxygen can give only a two-electron reduction and not a four-electron reduction leading to water. However, no solid proof of this concept has been provided and, on the other hand, it is known that supported Pd crystallites can undergo a fast surface reconstruction in the presence of H_2 and O_2, suggesting that during the synthesis of H_2O_2, particularly at high pressure, a reconstruction of Pd crystallites may occurs. If this is true, it is difficult to accept the idea that the preparation method could lead to the formation of Pd crystallites exposing only one selective crystal face.

One common observation in patents and fundamental studies is that an acidic support promotes the selectivity. H_2O_2 is more stable in acid medium, but a strong acid could create problems of corrosion and further treatment of the solution. One of the motivations for the use of acidic supports was thus the reduction of the required concentration of inorganic acids in solution and hence minimization of its impact on corrosion. However, re-adsorption of H_2O_2 (a weak acid) should be avoided on a strong acidic support, and thus the rate of consecutive reactions of H_2O_2 minimized. Among the solid acids used for the direct synthesis of H_2O_2, zirconia [87], sulfated zirconia [88] and fluorinated alumina [89] should be mentioned. However, low yields of hydrogen peroxide were obtained with these methods. In contrast, excellent yields have been reported using functionalized carbons with sulfonic acid groups or sulfonic acid functionalized polystyrene resins (PS-SO_3H) [90–92]. As commented in the previous section, the functionalization with sulfonic groups of carbon has a similar effect of promotion on the selectivity and performances. Catalysts prepared by anchoring Pd^{II} ions onto PS-SO_3H ion-exchange resins (1.34–1.49% w/w Pd) are highly effective for the direct synthesis of hydrogen peroxide with methanol as solvent at 40 °C. The hydrogen peroxide production rates are about $40\,g\text{-}H_2O_2\,g\text{-}Pd^{-1}\,h^{-1}$, which are good, but not two orders of magnitude higher than other published results, as claimed in patents and open literature [3, 91]. However, the completely different reaction conditions made reliable comparisons difficult, as mentioned in the previous section. Note, however, that the stability of organic resins at high pressure and in the presence of a strong oxidant (H_2O_2) could certainly be an issue for long-term performance. This is why companies preferred typically to use as the support graphite-type carbons (with good stability against oxidation), eventually functionalized with sulfonic groups or which already contain sulfur (carbon black), which reasonably create sulfonic groups during the carbon pre-treatment before anchoring the active component (mono- or bimetallic Pd).

The question is whether these sulfonic groups have only a role in changing the zero-point charge of the surface and improve the dispersion of Pd-based particles, or have an additional role. Fierro et al. [91] suggested on the basis of XPS studies that the high performances of Pd/PS-SO_3H systems result from the ability of the sulfonic acid groups of the resin to interact with and stabilize the Pd^{II} ions without further reduction to metallic palladium. This interaction creates a positive charge on the Pd particles, which enhances the selectivity for hydrogen peroxide.

There are no reliable data that indicate whether this conclusion is valid for all types of catalysts or is instead specific for the catalysts studied by Fierro et al. [91], because -SO_3H functional groups also show other functions to which the effect on selectivity could be related. Fierro et al. [3], in their review, demonstrated that the high selectivity for H_2O_2 can be achieved in the presence of cationic Pd species, but such proof has not been reported in literature, because the reaction mechanism is complex and changes in the surface state of Pd as the H_2O_2 synthesis reaction occurs. Therefore, there are still no unequivocal conclusions on the real nature of the active and selective sites in direct H_2O_2 synthesis.

A parameter that determines the performances, as outlined also in patents, is the mean diameter of Pd or doped-Pd particles. This is also one of the claims in Headwaters Nanokinetix Inc. patents. It seems that a maximum in the activity/selectivity as a function of the particle size is present. Figure 8.11 reports the effect of the Pd-particle size (Pd supported on alumina, prepared by deposition–precipitation method) in the direct synthesis of H_2O_2 in water at atmospheric pressure [77]. Low

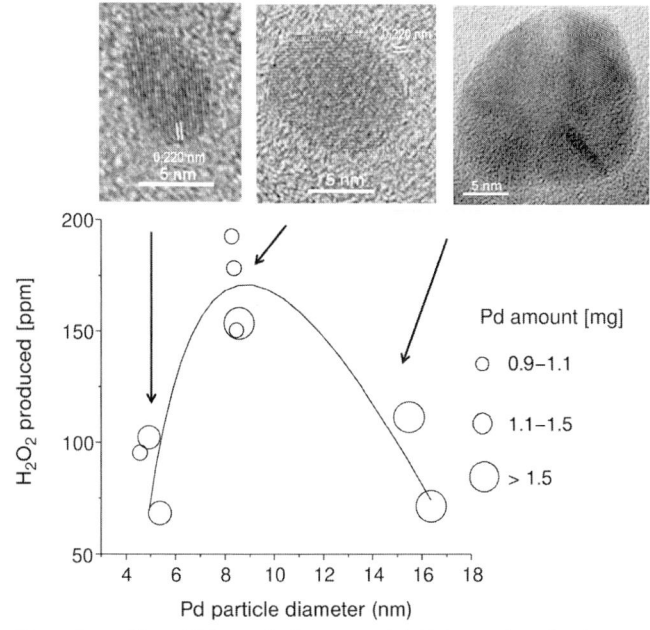

Figure 8.11 Effect of Pd particle diameter in Pd/Al$_2$O$_3$ catalysts for direct H$_2$O$_2$ synthesis (water, atmospheric pressure). (Adapted from [77]).

pressure is necessary, to minimize the possible changes of the Pd particles during the reaction.

Data indicate the presence of a maximum in the performances that is nearly independent on the metal loading. Transmission electron micrographs of the Pd particles indicate that the larger particles (around 15–16 nm) show an irregular, multifaced structure. A more regular structure is observed in the smaller round particles, but apparently with different preferential orientation depending on the dimensions. Note that in the claims of Headwaters Nanokinetix Inc. patents the existence of an optimal diameter for Pd particles is also indicated, to which the preferential exposure of the active/selective surface is related. Data in Figure 8.11 do not prove this statement, but are in line with this indication.

This concept is further demonstrated in Figure 8.12, which reports the performances of two Pd thin-film membranes with different morphology in H$_2$O$_2$ synthesis at room temperature and atmospheric pressure.

A more irregular and probably defective structure of Pd-crystallites leads to a much lower rate of H$_2$O$_2$ synthesis with respect to more regular crystals, even if the specific metal surface area is lower [79, 80]. It was also observed that the more irregular (defective) crystals show a higher rate of conversion of H$_2$O$_2$ into water. Figure 8.13 shows a simplified model to explain why steps and kinks on the Pd crystal surface could favor the dissociative chemisorption of O$_2$ that leads to the formation of water instead of H$_2$O$_2$, and the decomposition of H$_2$O$_2$ to form water, both of which are responsible for lowering selectivity.

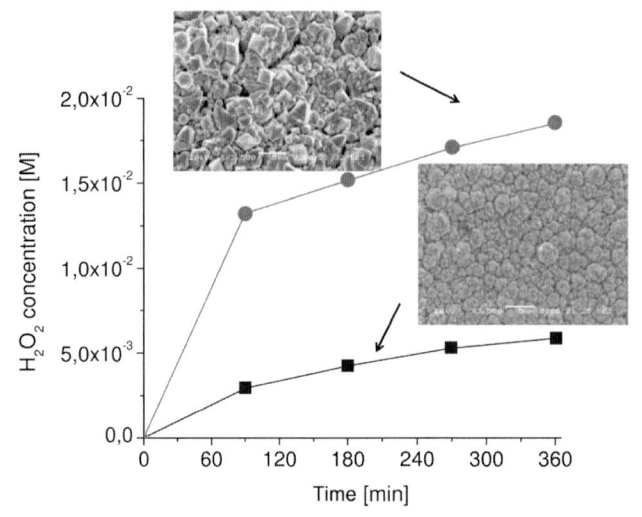

Figure 8.12 Effect of morphology of Pd thin-film membranes for direct H_2O_2 synthesis (water, atmospheric pressure). (Adapted from [79, 80]).

The key aspects determining the selectivity in H_2O_2 formation and in turn catalyst productivity are:

- The relative surface coverage of chemisorbed bimolecular oxygen vs. chemisorbed hydrogen, which depends on the surface crystalline structure and the O_2/H_2 concentration in the liquid/solid boundary layer; following the simplified approach shown in Figure 8.6a, when chemisorbed O_2 is surrounded in close vicinity by two or more chemisorbed H_2 species, a four-electron reaction takes place, leading to water, instead of a two-electron reaction that leads to H_2O_2.
- The relative rates of undissociative O_2 chemisorption vs. dissociative O_2 chemisorption, which depend in general terms on the presence of defective Pd sites on the surface (Figure 8.13).

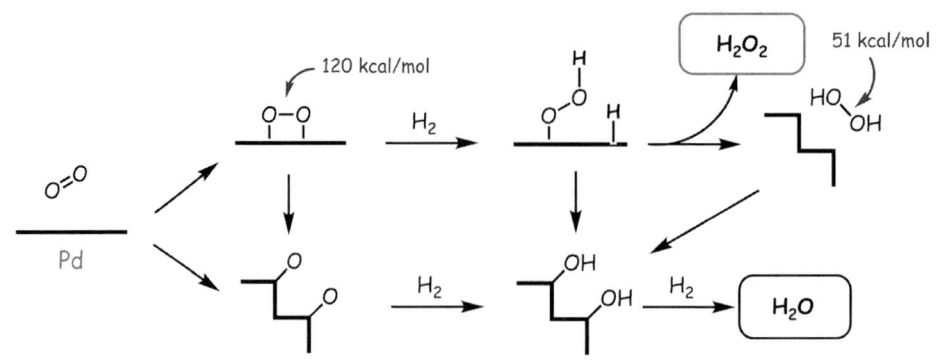

Figure 8.13 Simplified model to explain the role of defects in Pd in lowering selectivity [78].

- The effect of H_2 in the modification of Pd properties. This aspect has never been studied in relation to direct H_2O_2 synthesis, but could be important, particularly in explaining the role of Pd particle size and some of the modification effects of catalytic behavior with time on stream. In fact, it is known that H_2 may rapidly dissociate on Pd and diffuse in the bulk as H atoms. This is the principle the basis of the permeoselective behavior of Pd membranes for H_2 separation. The formation of interstitial H atoms leads to an expansion of the Pd lattice even before the formation of regular structures such as Pd-hydrate. In small Pd crystallites, H cannot diffuse in the bulk and thus remains mainly on the surface. In medium-size Pd crystallites the diffusion of H in the bulk reduces the surface concentration and at the same time induces an expansion of the lattice, leading to an increase of the Pd–Pd distance. Both these effects contribute in limiting the $4e^-$ vs. $2e^-$ reaction with O_2, and thus allow an increase in selectivity. This effect does not exclude the possibility of stabilization of specific, more effective crystal planes, due to an interaction with the support. For larger particles, the more irregular structure, and the lowering of the surface to bulk ratio, leads to a further lowering of performance. It is thus possible to explain the presence of a maximum in activity and selectivity (Figure 8.12), even if the mechanism discussed above is still rather speculative and further specific studies are necessary to clarify the reaction mechanism and the nature of the surface sites.

In general terms a better link of the reaction network to the nature of the active Pd surface is needed. An effort in this direction was made by Abate et al. [79] and is summarized in Figure 8.14. Reference is made to this work to discuss the details of the relationship between Pd species and selectivity in H_2O_2 formation, even if, notably, further results are necessary for a better understanding.

Based on this discussion, it is possible to clarify some aspects of the literature interpretations on the nature of the active species. Lunsford and coworkers [49, 93] published many papers indicating that the active species is colloidal palladium, which implies the easy dissolution of Pd in solution. An analysis of patents clearly reveals that this is not the case for active catalysts and various patents explicitly indicate that there is no leaching of Pd. On the other hand, a colloid would be difficult, if not impossible, to manage in a commercial process and its recovery would be not viable at the very low concentrations of dissolved metal employed. In addition, the presence of even traces of Pd in commercial H_2O_2 could be extremely dangerous in terms of the possibility of explosion. Finally, if the solid is a simple reservoir for Pd going into solution, a deactivation is expected with time-on-stream in continuous operations.

It is known that soluble Pd complexes also catalyze the direct synthesis of H_2O_2 and it is thus not unlikely that colloidal particles could be active in the reaction. However, the first issue in studying commercial catalysts for the direct heterogeneous H_2O_2 synthesis is to verify the absence of leaching of the metal. In conclusion, the indication that the active species in direct H_2O_2 synthesis is a colloidal Pd species is a confusing interpretation, which, however, is still reported in reviews on this subject [3, 49]. Note, however, that Lunsford and coworkers slightly corrected this hypothesis, indicating that, due to the high HCl concentrations they used, formation

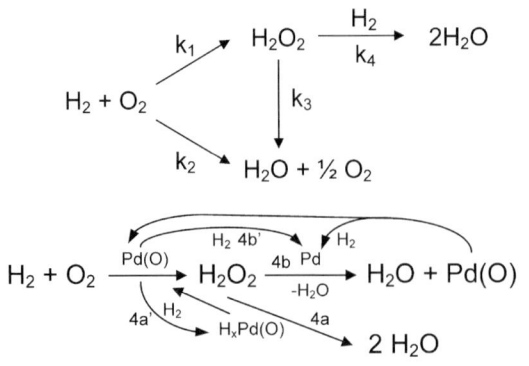

Figure 8.14 Reaction network and role of different Pd species in the direct synthesis of H_2O_2 [79].

of colloidal Pd in water occurs, but not in ethanol. In fact, the use of ethanol as a solvent allows productivities that are about two times higher [94].

A similar question is whether Pd or other metals such as Au are active/selective in the reaction. In this regard, recently [96–100], starting from the earlier observations of Hutchings et al. [95], several authors investigated supported gold catalysts for the direct H_2O_2 synthesis, even though Hutchings and coworkers in subsequent papers clearly showed that Au-Pd catalysts are superior to Au catalysts and that the microstructure of the metal alloy nanoparticles shows a core–shell morphology with Pd concentrated on the catalyst surface [101]. They showed also that when, due to redox treatments, the situation is inverted (Pd core and Au-rich external shell) the performances are lowered [102]. This evidences that gold, for which some activity was verified, shows definitively lower performances with respect to Pd. The role of gold is thus mainly due to the modification of the characteristics of Pd, possible in the directions discussed above. Patent analysis reported in the previous section also evidences that Au is usually used in low amounts (typically Pd/Au = 95 : 5) and its effect is quite similar to Pt, which is instead less active in the direct synthesis of H_2O_2 and catalyzes the H_2O_2 decomposition. The promotion effect of Au is not higher than that of Pt and, in fact, most companies (except BASF) use Pd-Pt catalysts instead of Pd-Au catalysts. The effect of Au, similarly to Pt, is therefore mainly due to its action as a structural promoter for the active surface of Pd, even if also electronic or other effects could contribute in determining the performances. Notably, theoretical investigations on direct synthesis of H_2O_2 have focused on the analysis of gold clusters, assuming a peculiar reactivity behavior of these active sites [103, 104]. The above observations evidence that the problem in the direct synthesis of H_2O_2 is different.

Finally, if the role of Au is essentially to act as a structural promoter for Pd, it is expected that the support influences this effect. The observation of Hutchings et al. [105] that Pd-Au/carbon catalysts show the presence of a homogeneous alloy (both Au and Pd on the surface) and improved reactivity with respect to Pd-Au/oxide (TiO_2, Al_2O_3 and Fe_2O_3), where a core–shell structure is observed, is thus not surprising. This result evidences that Au has a structural effect more than an electronic effect [106].

In view of the above considerations, the oxidation state of Pd has a significant effect and it was demonstrated that an oxidative pretreatment of the catalyst enhances performance [76–80], although clearly this is a transitory effect because the catalyst modifies with time on stream. Recently, these observations were rediscovered [106].

8.3.3
Role of the Solvent and of Promoters

The role of solvent and of promoters has been extensively discussed in open literature, but based on the discussion of patents reported in the previous section it may be evidenced that the critical aspects were not properly identified. We thus report some specific comments below, which are not intended to be a systematic discussion of the state-of-the-art:

- The use of organic solvents promotes by one order of magnitude the productivity, and this fact is essentially linked to the higher solubility of H_2 and O_2. However, organic solvents further limit the possible range of operations to avoid explosive mixtures. Operations with very high O_2 concentrations, which increase the selectivity, are possible only using water as solvent.

- Water as solvent is good when the use of the so-produced hydrogen peroxide solution is also compatible with the use of water and with the low concentration obtained (typically not more than 1–2 wt%), because the concentration operation is costly and risky. Therefore, for wastewater or some bleaching treatments the methodology could be applied, but for organic syntheses which typically require methanol as the solvent the synthesis in this latter medium is necessary.

- Analysis of solubility data indicates that increasing the chain length in alcohols increases the H_2 and O_2 solubility, and thus higher performances could be possible. The finding that ethanol is better than methanol [107] is thus not surprising. The use of isopropanol could be even better. However, it is academic project to study the reaction in ethanol, because ethanol is not suited to be used in the consecutive steps, for example, for propene oxide synthesis using TS-1 as catalyst. For this reason, nearly all patents using an organic medium focus on the use of methanol. Notably, in the literature it was suggested that, when using ethanol as solvent, acetate ions are formed and that these ions play a critical role in selectivity, blocking unselective Pd ensembles [108]. Acetate ions could be not formed from methanol, but the performances are equally good. Therefore, the role of acetate ions to promote selectivity does not seem to be of general validity, and can be associated with the specific investigated catalysts.

- Chlorine ions should be avoided for two main reasons: (i) corrosion and (ii) possible formation of dangerous Cl_2. Both factors are clearly critical for an industrial process and, for this reason, in patents H_2SO_4 or H_3PO_4 are used as acid additives. The tendency is to reduce them as much as possible, but compatibly with the need to obtain stable H_2O_2 solutions (in neutral solution H_2O_2 can easily decompose and it is thus not safe to produce such solutions; the claim reported in some patents and papers for the synthesis of H_2O_2 at neutral pH is thus of limited applied value). Actual typical values are 100–200 ppm of acid, which are compatible with conventional stainless steel. In addition, it is well known that Cl^- ions can form complexes with noble metals, favoring their leaching. Therefore, the use of Cl^- in the direct synthesis of H_2O_2 is definitively not recommended. Notably, many academic studies have been focused on the role of chloride ions in the direct synthesis of H_2O_2 [107, 109].

- NaBr or HBr is typically added in small amounts (few ppm), because an increase of productivity and, in part, of the selectivity is achieved. However, good performances (>70% selectivity) could be obtained also without these additives. Halide ions thus have some influence, but the effect is of secondary order, at least at the typical concentrations used for industrial development. It is thus surprising that a large number of publications (made last year; more were published in previous years), with several duplications, deal only with the effect of halides, often for not relevant concentrations [110]. In general, no specific advances have to be mentioned.

- Compressed CO_2 is an interesting, potentially clean solvent for the synthesis of H_2O_2, and also allows use of the same solvent for *in situ* clean selective oxidation reactions [48, 111–113]. CO_2-soluble catalysts have to be used. Even if potentially interesting, the results are still unsatisfactory for industrial development. Furthermore, there are safety aspects in decompressing CO_2 that should be better understood. The other alternative is to use CO_2 to expand organic solvents, to promote their properties (H_2 and particularly O_2 diffusivity). CO_2 can be the ballast to dilute the H_2-O_2 feed in substitution of N_2, and thus do not represent a cost. Preliminary results using CO_2-expanded methanol are very promising, with a nearly twice selectivity and a large increase of productivity compared with the same experiments performed using N_2 as the diluents [114]. These results were preliminary, and obtained at a maximum total pressure of 6.5 bar. However, when extended to a higher pressure range they could allow the synthesis of H_2O_2 under milder conditions (40 bar instead the ca. 100 bar reported in several patents), thereby decreasing the operational and fixed (reactor) costs.

8.4
Conclusion

Worldwide demand for H_2O_2 is increasing along with the desire to develop a process other than the conventional anthraquinone route, for the synthesis of hydrogen peroxide directly from H_2 and O_2. This chapter has discussed the state-of-the-art in patents and industrial development, evidencing that the process of catalytic direct synthesis of H_2O_2 will probably be commercialized within a few years. The main area of application will be for integration with the new process of propene oxide (PO) production from H_2O_2 and propene. Other potential large-scale chemical processes to which it could be integrated are caprolactam and phenol syntheses. For all these processes, a solution of at least 4–5% H_2O_2 in an organic solvent (methanol for PO) should be produced. This target has actually been met, but an improvement in the selectivity and productivity is still needed. Some of the patented results, however, already evidence that industrial targets of selectivity and productivity could be achieved using specific catalyst preparations, which are claimed to enhance the preferential formation of selective surface coordination of metal particles. Although, no definitive proof of this concepts is available, various indications that support this conclusion are presented. In water, the selectivity and productivity are definitively lower than in alcohols, but it is possible to use larger concentrations of O_2 in the feed. The target for aqueous solutions of H_2O_2 is mainly for wastewater treatment.

Intense activity on the fundamental side is also present in the literature. However, it is evidenced that, often, the aspects investigated were not the most relevant. Some suggested critical aspects to study are thus presented.

References

1 Jones, C.W. (1999) *Applications of Hydrogen Peroxide and Derivatives, RSC Clean Technology Monographs*, The Royal Society of Chemistry, Cambridge (UK).
2 (a) Hess, W.T. (1995) *Kirk-Othmer Encyclopedia of Chemical Technology*, 4th edn, Vol. 13, John Wiley & Sons, Inc., New York, pp. 961–995; (b) Eul, W., Moeller, A. and Steiner, N. (2001) Hydrogen Peroxide in *Kirk-Othmer Encyclopedia of Chemical Technology*, Online Edition (5th ed.), John Wiley & Sons, Inc., Hoboken, NJ.
3 Campos-Martin, J.M., Blanco-Brieva, G. and Fierro, J.L.G. (2006) *Angewandte Chemie – International Edition*, **45**, 6962.
4 (a) Delgado-Oyague, J.A., de Frutos, M.P. and Padilla-Polo, A. (1998) EP0839760, assigned to Repsol Q. (Spain); (b) de Frutos, M.P., Padilla-Polo, A. and Campos-Martin, J.M. (2001) EP1074548, assigned to Repsol Q. (Spain).
5 (a) Chen, Q. (2006) *Journal of Cleaner Production*, **14**, 708; (b) Chen, Q. (2008) *Chemical Engineering and Processing: Process Intensification*, **47**, 787.
6 Centi, G. and Perathoner, S. (2003) Selective Oxidation. Industrial, in *Encyclopedia of Catalysis* (Chief ed. I.T. Horváth), John Wiley & Sons, New York, Vol. 6, p. 239.
7 Chemical Week, June 27, p. 25 (2007) .
8 Perathoner, S. and Centi, G. (2005) *Topics in Catalysis*, **33**, 207.
9 Centi, G., Cavani, F. and Trifiro, F. (2001) *Selective Oxidation by Heterogeneous Catalysis*, Kluwer Academic/Plenum Publishers, New York.
10 Sheldon, R.A., Arends, I. and Hanefeld, U. (2007) *Green Chemistry and Catalysis*, Wiley-VCH, Weinheim.
11 Henkel, H. and Weber, W. (1914) US Patent 1,108,752.
12 Gosser, L.-W. and Schwartz, J.-A.T. (1989) US Patent 4,832,938, assigned to Du Pont de Nemour & Co. (S).
13 Paparatto, G. and De Alberti, G. (2003) US Patent 6,649,140, assigned to Eni and Enichem (Italy).
14 Fisher, M., Karbel, G., Stammer, A. *et al.* (2002) US Patent 6,375,920, assigned to BASF (Germany).
15 Zhou, B. and Lee, L.-K. (2001) US Patent 6,168,775, assigned to Hydrocarbon Techn. Inc. (US).
16 Haas, T., Stochniol, G. and Rollmann, J. (2006) US 7,005,528, assigned to Degussa AG (Germany).
17 Lewis, B. and Von Elbe, G. (1961) *Combustion, Flames and Explosions in Gases*, 2nd edn, Academic Press, New York, p. 22.

18 (a) Hutchins, H. (1998) WO Patent 9831629, assigned to Advanced Peroxide Technology (US); (b) Hutchins, H. (2003) WO Patent 082458, assigned to Princeton Advanced Techn. (US).
19 Brophy, J. (2005) *Focus on Catal*, **2**, 1.
20 Tonkovich, A.L., Jarosch, K.T.P. and Hesse, D.J. (2006) US Patent 7,029,647, assigned to Velocys (US).
21 Dada, E. and Lawal, A. (2007) US DOE Chemicals Industry of the future. Success Stories, http://www1.eere.energy.gov/industry/chemicals/pdfs/microchannel_h202_highlight.pdf.
22 *Chem & Eng News*, **82**, 39 (2004).
23 Brasse, C. and Jaeger, B. (2006) *Degussa Science Newsletter*, **17**, 4.
24 Zhou, B.(Sept. 2007) (Headwaters Technology), presentation for US EPA Presidential Green Chemistry Award.
25 Gopal, R. (2004) US Patent 6,712,949 assigned to The Electrosynthesis Co. (US).
26 Lehmann, T. Stenner, P. (2004) US Patent 6,685,818, assigned to Degussa (Germany).
27 Agladze, G.R., Tsurtsumia, G.S., Jung, B.-I. *et al.* (2007) *Journal of Applied Electrochemistry*, **37**, 375.
28 Zhou, J., Guo, H., Wang, X. *et al.* (2005) *Chemical Communications*, 1631.
29 Brill, W.F. (1987) US Patent 4, 661,337, assigned to Halcon SD Group (US).
30 Izumi, Y. Miyazaki, H. and Kawahara, S.-i. (1981) US Patent 4,279,883, assigned to Toyuyama Soda Ka Ka. (Japan).
31 Huckins, H.A. (2004) US Patent 6,752,978, assigned to Advanced Peroxide Techn. (US).
32 Germain, A. Pirard, J.-P. Delattre, V. *et al.* (1996) US Patent 5,500,202, assigned to Solvay (Belgium).
33 Dalton, A.I. and Skinner, R.W. (1982) US Patent 4,336,239, assigned to Air Products and Chem (US).
34 Rueter, M. (2006) US Patent 7,105,143, assigned to Headwaters Nanokinetix (US).
35 Zhou, B., Rueter, M.A., Lee, L.-K. and Pelrine, B.P. (2003) US Patent 6,576,214, assigned to Hydrocarbon Techn. Inc. (US).
36 Zhou, B., Rueter, M. and Parasher, S. (2006) US Patent 7,011,807, assigned to Headwaters Nanokinetix Inc. (US).
37 Paparatto, G., D'Aloisio, R., de Alberti, G. and Buzzoni, R. (2006) US patent 7,122,501, assigned to Eni and Enichem (Italy).
38 de Fruto Escrig, P. Campos Martin, J.M. Garcia Fierro, J.L. *et al.* (2007) US Patent 7,179,440, assigned to Repsol Quimica SA (Spain).
39 Choudhary, V.R., Samanta, C. and Jana, P. (2007) US Patent 7,288,240, assigned to Council of Scientific & Industrial Research – CSIR (India).
40 Haas, T., Stochniol, G. and Jahn, R. (2007) US Patent 7,241,908, assigned to Degussa AG (Germany).
41 Rueter, M., Zhou, B. and Parasher, S. (2006) US patent 7,144,565, assigned to Headwaters Nanokinetix Inc. (US).
42 Paparatto, G., De Alberti, G., D'Aloisio, R. and Bruzzoni, R. (2006) US Patent 7,101,526, assigned to ENI & Polimeri Europa (Italy).
43 Fischer, M. and Butz, T. (2006) US Patent 7,070,757, assigned to BASF AG. (Germany).
44 Devic, M. (2006) US Patent 7,060,244, assigned to Arkema (France).
45 Paparatto, G., Rivetti, F., Andrigo, P. *et al.* (2006) US Patent 7,048,905, assigned to Polimeri Europa and ENI (Italy).
46 Backman, E.J. and Hancu, D. (2003) US Patent 6,656,446, assigned to University of Pittsburgh (US).
47 Chen, Q. (2007) Direct synthesis of hydrogen peroxide from oxygen and hydrogen using carbon dioxide as an environmentally benign solvent and its application in green oxidation, PhD Thesis, University of Pittsburgh.
48 Hancu, D., Green, J. and Beckman, E.J. (2002) *Accounts of Chemical Research*, **35**, 757.
49 Lunsford, J.H. (2003) *Journal of Catalysis*, **216**, 455.

50 Yoshida, J.-i. and Okamoto, H. (2006) *Advanced Micro & Nanosystems*, **5** (Micro Process Engineering), 439.

51 Clerici, M.G. (2005) DGMK Tagungsbericht 2 (Proceedings of the DGMK/SCI-Conference "Oxidation and Functionalization: Classical and Alternative Routes and Sources" 2005), p. 165.

52 Grigoropoulou, G., Clark, J.H. and Elings, J.A. (2003) *Green Chemistry*, **5**, 1.

53 Zhao, J., Zhou, J., Su, J. *et al.* (2007) *AICHE Journal*, **53**, 3204.

54 Yamashita, H., Miura, Y., Mori, K. *et al.* (2007) *Pure and Applied Chemistry*, **79**, 2095.

55 Kuznetsova, N.I., Kirillova, N.V., Kuznetsova, L.I. *et al.* (2007) *Journal of Hazardous Materials*, **146**, 569.

56 Joshi, A.M., Delgass, W.N. and Thomson, K.T. (2007) *The Journal of Physical Chemistry C*, **111**, 7841.

57 (a) Lee, D.-K., Cho, I.-C., Lee, G.-S. *et al.* (2004) *Separation and Purification Technology*, **34**, 43; (b) Kim, S.-C., Park, H.-H. and Lee, D.-K. (2003) *Catalysis Today*, **87**, 51.

58 (a) Park, E.D., Hwang, Y.-S. and Lee, J.S. (2001) *Catalysis Communications*, **2**, 187; (b) Wang, Y. and Otsuka, K. (1995) *Journal of Catalysis*, **155**, 256.

59 Hölderich, W.F. and Dahlhoff, G. (2001) *Chemical Innovation*, **2**, 29.

60 Strukul, G.(ed.) (1992) *Catalytic Oxidations with Hydrogen Peroxide as Oxidant*, Kluwer, Dordrecht.

61 Centi, G. and Perathoner, S. (2003) *Catalysis Today*, **77**, 287.

62 The European Technology Platform for Sustainable Chemistry (Strategic Research Agenda and Implementation Action Plan documents) http://www.suschem.org/.

63 Cavani, F., Centi, G., Perathoner, S. and Trifirò, F. (2008) *Sustainable Industrial Chemical Processes – Principles Tools and Case Studies*, Wiley-VCH Verlag, Weinheim, in preparation.

64 Yamanaka, I. and Murayama, T. (2008) *Angewandte Chemie – International Edition*, **47**, 1900.

65 Yamanaka, I., Hashimoto, T., Ichihashi, R. and Otsuka, K. (2008) *Electrochimica Acta*, **53**, 4824.

66 Mededovic, S. and Locke, B.R. (2007) *Journal of Physics D – Applied Physics*, **40**, 7734.

67 Gupta, S.B. and Bluhm, H. (2007) *Water Science and Technology*, **55** (12, Oxidation Technologies for Water and Wastewater Treatment IV), 7.

68 Mittal, V., Kunz, H.R. and Fenton, J.M. (2006) *ECS Trans*, **1** (8, Durability and Reliability of Low-Temperature Fuel Cells Systems), 295.

69 Cai, R., Song, S., Ji, B. *et al.* (2005) *Catalysis Today*, **104**, 200.

70 (a) Yamanaka, I., Onizawa, T., Takenaka, S. and Otsuka, K. (2003) *Angewandte Chemie – International Edition*, **42**, 3653; (b) Yamanaka, I., Hashimoto, T. and Otsuka, K. (2002) *Chemical Letters*, **8**, 852.

71 (a) Dittmeyer, R. and Pashkova, A. Development of an inherently safe, cost-efficient and flexible process for direct synthesis of hydrogen peroxide from hydrogen and oxygen by means of catalytic membranes; (b) Pashkova, A., Svajda, K. and Dittmeyer, R. (2008) *Chemical Engineering Journal*, 139, 165; (c) Centi, G. and Dittmeyer, R., Final report EU NEOPS project G5RD-2002-00678.

72 Centi, G., Dittmeyer, R., Perathoner, S. and Reif, M. (2003) *Catalysis Today*, **79–80**, 139.

73 (a) Choudhary, V.R. Gaikwad, A.G. and Sansare, S.D. (2001) *Angewandte Chemie – International Edition*, **40**, 1776; (b) Choudhary, V.R. Sansare, S.D. and Gaikwad, A.G. (2002) US Patent 6,346,228, assigned to Council of Scientific and Industrial Research (India).

74 Bertsch-Frank, B., Hemme, I., Von Hippel, L. *et al.* (2001) EP 1,127,839, assigned to Degussa-Huls AG (Germany).

75 Vulpescu, G.D., Ruitenbeek, M., van Lieshout, L.L. *et al.* (2004) *Catalysis Communications*, **5**, 347.

76 Abate, S., Centi, G., Melada, S. et al. (2005) *Catalysis Today*, **104**, 323.
77 Melada, S., Pinna, F., Strukul, G. et al. (2005) *Journal of Catalysis*, **235**, 241.
78 Melada, S., Pinna, F., Strukul, G. et al. (2006) *Journal of Catalysis*, **237**, 213.
79 Abate, S. Centi, G. Perathoner, S. et al. (2006) *Topics in Catalysis*, **38**, 181.
80 Abate, S., Melada, S., Centi, G. et al. (2006) *Catalysis Today*, **117**, 193.
81 (a) Abate, S., Perathoner, S., Genovese, C. and Centi, G. (2006) *Desalination*, **200**, 760; (b) Abate, S., Centi, G., Perathoner, S. and Frusteri, F. (2006) *Catalysis Today*, **118**, 189.
82 (a) Niwa, S. Eswaramoorthy, M. Nair, J. et al. (2002) *Science*, **295**, 105; (b) Orita, H. and Itoh, N. (2004) *Applied Catalysis A – General*, **2588**, 17; (c) Sato, K., Hanaoka, T.-a., Niwa, S.-i. et al. (2005) *Catalysis Today*, **104**, 260.
83 Mizukami, F. (2005) *AIST Today*, **18** (Autumn), 4.
84 Vulpescu, G.D., Ruitenbeek, M., van Lieshout, L.L. et al. (2004) *Catalysis Communications*, **5**, 347.
85 Sato, K., Hanaoka, T.-a., Hamakawa, S. et al. (2006) *Catalysis Today*, **118**, 57.
86 Voloshin, Y., Halder, R. and Lawal, A. (2007) *Catalysis Today*, **125**, 40.
87 Choudhary, V.R. and Samanta, C. (2006) *Journal of Catalysis*, **238**, 28.
88 Melada, S., Rioda, R., Menegazzo, F. et al. (2006) *Journal of Catalysis*, **239**, 422.
89 Burch, R. and Ellis, P.R. (2003) *Applied Catalysis B – Environmental*, **42**, 203.
90 (a) de Frutos, M.P., Campos-Martin, J.M., Fierro, J.L.G. et al. (2003) EP1344747, assigned to Repsol Q. (Spain); (b) de Frutos, M.P., Padilla, A., Riesco, J.M. et al. (2004) EP1443020, assigned to Repsol Q. (Spain).
91 Blanco-Brieva, G., Cano-Serrano, E., Campos-Martin, J.M. and Fierro, J.L.G. (2004) *Chemical Communications*, 1184.
92 Burato, C., Centomo, P., Rizzoli, M. et al. (2006) *Advanced Synthesis and Catalysis*, **348**, 255.
93 (a) Dissanayake, D.P. and Lunsford, J.H. (2002) *Journal of Catalysis*, **206**, 173; (b) Dissanayake, D.P. and Lunsford, J.H. (2003) *Journal of Catalysis*, **214**, 113; (c) Chinta, S. and Lunsford, J.H. (2004) *Journal of Catalysis*, **225**, 249.
94 Han, Y.-F. and Lunsford, J.H. (2005) *Catalysis Letters*, **99**, 13.
95 Landon, P., Collier, P.J., Papworth, A.J. et al. (2002) *Chemical Communications*, 2058.
96 Yildiz, M. and Akin, A.N. (2007) *Turkish Journal of Chemistry*, **31**, 479.
97 Han, Y.-F., Zhong, Z., Ramesh, K. et al. (2007) *J Phys Chem C*, **111**, 8410.
98 Ma, S., Li, G. and Wang, X. (2006) *Chemical Letters*, **35**, 428.
99 Ishihara, T., Ohura, Y., Yoshida, S., Hata, Y. et al. (2005) *Applied Catalysis A – General*, **291**, 215.
100 Okumura, M., Kitagawa, Y., Yamagcuhi, K. et al. (2003) *Chemical Letters*, **32**, 822.
101 (a) Edwards, J.K., Solsona, B., Landon, P. et al. (2005) *Journal of Materials Chemistry*, **15**, 4595; (b) Edwards, J.K., Solsona, B.E., Landon, P. et al. (2005) *Journal of Catalysis*, **236**, 69; (c) Solsona, B.E., Edwards, J.K., Landon, P. et al. (2006) *Chemistry of Materials*, **18**, 2689.
102 Herzing, A.A., Carley, A.F., Edwards, J.K. et al. (2008) *Chemistry of Materials*, **20**, 1492.
103 Wang, F., Zhang, D., Sun, H. and Ding, Y. (2007) *J Phys Chem C*, **111**, 11590.
104 (a) Joshi, A.M., Delgass, W.N. and Thomson, K.T. (2007) *J Phys Chem C*, **111**, 7384; (b) Joshi, A.M. Delgass, W.N. and Thomson, K.T. (2005) *The Journal of Physical Chemistry. B*, **109**, 22392; (c) Wells, D.H., Delgass, W.N. and Thomson, K.T. (2004) *Journal of Catalysis*, **225**, 69.
105 (a) Edwards, J.K., Thomas, A., Carley, A.F. et al. (2008) *Green Chemistry*, **10**, 388; (b) Edwards, J.K. Carley, A.F. Herzing, A.A. et al. (2008) *Faraday Discussions*, **138**, (Nanoalloys) 225.
106 (a) Choudhary, V.R. and Jana, P. (2008) *Catalysis Communications*, **9**, 1624;

(b) Choudhary, V.R., Samanta, C. and Choudhary, T.V. (2006) *Applied Catalysis A – General*, **308**, 128.

107 (a) Liu, Q., Bauer, J.C., Schaak, R.E. and Lunsford, J.H. (2008) *Applied Catalysis A – General*, **339**, 130; (b) Liu, Q. and Lunsford, J.H. (2006) *Applied Catalysis A – General*, **314**, 94.

108 Han, Y.-F. and Lunsford, J.H. (2005) *Journal of Catalysis*, **230**, 313.

109 (a) Liu, Q., Bauer, J.C., Schaak, R.E. and Lunsford, J.H. (2008) *Applied Catalysis A – General*, **339**, 130; (b) Liu, Q. and Lunsford, J.H. (2006) *Journal of Catalysis*, **239**, 237.

110 (a) Samanta, C. and Choudhary, V.R. (2008) *Chemical Engineering Journal (Amsterdam, Netherlands)*, **136**, 126; (b) Samanta, C. and Choudhary, V.R. (2007) *Catalysis Communications*, **8**, 2222; (c) Choudhary, V.R. Samanta, C. and Choudhary, T.V. (2007) *Catalysis Communications*, **8**, 1310; (d) Samanta, C. and Choudhary, V.R. (2007) *Applied Catalysis A – General*, **330**, 23; (e) Choudhary, V.R. and Jana, P. (2007) *Applied Catalysis A-General*, **329**, 79; (f) Choudhary, V.R. Samanta, C. and Jana, P. (2007) *Applied Catalysis A – General*, **332**, 70; (g) Samanta, C. and Choudhary, V.R. (2007) *Applied Catalysis A – General*, **326**, 28; (h) Choudhary, V.R. Samanta, C. and Jana, P. (2007) *Applied Catalysis A – General*, **317**, 234–243; (i) Choudhary, V.R., Ingole, Y.V., Samanta, C. and Jana, P. (2007) *IEC Research*, **46**, 8566; (j) Choudhary, V.R. and Jana, P. (2007) *Journal of Catalysis*, **246**, 434.

111 (a) Chen, Q. and Beckman, E.J. (2007) *Green Chemistry*, **9**, 802; (b) Beckman, E.J. (2003) *Green Chemistry*, **5**, 332.

112 (a) Hancu, D. Green, J. and Beckman, E.J. (2002) *IEC Research* **41**, 4466; (b) Hancu, D. and Beckman, E.J. (2001) *Green Chemistry*, **3**, 80.

113 Jin, H. and Subramaniam, B. (2003) *Chemical Engineering Science*, **58**, 1897.

114 Abate, S., Barbera, K., Lanzafame, P. *et al.* (2008) *Preprints – American Chemical Society, Division of Petroleum Chemistry*, **53**, 227.

9
Recent Achievements and Challenges for a Greener Chemical Industry
Fabrizio Cavani and Nicola Ballarini

9.1
Introduction: Old and New Challenges for Oxidation Catalysis in Industry

The chemical industry is experiencing important changes, the driving force for which is not only the need to improve competitiveness and consolidate market positions while complying with regulations and rules for the protection of human health and of the environment, but also the increased awareness that in the near future all the raw materials and energy chain supply will have to be reconsidered.

In this context, in recent decades a pioneering role has been played by oxidation catalysis, which has been undergoing substantial modifications, both in process technology and in catalyst design, aimed at improving the performance of chemical processes especially in terms of selectivity, ameliorating the energy efficiency and decreasing the costs associated with waste stream decontamination and pollutants abatement [1].

Emblematic examples of achievements in this field are:

1. The shift from air-based, once-through processes to oxygen-based recycle processes, and the corresponding change from reactant-lean to oxidant-lean processes. This not only considerably reduces the emissions and makes purge streams more concentrated and hence more easily combusted but also may lead to improved selectivity and productivity. Examples are the oxychlorination of ethylene to 1,2-dichloroethane and the epoxidation of ethylene.

2. The synthesis of intermediates and monomers from alkanes by means of oxidative processes, in part replacing alkenes and aromatics as the traditional building blocks for the chemical industry [2]. Besides the well-known oxidation of *n*-butane to maleic anhydride, examples of processes implemented at the industrial level are (i) the direct oxidation of ethane to acetic acid, developed by Sabic; (ii) the ammoxidation of propane to acrylonitrile, developed by INEOS (former BP) and by Mitsubishi, and recently announced by Asahi to soon become commercial; (iii) the partial oxidation of methane to syngas (a demonstration unit is being built by ENI). Many other reactions are currently being investigated, for example, (i) the

oxidation of propane to acrylic acid, (ii) the oxidation of isobutane to methacrylic acid, (iii) the oxidative dehydrogenation of light alkanes to the corresponding olefins, (iv) the oxychlorination of ethane to vinyl chloride, (v) the oxidation of n-hexane and cyclohexane to adipic acid and (vi) the oxidation of n-butane to 1,4-butanediol. In some cases the yield and selectivity so far achieved are still far from being of industrial interest; however, research efforts have led to considerable steps forward.

3. The development of new types of heterogeneous catalytic systems for liquid-phase oxidations that made possible the use of environmentally friendly oxidants, that is, hydrogen peroxide and air, in place of oxidants like hydroperoxides (that lead to the co-formation of the product of reduction), HNO_3 and Cr^{6+}. Noteworthy examples are titanosilicates, and in general crystalline silicates and silico-aluminates incorporating metal ions such as V^{4+}, Sn^{4+} and others as well. The recently implemented processes for the synthesis of propene oxide by direct epoxidation of propene and of cyclohexanone oxime by ammoximation of cyclohexanone, both using hydrogen peroxide as the oxidant and avoiding the formation of co-products or waste compounds, have followed the main road paved by the previously developed hydroxylation of phenol to diphenols – all of these processes made possible because of the discovery of the TS-1 catalyst by ENI researchers.

However, new challenges are now driving researchers towards the exploration of new synthetic catalytic routes that use renewable raw materials in place of building blocks derived from oil, natural gas and carbon, adopt the intensification concept, decreasing risks associated with oxidation processes and lowering investment costs for new technologies, and make use of new types of catalysts not only for new synthetic pathways, but also for established industrial processes, thus permitting the overcoming of apparently unsurpassable yields.

This chapter describes a few examples of recent achievements in catalytic selective oxidation that illustrate current trends, perspectives of development and future directions in this field.

9.2
Recent Successful Examples of Alkanes Oxidation

9.2.1
Oxidation of Ethane to Acetic Acid

The major conventional processes for the production of acetic acid include the carbonylation of methanol (originally developed by Monsanto, and now carried out by several companies, such as Celanese-ACID OPTIMIZATION, BP-CATIVA, etc.), the liquid-phase oxidation of acetaldehyde, still carried out by a few companies, and the liquid-phase oxidation of n-butane and naphtha. More recent developments include the gas-phase oxidation of ethylene, developed by Showa Denko K.K., and the liquid-phase oxidation of butenes, developed by Wacker [2a].

The synthesis of acetic acid by direct gas-phase oxidation of ethane:

$$C_2H_6 + 3/2 O_2 \rightarrow CH_3COOH + H_2O$$

is economically viable only in a location with an advantageous ethane position. In this case, the competition would not be against the cost of ethylene, but rather against the low cost methanol/syngas, the raw materials of the carbonylation technology. Investment in world-scale state-of-the-art methanol carbonylation would require around 110 million US dollars, while a direct ethane conversion at a scale of 200 000 tons y^{-1} would be slightly higher [2a]. Nevertheless, the ethane oxidation technology may permit the resolution of technical problems (mainly related to corrosion) of the methanol carbonylation technology.

Different classes of catalysts have been claimed for the oxidation of ethane to acetic acid [3], but the catalyst that gives the best performance is made of a mixed oxide of Mo/V/Nb (plus other components in minor amounts). This compound was first described in a paper by Thorsteinson et al. [3a] – a paper that is considered nowadays a milestone in the field of the selective oxidation of alkanes, in view of the number of active phases that have been developed starting from catalysts described therein. Several patents were also issued by Union Carbide [3a–f], now Dow Chemical, regarding this system and the ETHOXENE process. The activity in ethane oxidation was attributed to the development of a crystalline phase characterized by a broad X-ray diffraction reflection at $d = 4.0$ Å. The best composition was claimed to be $Mo_{0.73}V_{0.18}Nb_{0.09}O_x$, which reached 10% conversion of ethane at 286 °C with almost total selectivity to ethylene; the selectivity decreased with increasing temperature, due to the formation of carbon oxides. The main peculiarity of this catalyst is its capability to activate the paraffin at low temperatures (<250 °C). The process and the catalyst claimed were developed for ethane oxidative dehydrogenation, and acetic acid was only a minor by-product of the reaction; however, the use of pressures above atmospheric enhanced the selectivity of acetic acid.

Catalysts similar to those claimed by Union Carbide were later studied by Bordes and coworkers [4], and by Burch and coworkers [5]. Merzouki et al. [4a, b] proposed that the Mo/V/Nb/O catalyst is made up of $(VNbMo)_5O_{14}$-type microdomains in a MoO_3 matrix. At 200 °C, a selectivity of 45% to acetic acid and 45% to ethylene was obtained at 25% ethane conversion; an increase of temperature caused a loss in selectivity to acetic acid in favor of that to ethylene. Burch and Swarnakar [5a] compared the reactivity of Mo/V/O and Mo/V/Nb/O systems. The former contained MoO_3, $Mo_6V_9O_{40}$ and $Mo_4V_6O_{25}$ crystalline compounds, while the latter also contained $Mo_3Nb_2O_{11}$, the most intense diffraction line of which occurred at 4.01 Å. The addition of Nb increased both activity and selectivity, and the formation of $Mo_3Nb_2O_{11}$ was proposed to account for the increase in performance. The product distribution was independent of the conversion, indicating the absence of consecutive reactions.

Researchers from Union Carbide proposed a mechanism in which ethane is first adsorbed on Mo^{6+} or V^{5+} [5c] sites to form an ethoxide species; the latter transforms into ethylene via β-elimination. However, the surface ethoxide can also be oxidized further and by α-elimination form acetaldehyde and then a surface acetate, which

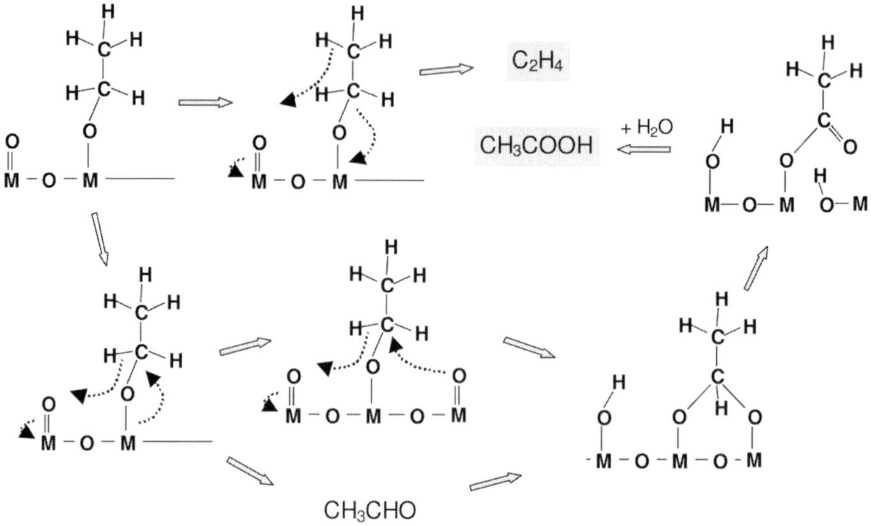

Figure 9.1 Mechanism proposed in the literature for the oxidation of ethane to acetic acid [2b].

upon hydrolysis yields acetic acid [5c] (Figure 9.1 [2b]). The α-elimination step is favored at lower temperatures than the β-elimination. Acetic acid is more stable than ethylene towards consecutive reactions of combustion, and for this reason an increase in the total pressure results in an increased formation of acetic acid (Figure 9.1) [2b].

In patents by Saudi Basic, similar systems (Mo/V/Nb/P/O; Mo/V/La/Pd/O; Mo/V/Nb/Pd/O) were claimed to perform excellently in the oxidation of ethane to acetic acid under pressure (10–20 atm), and in the acetoxylation of ethane/ethylene to vinyl acetate [6]. In these patents, the feed composition is not very rich in ethane, and the limiting reactant is the hydrocarbon. This allows operation at high ethane conversion, while maintaining high selectivity to acetic acid. The reactor for acetic acid synthesis can be integrated with an ethylene acetoxylation reactor (Figure 9.2); in the first reactor, $MoV_{0.396}Nb_{0.128}Pd_{0.00019}O_x$ catalyzes the transformation of ethane into acetic acid and ethylene, with approx. 45% alkane conversion and 40% selectivity to each of the two products, while in the second reactor the two molecules react in the presence of O_2 to produce vinylacetate. No need for intermediate separation is claimed, since N_2, H_2O, CO_2 and ethane are diluents for the second stage. Analogous patents, claiming processes for integrated vinylacetate production, were also issued by BP [7a] and by Aventis [8a].

The catalyst claimed by Saudi Basic is in large part amorphous, but shows a few diffraction lines relative to a crystalline compound characterized by d values at 4.03 (100% I), 3.57, 2.01 and 1.86 Å. Doping the compound with P improves the yield to acetic acid; this was attributed to an enhancement of surface acidity, which facilitates the ethylene adsorption and the acetic acid desorption. A detailed investigation of catalyst composition confirmed that the former model originally proposed by Merzouki et al. [4a, b] was the most likely [4c]. Therefore, the excellent properties

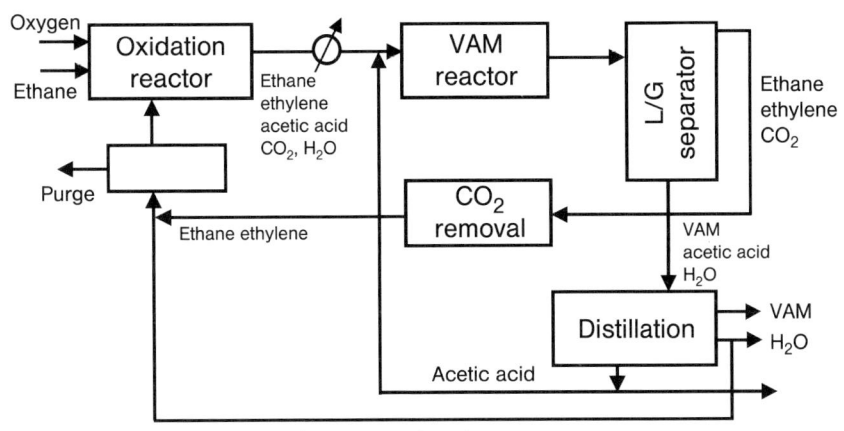

Figure 9.2 Simplified flow-sheet of the integrated Saudi Basic SABOX process for vinylacetate (VAM) production from ethane [6b].

of the Mo/V/Nb/O system are due to the great flexibility of the Mo_5O_{14}-like structure; this is able to incorporate other elements (V, Nb, and even small amounts of Pd), which contribute to improve the activity and selectivity.

Both catalyst and process development were improved by Hoechst (then Aventis), which patented Mo/Pd/(Re)/O systems, also containing several promoters [8, 9]. The same authors published a comprehensive study of the $MoV_{0.25}Nb_{0.12}Pd_{0.0005}O_x$ catalyst [10]. The high dispersion of Pd(II), maintained under reaction conditions through the incorporation into the Mo_5O_{14}-like phase, provided the active sites for the consecutive transformation of ethylene into acetic acid, via a Wacker-like mechanism, in which water played an important role. This reaction contributed to the high selectivity to acetic acid, but the principal reaction pathway leading to acetic acid formation was the direct one, occurring on ethane. Other authors instead found that the main role of Pd is to improve the catalytic activity [11]. Linke *et al.* [10c] simulated the performance of the catalyst in fixed-bed and in fluidized-bed reactors, and studied the effect of main reaction parameters. In fluidized-bed, a lower selectivity to acetic acid was achieved than in fixed-bed, mainly due to the slow mass transfer of oxygen between the gas bubbles and the emulsion phase. In one patent issued by Hoechst [9b], a cascade reactor with an oxygen feed between reactor levels was proposed, which provided higher ethane conversion rates. Recent patents issued by BP claim Mo/V/O systems doped with Sn, or containing Au instead of Pd [12].

The use of hydrothermal-like synthesis for the preparation of Mo/V/Al/O catalysts, starting from Anderson-type polyoxometalates, leads to monophasic crystalline compounds, also containing other trivalent elements (Fe, Ga), and dopants [13]. For example, the incorporation of Ti in Mo/V/Al/O gives rise to the formation of a compound having stoichiometry $Mo_6V_2Al_1Ti_{0.5}O_x$, in which the addition of Ti improves the activity with respect to the undoped system.

Other classes of catalysts investigated include supported Keggin-type P/Mo/V/O polyoxometalates [14–16], and titania-supported vanadium oxide or V/P/(Mo)/O

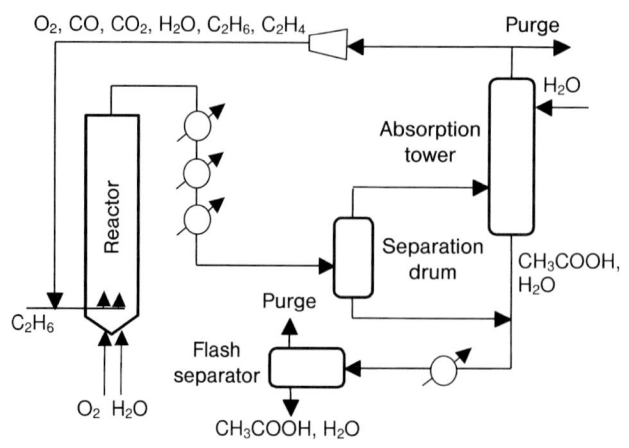

Figure 9.3 Simplified flow-sheet of the Standard Oil process for the oxidation of ethane to acetic acid [21].

[15b, 17, 18]. The performance of bulk vanadyl pyrophosphate $(VO)_2P_2O_7$, the catalyst used industrially for n-butane oxidation to maleic anhydride, has also been described in some papers [19, 20], and claimed by The Standard Oil Co at the beginning of 1990s [21]. The latter company described the use of a fluidized-bed reactor, with feed of fresh ethane and oxygen and recycle of most of the reactor effluents, after separation of acetic acid. In this way a high concentration of CO_2 was maintained in the reactor to better control the temperature rise caused by the exothermic reactions, and also to obviate the separation of carbon oxides from ethylene, including the costly cryogenic separation of CO. A flow sheet of the process is illustrated in Figure 9.3. In this case, the catalyst described had the empirical formula $Mo_{0.37}Re_{0.25}V_{0.26}Nb_{0.07}Sb_{0.03}Ca_{0.02}O_x$ and was analogous to that reported in ref. [7b].

Table 9.1 summarizes catalyst compositions and corresponding performances. The oxidation of ethane to acetic acid is now commercial; an industrial plant is installed, with the technology developed by Saudi Basic. Elements that have contributed to the successful development of the process are (1) the discovery of a catalytically active compound, the multifunctional properties of which can be modified and tuned to be adapted to reaction conditions through incorporation of various elements; (2) the stability of the main products, ethylene and acetic acid, which do not undergo extensive consecutive degradation reactions; (3) the possibility of recycling the unconverted reactant and the major by-product, ethylene; (4) the use of reaction conditions that minimize the formation of CO; and (5) an acceptable overall process yield.

9.2.2
Ammoxidation of Propane to Acrylonitrile

In current process technology for the manufacture of acrylonitrile, propene feedstock cost represents about 67% of the cost of production. The price differential between

Table 9.1 Summary of the performance of selected catalysts in ethane oxidation.

Catalyst	Conditions	Conversion of ethane (%)	Selectivity for acetic acid (%)	Selectivity for ethylene (%)	Reference
$Mo_{0.73}V_{0.18}Nb_{0.09}O_x$	$T\ 200\,°C$, $P\ 1\,atm$.	2.3	100	0	[4a]
$Mo_{0.75}V_{0.22}Nb_{0.03}O_x$	86% C_2, 6% O_2, 8% H_2O, $T\ 323\,°C$, $P\ 8.3\,atm$	7.5	18.4	67.6	[3b]
$Mo_{2.5}V_{1.0}Nb_{0.32}O_x$	15% C_2, 85% air; $T\ 260\,°C$, $P\ 14\,atm$	65.0	30.5	26.9	[6a]
$Mo_{2.5}V_{1.0}Nb_{0.32}P_{0.042}O_x$		53.3	49.9	10.5	[6a]
$P_{0.015}Mo_{0.153}V_{0.014}Ti_{1.0}O_x$	62% C_2, 17% O_2, 10% N_2, 12% H_2O, $T\ 275\,°C$, $P\ 6.3\,atm$	6.5	38	11	[18c]
V/TiO_2	$T\ 225\,°C$, $P\ 1\,atm$	0.5	73	8	[17a]
$(VO)_2P_2O_7$	55% C_2, 11% O_2, 34% N_2, $T\ 333\,°C$, $P\ 14\,atm$	5.7	14.8	48.0	[21b]
$V_1P_1Re_{0.016}O_x$		4.4	27.1	8.2	[21b]
$Mo_{0.37}Re_{0.25}V_{0.26}Nb_{0.07}Sb_{0.03}Ca_{0.02}O_x$	$T\ 227\,°C$, $P\ 28\,atm$	14	78	12	[7b]
$Mo_1V_{0.628}Pd_{0.000288}La_{0.00001}O_x$	Only by-product was CO_2. $T\ 260\,°C$, $P\ 14\,atm$, 15% C_2, 85% air, GHSV $1100\,h^{-1}$	58.2	67.9	0	[6c]
$Mo_6V_2Al_1Ti_{0.5}O_x$	$T\ 380\,°C$, $P\ 1\,atm$, 15% C_2, 5% O_2, 10% H_2O	19.8	6.5	54.0	[13a]
$H_5PMo_{10}V_2O_{40}$ 30 wt% on SiO_2	$T\ 400\,°C$, $P\ 1\,atm$, 23% C_2, 11% O_2, 20% H_2O	12	5	50	[16]

propene and propane depends on many factors, but it may be estimated to be on average 360\$ ton^{-1} in 2007. This price differential makes competitive a propane ammoxidation process using the current available catalysts and in fact already some plants started to be revamped. BP built a propane demonstration unit that is integrated with its Green Lake, Texas, facilities. However, INEOS, which acquired Innovene in 2005, the olefin and derivatives business formerly owned by BP, is now silent on possible developments in this technology. Mitsubishi Chemical with BOC has tested a propane process at Mizushima, Japan; however, Mitsubishi also appears not to be advancing their propane ammoxidation technology. Asahi Kasei has commercialized its propane-based process by converting an existing 70 000 tons y^{-1} line to propane in Ulsan, South Korea, at its subsidiary Tongsuh Petrochemica; the plant went on stream at the beginning of 2007. Asahi Kasei is planning a new acrylonitrile plant from propane, with a capacity of 200 000 tons y^{-1}.

The reaction conditions claimed by the various companies that investigated the ammoxidation of propane are different. Sometimes propane-rich conditions have been reported, that is, in earlier patents from Standard Oil (then BP, now INEOS), while in other cases propane-lean conditions are preferred [22]. In the former case, the conversion of propane is low, and therefore the unconverted paraffin is recycled. Mitsubishi was the first to claim the use of hydrocarbon-lean conditions, for example, conditions in which very high propane conversion can be reached [23]. In more recent patents, BP also claims the use of analogous conditions, with oxygen-rich feed and propane as the limiting reactant. However, the lower activity of antimonates makes necessary the use of temperatures that are approximately 50 °C higher than those employed with the Mitsubishi catalyst. Indeed, the economics of the process is greatly affected by the value of the by-products (HCN and acetonitrile), by the degree of ammonia conversion and the cost for its recovery, by the costs for separation and eventual recycle of unconverted propane, by the stream costs, especially when oxygen is used instead of air in a recycle process, and by the costs for disposal of liquid effluents [2a].

Several reviews have discussed the catalyst and catalytic reaction chemistry in propane ammoxidation [24]. Two principal catalytic systems have been proposed in the literature so far, based on V-antimonates with rutile structure and on multicomponent molybdates, Mo/V/Nb/Te/O. Amongst the former is the Al/Sb/V/W/O system that gives the highest acrylonitrile yield (about 39%) [25]. However, the best performing system discovered so far employs the Mo/V/Nb/Te/O catalyst, which has shown to be both active and selective, giving acrylonitrile yields of 62% [26]. Mitsubishi has developed the base catalyst composition: $MoV_{0.3}Te_{0.23}Nb_{0.12}O_x$. The performance can be further increased by the addition of dopants, such as Sb, B or Ce. Key characteristics of this catalyst are not only its composition, but also the procedure of synthesis and activation and the modality of doping. This catalyst can be prepared by hydrothermal synthesis, which results in nucleation and growth of the so-called M1 and M2 phases with well-defined crystal morphologies [26–28].

Asahi [29] has modified the composition of the Mitsubishi catalyst, by incorporation of Sb in place of Te in the M1 phase; this catalyst is more stable and provides a longer catalyst lifetime. The main features of the Asahi process and catalysts are:

1. For the catalyst based on $Mo_{1.0}V_{0.33}Nb_{0.11}Te_{0.22}O_x$, a precise phase composition is claimed, as defined by the ratio of the X-ray diffraction peak intensity at 2θ $27.3°$ and $28.2°$ ($CuK\alpha$), which minimizes the combustion of ammonia to N_2 [29d,j]. For instance, when the degree of ammonia decomposition is only 29.4%, the yield to acrylonitrile is 58.9% at 93.1% propane conversion, with a feed composition: propane/ammonia/oxygen/inert equal to $1.0:1.2:3.0:14.8$, at $420°C$ and W/F $0.5\,g\,s\,cc^{-1}$. When the same active phase is diluted in silica by means of the spray-drying procedure, the best yield to acrylonitrile achieved is 52.7% at $430°C$ [29e], but the addition of a dopant (Yb, Er, Dy, Nd, etc.) increases the yield up to 55–56%. Dopants also allow the feed of a lower ammonia/propane ratio, for example, 0.8 instead of the stoichiometric 1.0, while reaching in high yield to acrylonitrile with respect to ammonia, and minimizing ammonia combustion [29h].

2. For the catalyst based on Mo/V/Nb/Sb/O, the composition giving the highest acrylonitrile yield is $Mo_1V_{0.23}Sb_{0.26}Nb_{0.09}O_x/SiO_2$ (45 wt%), with 49.5% propane conversion and 68% acrylonitrile selectivity at $420°C$ and with a feed composition propane/ammonia/oxygen/inert equal to $1:0.7:1.7:5.3$ [29g]. The positive effect achieved by addition of a metal oxide having the rutile structure, for example, TiO_2, SnO_2 or $MeSbO_4$, is also claimed [29k]. Notably, with such systems the conversion of propane is lower than that obtained with the Mo/V/Nb/Te/O catalysts; in fact, a lower ammonia/propane feed ratio is used.

A third catalytic system was proposed more recently and based on vanadium aluminum oxynitrides (VALON) [30]. The maximum acrylonitrile yield reported was about 30%, but with acrylonitrile productivity four times higher than for V/Sb/W/Al/O catalysts and one order of magnitude than for Mo/V/Nb/Te/O. Other companies have studied and developed proprietary formulations but, in general, catalytic systems belong either to the antimonates family (Rhodia, BASF, Nitto, Monsanto) [31–33] or to the molybdates family.

All catalysts claimed are "multifunctional" systems. Indeed, the formation of acrylonitrile from propane occurs mainly via the intermediate formation of propene; the latter is then transformed into acrylonitrile via the allylic intermediate (Figure 9.4).

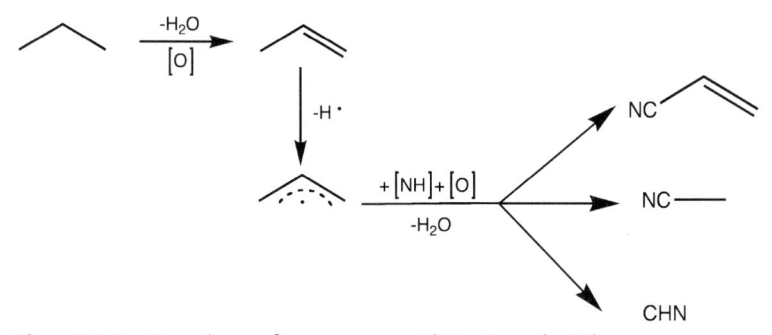

Figure 9.4 Reaction scheme of propane ammoxidation to acrylonitrile.

Therefore, the catalyst possesses different types of active sites: one site that can activate the paraffin and oxide hydrogenate it to the olefin, and one that (amm) oxidizes the adsorbed olefin intermediate.

The multifunctionality is achieved through either the combination of two different compounds (phase-cooperation) or the presence of different elements inside a single crystalline structure. In antimonates-based systems, cooperation between the metal antimonate (having a rutile crystalline structure), employed for propane oxidative dehydrogenation and propene activation, and the dispersed antimony oxide, active in allylic ammoxidation, is made more efficient through the dispersion of the latter compound over the former. In metal molybdates, one single crystalline structure contains both the element active in the oxidative dehydrogenation of the hydrocarbon (vanadium) and those active in the transformation of the olefin and in the allylic insertion of the NH^{2-} species (tellurium and molybdenum).

Many researchers have investigated the chemical-physical and structural features of the M1 (orthorhombic) and M2 (hexagonal) phases in Mo/V/Nb/Te/O [34–38]. The M1 phase alone is capable of selectively transforming propane, but the presence of the M2 phase is claimed to improve the selectivity under more demanding conditions, for example, at high conversion [36]. Under demanding conditions symbiosis between the M1 and M2 phases occurs, with the latter serving as a co-catalyst to the former, transforming desorbed propene into acrylonitrile. A maximum acrylonitrile yield of 61.8% (86% conversion, 72% selectivity at 420 °C) was achieved with a nominal catalyst composition $Mo_{0.6}V_{0.187}Te_{0.14}Nb_{0.085}O_x$, comprised of 60% M1, 40% M2 and a trace of $TeMo_5O_{16}$ [37].

Figure 9.5 illustrates a simplified scheme of the Mitsubishi process. The process includes the recycle of unconverted propane, and the BOC-PSA technology for rejection of N_2; the latter is present in the feed, which contains oxygen-enriched air, and is also generated in the reactor by ammonia combustion. The unconverted hydrocarbon is recovered and recycled to the reactor [23f,g]. One Mitsubishi patent claims the differentiation of ammonia along the catalytic bed [23d]. This might

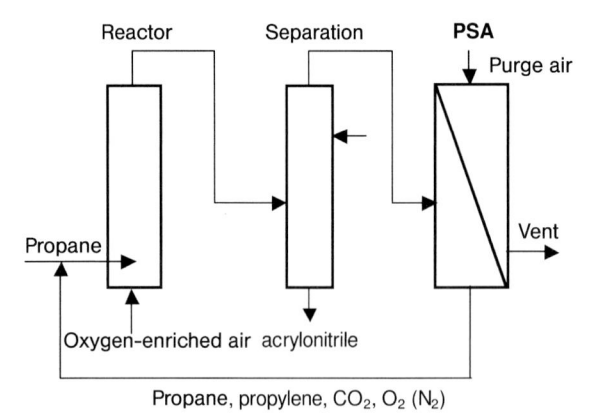

Figure 9.5 Simplified flow-sheet of the Mitsubishi/BOC process for propane ammoxidation.

become necessary if the catalyst were particularly active in ammonia combustion to N_2; in this case, in fact, when all reactants are fed together at the reactor inlet, the ammonia concentration rapidly declines along the bed, causing the preferential formation of propene and carbon oxides in the final part of the reactor. Differentiation of ammonia along the reactor may in part overcome the problem of ammonia combustion, and favor the transformation of the alkane and of the alkene intermediately formed into acrylonitrile.

Several papers deal with the study of metal antimonates having the rutile structure [39–49]. In particular, V/Sb/O and Fe/Sb/O systems have been the object of many investigations, aimed at understanding the nature of these mixed oxides, and at identifying the active species. Indeed, the preparation of a truly stoichiometric metal antimonate, $MeSbO_4$, is a difficult task. The method of preparation affects the nature of the catalysts prepared, but in general non-stoichiometry is a particular feature of these systems. The most striking case is V/Sb/O, for which the following composition has been reported for the catalyst having V/Sb atomic ratio equal to 1/1: $V_{0.92}Sb_{0.92}O_4$ (quasi-$VSbO_4$) [41]. This cation-deficient structure, having 0.04 cationic positions unoccupied for each O^{2-} anion, contains Sb^{5+}, while vanadium is present both as V^{4+} and as V^{3+}. Electroneutrality is obtained for the composition $V_{0.28}^{3+}V_{0.64}^{4+}Sb_{0.92}^{5+}O_4$ [43a].

Table 9.2 summarizes the performance of some V/Sb/O-based catalysts [40b]. BP claimed the same catalyst type for propane-lean conditions, with high conversion of the paraffin, and propane-rich conditions. In these catalysts, the main component is the quasi-$VSbO_4$ rutile, which transforms the alkane into the unsaturated intermediate. The latter may either desorb to yield propene or be transformed into acrylonitrile over the SbO_x "overlayers," the amount of which is a function of the Sb in excess with respect to the rutile formula [24c].

In systems developed by Rhodia [31] the main component is SnO_2 (cassiterite), which is inactive in the reaction of propane ammoxidation, while it acts as the carrier for the active components: V/Sb/O and SbO_x. Tin oxide facilitates the dispersion of the active components, and yields a multifunctional catalyst in which the various species can effectively cooperate in the reaction.

Rutile-type Cr antimonate, $CrSbO_4$, is fairly active and selective in propane ammoxidation; in catalysts studied by Snamprogetti, adding V considerably improves

Table 9.2 Performance of some V/Sb/O-based catalysts described in the literature.

Catalyst formula	T (°C)	$C_3/NH_3/O_2/H_2O$/inert (mol.%)	C_3H_8 conversion (%)	Selectivity for acrylonitrile (%)	Reference
$VSb_5W_{0.5}Te_{0.5}Sn_{0.5}O_x$–$SiO_2$	500	6.5:13:12.9:19.4:48.4	68.8	56.7	[22c]
$VSb_{1.4}Sn_{0.2}Ti_{0.2}O_x$	460	51:10.2:28.6:10.2:0	14.5	61.9	[22b]
$VSb_{1.4}Sn_{0.2}Ti_{0.1}O_x$	480	6.4:7.7:18.6:0:67.3	40.3	47.5	[22j]
$VSb_5Bi_{0.5}Fe_5O_x$–Al_2O_3	440	7.5:15:15:20:42.5	39	77	[31d]
$VSb_5Sn_5O_x$	450	8:8:20:0:64	30	49	[31b]

catalyst activity [46]. The atomic ratio (Cr + V)/Sb is the main compositional parameter affecting the nature of these crystalline compounds. When almost equiatomic ratios are used (e.g., in Cr/V/Sb 1:1:1), the corresponding XRD pattern shows the presence of a single, well-crystallized rutile-type phase, $CrVSbO_6$, which in practice corresponds to an equimolar solid solution between $CrSbO_4$ and VO_2 (both characterized by the rutile-type structure). In Cr/V/Sb/O samples with atomic ratios of Cr/V/Sb (1:x:1), the formula $Cr_1V_xSb_1O_{4+2x}$ was extrapolated, with composition ranging from $CrSbO_4$ to $CrVSbO_6$. In these catalysts, V is present mainly as V^{4+}, and samples are extremely active in propane (amm)oxidation but poorly selective to acrylonitrile, leading to the prevailing formation of propene and carbon oxides. When samples are instead prepared with an atomic ratio (Cr + V)/Sb < 1, V is present mainly as V^{3+} (Figure 9.6). The presence of excess Sb leads to a considerable improvement of the selectivity to acrylonitrile. The main peculiarity of these Cr-based antimonates is the ability to host excess Sb in the structure, thereby developing nonstoichiometric compounds, especially in samples with lower V content. This renders the transformation of the intermediately formed propene into acrylonitrile more efficient. Also in these systems, however, an excess Sb is necessary with respect to the amount incorporated in the rutile mixed oxide, to reach good selectivity to acrylonitrile.

An important dopant for rutile-type mixed oxides is Nb oxide [46a]. Bañares *et al.* [49] found that when used as the support for V/Sb/O, Nb_2O_5 formed new phases by reaction with V and Sb under catalytic reaction conditions; these phases, of unclear nature, affected the catalytic performance in propane ammoxidation. When instead Nb was added as a promoter for the alumina-supported V/Sb/O system, the interaction between the active components led to an improvement of catalytic performance with respect to the undoped V/Sb/O. Nb also forms rutile-type mixed

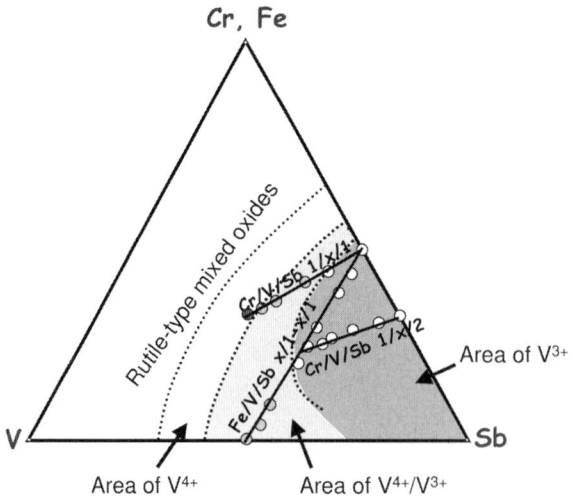

Figure 9.6 Areas of existence of different V species in a composition diagram for the Cr(Fe)/V/Sb/O rutile-type system [46d].

oxides with Cr, V and Sb, and modifies the properties of Sb cations in allylic ammoxidation [46a].

Therefore, the incorporation of other elements in the rutile lattice may substantially improve the performance of V antimonate [43f]. This is also the case for Al^{3+}, Ti^{4+} and W^{4+} [43d,e]. A yield to acrylonitrile of almost 40% has been reported for an Al/Sb/V/W/O catalyst [25]. It was demonstrated that, if compared with the pure rutile V/Sb/O phase, the better catalytic properties of (Al,V)SbO$_4$ and of *quasi*-Sb(V, W)O$_4$ can be rationalized in terms of the "site-isolation" theory. According to this theory, which was originally formulated by Callahan and Grasselli [50], a catalyst selective for partial oxidation can be developed by creating structural isolation of the active site.

9.3
New Oxidation Technologies: Oxidative Desulfurization (ODS) of Gas Oil

Diesel regulations in Europe and North America demand a sulfur reduction down to either 10 or 15 wppm by 2009. Refiners will have to meet Ultra-Low Sulfur Diesel (ULSD) regulations at a significant cost; in most cases, either a new high-pressure hydrotreating unit or a revamping of medium-pressure hydrotreating plant will be required. Operation at harder conditions, however, implies a penalty in fuel characteristics, and high investment and operative costs, due to the high hydrogen consumption. Therefore, the need to cut the costs offers opportunities to non-hydrogen-consuming processes, operating at low severity in terms of temperature and pressure.

Innovative processes for ultra-deep desulfurization are:
1. Selective adsorption, that is, sulfur removal by selective interaction between compounds and transition metal based adsorbents.
2. Biodesulfurization by specialized bacteria.
3. Oxidation and extraction, that is, liquid-phase oxidation with organic peroxides followed by separation of the oxidized sulfur (oxidative desulfurization, ODS) [51–58].

In recent years, a few companies have been developing ODS processes to reduce the sulfur level in transportation fuels. With regard to gas oil, Sulphco Inc., Unipure Corporation and Lyondell LLC were active in promoting their ODS processes. The key feature of this technology is its complementarities with the hydrodesulfurization. In fact, compounds less reactive towards hydrogen are those that show the greater reactivity towards oxygen; this is the case of dibenzothiophenes, and especially of 4-methyldibenzothiophene and 4,6-dimethyldibenzothiophene, which are very difficult to convert by HDS due to their steric hindrance. These compounds can be oxidized under mild conditions by peroxides, and be transformed into sulfoxides and sulfones; the latter are highly polar and, having physical properties significantly different from those of the hydrocarbons in the gas oil, can be easily removed (Figure 9.7).

Figure 9.7 Oxidative desulfurization of benzothiophenes.

Therefore, an ODS unit might effectively treat the 300–500 ppm S-containing outlet stream of a hydrodesulfurization unit; in the latter, all those compounds that are less reactive towards oxygen, for example, mercaptans and thiophenes, are removed. The main problem in ODS processes is the use of expensive or corrosive chemicals and the expensive liquid–liquid extraction unit for the sulfones separation. For example, organic hydroperoxides are in general preferred over H_2O_2, to avoid the use of a solvent for the oxidation section. The key to a successful implementation of this technology is effectively integrating the ODS unit with existing hydrotreating units. Figure 9.8 shows a general scheme of the ODS process; it includes three sections, (1) a peroxide supply section, including peroxide storage and handling facilities, (2) a sulfone generation section and (3) a sulfone separation section, achieved by means of either solvent extraction or with adsorption on a solid. The use of a specific adsorbent, for example, silica gel, may be advantageous, because solvent extraction suffers from being capital-intensive, difficult to operate and from the high diesel yield loss. After saturation, the adsorbent is regenerated in a cyclic operation by means of a suitable solvent, for example, an aliphatic or aromatic hydrocarbon. Finally, the adsorbent is separated from the sulfones by means of distillation [59a].

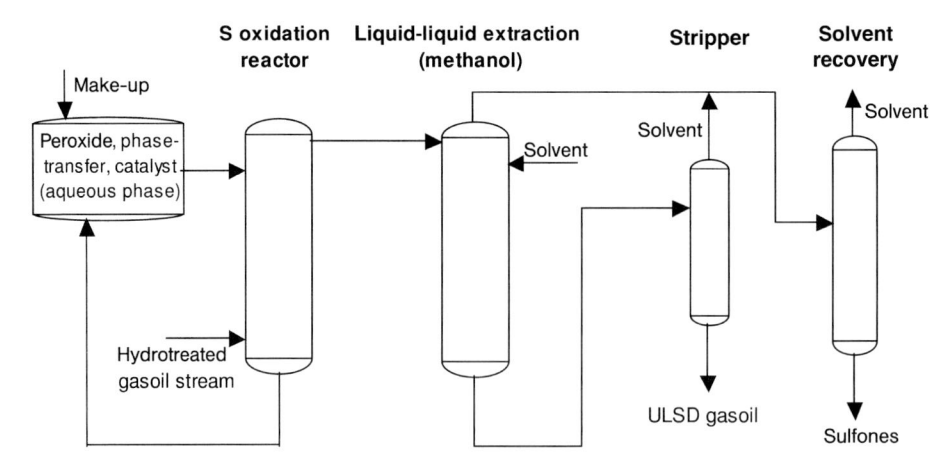

Figure 9.8 Simplified scheme of an ODS process.

9.3 New Oxidation Technologies: Oxidative Desulfurization (ODS) of Gas Oil

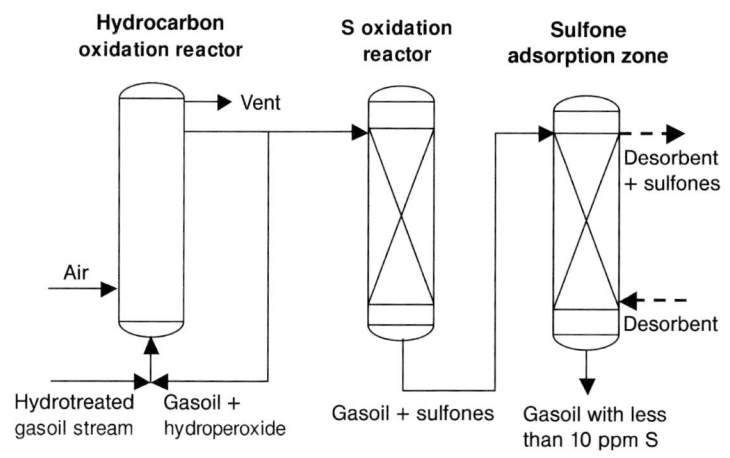

Figure 9.9 Simplified flow-sheet of the ENI-UOP ODS process.

However, the high cost of hydrogen peroxide and organic hydroperoxides makes economic comparison unfavorable. Therefore, ENI SpA and UOP LLC have jointly developed a new ODS process for the production of ULSD [59], in which the hydroperoxide is produced *in situ* (Figure 9.9). In the process, the first step is a circulating reaction loop in which the hydrotreated hydrocarbon stream is mixed with air to produce a hydroperoxide-containing stream in the presence of an organic initiator, at 130 °C and 70 atm; no catalyst is necessary in this step. The stream, containing approximately 2000 wppm O as peroxide, and in which a fraction of S has been already oxidized to sulfone, is then fed to the sulfur oxidation section, where a suitable heterogeneous catalyst is loaded, to complete the reaction and reach a level of unconverted S lower than 10 ppm.

The sulfur oxidation is carried out at pressure higher than 8 atm and below 180 °C, with a proprietary supported-Mo oxide-based catalyst, for example, an alpha alumina-supported $MgMoO_4$ catalyst, operating at 110 °C and 17 atm [59c]. All the products produced by oxidation side reactions and by hydroperoxide reduction are separated from the gas oil stream together with the sulfones. This operation may result in diesel yield loss; therefore, the valorization or upgrade of this oxidized stream affects the process economics. This stream can be blended into the heating oil pool or treated in a hydrocracking unit to recover valuable products.

Several homogeneous and heterogeneous catalysts are active and selective for the ODS of benzothiophenes. Heteropolycompounds (HPA) are amongst the most efficient systems for this reaction. In the literature, $H_3PW_{12}O_{40}$ [60] was studied using H_2O_2 as the oxidant in a two-phase system in the presence of a phase-transfer agent. Even if this system proved to be quite efficient, it presents some drawbacks: it is difficult to handle at large scale, catalyst productivity is low referred to the HPA, H_2O_2 is an expensive chemical to perform this reaction and its selectivity is not very high.

Various Keggin-type polyoxometalates (POMs) were tested as catalysts for the ODS of gas oil with *t*-butyl hydroperoxide. Alumina-supported phosphomolybdic acid ($H_3PMo_{12}O_{40}$) proved to be quite active, yielding sulfur conversion higher than

70% [58]. In a batch reactor, the alumina-supported V-substituted POM $H_5PMo_{10}V_2O_{40}$ reached complete conversion of the S-containing compound within 2 hours. Tests performed in continuous reactor showed that both POMs gave conversion of S well above the 95% with a life exceeding 1600 hours time-on-stream, and with an average residual S content in the outlet stream lower than 10 ppm, when a hydrotreated diesel with 300 wt ppm S was fed to the reactor. The oxidant was fed with a TBHP/S molar ratio close to 10. The V-containing POM, however, also catalyzed the hydroperoxide decomposition, which converted almost completely; residual TBHP was in this case less than 100 ppm, whereas with the supported $H_3PMo_{12}O_{40}$ the residual amount of oxidant was higher than 800 ppm. These two heteropolyacids could be easily regenerated at the end of their catalytic life for heat treatment under air flow at mild conditions (250 °C).

9.4
Process Intensification in Catalytic Oxidation

Recently, microreactor technology has received a great deal of attention, because it is becoming clear that many of the traditional methods for synthesizing chemicals are not sustainable and must be changed. Microreactors offer a solution because they are inherently less wasteful than traditional methods and because they provide a superior reaction control (Figure 9.10) [61]. The unique heat transfer properties of microreactors allow reactions to be efficiently controlled, and this is clearly very important for fast and exothermic reactions, like oxidations. Furthermore, the small volumes used by microreactors enable the safe use of dangerous reactants. Microchannel reactors are intrinsically safe, since the major reasons for explosions are not found in microreactors: (i) thermal runaway and (ii) uncontrolled self-accelerating radical

Advantages	
Increased safety	Low pressure drop
High surface-to-volume ratio	Low operating and maintenance costs
Better mass and heat transfer	Minimal environmental impact
High volumetric productivity	High volumetric productivity

Microreactors

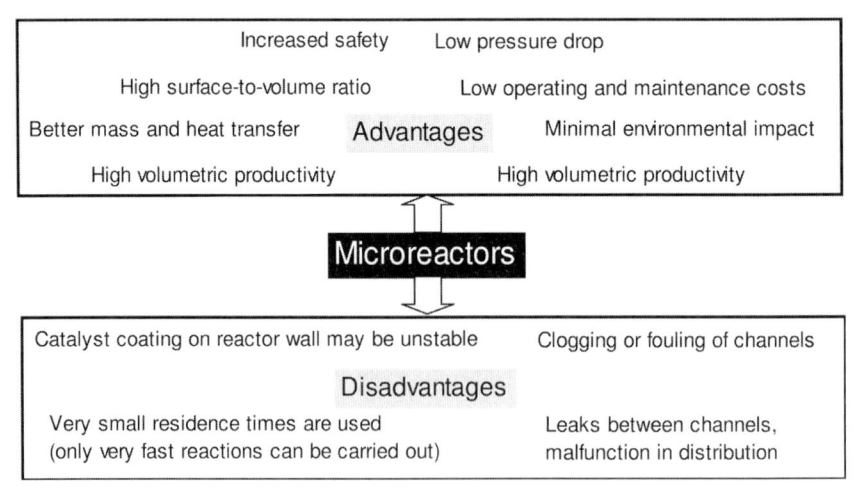

Disadvantages	
Catalyst coating on reactor wall may be unstable	Clogging or fouling of channels
Very small residence times are used (only very fast reactions can be carried out)	Leaks between channels, malfunction in distribution

Figure 9.10 Advantages and disadvantages of microreactors.

chain propagation. In fact, the flame arrestor effect that quenches chain growth by termination at the walls is favored because of the short diffusion distances.

Several examples of oxidation reactions, both in the liquid and in the gas phase, have been investigated in microreactors. Often the use of the microstructured device allows a better selectivity to the product of partial oxidation, because of a better temperature control on the catalyst surface (see, for instance, several examples in reviews [61a,b]). Indeed, several gas-phase oxidations can be completed in milliseconds, at significantly high temperatures.

One successful example of industrial production carried out in microreactors is the BASF process for alcohol oxidation [62]. Another relevant example is a new plant for the production of over 100 000 tons y^{-1} of hydrogen peroxide (HP) planned by UOP, now under the stage of pilot processing and basic engineering. UOP's interest was to have this synthesis route in the framework of propene oxide manufacture. Based on microstructured mixing units, the new process has been studied together with IMM (Mainz, D), and realizes the direct contacting of hydrogen and oxygen (without inert gas) in the presence of a heterogeneous catalyst based on noble metal. The latter has to be maintained in an oxidized state to be active and selective to HP. A selectivity as high as 80–85% at 90% conversion and a productivity of $2\,g_{HP}\,(g_{cat}\,h)^{-1}$ was reported at an oxygen/hydrogen ratio of 1.5–3, at 50 °C and 20 bar. The maximum HP concentration obtained, however, is only 1.7% [63]. Also, recent patents from Velocys claim microreactor devices for the synthesis of HP by direct reaction between hydrogen and oxygen [64].

A consortium including Degussa, Uhde and academic teams has developed an innovative approach for the gas-phase epoxidation of propene with HP, catalyzed by TS-1. A pilot unit has been designed and engineered at the Degussa site in Hanau-Wolfgang [65]. The joint project DEMiS (Demonstration project for the Evaluation of Microreaction technology in industrial Systems) uses microstructured devices for the gas-phase epoxidation of propene with vapors of HP (Figure 9.11). The use of the gas phase avoids the necessity of using a solvent for the reaction, and allows operation at high space–time yields. A gas-phase process involving HP is extremely challenging, since safe and stable evaporation of HP is required, and precautions have to be taken to minimize the risk of explosions of the gas mixture. Using a microreactor solved these problems. At 140 °C, 1 bar, 5% HP and 15% propene, a catalyst activity greater than $80\,mol_{PO}\,(kg_{cat}\,h)^{-1}$ was obtained on the laboratory scale; activity of this magnitude in a conventional reactor would cause hot spots and increased decomposition of the HP. For a propene conversion of 5–20%, the selectivity to PO obtained was >90%; the selectivity based on HP was about 25%, but it could be improved up to 60% by using a molar excess of propene of about 6.6. The microreaction technology allows a better temperature control, and a safer provision and handling of HP both within and outside the range of explosive mixtures with propene and PO. To minimize the decomposition of HP during the evaporation, a microstructured falling-film evaporator is used, placed below the mixing device and reaction zone. The catalyst was coated onto the reactor module, forming a layer several hundred micrometers thick; one problem met was the slow accumulation of deposits on catalysts, deriving from by-products formation.

Figure 9.11 The DEMiS reactor for propene epoxidation with HP vapors (courtesy of Uhde).

This route is emblematic of the new intensification approach currently used in the chemical industry, which is aimed at finally developing a safer, less-capital and energy intensive and more flexible production of bulk chemicals.

9.5
An Alternative Approach: Anaerobic Oxidation with Metal Oxides in a Cycle Process (from an Oxidation Catalyst to a Reusable Stoichiometric Oxidant)

In oxidation reactions involving redox-type interaction between the catalyst and the reactant, the selectivity to the product of partial oxidation is a function of several factors, including [2c,d]:

1. The redox properties of the metal cation used for the activation of the organic substrate. These depend not only a function on intrinsic characteristics of the metal ions in the solid, but also on the reaction conditions (i.e., temperature and

9.5 An Alternative Approach: Anaerobic Oxidation with Metal Oxides in a Cycle Process

feedstock composition), since the latter affect the characteristics of the catalyst under working conditions.

2. The presence of consecutive reactions involving the desired product, and the reactivity of the product towards unselective oxidizing attacks occurring at the catalyst surface. For example, the development of electrophilic adsorbed oxygen species can be detrimental for selectivity.

3. The presence of gas-phase homogeneous reactions of combustion.

All these problems can be potentially solved by carrying out the reaction with decoupling of the redox steps in two separate vessels (cycle operation): one for the contact between the catalyst (which acts as a true oxidizing agent) and the organic substrate, and one for the reoxidation of the reduced catalyst by contact with molecular oxygen (Figure 9.12) [66]. This approach, also referred to as *anaerobic oxidation*, is possible because the lattice oxygen of the catalyst is used in a stoichiometric reaction with a reactant to yield the oxygenated product [67]. Eventually, a Circulating Fluid Bed Reactor allows the circulation of the catalyst from the reactor zone, where it is put in contact with the hydrocarbon, to the reoxidation vessel. The redox solid has to be fabricated in the form of a free-flowing powder, allowing pneumatic transport from one reactor to the other.

Reductant (hydrocarbon, H_2)

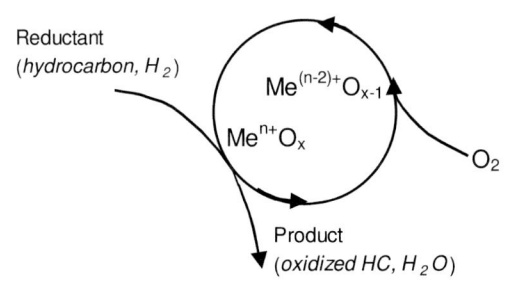

$Me^{(n-2)+}O_{x-1}$

$Me^{n+}O_x$

O_2

Product (oxidized HC, H_2O)

Figure 9.12 The redox mechanism [67].

The following are the advantages for the anaerobic oxidation, as compared to the conventional co-feed operation:

1. Possibility of using greater concentrations of the hydrocarbon than in the co-feed process. Higher concentration means higher productivity, or smaller equipment size, and the handling of smaller volumes of gas. Product recovery costs are lower, since the regeneration off-gas stream is kept separate from the product gas stream. Moreover, a better temperature control in the catalytic bed is achieved, because of the thermal characteristics of hydrocarbons as compared to nitrogen or other ballasts.

2. Improved safety, since in no section of the plant are the hydrocarbon and oxygen fed together.

3. Higher selectivity to the product of partial oxidation for several reasons: (i) minimization of reactions occurring between the hydrocarbon and reactive adsorbed

oxygen species that may develop at the catalyst surface in the presence of O_2 and (ii) a catalyst that on average is more reduced than the steady-state catalyst working in co-feed operation. This in general may represent an advantage in selective oxidation reactions; however, it is necessary to avoid over-reduction of the catalyst, since this would cause formation of coke.

4. Possibility of optimizing independently in each step (reaction and regeneration) the reaction parameters, such as the reactant concentration, the gas residence time and the solid residence time and the temperature.

The properties a catalyst must have to be suitable for use in a cycle operation are:

1. It must be able to readily furnish ionic oxygen to the organic substrate at an acceptable rate, and in amount sufficient to allow recirculation of reasonable masses of catalyst per unit time (otherwise, energy costs for transport would increase excessively). In addition, the solid must maintain structural integrity through many redox cycles.

2. It has to easily recover the original oxidation state by contact with air. Usually the reoxidation of a reduced catalyst is the rate-determining step, and therefore residence times in the regenerator are much higher than in the synthesis reactor.

3. It must have mechanical properties sufficiently high to allow resistance under conditions of fast fluidization or pneumatic transport.

The cycle approach for oxidation has been adopted at an industrial level for the Wacker–Chemie process for acetaldehyde production, in which ethylene is first put in contact with the oxidized catalyst solution, containing palladium chloride, and in the second step the solution containing the reduced catalyst is sent to a regeneration reactor containing cupric chloride and inside which also air is fed. The regenerated catalyst solution is returned to the first oxidation stage. Another industrial application is the Lummus process for the anaerobic ammoxidation of o-xylene to o-phthalonitrile [68]. Du Pont has developed the oxidation of n-butane to maleic anhydride catalyzed by V/P/O, in a CFBR reactor, and built a demonstration unit in Spain [69]; however, a few years ago the plant was shut down, due to the bad economics.

However, from patent literature it can be inferred that for practically all oxidation reactions the two-step approach has been claimed as that one potentially giving considerable advantages with respect to the co-feed operation. For example, alternate feeding of propene and oxygen has been claimed for the direct synthesis of propene oxide, over supported Ag-based catalysts [70a]. The catalyst is first contacted with the olefin in the absence of oxygen, and is then rapidly contacted with oxygen, for 20 s. The epoxide is obtained with selectivity higher than 55%; then the oxygen is discontinued for a few minutes, and the cycle of oxygen pulse is repeated.

The anaerobic oxidative coupling of methane to ethylene was studied by ARCO Chemical [70b]. The reaction occurs at a very high temperature of 850–900 °C; with a $Li/B/Mn/Mg/O$-SiO_2 catalyst, a conversion of 22% was achieved with selectivity to C_2 compounds of about 60% under anaerobic conditions. Monsanto [71] studied the oxidative dimerization of toluene to stilbene. Using a K/Bi/O catalyst at 575 °C, the anaerobic oxidation gave 46% conversion with 81% selectivity to stilbene, while under co-feed conditions the conversion was 38% and the selectivity 73%.

A few examples of anaerobic oxidations reported recently in the literature are described more in detail below.

9.5.1
Anaerobic Oxidation of Propene to Acrolein in a CFBR Reactor

Elf Atochem (now Arkema) and Du Pont have claimed a cycle process for the oxidation of propene to acrolein [70a]. In a first transport-bed reactor (a riser, where the catalyst is transported upwards by the gas) propene is put in contact with the catalyst, a Bi/Mo/W/Co/Fe mixed oxide, while in the second one, a fluidized-bed reactor, the catalyst that in the first step has been reduced by the olefin is reoxidized with air. The catalyst is continuously transported from the first reactor to the regenerator, by means of a CFBR. The reduced solids from the riser reactor are separated from the product gas, acrolein is recovered from the effluent gases, and the remaining gases can be vented or recirculated to the riser. The solids are then stripped for any gas, and conveyed to the regenerator. The stripped gases are mixed with the reactor effluent gases; off-gases from the regenerator are vented. The feedstock to the riser is made of propene and inerts (steam and nitrogen); the propene concentration is in the range 2 to 20%. Alternatively, part of the reoxidation of the reduced catalyst can be done in the riser itself (and then completed in the regenerator), and for this purpose a small amount of oxygen can be fed to the riser together with propene and the ballast. This may reduce the amount of catalyst that has to be circulated per unit time.

The catalytically active phase described by Elf Atochem/Du Pont for anaerobic oxidation is the same as used for propene oxidation under co-feed conditions [72b], a mixed molybdate of Fe, Co and Bi, with further components present in minor amounts (K, Cs, W, Mn, Cr). The following active phase composition is reported: $Mo_{12}Co_{3.5}Bi_{1.1}Fe_{0.8}W_{0.5}Si_{1.4}K_{0.05}O_x$. This active phase is able to furnish the organic substrate with a large amount of bulk ionic oxygen, without undergoing irreversible structural collapse. Another feature is the great anion mobility in the structure, which makes possible fast ion transport through the lattice. The mechanical and morphological properties of the catalyst for pneumatic transport are conferred by 40–50 wt% of silica gel.

The best per pass yield to $C_2 + C_3$ products (aldehydes plus acids with two and three C atoms) with the said catalyst was obtained at a propene conversion of 61.3% (selectivity to acrolein 83.7%), at the reaction temperature of 355 °C, with the following feed composition: $C_3H_6/H_2O/N_2$ 11.6 : 10.0 : 78.4 (mol.%), with a gas contact time of 2.4 s. A decrease in solids circulation rate, while keeping gas residence time constant, led to a considerable decrease in propene conversion, while selectivity to $C_2 + C_3$ oxygenated products was not much affected by circulation rate. With a less concentrated feed, the amount of solid to be circulated for a defined olefin conversion is lower, but productivity also becomes lower. Other catalysts based on Bi/Mo/O or on V/Mo/W/Cu/O [72c] afforded conversions >70% and selectivity >90%; industrial processes that operate with a co-feed of all reactants yield at best 82–83% selectivity, with 95% propene conversion.

9.5.2
Anaerobic Synthesis of 2-Methyl-1,4-Naphthoquinone (Menadione)

In the industrial production of menadione (2-methyl-1,4-naphthoquinone, the intermediate for the synthesis of vitamins of the K group), the conventional route using 2-methylnaphthalene and chromium oxide in sulfuric acid as the oxidant co-produces a large amount of inorganic salts. Therefore, many efforts are being made aiming for the development of a catalytic route, and several systems have been investigated that make use of either HP [73] or O_2 [74] as the oxidants for 2-methylnaphthalene or 2-methyl-1-naphthol (Figure 9.13).

Matveev [74] first described the use of aqueous solutions of Keggin-type P/Mo/V polyoxiometalates (POM) for the selective, stoichiometric oxidation of 2-methyl-1-naphthol, in a liquid bi-phase system, at moderate reaction conditions (Vikasib technology). The redox-decoupled procedure guarantees a high selectivity to the desired product, since in the absence of oxygen the O^{2-}-insertion step is favored with respect to unselective radical reactions. After the reaction, the aqueous phase containing the reduced POM is re-oxidized with air, at high temperature and pressure. The authors investigated the effect of the POM composition on the productivity and selectivity. The best performance in the synthesis step was obtained with the POM having composition $H_5PMo_{10}V_2O_{40}$; this compound is also the one that can be more easily re-oxidized with oxygen. The reduction level for the POM corresponds to the total reduction of the V^{5+} ion in the Keggin anion to V^{4+} (α = degree of V reduction = 1).

Although the reaction is very selective, the re-oxidation step remains a crucial point that limits the overall productivity of the process. In this regard, it was found that it is possible to obtain a high selectivity to menadione even when starting from the partially oxidized catalyst, instead of the fully oxidized one. In other words, there is no need to re-oxidize the reduced POM ($\alpha = 0$) completely, but a partial reoxidation ($\alpha = 0.25$) is sufficient to obtain a POM that can be used for the successive synthetic step by contact with 2-methyl-1-naphthol [75]. This is an advantage compared to the

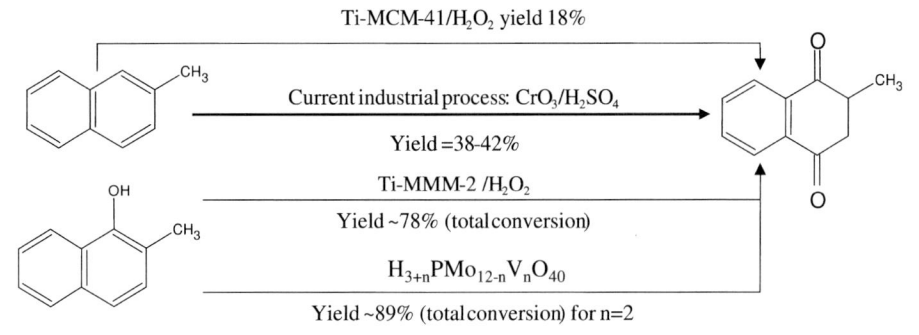

Figure 9.13 Various approaches for the synthesis of menadione.

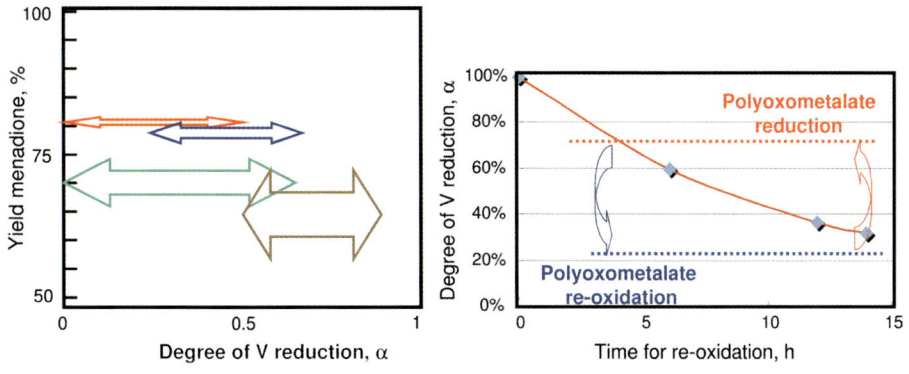

Figure 9.14 Modified redox approach for the synthesis of menadione [75].

use of a fully re-oxidized POM as a stoichiometric oxidant for methylnaphthol during subsequent cycles, because the partial reoxidation (from $\alpha = 1$ to $\alpha = 0.25$) takes much less time than a total reoxidation (from $\alpha = 1$ to $\alpha = 0$), and can be done under milder conditions. Even more conveniently, a good selectivity to menadione is obtained during the synthetic step when a maximum average reduction level corresponding to $\alpha = 0.60$–0.75 is reached. Therefore, the best performance is obtained by shuttling the V average oxidation state in the POM solution between $\alpha \approx 0.25$ and $\alpha \approx 0.75$ (Figure 9.14), with a vanadium/methylnaphthol initial molar ratio equal to 8. This combines a high selectivity to menadione during the synthetic step with the shorter time needed for the reoxidation of the POM with air. Notably, a continuous variation of the catalytic performance (in terms of yield to menadione) in the range of α between 0 and 1 implies that the POM redox behavior is not due to the contribution of each molecular-like Keggin unit, but is rather closer to that of a solid showing collective properties.

9.5.3
Anaerobic Oxidative Dehydrogenation of Propane to Propene

New applications of the cycle concept might be for those oxidation processes that have not yet reached the commercial stage because of the low selectivity to the desired product. One example is the oxidative dehydrogenation (ODH) of alkanes to the corresponding olefins [24d,76]. Indeed, looking at the literature data, numerous catalytic systems have been described for which the selectivity to the olefin, if extrapolated to a very low partial pressure of oxygen, is very high (even higher than 90%). This suggests that the anaerobic reaction between the catalyst and the hydrocarbon may be very selective to the formation of the olefin.

Several papers describe the anaerobic ODH of alkanes to olefin with catalysts based on metal oxides [77–83]. In many cases, an improved selectivity to the olefin as compared to the co-feed operation has been achieved; in other cases, no effective improvement has been reported. For instance, catalysts made of V_2O_5-SiO_2 showed a remarkably improved selectivity to propene with respect to the co-feed conditions for propane ODH, whereas the same remarkable effect was not observed with catalysts

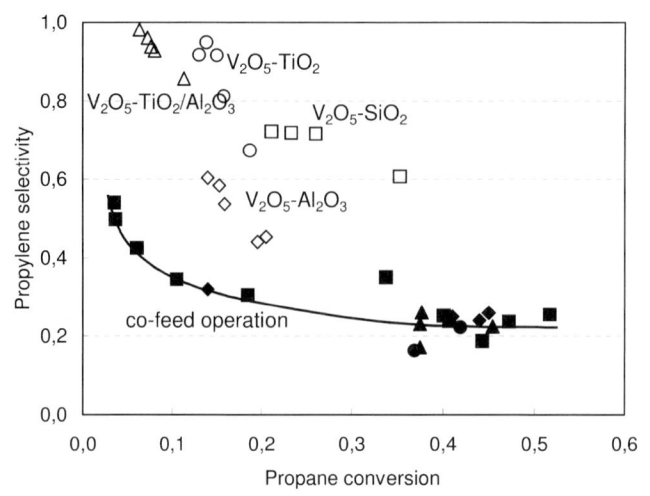

Figure 9.15 Comparison of selectivity to propene obtained by cyclic feed (redox mode) (open symbols) and co-feed (filled symbols), for supported vanadium oxide catalysts [76m]. V_2O_5-Al_2O_3 (◇, ◆), V_2O_5-TiO_2 (○, ●), V_2O_5-SiO_2 (□, ■), V_2O_5-TiO_2/Al_2O_3 (△, ▲).

in which vanadium oxide was deposited on alumina or titania (Figure 9.15) [76m]. Indeed, the very high selectivity to propene obtained at low propane conversion in the redox mode was due to the relevant contribution of dehydrogenation to olefin formation over the strongly reduced catalyst.

One condition for obtaining an improved selectivity with the cycle operation is the presence of catalysts in which V sites are highly dispersed; this was achieved by preparing vanadium oxide-silica co-gels, with a V_2O_5 content lower than 10% [77b,e]. The improvement in selectivity to propene was around 30 percentage points (60% selectivity in cyclic mode versus 30% in co-feed mode) for a propane conversion of 35%.

The same concept has been used also with an alkane dehydrogenation catalyst [82, 84], to shift the equilibrium of the reaction by using a catalyst selective in H_2 combustion. This can be carried out by feeding O_2 together with the alkane; a (post) transition metal or its oxide can be used to selectively oxidize hydrogen [85, 86]. This approach has the advantage of the energy released by exothermic oxidation, which is needed to aid dehydrogenation. However, by mixing oxygen, hydrogen and hydrocarbons at high temperatures, a dangerous mixture may be obtained. The separation of reactants using dense ion-conducting oxides or oxygen-selective membranes reduces the risk of explosion; however, as an alternative approach, the cyclic concept may be applied. Specifically, a solid oxygen carrier (SOC) is used to selectively remove H_2 from the reactor and improve product yield [82c,84]. In the second step, oxygen is flowed to re-oxidize the reduced SOC; the latter is typically a metal oxide that can easily sustain the redox cycle.

The reduction of SOC by H_2 can be either endo- or exothermic, depending on the type of oxide; in the latter case, the reduction provides the energy gain to support the endothermic dehydrogenation; otherwise, during the re-oxidation step the heat released is stored in the catalytic bed and carried over to the next cycle. The authors reported that the best SOC was $Ce_{0.9}W_{0.1}O_x$ (selectivity to H_2 oxidation higher than 97%), which – when coupled with a dehydrogenation catalyst – made it possible to enhance the conversion and selectivity to propene compared to a conventional dehydrogenation reactor [82].

9.5.4
Production of Hydrogen from Methane with Oxide Materials and Inherent Segregation of Carbon Dioxide

Selected oxides, once reduced with hydrocarbons, are capable of being re-oxidized by splitting H_2O into H_2 and [O], which in turn is incorporated in the solid [87]. This scheme can be exploited by separating the two reactions spatially or temporally, with the obvious advantage of obtaining a stream of almost pure H_2 (virtually carbon free) when the solid is oxidized with water, and a concentrated stream of CO_2/H_2O when the solid is reduced by methane or by any other reducing agent. For instance, reduction by the hydrocarbon and water splitting can take place in two vessels with a moving catalyst in a CFB reactor.

ENI is exploiting this principle for the development of an innovative technology (one-step hydrogen) for H_2 production from fossil fuels with an inherent confinement of CO_2 in a concentrated stream, ready to be buried [88].

The reaction stoichiometries are:

$$4MeO_y + CH_4 \rightarrow 4MeO_{(y-1)} + 2H_2O + CO_2$$

$$4MeO_{(y-1)} + 4H_2O \rightarrow 4MeO_y + 4H_2$$

overall:

$$CH_4 + 2H_2O \rightarrow 4H_2 + CO_2$$

The use of iron oxides in a reversible, cyclic, reduction-and-oxidation loop for hydrogen production had been described previously by Otsuka [89]. A cyclic chemical combustion on iron-oxide-based solids has also been proposed for the combustion of methane with inherent separation of CO_2 [90]. The storage of H_2 from methane mediated by iron oxides has been described recently [89b,91]. However, pure iron oxide cannot be used as an oxygen carrier, because it is very difficult to synthesize in the form of microspheres and because it cannot be easily reduced at high temperature with methane in a controlled way. In fact, the iron oxide particles become overreduced on the external surface, exposing metallic iron and promoting carbon deposition while maintaining an oxidized core. This prevents a complete exploitation of the available oxygen. Therefore, deposition of iron oxide on a suitable carrier may be desirable, not only to obtain particles having the desired morphology characteristics, but also to promote the formation of a dispersed iron oxide, the reduction of which can be better controlled.

ENI has been developing a material with fluidizability characteristics, made of either ceria-supported or alumina-supported iron oxide [88]. Alumina is, however, pre-treated to avoid the formation of $FeAl_2O_4$, a species that is not capable of being re-oxidized by water. For instance, microspheroidal alumina can be treated to form a protective coating made of a spinel phase, for example, either $MgAl_2O_4$ or $ZnAl_2O_4$. Addition of promoters, Cr and Ce, also improves hydrogen productivity and oxygen efficiency.

The progressive reduction of Fe_2O_3 by methane occurs by means of successive steps:

$$12Fe_2O_3 + CH_4 \rightarrow 8Fe_3O_4 + CO_2 + 2H_2O \text{ (very fast)}$$

$$4Fe_3O_4 + CH_4 \rightarrow 12FeO + CO_2 + 2H_2O$$

$$Fe_3O_4 + CH_4 \rightarrow 3FeO + CO + 2H_2$$

(Indeed, wustite is a non-stoichiometric compound: $Fe_{1-\delta}O$.)

$$FeO + CH_4 \rightarrow Fe + CO + 2H_2$$

As soon as metallic iron is formed, the cracking of methane to carbon (a polymorph of the graphite structure) and H_2 begins, with development of the Fe_3C orthorhombic structure; the latter eventually decomposes into αFe and filamentous carbon. Ceria, when used as the support, also undergoes reduction to CeO_{2-y}, probably because of the considerable amount of hydrogen coming from methane cracking. During reoxidation with water, the solid follows the reverse path with respect to the reduction, with the only exception of the iron carbide, which is directly transformed into wustite.

The conceptual reactor design proposed by ENI includes three steps (Figure 9.16) [88d]:

1. In a first reactor (R1), the oxidized material is put in contact with 100% methane, to reduce hematite to wustite; the formation of metallic iron has to be preferentially avoided. Carbon dioxide and water are the reaction products. This reaction is endothermic.

$$4MeO_y \rightarrow 4MeO_{y-1} + 2O_2 \quad \Delta H° > 0$$

$$CH_4 + 2O_2 \rightarrow CO_2 + 2H_2O \quad \Delta H° = -192 \text{ kcal mol}^{-1}$$

overall:

$$4MeO_y + CH_4 \rightarrow 4MeO_{(y-1)} + 2H_2O + CO_2$$

For the spontaneous reduction of hematite:

$$Fe_2O_3 \rightarrow 2FeO + \tfrac{1}{2}O_2 \quad \Delta H° = +70 \text{ kcal mol}^{-1}$$

and therefore the overall reaction occurring in R1 is:

$$4Fe_2O_3 + CH_4 \rightarrow 8FeO + 2H_2O + CO_2 \quad \Delta H° = +88 \text{ kcal mol}^{-1}$$

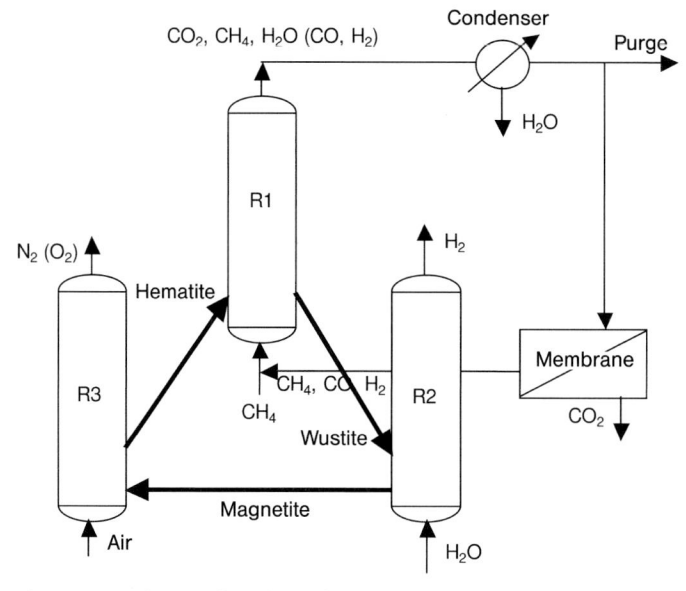

Figure 9.16 Schematic flow-sheet of the ENI process for H_2 production with inherent CO_2 sequestration [88d].

2. In a second reactor (R2), the reduced oxide is reoxidized in part with water (20% steam in N_2), to produce hydrogen and a partially reoxidized material (e.g., a solid in prevalence containing magnetite).

3. In a third reactor (R3, thermal support unit), the solid obtained in R2 is fully oxidized to hematite with air. This third zone is aimed at closing the thermal balance of the process; the heat released is used to carry out the endothermic step. In practice, this would correspond to the burning of a part of the hydrogen produced, to supply the heat for this step. This third unit can be avoided if the endothermicity of the process is accepted, and the reoxidation with water is completed in R2, so maximizing the amount of H_2 produced per unit weight of catalyst.

Evidently, the production of the reduction gases of the solid (CO_2 and H_2O) and of H_2 in separate zones considerably reduces separation and purification costs of hydrogen.

Since:

$$MeO_{(y-1)} + \tfrac{1}{2}O_2 \rightarrow MeO_y \qquad \Delta H° < 0$$

and

$$H_2O \rightarrow H_2 + \tfrac{1}{2}O_2 \qquad \Delta H° = +58\,\text{kcal mol}^{-1}$$

(while the combustion of H_2 is exothermic by $58\,\text{kcal mol}^{-1}$), it turns out that the burning of approximately the 17% of the hydrogen produced per each mole of methane suffices to obtain thermal neutrality. The same can be achieved by carrying out a partial reoxidation of the reduced oxide in R2, and by leading the reoxidation to completion with air in R3. For instance, if the oxidation of FeO with H_2O in R2 is limited to produce Fe_3O_4 ($\Delta H°$ $-13\,\text{kcal}\,\text{mol}^{-1}$ per mole of H_2O), and then the oxidation of Fe_3O_4 to Fe_2O_3 is carried out with air in R3 ($\Delta H°$ $-34\,\text{kcal}\,\text{mol}^{-1}$ Fe_3O_4), for an overall reoxidation stoichiometry:

$$8\text{FeO} + \tfrac{8}{3}H_2O + \tfrac{2}{3}O_2 \rightarrow 4Fe_2O_3 + \tfrac{8}{3}H_2 \text{ (in R2 + R3)}$$

and a process stoichiometry:

$$CH_4 + \tfrac{2}{3}H_2O + \tfrac{2}{3}O_2 \rightarrow CO_2 + \tfrac{8}{3}H_2 \text{ (in R1 + R2 + R3)}$$

the reaction heat supplied by the exothermal reactions in R2 and R3 more than compensates for the endothermic reaction occurring in R1. In this case, the productivity of H_2 is $^2/_3$ that obtained when FeO is fully reoxidized to Fe_2O_3 with H_2O.

The process is called "one-step hydrogen" because two of the three sections (including the syngas production step) in which a CO_2-free hydrogen production can be divided (i.e., steam reforming, water–gas shift and CO_2 adsorption) are removed in this process. It is evident that the amount of oxygen exchanged per unit of mass is of paramount importance; it dictates the hydrogen production and governs the heat released during oxidation. If it exceeds an optimum value, the adiabatic temperature rise can be excessive, while an optimal profile has to be carefully maintained through the cycle to reach the highest efficiency.

9.6
Current and Developing Processes for the Transformation of Bioplatform Molecules into Chemicals by Catalytic Oxidation

The present utilization of carbohydrates as a feedstock for the chemical industry is modest, when considering their ready availability, low cost and huge potential [92]. The bulk of the annually renewable carbohydrate biomass consists of polysaccharides, but their non-food utilization is still modest. The low-molecular-weight carbohydrates, that is, the constituent units of these polysaccharides, are potential raw materials for several commodity chemicals; in fact, glucose (available from cornstarch, bagasse, molasses, wood), fructose (inulin), xylose (hemicelluloses) or the disaccharide sucrose (world production 140 Mtons year^{-1}) are inexpensive and available on a scale of several ten thousands.

A few examples of chemicals produced by oxidation of sugars include:

1. D-Gluconic acid: the production of this compound is estimated to be around 100 000 ton year^{-1}. It is obtained by either chemical (e.g., hypochlorite, Pt/O_2), electrochemical or enzymatic oxidation of D-glucose. Aside from its use in the food processing industry as an acidulant, it has important non-food uses as a sequestrant in cleaners, as a metal-etching agent and as a latent acid catalyst in textile printing. Charcoal- and titania-supported gold gives better performance than

previously reported Pt- and Pd-based catalysts in activity and selectivity, which is nearly 100% to gluconic acid. However, long-term stability is still too low for industrial application [93, 94].

2. Furoic acid (furan-2-carboxylic acid, or pyromucic acid) is used as a bactericide, and the furoate esters are used as flavoring agents, as antibiotic and corticosteroid intermediates. It is obtained by the enzymatic or chemical/catalytic aerobial oxidation of furfural (2-furalaldehyde); the latter is the only unsaturated large-volume organic chemical prepared from carbohydrates today. D-Xylose and L-arabinose, the pentoses contained in the xylan-rich portion of hemicelluloses from agricultural and forestry wastes, under the conditions used for hydrolysis undergo dehydration to furfural.

In 2004, a conceptual milestone was set by the US Department of Energy by identifying from a list of 300 candidates the top sugar-derived building blocks (Figure 9.17) [95]. These "bioplatform chemicals" have the potential to be converted into several chemicals and materials by chemical transformations, and it is evident that catalytic oxidation may play an important role for the full industrial utilization of these building blocks.

Examples of oxidative transformations for the synthesis of chemicals starting from sugars and sugars-derived compounds include the following (Figure 9.18):

1. Oxidized sugars are materials having very large worldwide consumption. Although malic, tartaric, aspartic and glutamic acid are sugar-derived dicarboxylic acids with industrial significance as building blocks, those dicarboxylic acids that utilize the entire carbon chain of the starting sugar may also become of importance. For instance, glucaric acid can be produced by oxidation with nitric acid of glucose (e.g., from starch), but interest exists in finding heterogeneous systems that catalyze the oxidation with air or diluted hydrogen peroxide. Pt-catalyzed oxidation with oxygen seems to be a promising approach. Glucaric acid can serve as a starting point for the synthesis of a wide range of products with high-volume markets, such as polyhydroxypolyamides (new nylons). Esters of glucaric acid are also potential starting materials of new types of hyperbranched polyesters. Other examples of dicarboxylic acids include galactaric acid from lactose, xylaric acid from xylose and acids derived from disaccharides such as sucrose and isomaltulose. The same approach could also be applicable to the oxidation of other inexpensive sugars, such as xylose or arabinose. In many cases, the technical barriers for their large-scale production are mainly the development of efficient catalytic selective oxidation technologies, to eliminate the need for nitric acid as the stoichiometric oxidant; in this context, Pt-catalyzed oxidation with O_2 is the most promising approach. For instance, sucrose is cheaper than its component sugars D-glucose and D-fructose, and is available in a huge amount and in great purity; the primary hydroxyl groups can be transformed into the corresponding carboxylic groups, and the compounds obtained are of great industrial relevance. Oxidation of sucrose with Pt/C and O_2 at 80–100 °C yields the preferred formation of sucrose-6,6-dicarboxylic acid; the latter can be prevented from further oxidation with continuous electrodyalitic removal [96, 97]. Great

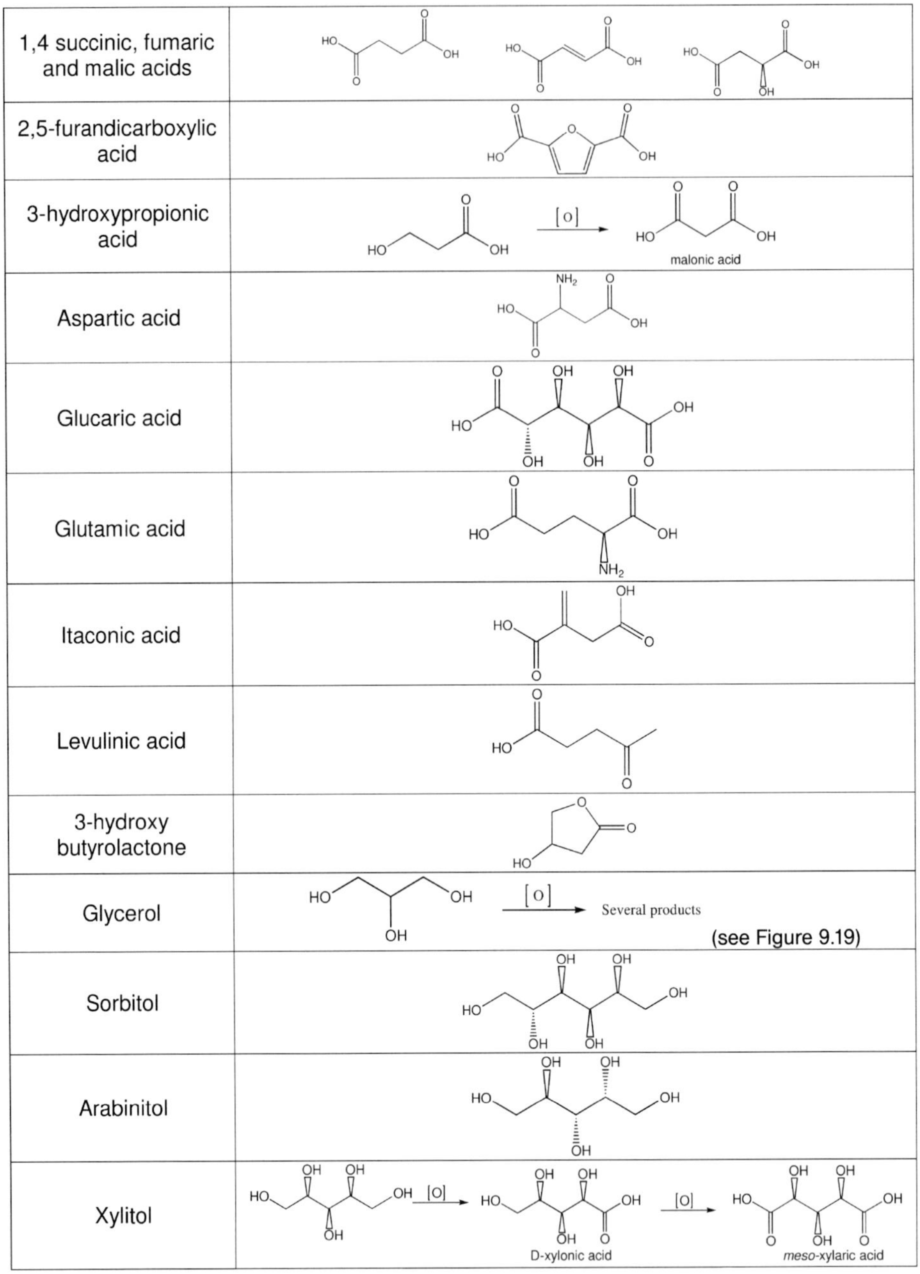

Building block	Structure
1,4 succinic, fumaric and malic acids	(succinic acid, fumaric acid, malic acid structures)
2,5-furandicarboxylic acid	(2,5-furandicarboxylic acid structure)
3-hydroxypropionic acid	(3-hydroxypropionic acid) —[O]→ malonic acid
Aspartic acid	(aspartic acid structure)
Glucaric acid	(glucaric acid structure)
Glutamic acid	(glutamic acid structure)
Itaconic acid	(itaconic acid structure)
Levulinic acid	(levulinic acid structure)
3-hydroxy butyrolactone	(3-hydroxybutyrolactone structure)
Glycerol	(glycerol) —[O]→ Several products (see Figure 9.19)
Sorbitol	(sorbitol structure)
Arabinitol	(arabinitol structure)
Xylitol	(xylitol) —[O]→ D-xylonic acid —[O]→ meso-xylaric acid

Figure 9.17 Top bioplatform building blocks, and some examples of chemicals that can be obtained by oxidation [95].

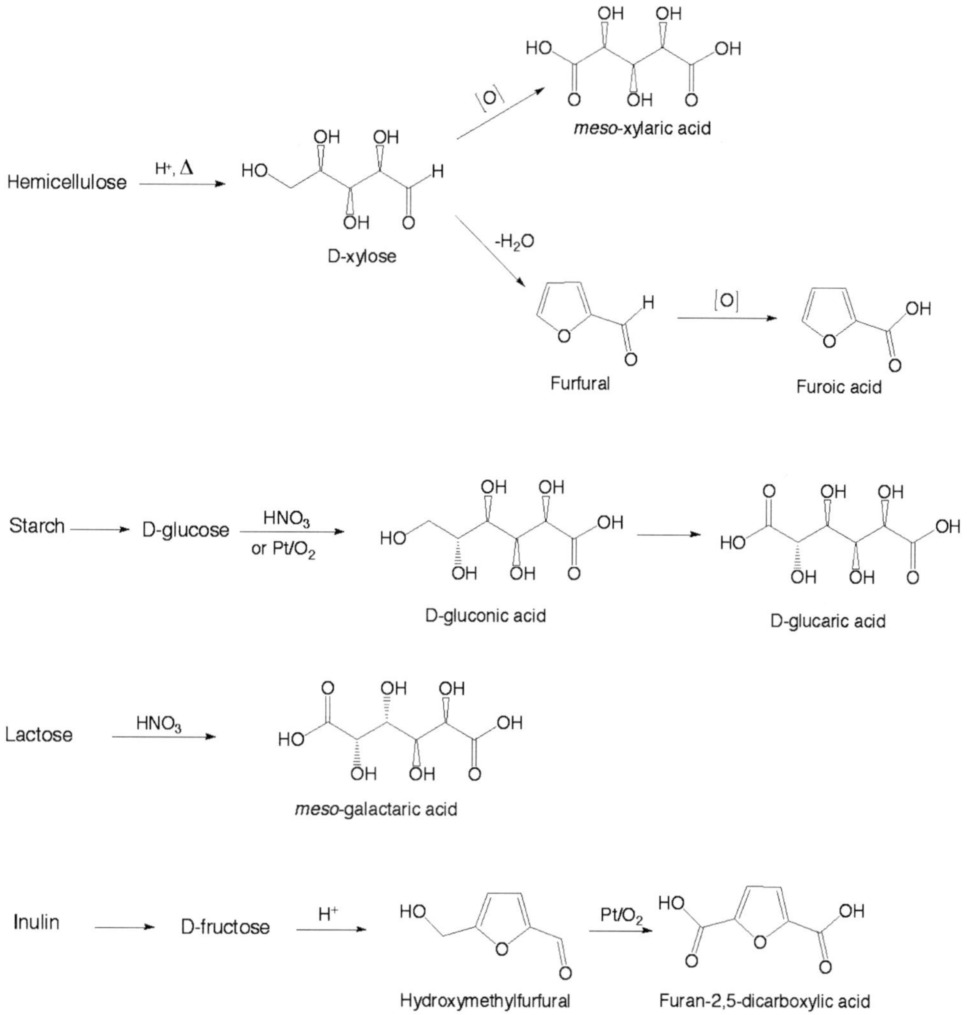

Figure 9.18 Chemicals obtained from monosaccharides by oxidation.

industrial potential can also be attributed to glucuronyloxymethyl furoic acid, obtained by acid treatment and catalytic oxidation of sucrose-derived isomaltulose.

2. Xylitol and L-arabinitol (five-carbon sugar alcohols that can be obtained by hydrogenation of the corresponding pentoses D-xylose and L-arabinose, respectively, with the same technology by which sorbitol is obtained from glucose) can be the starting compounds for the synthesis of various acids, such as xylaric and xylonic acids, arabonic and arabinoic acids (the dicarboxylic acid and the carboxylic acid, respectively). These compounds have the potential for the preparation of polyesters and polyamides. Also in this case, catalysts that afford the use of oxygen as the oxidizing agent need to be developed.

3. Furan-2,5-dicarboxylic acid also has tremendous industrial potential, because it could replace oil-derived diacids such as adipic or terephthalic acid as monomers for polyesters and polyamides [98, 99]. This diacid can be synthesized by Pt-catalyzed oxidation with O_2 of 5-hydroxymethylfurfural; the latter is obtained by acid-catalyzed dehydration of D-fructose or fructosans (inulin); the latter, however, are too expensive as starting materials, and yields from glucose-based waste raw materials are no higher than 40%. Therefore, the potential attractive option of furan-2,5-dicarboxylic acid will develop only after an efficient generation of 5-hydroxymethylfurfural from forestry waste materials has been developed. The same compound is also the starting material for the synthesis of other interesting chemicals obtained by oxidative processes, such as 5-hydroxymethylfuroic acid, 5-formylfuran-2-carboxylic acid and the 1,6-dialdehyde.

4. Levulinic acid is formed by the treatment of six-carbon sugar carbohydrates from starch or lignocellulosics with acids, or by acid treatment plus a reductive step of five-carbon sugars derived from hemicellulose. Levulinic acid can serve as a building block for the synthesis of many derivatives; of interest may be the selective oxidation to succinic and acrylic acid. β-Acetylacrylic acid could be used in the production of new acrylate polymers.

9.6.1
Glycerol: A Versatile Building Block

The worldwide production of glycerol is estimated to be around 750 000 ton y^{-1}, almost entirely made from natural triglycerides; nowadays, in fact, only 12% of the worldwide production is made synthetically. The increased availability of glycerine, the aqueous solution of glycerol obtained as the co-product in triglycerides transesterification for biodiesel production (10% w/w related to the vegetable oil of the starting material), has caused the decrease of glycerol cost. With an estimated capacity of biodiesel of 12 M metric tons per year in 2010, glycerol is becoming a bulk renewable feedstock. The cost of raw (crude) glycerol was between 250 and 300€ ton^{-1} before 2003, but it has been falling in recent years to well below 50€ ton^{-1}; most biodiesel producers now attach zero value to the crude glycerol. The cost of refined glycerol has been fluctuating remarkably in recent years, from a value ranging between 800 and 1200€ ton^{-1} before 2003 to a minimum of approx 400–500€ ton^{-1} in 2006 (which corresponds to fuel value), to almost 1000€ ton^{-1} in 2007.

The volumes of "new" glycerol are huge and increasing; announced new biodiesel plants will add over 2 M tons of glycerol to the current production in the USA and EU by 2010; in 2005 the total worldwide yearly demand for glycerol was estimated to be less than 1 M tons. New players, including very large agribusiness companies such as Cargill and Archer Daniels Midland, will dominate the production of glycerol.

Nowadays, glycerol finds application only for niche markets; therefore the co-production of glycerine is becoming a burden for biodiesel producers and a bottleneck limiting the market of this biofuel. The economics of biodiesel depend heavily on using its co-product; a high-value use for glycerol could reduce the cost of biodiesel by as much as 5–10€cent L^{-1}. However, there is not enough demand for glycerol to make use of all the co-product expected. Therefore, intensive research is aimed at the valorization of glycerol by catalytic transformation into chemicals; the focus is on developing new catalytic routes to monomers or intermediates for the production of commodity chemicals, thus expanding the market beyond the limited current applications of glycerol. A relevant example in this direction is the Du Pont process for the enzymatic transformation of glycerol into 1,3-propandiol, the co-monomer for the high-performance polyester (polytrimethyleneterephthalate) Sorona.

The transformation of glycerol by means of catalytic oxidation has been investigated for many years. The selective oxidation of glycerol leads to a broad family of derivatives [100]; there is a wide literature on the liquid-phase oxidation with O_2, catalyzed by supported Pt/Bi [101–105], and by gold-based catalysts [105b,c, 106]. The type of compounds obtained and the corresponding yields are greatly affected by factors such as the pH of the solution, the nature of the main active component and the presence of Bi as promoter (Figure 9.19). The effect of each parameter and of the catalyst composition on yields and selectivities has been reviewed recently [100].

Gas-phase oxidation of glycerol has been less investigated than liquid-phase oxidation; it occurs via a two-step catalyzed reaction involving first the dehydration of glycerol into acrolein, catalyzed by an acid, and then its oxidation. The same reactions can be conducted in two distinct reactors, in which the first step can be carried out with an acid catalyst such as phosphoric acid over alumina [107]. Then acrolein is oxidized to acrylic acid with a conventional alumina-supported Mo/V/Cu/O catalyst.

Arkema has recently issued patents on the gas-phase dehydration of glycerol to acrolein and on the oxidation of glycerol to acrylic acid [107]. The dehydration can be carried out at 300 °C, with a ZrO_2/WO_3 catalyst; the conversion of glycerol is total, and the yield to acrolein is 72%, when a feed made of an aqueous solution of glycerol (20%) is used. The addition of O_2 in the feed stream has the effect of yielding also the formation of acrylic acid (4.5%), and of lowering the yield to acrolein.

In a two-bed approach, a multimetal Mo vanadate, of composition $Mo_{12}V_{4.8}Sr_{0.5}W_{2.4}Cu_{2.2}O_x$ [72c], developed for the anaerobic oxidation of propene to acrolein in a cycle process, can be loaded in a second catalytic bed, downstream of the first one made of Zr/W/O. At 280 °C and in the presence of O_2, the conversion of glycerol is total, and the yield to acrylic acid is 74.9%, with less than 1% of acrolein [108b]. A multi-bed catalytic reactor is also claimed for the production of acrylic acid in which propene, glycerol and air are co-fed to the reactor. The first bed serves for the dehydration of glycerol to acrolein, the second bed for the oxidation of propene to acrolein, and the third bed for the oxidation of acrolein to acrylic acid [108c, e].

9.7 Conclusion

Promising fields of development for a greener and more efficient chemical industry in the area of catalytic selective oxidation are several, ranging from the use of

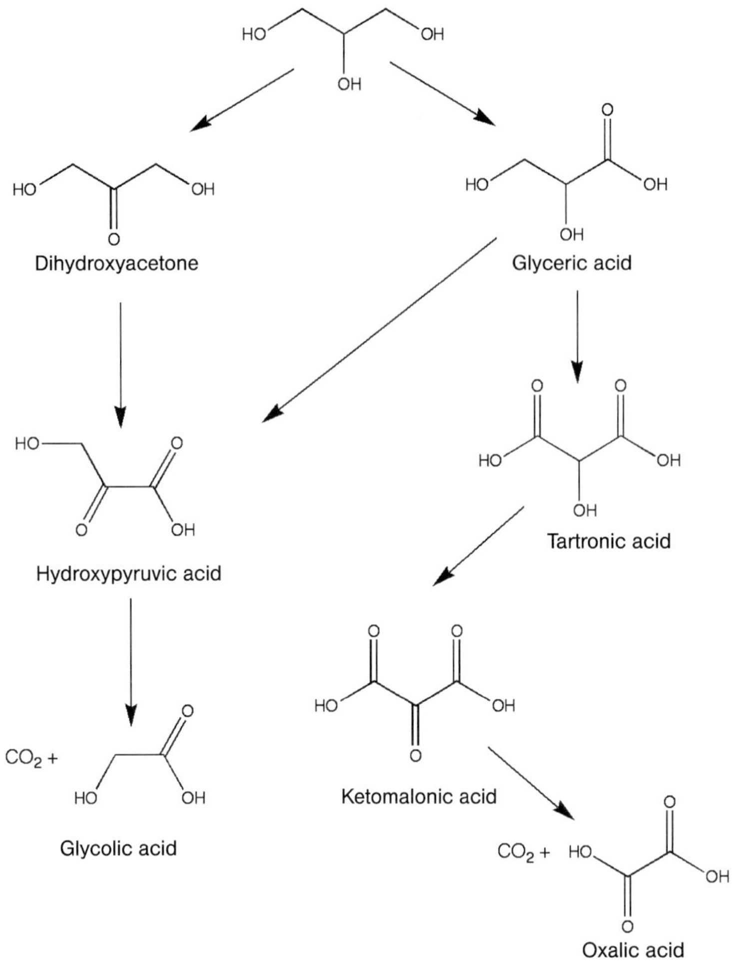

Figure 9.19 Products obtained in the liquid-phase oxidation of glycerol with O_2 catalyzed by supported Pt and Au catalysts.

alternative feedstocks, for example, bioplatform molecules and alkanes in place of traditional building blocks derived from oil, to the use of new technologies, such as miniaturized devices, and of alternative approaches for contacting reactants, for example, anaerobic oxidation, to new types of reactions and catalysts. The several examples available in the scientific and patent literature show that margins exist not only for the improvement of technologies currently employed but also for rethinking the industrial chemical production through an innovative and alternative vision. It is also evident that a huge effort is needed to develop new technologies that are environmentally more friendly than those they are going to replace, while being at the same time economically sustainable. To be successful, this effort requires for the knowledge developed inside each research area to be transferred into areas that may not be strictly scientifically adjacent, but which may benefit from innovative approaches used by other disciplines. This is particularly true in the field of catalytic selective oxidation, a scientific area in which major recent advancements have been possible thanks to the cooperation between scientists active in the fields of chemical engineering, materials chemistry and chemical/physics, synthetic chemistry and catalytic technology.

References

1. (a) Thomas, J.M. and Raja, R. (2006) *Catal Today*, **117**, 22; (b) ten Brink, G.J., Arends, I.W.C.E. and Sheldon, R.A. (2000) *Science*, **287**, 1636; (c) Lenoir, D. (2006) *Angewandte Chemie – International Edition*, **45**, 3206; (d) Ratnasamy, P., Raja, R. and Srinivas, D. (2005) *Philosophical Transactions of the Royal Society A*, **363**, 1001; (e) Arends, I.W.C.E. and Sheldon, R.A. (2001) *Applied Catalysis A-General*, **212**, 175; (f) Mizuno, N., Hikichi, S., Yamaguchi, et al. (2006) *Catalysis Today*, **117**, 32; (g) Kamata, K., Yonehara, K., Sumida, Y. et al. (2003) *Science*, **300**, 964; (h) Arakawa, H. et al. (2001) *Chemical Reviews*, **101**, 953; (i) Hoelderich, W. (2000) *Applied Catalysis A – General*, **194/195**, 487; (j) Noyori, R., Aoki, M. and Sato, K. (2003) *Chemical Communications*, 1977; (k) Min, B.K. and Friend, C.M. (2007) *Chemical Reviews*, **107**, 2709;(l) Clerici, M.G., Ricci, M. and Strukul, G. (2007) *Metal-catalysis in Industrial Organic processes*, Royal Society of Chemistry, Cambridge, pp. 23; (m) Brégeault, J.M. (2003) *Dalton Transactions*, 3289.

2. (a) Morgan, M. (2002) *Hydrocarbon Engineering*, 14; (b) Bhasin, M.M. (2003) *Topics in Catalysis*, **23**, 145; (c) Centi, G., Cavani, F. and Trifirò, F. (2001) *Selective Oxidation by Heterogeneous Catalysis Recent Developments* (eds M.V. Twigg and M.S. Spencer), Plenum Publishing Corporation, New York & London, Series: Fundamental and Applied Catalysis; (d) Arpentinier, P., Cavani, F. and Trifirò, F. (2001) *The Technology of Catalytic Oxidations*, Editions Technip, Paris; (e) Fokin, A.A. and Schreiner, P.R. (2003) *Advanced Synthesis and Catalysis*, **345**, 1035; (f) Brazdil, J.F. (2006) *Topics in Catalysis*, **38**, 289; (g) Grasselli, R.K. (2005) *Catalysis Today*, **99**, 23; (h) Costine, A. and Hodnett, B.K. (2005) *Applied Catalysis A – General*, **290**, 9; (i) Misono, M. (2002) *Topics in Catalysis*, **21**, 89; (j) Bordes, E. (2000) *Comptes Rendus de l'Académie des Sciences Series IIc, Chimie*, **3**, 725; (k) Sinev, M.Yu. (2003) *Journal of Catalysis*, **216**, 468.

3. (a) Thorsteinson, E.M., Wilson, T.P., Young, F.G. and Kasai, P.H. (1978) *Journal of Catalysis*, **52**, 116; (b) Young, F.G. and Thorsteinson, E.M. (1981) US Patent 4,250,346, assigned to Union Carbide Co; (c) McCain, J.H. and Charleston, W.V. (1985) US Patent 4,524,236, assigned to Union Carbide Co; (d) McCain, J.H. (1986) US Patent 4,568,790, assigned to Union Carbide Co; (e) Manyik, R.M., Brockwell, J.L. and Kendall, J.E. (1990) US Patent 4,899,003, assigned to Union Carbide Chemical and Plastics Co. (f) Manyik, R.M. (1986) US Patent 4,596,787, assigned to Union Carbide Co.

4. (a) Merzouki, M., Taouk, B., Monceaux, L. et al. (1992) *Studies in Surface Science and Catalysis*, **72**, 165; (b) Merzouki, M., Taouk, B., Tessier, L. et al. (1993) *Studies in Surface Science and Catalysis*, **75**, 753; (c) Roussel, M., Bouchard, M., Bordes-Richard, E. et al. (2005) *Catalysis Today*, **99**, 77.

5. (a) Burch, R., and Swarnakar, R. (1991) *Applied Catalysis*, **70**, 129; (b) Ruth, K., Kieffer, R. and Burch, R. (1998) *Journal of*

Catalysis, **175**, 16; (c) Ruth, K., Burch, R. and Kieffer, R. (1998) *Journal of Catalysis*, **175**, 27; (d) Oyama, S.T. (1991) *Journal of Catalysis*, **128**, 210.

6 (a) Karim, K., Al-Hazmi, M.H. and Mamedov, E. (2000) US Patent, 6,013,597, and WO Patent Appl. 99/13980, assigned to Saudi Basic Ind.; (b) Karim, K. and Adris, A.E.M. (2000) US Patent 6,143,921, assigned to Saudi Basic Ind; (c) Karim, K., Al-Hazmi, M.H. and Khan, A. (2000) US Patent 6,087,297; (d) (2000) US Patent 6,060,421, assigned to Saudi Basic Ind.

7 (a) Jobson, S. and Watson, D.J. (1998) WO Patent Appl 98/05620, assigned to BP Chemicals; (b) Hallett, C. (1991) Eur Patent 480,594, assigned to BP Chemicals.

8 (a) Zeyss, S., Dingerdissen, U. and Fritch, J.WO Patent Appl. 2001/090042 and 090043, assigned to Aventis R&T; (b) Zeyss, S., Dingerdissen, U. and Fritch, J.WO Patent Appl. 2001/090042 and 2001/090043, assigned to Aventis R&T; (c) Borchert, H., Dingerdissen, U. and Weiguny, J. (1997) WO Patent Appl. 97/44299, assigned to Hoechst; (d) Borchert, H. and Dingerdissen, U. (2000) Eur Patent 1,025,075, assigned to Hoechst AG; (e) Zeyss, S., Dingerdissen, U., Baerns, M. *et al.* WO Patent Appl. 2002/038526, assigned to Aventis R&T; (f) Zeyss, S. and Dingerdissen, U.WO Patent Appl. 2001/090039, assigned to Aventis R&T.

9 (a) Borchert, H. and Dingerdissen, U. (1998) WO Patent Appl. 98/05619, assigned to Hoechst; (b) Borchert, H., Dingerdissen, U. and Roesky, R. (1998) WO Patent Appl. 98/47850 and 98/47851, assigned to Hoechst.

10 (a) Linke, D., Wolf, D., Baerns, M. *et al.* (2002) *Journal of Catalysis*, **205**, 16; (b) Linke, D., Wolf, D., Baerns, M., Zeyβ, S. and Dingerdissen, U. (2002) *Journal of Catalysis*, **205**, 32; (c) Linke, D., Wolf, D., Baerns, M. *et al.* (2002) *Chemical Engineering Science*, **57**, 39.

11 (a) Bergh, S., Cong, P., Ehnebuske, B. *et al.* (2003) *Topics in Catalysis*, **23**, 65; (b) Bergh, S., Guan, S., Hagemeyer, A. *et al.* (2003) *Applied Catalysis A – General*, **254**, 67.

12 (a) Ellis, B.WO Patent Appl. 2003/033138, assigned to BP Chemicals; (b) Ellis, B.WO Patent Appl. 2004/033090, assigned to BP Chemicals; (c) Brazdil, J.F., George, R.J. and Rosen, B.I.WO Patent Appl. 2005/018804, assigned to BP Chemicals.

13 (a) Chen, N.F., Oshihara, K. and Ueda, W. (2001) *Catalysis Today*, **64**, 121; (b) Oshihara, K., Nakamura, Y., Sakuma, M. and Ueda, W. (2001) *Catalysis Today*, **71**, 153; (c) Ueda, W. and Oshihara, K. (2000) *Applied Catalysis A – General*, **200**, 135.

14 (a) Min, J.S. and Mizuno, N. (2001) *Catalysis Today*, **71**, 89; (b) Min, J.S. and Mizuno, N. (2001) *Catalysis Today*, **66**, 47; (c) Li, W. and Ueda, W. (1997) *Studies in Surface Science and Catalysis*, **110**, 433.

15 (a) Bordes, E., Gubelman, M. and Tessier, L. (2000) US Patent 6,114,274, assigned to Rhone-Poulenc Chimie; (b) Enache, D.I., Bordes-Richard, E., Ensuque, A. and Bozon-Verduraz, F. (2004) *Applied Catalysis A – General*, **278**, 103.

16 Sopa, M., Wącław-Held, A., Grossy, M. *et al.* (2005) *Applied Catalysis A – General*, **285**, 119.

17 (a) Tessier, L., Bordes, E. and Gubelmann-Bonneau, M. (1995) *Catalysis Today*, **24**, 335; (b) Blaise, F., Bordes, E., Gubelmann, M. and Tessier, L. (1994) Eur Patent 627,401, assigned to Rhone-Poulenc.

18 (a) Volta, J.C. (2001) *Topics in Catalysis*, **15**, 121; (b) Roy, M., Ponceblanc, H. and Volta, J.C. (2000) *Topics in Catalysis*, **11**, 101; (c) Aubry, A., Gubelmann, M. and Le Govic, A.M. (1998) US Patent 5,750,777, assigned to Rhone-Poulenc; (d) Roy, M., Gubelmann-Bonneau, M., Ponceblanc, H. and Volta, J.C. (1996) *Catalysis Letters*, **42**, 93.

19 Ciambelli, P., Galli, P., Lisi, L. et al. (2000) *Applied Catalysis A – General*, **203**, 133.
20 Fakeeha, A.H., Fahmy, Y.M., Soliman, M.A. and Alwahabi, S.M. (2000) *Journal of Chemical Technology and Biotechnology*, **75**, 1160.
21 (a) Blum, P.R. and Pepera, M.A. (1992) Eur Patent 518,548, assigned to The Standard Oil Co; (b) Blum, P.R. and Pepera, M.A. (1994) US Patent 5,300,682, assigned to The Standard Oil Co; (c) Benkalowycz, N.C., Wagner, D.R. and Blum, P.R. (1993) Eur Patent 546,677, assigned to The Standard Oil Co.
22 (a) Guttmann, A.T., Grasselli, R.K. and Brazdil, J.F. (1988) US Patent 4,746,641; (b) Lynch, C.S., Glaeser, L.C., Brazdil, J.F. and Toft, M.A. (1992) US Patent 5,094,989; (c) Guttmann, A.T., Grasselli, R.K. and Brazdil, J.F. (1988) US Patent 4,788,317; (d) Bartek, J.P., and Guttmann, A.T. (1989) US Patent 4,797,381; (e) Glaeser, L.C., Brazdil, J.F. and Toft, M.A. (1989) US Patent 4,837,191; (f) Seely, M.J., Friedrich, M.S. and Suresh, D.D. (1990) US Patent 4,978,764; (g) Suresh, D.D., Seeley, M.J., Nappier, J.R. and Friedrich, M.S. (1992) US Patent 5,171,876; (h) Brazdil, J.F., Glaeser, L.C. and Toft, M.A. (1992) US Patent 5,079,207; (i) Bartek, J.P., Ebner, A.M. and Brazdil, J.R. (1993) US Patent 5,198,580; (j) Brazdil, J.F. and Cavalcanti, F.A.P. (1996) US Patent 5,576,469; (k) Brazdil, J.F. and Cavalcanti, F.A.P. (1996) US Patent 5,498,588; all patents assigned to The Standard Oil Co.
23 (a) Ushikubo, T., Oshima, K., Ihara, T. and Amatsu, H. (1996) US Patent 5,534,650; (b) Ushikubo, T., Oshima, K., Kayo, A. et al. (1992) Eur Patent 529,853; (c) Ushikubo, T., Oshima, K., Umezawa, T. and Kiyono, K. (1992) Eur Patent 512,846; (d) Ushikubo, T., Oshima, K., Ihara, T. and Amatsu, H. (1996) US Patent 5,534,650; (e) Ushikubo, T., Koyasu, Y. and Nakamura, H. (1996) US Patent 767,164; all patents assigned to Mitsubishi Chemical Co; (f) Ramachandran, R., Maclean, D.L. and Satchell, D.P. (1989) US Patent 4,849,538, assigned to The BOC Group; (g) Ramachandran, R. and Dao, L. (1994) Eur Patent 646,558, assigned to The BOC Group.
24 (a) Centi, G. and Perathoner, S. (1998) *Chemotherapy*, **28**, 13; (b) Andersson, A., Hansen, S. and Wickman, A. (2001) *Topics in Catalysis*, **15**, 103; (c) Centi, G., Perathoner, S. and Trifiro, F. (1997) *Applied Catalysis A – General*, **157**, 143; (d) Cavani, F. and Trifirò, F. (2003) *Basic Principles in Applied Catalysis* (ed. M. Baerns), Series in Chemical Physics 75, 21, Springer, Berlin; (e) Ballarini, N., Cavani, F. and Trifirò, F. (2005) Proceedings DGMK Conference "Oxidation and Functionalization of alkanes: Classical and Alternative Routes and Sources", DGMK Tagungsbericht 2005-2, pp. 19; (f) Sokolovskii, V.D., Davydov, A.A. and Ovsitser, O.Yu. (1995) *Catalysis Reviews: Science and Engineering*, **37**, 425; (g) Moro-Oka, Y. and Ueda, W. (1994) *Catalysis*, **11**, 223; (h) Prada, S.R. and Grange, P. (2003) *Oil Gas European Magazine*, **29**, 145.
25 Nilsson, J., Landa-Cánovas, A.R., Hansen, S. and Andersson, A. (1999) *Journal of Catalysis*, **186**, 442.
26 (a) Grasselli, R.K., Burrington, J.D., Buttrey, D.J. et al. (2003) *Topics in Catalysis*, **23**, 5; (b) Ueda, W. and Oshihara, K. (2000) *Applied Catalysis A – General*, **200**, 135; (c) Oshihara, K., Hisano, T. and Ueda, W. (2001) *Topics in Catalysis*, **15**, 153.
27 Botella, P., Lopez Nieto, J.M. and Solsona, B. (2002) *Catalysis Letters*, **78**, 383.
28 Aouine, M., Dubois, J.L. and Millet, J.M.M. (2001) *Chemical Communications*, **13**, 1180.
29 (a) Hamada, K. and Komada, S. (1999) US Patent 5,907,052; (b) Komada, S., Hinago, H., Kaneta, M. and Watanabe, M. (1998) Eur Patent 895,809; (c) Midorikawa, H., Sugiyama, N. and Hinago, H. (1999) US Patent 5,973,186;

(d) Hinago, H. and Komada, S. (2000) US Patent 6,063,728; (e) Komada, S., Kaneta, M. (2000) US Patent 6,143,690; (f) Hinago, H. and Yano, H. (2003) US Patent 6,610,629; (g) Hinago, H. and Watanabe, M. (2006) US Patent 7,109,144; (h) Komada, S. and Hamada, K. (2000) US Patent 6,043,186; (i) Midorikawa, H., Sugiyama, N. and Hinago, H. (2000) US Patent 6,080,882; (j) Hinago, H. and Komada, S. (2000) US Patent 6,143,916; (k) Komada, S., Hinago, H., Nagano, O. and Watanabe, M. (2006) US Patent 7,087,551; all assigned to Asahi Kasei Kabushiki Kaisha.

30 (a) Florea, M., Prada Silvy, R. and Grange, P. (2003) *Catalysis Letters*, **87**, 63; (b) Prada Silvy, R., Florea, M., Blangenois, N. and Grange, P. (2003) *AICHE Journal*, **49**, 2228; (c) Florea, M., Prada Silvy, R. and Grange, P. (2005) *Applied Catalysis A – General*, **286**, 1; (d) Olea, M., Florea, M., Sack, I. *et al.* (2005) *Journal of Catalysis*, **232**, 152.

31 (a) Albonetti, S., Blanchard, G., Burattin, P. *et al.* (1996) Eur Patent 723,934; (b) Albonetti, S., Blanchard, G., Burattin, P. *et al.* (1997) Eur Patent 932,662; (c) Blanchard, G., Burattin, P., Cavani, F. *et al.* (1997) WO Patent Appl. 97/23,287; (d) Blanchard, G. and Ferre, G. (1994) US Patent 5,336,804, all patents assigned to Rhodia; (e) Albonetti, S., Blanchard, G., Burattin, P. *et al.* (1997) *Catalysis Letters*, **45**, 119; (f) Albonetti, S., Blanchard, G., Burattin, P. *et al.* (1998) *Catalysis Today*, **42**, 283.

32 Mimura, Y., Ohyachi, K. and Matsuura, I. (1999) *Science and Technology in Catalysis 1998*, Kodansha, Tokyo, pp. 69.

33 Bowker, M., Kerwin, P. and Eichhorn, H.-D. (1997) UK Patent 2,302,291, assigned to BASF.

34 Ueda, W., Chen, N.F. and Oshihara, K. (1999) *Chemical Communications*, 517; (b) Oshihara, K., Nakamura, Y., Sakuma, M. and Ueda, W. (2001) *Catalysis Today*, **71**, 153; (c) Ueda, W., Oshihara, K., Vitry, D. *et al.* (2002) *Catalysis Surveys from Japan*, **6**, 33; (d) Vitry, D., Morikawa, Y., Dubois, J.-L. and Ueda, W. (2003) *Topics in Catalysis*, **23**, 47; (e) Vitry, D., Morikawa, Y., Dubois, J.-L. and Ueda, W. (2003) *Applied Catalysis A – General*, **251**, 411; (f) Vitry, D., Dubois, J.-L. and Ueda, W. (2004) *Journal of Molecular Catalysis A-Chemical*, **220**, 67; (g) Watanabe, N. and Ueda, W. (2006) *Industrial & Engineering Chemistry, Research*, **45**, 607.

35 Roussel, M., Bouchard, M., Bordes-Richard, E. *et al.* (2005) *Catalysis Today*, **99**, 77.

36 (a) Holmberg, J., Grasselli, R.K. and Andersson, A. (2004) *Applied Catalysis A – General*, **270**, 121; (b) Holmberg, J., Haeggblad, R. and Andersson, A. (2006) *Journal of Catalysis*, **243**, 350.

37 (a) DeSanto, P., Jr, Buttrey, D.J., Grasselli, R.K. *et al.* (2003) *Topics in Catalysis*, **23**, 23; (b) DeSanto, P., Jr, Buttrey, D.J., Grasselli, R.K. *et al.* (2004) *Zeitschrift für Kristallographie*, **219**, 152; (c) Grasselli, R.K., Buttrey, D.J., DeSanto, P. Jr *et al.* (2004) *Catalysis Today*, **91–92**, 251.

38 Millet, J.M.M., Roussel, H., Pigamo, A. *et al.* (2002) *Applied Catalysis A – General*, **232**, 77.

39 (a) Berry, F.J., Brett, M.E. and Patterson, W.R. (1983) *Journal of the Chemical Society – Dalton Transactions*, 9 and 13; (b) Berry, F.J., Brett, M.E. and Patterson, W.R. (1982) *Journal of the Chemical Society. Chemical Communications*, 695; (c) Berry, F.J., Brett, M.E., Marbrow, R.A. and Patterson, W.R. (1984) *Journal of the Chemical Society – Dalton Transactions*, 985; (d) Berry, F.J., Holden, J.G. and Loretto, M.H. (1987) *Journal of the Chemical Society – Faraday Transactions I*, **83**, 615; (e) Berry, F.J., Holden, J.G. and Loretto, M.H. (1986) *Solid State Communications*, **59**, 397.

40 (a) Teller, R.G., Antonio, M.R., Brazdil, J.F. and Grasselli, R.K. (1986) *Journal of Solid State Chemistry*, **64**, 249; (b) Grasselli, R.K. (1999) *Catalysis Today*, **49**, 141.

41 Birchall, T. and Sleight, A.E. (1976) *Inorganic Chemistry*, **15**, 868.

42 (a) Centi, G. and Mazzoli, P. (1996) *Catalysis Today*, **28**, 351; (b) Catani, R. and Centi, G. (1991) *Journal of the Chemical Society. Chemical Communications*, 1081.

43 (a) Hansen, S., Ståhl, K., Nilsson, R. and Andersson, A. (1993) *Journal of Solid State Chemistry*, **102**, 340; (b) Landa-Canovas, A., Nilsson, J., Hansen, S. et al. (1995) *Journal of Solid State Chemistry*, **116**, 369; (c) Nilsson, R., Lindblad, T. and Andersson, A. (1994) *Journal of Catalysis*, **148**, 501. (d) Nilsson, J., Landa-Canovas, A.R., Hansen, S. and Andersson, A. (1996) *Journal of Catalysis*, **160**, 244; (e) Wickman, A., Wallenberg, L.R. and Andersson, A. (2000) *Journal of Catalysis*, **194**, 153; (f) Andersson, A., Hansen, S. and Wickman, A. (2001) *Topics in Catalysis*, **15**, 103.

44 (a) Allen, M.D. and Bowker, M. (1995) *Catalysis Letters*, **33**, 269; (b) Bowker, M., Bricknell, C.R. and Kerwin, P. (1996) *Applied Catalysis A – General*, **136**, 205; (c) Poulston, S., Price, N.J., Weeks, C. et al. (1998) *Journal of Catalysis*, **178**, 658.

45 Magagula, Z. and van Steen, E. (1999) *Catalysis Today*, **49**, 155.

46 (a) Ballarini, N., Cavani, F., Cimini, M. et al. (2006) *Journal of Catalysis*, **241**, 255; (b) Cimini, M., Millet, J.M.M. and Cavani, F. (2004) *Journal of Solid State Chemistry*, **177**, 1045; (c) Cimini, M., Millet, J.M.M., Ballarini, N. et al. (2004) *Catalysis Today*, **91**, 259; (d) Ballarini, N., Cavani, F., Cimini, M. et al. (2003) *Applied Catalysis A-General*, **251**, 49; (e) Ballarini, N., Cavani, F., Giunchi, C. et al. (2001) *Topics in Catalysis*, **15**, 111; (f) Ballarini, N., Cavani, F., Ghisletti, D. et al. (2003) *Catalysis Today*, **78**, 237.

47 (a) Roussel, H., Mehlomakulu, B., Belhadj, F. et al. (2002) *Journal of Catalysis*, **205**, 97; (b) Nguyen, D.L., Ben Taarit, Y. and Millet, J.M.M. (2003) *Catalysis Letters*, **90**, 65; (c) Millet, J.M.M., Marcu, J.C. and Herrmann, J.M. (2005) *Journal of Molecular Catalysis A – Chemical*, **226**, 111.

48 Xiong, G., Sullivan, V.S., Stair, P.C. et al. (2005) *Journal of Catalysis*, **230**, 317.

49 (a) Guerrero-Perez, M.O., Fierro, J.L.G. and Bañares, M.A. (2006) *Topics in Catalysis*, **41**, 43; (b) Guerrero-Perez, M.O., Fierro, J.L.G. and Bañares, M.A. (2003) *Catalysis Today*, **78**, 387; (c) Guerrero-Perez, M.O., Fierro, J.L.G. and Bañares, M.A. (2003) *Physical Chemistry Chemical Physics*, **5**, 4032; (d) Guerrero-Perez, M.O., Fierro, J.L.G. and Banares, M.A. (2006) *Catalysis Today*, **118**, 366; (e) Guerrero-Perez, M.O., Martinez-Huerta, M.V., Fierro, J.L.G. and Banares, M.A. (2006) *Applied Catalysis A – General*, **298**, 1; (f) Guerrero-Perez, M.O. and Banares, M.A. (2007) *The Journal of Physical Chemistry*, **111**, 1315.

50 Callahan, J.L. and Grasselli, R.K. (1963) *AICHE Journal*, **9**, 755.

51 Martinie, G.M., Al-Shahrani, F.M. and Dabbousi, B.O.US Patent Appl. 2007/0051667.

52 Otsuki, S., Nonaka, T., Takashima, N. et al. (2000) *Energy & Fuels*, **14**, 1232.

53 Hulea, V., Moreau, P. and Di Renzo, F. (1996) *Journal of Molecular Catalysis A – Chemical*, **111**, 325.

54 Baucherel, X. and Sheldon, R.A. (2002) WO Patent Appl. 02/100810, assigned to Imperial Chem, Ind.

55 Zinnen, H.A., and Cabrera, C.A. (2005) WO Patent Appl. 05/019386, assigned to UOP LLC.

56 (a) Corma, A., Domine, M.E. and Martínez, C. (2002) WO Patent Appl. 02/083819, assigned to CSIC-UPV; (b) Corma, A., Domine, M.E. and Martínez, C. (2003) WO Patent Appl. 03/044129, assigned to CSIC-UPV; (c) Chica, A., Corma, A. and Domine, M.E. (2006) *Journal of Catalysis*, **242**, 299; (d) Concepcion, P., Corma, A., López-Nieto, J.M. and Pérez-Pariente, J. (1996) *Applied Catalysis A – General*, **143**, 17.

57 Hulea, V., Fajula, F. and Bousquet, J. (2001) *Journal of Catalysis*, **198**, 179.

58 De Angelis, A., Pollesel, P., Molinari, D. et al. (2007) *Pure and Applied Chemistry*, **79**, 1887.

59 (a) Schultz, M.A., Gatan, R.M., Brandvold, T.A. and Gosling, C.D. (2007) US Patent 7,186,328, assigned to UOP LLC; (b) Molinari, D., Baldiraghi, F., Gosling, C. and Gatan, R. (2005) Proceedings of the DGMK/SCI Conference, "Oxidation and Functionalization: Classical and Alternative Routes and Sources", Milan, pp. 259; (c) Gosling, C.D., Gatan, R.M. and Barger, P.T. (2007) US Patent 7,297,253, assigned to UOP LLC; (d) Zinnen, H.A. and Cabrera, C.A.US Patent Appl. 2005/040078A1; assigned to UOP LLC.

60 (a) Yazu, K. et al. (2001) *Energy & Fuels*, **15**, 1535; (b) Te, M., Fairbridge, C. and Ring, Z. (2001) *Applied Catalysis A – General*, **219**, 267; (c) Li, C., Jiang, Z., Gao, J. et al. (2004) *Chemistry – A European Journal*, **10**, 2277; (d) Collins, F.M., Lucy, A.R. and Sharp, C. (1997) *Journal of Molecular Catalysis A – Chemical*, **117**, 397.

61 (a) Mason, B.P., Price, K.E., Steinbacher, J.L. et al. (2007) *Chemical Reviews*, **107**, 2300; (b) Jähnisch, K., Hessel, V., Löwe, H. and Baerns, M. (2004) *Angewandte Chemie – International Edition*, **43**, 406; (c) Oroskar, A.R., Vanden Bussche, K. and Abdo, S.F. (2001) Microreaction Technology – IMRET 5. in Proceedings of the 5th International Conference on Microreaction Technology (eds M. Matlosz, W. Ehrfeld and J.P. Baselt), Springer, Berlin, pp. 153; (d) Haswell, S.J. and Watts, P. (2003) *Green Chemistry*, **5**, 240; (e) Klemm, E., Döring, H., Geisselmann, A. and Schirrmeister, S. (2007) *Chemical Engineering & Technology*, **30**, 1615.

62 (a) Wörz, O., Jäckel, K.P., Richter, T. and Wolf, A. (2001) *Chemical Engineering & Technology*, **24**, 138; (b) Wörz, O., Jäckel, K.P., Richter, T. and Wolf, A. (2000) *ChemIngTech*, **72**, 460.

63 Pennemann, H., Hessel, V. and Löwe, H. (2004) *Chemical Engineering Science*, **59**, 4789.

64 Tonkovich, A.L., Jarosch, K.T.P. and Hesse, D.J. (2006) US Patent 7,029,647, assigned to Velocys.

65 (a) Plettig, M., Döring, H., Dietzsch, E. et al. (2005) Proceed DGMK/SCI Conference on "Oxidation and Functionalization: Classical and Alternative Routes and Sources", Milan, pp. 177; (b) Markowz, G., Schirrmeister, St., Albrecht, J. et al. (2005) *Chemical Engineering & Technology*, **28**, 459; (c) Klemm, E., Dietzsch, E., Schwarz, T. et al. (2008) *Industrial & Engineering Chemistry Research*, **47**, 2086.

66 Seiler, H. and Emig, G. (2004) *Basic Principles in Applied Catalysis*, Springer, Berlin, pp. 505.

67 Mars, P. and van Krevelen, D.W. (1954) *Chemical Engineering Science – Special Supplement*, **3**, 41.

68 Sze, M.C. and Gelbein, A.P. (February 1976) *Hydrocarbon Processing*, **3**, 103.

69 (a) Contractor, R.M., Bergna, H.E., Horowitz, H.S. et al. (1987) *Catalysis Today*, **1**, 49; (b) Contractor, R.M., Bergna, H.E., Horowitz, H.S. et al. (1988) *Advances in Chemical Conversions for Mitigating Carbon Dioxide*, **38**, 645.

70 (a) Gaffney, A.M., Jones, C.A., Pitchai, R. and Kahn, A.P. (1997) US Patent 5,698,719, assigned to ARCO Chem Tech; (b) Gaffney, A.M., Jones, A.C., Leonard, J.J. and Sofranko, J.A. (1988) *Journal of Catalysis*, **114**, 422.

71 Tremond, S.J. and Williamson, A.N. (1981) US Patent 4,254,293.

72 (a) Contractor, R.M., Anderson, M.W., Campos, D. et al. (1999) WO Patent Appl. 99/03809, assigned to Elf Atochem and Du Pont; (b) Bergna, H. (1988) US Patent 4,769,477, assigned to Du Pont; (c) Contractor, R.M., Andersen, M.W., Campos, D. et al. (2001) US Patent 6,310,240, assigned to E.I. du Pont de Nemours & Co, and Atofina.

73 (a) Gilbert, L. and Mercier, C. (1993) *Studies in Surface Science and Catalysis*, **78**, 51; (b) Anunziata, O.A., Beltramone, A.R.

and Cussa, J. (2004) *Applied Catalysis A – General*, **270**, 77; (c) Narayanan, S., Murthy, K., Reddy, K.M. and Premchonder, N. (2002) *Applied Catalysis A – General*, **228**, 161; (d) Kholdeeva, O.A., Zalomaeva, O.V., Sorokin, A.B. et al. (2007) *Catalysis Today*, **121**, 58; (e) Kholdeeva, O.A., Zalomaeva, O.V., Shmakov, A.N. et al. (2005) *Journal of Catalysis*, **236**, 62.

74 (a) Matveev, K.I., Odyakov, V.F. and Zhizhina, E.G. (1996) *Journal of Molecular Catalysis A – Chemical*, **114**, 151; (b) Matveev, K.I., Zhizhina, E.G. and Odyakov, V.F. (1995) *Reaction Kinetics and Catalysis Letters*, **55**, 47.

75 Monteleone, F., Cavani, F., Felloni, C. and Trabace, R. WO Patent Appl. 2004/014832, assigned to Vanetta SpA.

76 (a) Kung, H.H. (1994) *Advances in Catalysis*, **40**, 1; (b) Cavani, F. and Trifirò, F. (1995) *Catalysis Today*, **24**, 307; (c) Mamedov, E.A. and Cortes-Corberan, V. (1995) *Applied Catalysis A – General*, **127**, 1; (d) Albonetti, S., Cavani, F. and Trifiro, F. (1996) *Catalysis Reviews: Science and Engineering*, **38**, 413; (e) Cavani, F. and Trifirò, F. (1997) *Catalysis Today*, **36**, 431; (f) Blasco, T. and Lopez Nieto, J.M. (1997) *Applied Catalysis A – General*, **157**, 117; (g) Baerns, M. and Buyevskaya, O. (1998) *Catalysis Today*, **45**, 13; (h) Grasselli, R.K. (1999) *Catalysis Today*, **49**, 141; (i) Bañares, M.A. (1999) *Catalysis Today*, **51**, 319; (j) Cavani, F. and Trifirò, F. (1999) *Catalysis Today*, **51**, 561; (k) Grzybowska-Swierkosz, B. (2002) *Topics in Catalysis*, **21**, 35; (l) Grabowski, R. (2006) *Catalysis Reviews*, **48**, 199; (m) Cavani, F., Ballarini, N. and Cericola, A. (2007) *Catalysis Today*, **127**, 113.

77 (a) Ballarini, N., Cavani, F., Cericola, A. et al. (2004) *Catalysis Today*, **91–92**, 99; (b) Ballarini, N., Cavani, F., Ferrari, M. et al. (2003) *Journal of Catalysis*, **213**, 95; (c) Ballarini, N., Cavani, F., Cortelli, C. et al. (2003) *Catalysis Today*, **78**, 353; (d) Ballarini, N., Cavani, F., Cericola, A. et al. (2004) *Studies in Surface Science and Catalysis*, **147**, 649; (e) Ballarini, N., Calestani, G., Catani, R. et al. (2005) *Studies in Surface Science and Catalysis*, **155**, 81; (f) Ballarini, N., Battisti, A., Cavani, F. et al. (2006) *Applied Catalysis A – General*, **307**, 148.

78 (a) Creaser, D., Andersson, B., Hudgins, R.R. and Silveston, P.L. (1999) *Chemical Engineering Science*, **54**, 4365; (b) Creaser, D., Andersson, B., Hudgins, R.R. and Silveston, P.L. (1999) *Applied Catalysis A – General*, **187**, 147; (c) Creaser, D., Andersson, B., Hudgins, R.R. and Silveston, P.L. (1999) *Journal of Catalysis*, **182**, 264.

79 Kondratenko, E.V., Cherian, M. and Baerns, M. (2005) *Catalysis Today*, **99**, 59.

80 Grabowski, R., Pietrzyk, S., Słoczynski, J. et al. (2002) *Applied Catalysis A – General*, **232**, 277.

81 Vrieland, G.E. and Murchison, C.B. (1996) *Applied Catalysis A – General*, **134**, 101.

82 (a) de Graaf, E.A., Rothenberg, G., Kooyman, P.J. et al. (2005) *Applied Catalysis A – General*, **278**, 187; (b) de Graaf, E.A., Andreini, A., Hensen, E.J.M. and Bliek, A. (2004) *Applied Catalysis A – General*, **262**, 201; (c) van der Zande, L.M., de Graaf, A. and Rothenberg, G. (2002) *Advanced Synthesis and Catalysis*, **344**, 884.

83 (a) Sugiyama, S., Hashimoto, T., Tanabe, Y. et al. (2005) *Journal of Molecular Catalysis A – Chemical*, **227**, 255; (b) Sugiyama, S., Hashimoto, T., Shigemoto, N. and Hayashi, H. (2003) *Catalysis Letters*, **89**, 229.

84 (a) Grasselli, R.K., Stern, D.L. and Tsikoyiannis, J.G. (1999) *Applied Catalysis A – General*, **189**, 1; (b) Grasselli, R.K., Stern, D.L. and Tsikoyiannis, J.G. (1999) *Applied Catalysis A – General*, **189**, 9.

85 Lin, C.H., Lee, K.C. and Wan, B.Z. (1997) *Applied Catalysis A – General*, **164**, 59.

86 (a) Låte, L., Rundereim, J.-I. and Blekkan, E.A. (2004) *Applied Catalysis A – General*, **262**, 53; (b) Låte, L., Thelin, W. and

Blekkan, E.A. (2004) *Applied Catalysis A – General*, **262**, 63.
87 Kodama, T. and Gokon, N. (2007) *Chemical Reviews*, **107**, 4048.
88 (a) Cornaro, U. and Sanfilippo, D. US Patent 2004/0152790 A1, assigned to ENI SpA and Snamprogetti SpA; (b) Sanfilippo, D., Paggini, A., Piccoli, et al. (2001) Eur Patent 1,134,187, assigned to Snamprogetti; (c) Gemmi, M., Merlini, M. et al. (2005) *Journal of Applied Crystallography*, **38**, 353; (d) Sanfilippo, D., Miracca, I., Cornaro, U. et al. (2004) *Studies in Surface Science and Catalysis*, **147**, 91.
89 (a) Otsuka, K., Mito, A. and Takenaka, S. (2001) *International Journal of Hydrogen Energy*, **26**, 191; (b) Takenaka, S., Serizawa, M. and Otsuka, K. (2004) *Journal of Catalysis*, **222**, 520.
90 Mattisson, T., Jardnas, A. and Lyngfelt, A. (2003) *Energy & Fuels*, **17**, 643.
91 Svoboda, K., Slowinski, G., Rogut, J. and Baxter, D. (2007) *Energy Conversion and Management*, **48**, 3063.
92 (a) Corma, A., Iborra, S. and Velty, A. (2007) *Chemical Reviews*, **107**, 2411; (b) Gallezot, P. (2007) *Green Chemistry*, **9**, 295; (c) Metzger, J.O. (2006) *Angewandte Chemie – International Edition*, **45**, 696; (d) Lichtenthaler, F.W. and Peters, S. (2004) *Comptes Rendus Chimie*, **7**, 65; (e) Stevens, C.V. and Verhé, R.G. (eds) (2004) *Renewable Resources Scope and Modification for Non-Food Applications*, John Wiley & Sons, Chichester; (f) Kamm, B., Gruber, P.R. and Kamm, M. (eds) (2006) *Biorefineries – Industrial Processes and products*, Wiley-VCH, Weinheim; (g) Centi, G. and van Santen, R.A. (eds) (2007) *Catalysis for Renewables*, Wiley-VCH, Weinheim; (h) Graziani, M. and Fornasiero, P. (eds) (2007) *Renewable Resources and Renewable Energy – A Global Challenge*, CRC Taylor & Francis, Boca Raton.
93 Biella, S., Prati, L. and Rossi, M. (2002) *Journal of Catalysis*, **206**, 242.
94 Willke, T., Prüße, U. and Vorlop, K.D. (2006) *Biorefineries – Industrial Processes and Products* (eds B. Kamm, P.R. Gruber and M. Kamm), Wiley-VCH, Weinheim, pp. 385.
95 Werpy, T. and Petersen, G. (eds) (2004) *Top Value Added Chemicals from Biomass Volume I – Results of Screening for Potential Candidates from Sugars and Synthesis Gas*, Pacific Northwest National Laboratory (PNNL), National Renewable Energy Laboratory (NREL), Office of Biomass Program (EERE).
96 Kunz, M., Schwarz, A., and Kowalczyc, J. (1997) DE Patent 19,542,287.
97 Edye, L.A., Meehan, G.V. and Richards, G.N. (1991) *Carbohydrate Chemistry*, **10**, 11 and **13**, 273.
98 Moreau, C., Belgacem, M.N. and Gandini, A. (2004) *Topics in Catalysis*, **27**, 11.
99 Gallezot, P. (2007) *Catalysis for Renewables* (eds G. Centi and R.A. van Santen), Wiley-VCH, Weinheim.
100 Sels, B., D'Hondt, E. and Jacobs, P. (2007) *Catalysis for Renewables* (eds G. Centi and R.A. Van Santen), Wiley-VCH, Weinheim, pp. 223.
101 (a) Garcia, R., Besson, M. and Gallezot, P. (1995) *Applied Catalysis A – General*, **70**, 1027; (b) Garcia, R. and Gallezot, P. (1997) *Catalysis Today*, **37**, 405.
102 (a) Mallat, T. and Baiker, A. (1994) *Catalysis Today*, **19**, 247; (b) Mallat, T. and Baiker, A. (2004) *Chemical Reviews*, **104**, 3037.
103 Kimura, H. (1993) *Applied Catalysis A-General*, **105**, 147.
104 Abbadi, A. and van Bekkum, H. (1996) *Applied Catalysis A – General*, **148**, 113.
105 (a) Carrettin, S., McMorn, P., Johnston, P. et al. (2003) *Physical Chemistry Chemical Physics*, **5**, 1329; (b) Carrettin, S., McMorn, P., Johnston, P. et al. (2002) *Chemical Communications*, 696; (c) Carrettin, S., McMorn, P., Johnston, P. et al. (2004) *Topics in Catalysis*, **27**, 131.
106 (a) Porta, F. and Prati, L. (2004) *Journal of Catalysis*, **224**, 397; (b) Dimitratos, N.,

Porta, F. and Prati, L. (2005) *Applied Catalysis A – General*, **291**, 210; (c) Dimitratos, N., Messi, C., Porta, F. *et al.* (2006) *Journal of Molecular Catalysis A-Chemical*, **256**, 21.

107 Bub, G., Mosler, J., Sabbach, A. *et al.* WO Patent Appl. 2006/092272.

108 (a) Dubois, J.L., Duquenne, C., Hoelderich, W. and Kervennal, J. (2006) Eur Patent 1,848,681; (b) Dubois, J.L., Duquenne, C. and Hoelderich, W. (2006) Eur Patent 1,874,720; (c) Dubois, J.L. WO Patent Appl. 2007/090990; (d) Dubois, J.L.WO Patent Appl. 2008/007002; Dubois, J.L., Duquenne, C. and Hoelderich, W. (2006) Eur Patent 1,853,541; (e) Dubois, J.L. WO Patent Appl. 2007/090991; (f) Dubois, J.L. WO Patent Appl. 2008/007002; all assigned to Arkema France.

Index

a

A Pd(OAc)$_2$/phenanthroline catalytic system 63
acetic acid 12
acetic acid production
– carbonylation of methanol 290
– direct gas-phase oxidation of ethane 291
– gas-phase oxidation of ethylen 290
acetoxylation of ethane/ethylene to vinyl acetate 292
active oxygen species, nature 15
active site, physical ingredients 34
active surface 10
Advanced Peroxide Technology 257
Ag/HT(hydrotalcite) 167
Ag$_5$[PV$_2$Mo$_{10}$O$_{40}$] 200
air calcination 21
aliphatic oxygenates 110
alkoxides 11
alkylated anthraquinone process 253
alkylated anthraquinone process of H$_2$O$_2$ production, drawbacks 254
alkylated polyethyleneimine-POM hybrid heterogeneous catalyst 203
aluminophosphate molecular sieves (AlPO) 178
α-amino acids, source of Schiff-base ligands 45
ammoxidation
– propane to acrylonitrile 289
– with N$_2$O oxidant 230
ammoxidation of propane to acrylonitrile
– Asahi process and catalysts 297
– catalyst systems 296
– mechanism of catalysis (reaction scheme) 297
– metal antimonates cataylst system 299
– Mitsubishi process 298
– reaction conditions 296
– rutile type Cr antimonate catalyst 299
– vanadium aluminum oxynitrides (VALON) catalyst system 297
amorphous oxides 16
anaerobic ammoxidation, Lummus process 308
anaerobic oxidation (cycle process)
– advantages 307
– definition 307
– oxidation propene to acrolein 309
– properties of catalyst 308
anaerobic oxidative coupling 308
anatase-type Ti species 139
antimony, lattice oxygen modifier 14
asymmetric catalysis on surfaces chiral reaction 44
atomic oxygen (O*) 61
Au, CO oxidation 79
13-Au atom clusters 91
– catalytic activity 91
Au catalysts 78
– alkane oxidation 95
– complete oxidation of VOCs 93
– cyclohexan oxidation 116
– diol oxidation 112
– particle size distribution 79
– preparation methods 116
Au colloids 78
Au mono- and bilayer structures rates of CO oxidation 90
Au nanoparticles (AuNPs) 77ff, 115f, 116
Au NPs
– effect of size of Au nanoparticles 101
– gas-phase alkene epoxidation 101
– propylene epoxidation reaction pathway 103
– removing CO from H$_2$ 84

Au on solid polymers 78
Au submicron tube 78
Au supported on titanosilicates (Au/TS-1) 103
– mechanism of propylene epoxidation 106
Au/3D Ti-SiO$_2$
– effect of pore diameters on propylene epoxidation 104
– effect ofpromotors on propylene epoxidation 104
Au/AC, mechanism of glucose oxidation 111
Au/activated carbon (AC) 107
Au/Al$_2$O$_3$, Au/TiO$_2$, glucose oxidation 115
Au/C, selectiveoxidation of cyclohexene 116
Au/CeO$_2$, oxidation allylic alcohols 108
Au/CeO$_2$ catalyst 164
Au/CeO$_2$, Au/C, mechanism of alcohol oxidation 110
Au/CeO$_2$-Co$_3$O$_4$ 86
Au/Fe$_2$O$_3$ catalysts 57, 86
Au/metal oxides oxidation of mono-alcohols 107
Au/MnO$_2$-TiO$_2$ 86
Au/TiO$_2$ 80
– FT-IR spectra for CO adsorption 89
– gas-phase epoxidation of propylene 101
– preparation method on catalytic performance 101
Au/TiO$_2$, TOFs of CO oxidation 82
Au/TiO$_2$ catalysts
– preparation 80
– turnover frequencies of CO oxidation 80
Au-catalyzed alcohol oxidation 108
Au-catalyzed glycerol oxidation 114
Au-catalyzed liquid-phase oxidation 112
Au-Co/CeO$_2$/TiO$_2$/SnO$_2$ 86
Au-Pd/TiO$_2$ 166
AuPt/A zeolite 86

b

Baeyer–Villiger oxidation 150, 157
Baeyer–Villiger reaction 175
– HT catalysator 175
BASF process for alcohol oxidation 305
B-containing Ti-MWW 140
benzene-to-phenol transformation 63
B-free Ti-MWW 140, 141
1,1′-binaphthol (BINOL), chiral source for asymmetric catalysis 45
Biodesulfurization 301
bioethanol transformation, Au catalysts 108
biomass derived feedstock, transformation 108
biomimetic organometallic catalysts 186
bulk catalysts 12
butane 2
butane oxidation 5
– wiring diagram 30

c

C–H activation 7, 14, 32, 34. 157, 177
C–H bonds 3
cages of zeolite 84
calcination reactions 16
carbohydrates, feedstock for chemical industry 316
carbon sugar alcohols 319
carbonyl compounds 157
carboxidation 232, 246
– bicyclic alkenes 238
– cyclic alkenes 234
– cyclodienes 237
– heterocyclic 238
– linear alkenes 232
– mechanism 232
– polybutadien rubber 241
– polyethylene 240
catalyst deactivation 230
catalyst dynamics 27
catalyst productivity 275
catalysts, polynuclear sites 186
catalytic cycle 16
catalytic dehydrogenation, ethylbenzene to styrene 28
catalytic ensemble structures 43
catalytic liquid phase oxidations with N$_2$O 231
catalytic selective oxidation 323
catalytic systems, MCM-41-supported Cu catalysts 63
catalyzed selective oxidation 1
CeO$_2$-supported Cu$^+$-cluster 52
chemisorption of oxygen 18
chiral dimer structure 45
chiral self-assembly 51
chiral self-dimerization 45
chiral self-dimerization of metal 44
chlorinated organic compounds 77
chlorohydrin route 128
Chromium-containing silicalite-2 150
cluster, active site 3
cold start emission 95
combustion 77
competitive oxidation 136
concave oxide surfaces 43
correlation alkane oxidation and CO oxidation 95
CrAPO-11 150

CrAPO-5 150
CrS-1 150
CrS-2 150
Cu and Ce catalysts, catalytic performances 55
Cu nanocluster catalysts, syntheses 52
Cu/Ce-noCTAB, catalytic properties 57
cumene process 58, 223
cumene recycling process 100
CVD catalyst 65
cycle approach for oxidation 308
cyclic reversibility 16
cytochrome P-450 186

d
3D acidity 43
Del(delaminated)-Ti-MWW 143
density functional theory (DFT), CO oxidation over Au catalysts 90
DFT calculations 68
– benzene–O_2 reaction 69
diameter of Au particles, influence of CO oxidation 83
dicarboxylic acids, synthesis from carbohydrates 317
di-copper-substituted γ-Keggin silicotungstate $[\gamma-H_2SiW_{10}O_{36}Cu_2(\mu-1,1-N_3)_2]^{4-}$ 193
diffusion processes, quantification 16
Dihydroxy compounds 61
di-iron-substituted silicotungstate $[\gamma-SiW_{10}\{Fe(OH_2)\}_2O_{38}]^{6-}$ 192
di-metal-substituted POMs 192
dioxygen 125, 217
– difficulties in oxidation reactions 217
di-oxygen molecule 10, 11
direct catalytic synthesis of H_2O_2
– catalyse with nobel elements 263
– catalyst productivity 265
– electrochemical synthesis 271
– problems of chloride and other halide ions 281
– influence of Pd and other metals metal in catalytic selctivity 280
– influence of solvent 264, 281
– key aspects of selectivity and productivity 278
– nature of active species 279
– nature of catalyst 275
– patents 264
– role of acidic supports 275
– role of diameter of catalyst particles 276
– role of promotors 281
– safe operations 271
– solvent selection parameter (SSP) 265
– use of catalytic membranes 271

direct epoxidation of propylene 101
direct H_2O_2 synthesis, microchanneled reactors 258
direct hydroxylation of benzene 60
direct oxidation
– ethane to acetic acid 289
– n-butane to 1,4-butanediol 290
direct oxidation of ethane to acetic acid
– catalysts 291
direct oxidation of ethene to acetic acid
– mechanism 291
direct synthesis process for hydrogen peroxide (DSHP) 260
distinctive lipophiloselectivity 203
divacant Keggin-type silicodecatungstate 190
di-vanadium-substituted silicotungstate $[SiW_{10}O_{38}V_2(\mu-OH)_2]^{4-}$ 193
DOW-BASF 261
DP method 80

e
electroless plating deposition – EPD 274
electronegativity, M in M=O groups 18
electrophilic charged di-oxygen species 15
electrophilic oxygen 10, 17ff
electrophilic reaction pathway 18
electrophilicity 18
enantioselectivity 44
– influence of Schiff-base ligends 49
epoxidation 166
– with N_2O oxidant 230
epoxidation of stilbene, solvent effect 116
ethane, selective oxidation 222
ETHOXENE process 291
ethylbenzene dehydrogenation 5
ethylbenzene to styrene oxidation 4, 5
ethylene epoxidation 19
EXAFS analysis, Cu–O, Cu–Cu 54

f
[Fe, Al]MFI catalysts 59
[Fe, Al]MFI zeolite, UV/VIS diffuse reflectance spectroscopy 59
Fe K-edge XANES 59
Fe/ZSM-5, catalyst 58
Fe-containing zeolites, catalytic hydroxylation 229
ferrosilicates 151
FeZSM-5 zeolites
– ammoxidation 230
– Mössbauer spectra 225
– oxidation catalyst 223
FMC 258
furan-2,5-dicarboxylic acid 320

g

gel-permeation chromatography (GPC) 241
glucose oxidation 115
glycerol
– gas phase dehydration 321
– gas phase oxidation 321
– selctive oxidation to organic products 321
gold catalysts 5
gold nanoparticles 164
GPC curves, molecular weight distribution 241
green chemistry 218, 270
green oxidations 194
green, sustainable chemical processes 107

h

H_2O_2, oxidation pbenzene to phenol 60
H_2O_2 production, outlook 254
H_2-TPR XAFS studies 62
$H_3[PW_{12}O_{40}]$ in CPC 187
HAP-γ-Fe_2O_3 160
heterogeneous catalysts 157
heterogeneous catalytic reactions 107
heterogeneous epoxidation of olefins 171
heterogeneous NIPAM-POM hybrid catalyst 202
heterogeneous POM catalysts 194
heterogeneous POM-based catalysts 211
heteropolyacid systems, bulk catalyst 12
heteropolyoxometalates 187ff, 210
high-performance catalysts, structural studies 27
homogeneous C–H bond activation 14
homogeneous catalysts 43, 157
homogeneous catalytic oxidation 2
homogeneous catalytic system 63
homogeneous nucleation mechanism 133
HPPO(hydrogen peroxid propene oxide) process 260
$[\gamma-H_4SiW_{10}O_{36}]^{4-}$, DTF calculations 191
HT (hydroltalcite), olefin oxidation) 171
HT(hydrolatcite) catalysts 170
– mechanism olefin epoxidation 171
HTS (hydrothermally syntesized)-Ti-MWW 141
Huror-Dow process 253
hydoxyapatites
– properties 158
– structure 158
hydrodesulfurization (HDS) 301
hydrogen peroxide (H_2O_2)
– applications 255
– kinds of synthesis 253

hydrogen peroxide(H_2O_2)
– direct synthesis 257
– direct synthesis by HTI 262
– electrocatalytic synthesis 262
– propene oxidation 260
hydrogen production from methane, inherent segregation of CO_2 373
hydrotalcite (HT) 167, 169
hydroxylation, aromatic compounds 223
hyperoxo species 15

i

immobilization of POMs 194
IMP method 80
IM-SiO_2 209
in situ structural studies, HPA salt 26
in situ techniques 19
in situ XPS 35
inorganic–organic hybrid POMs 192
intensification concept 290
γ-isomer Keggin-type silicodecatungstate $[\gamma-SiW_{10}O_{36}]^{8-}$ 192
isopolyoxometalates 187 ff, 210
isotope experiments, structure O_2 adsorption and reaction 89
isotopic exchange reaction 19
isotopic scrambling experiments 16

j

junction perimeter hypothesis 90

k

$K_2[\{WO(O_2)_2(H_2O)_2\}(\mu-O)]$, allylic alcohols epoxidation 190
Keggin anions 187
ε-Keggin POMs $[Mo_{12}O_{39}(\mu-OH)_{10}H_2\{X(H_2O)_3\}_4]$ ($X = Co^{2+}$, Mn^{2+} and Cu^{2+}) 200
Keggin units 16
Keggin-type P/Mo/V polyoxometalates, menadion synthesis 310
Keggin-type polyoxometalates, catalyst for ODS 303
kinetic of selectivity of partial oxidations, bond energy of molecules 8
kinetic selectivity 8

l

lacunary POMs 187, 210
Langmuir–Hinshelwood mechanism 87
lattice defects 15
lattice oxygen 1, 2, 3, 11, 14, 15ff, 22, 23, 34, 54, 68
lattice oxygen dynamics 3

layered double hydroxides 210
levulinic acid 320
liquid-phase oxidation 63, 290
liquid-phase oxidation processes 185
liquid-phase oxidation reactions 157
liquid-phase selective oxidations 106

m
main valence band transitions 25
Mars–van Krevelen type (MvK) reaction "mechanism" 1
^{19}MAS NMR 137
membrane catalyst (H_2O_2 synthesis), Pd-Ag interlayer 274
membrane reactor system, oxidation of benzene 61
menadione (2-methyl-1,4-naphthoquinone) synthesis 310
mesoporous SiO_2 84
metallic Cu clusters, properties 52
metallosilicate zeolites 151
metallosilicates, post-synthesis method 137
metallosilicates with MFI structure 145
metal–oxygen
– bonding 17
– metal–oxygen interaction 17
metal-to-oxygen bond, polarity 17
methan oxidation, formation condensation products 222
methane oxidation by N_2O 220, 221
methanol dehydrogenation 19
methyl ethyl ketone, C–C bond breaking 14
MFI-type zeolites, oxidation catalyst 58
micropores 43
microreactor, synthesis of hydrogen peroxide 305
microreactor technology 304
microreactors, oxidation reactions 305
mixed-addenda POMs $[PV_2Mo_{10}O_{40}]^{5-}$, $[PV_8Mo_4O_{40}]^{11-}$ 205
MMO catalyst systems 21
MMO systems 12, 15ff, 22, 26
Mo oxide systems 217
Mo/Pd/(Re)/O systems, ethylene/ethene oxidation to acetic acid 293
Mo/V/Nb/O catalyst, direct oxidation ethane to acetic acid 291, 292
Mo/V/Nb/Te/O catalyst, ammoxidation of propane 296
model catalysis 5
molecular O_2, benzene oxidation 62
monatomic oxygen donors 218
monolayer oxide catalysts 2
mono-oxo species 15
mono-ruthenium-substituted POM TBA_4H $[SiW_{11}O_{39}Ru(H_2O)]$ 200
MoO_2-supported Cu^0-cluster 52
multernary compounds 22
MWW aluminosilicate 137

n
2-naphthol, asymmetric oxidative coupling 44
N_2O 10, 245ff
– properties 218
N_2O epoxidation of cholesteryl benzoate 231
N_2O oxidant, specifity compared to O_2 227
nanoscience 5
nanostructuring 16, 34
narrow band-gap semiconductors, redox catalysts 12
NH_3 66
N-interstitial Re cluster 70
$[Ni(tacn)_2]_2[\gamma\text{-}SiW_{10}O_{38}V_2(\mu\text{-}OH)_2]$, catalytic activity 204
nitrous oxide (N_2O) 218
– catalytiv reactions 246
non catalytic N_2O oxidation 232
non-dalton composition 16
novel titanosilicate 143
nucleophilic (O^{2-}) oxo-anion 15
nucleophilic oxygen 17, 21
nucleophilicity 18

o
O^- radical 221
olefins, cis-dihydroxylation 173
olid oxygen carrier (SOC) 312
one-step hydrogen 316
one-step phenol process (AlphOx) 224
on-site production of chemicals 270
organic oxygenates 77, 97
organic polycation components 210
Os catalyst, cis-dihydroxylation of olefins 173
over-oxidation 17
oxidant-lean processes 289
oxidation
– pathway of activation 8
– propane to acrylic acid 12
– reaction path 8
– reaction scenario 8
oxidation catalysis 15, 34, 43, 126, 151, 186, 289
oxidation catalysts, tunable electronic structure 22
oxidation of hydrocarbons 77
oxidation of sugers
– furoic acid 317
– gluconic acid 316

oxidation processes 27, 69, 77, 217, 257
– N$_2$O emisson 244
oxidation reactions
– selectivity to the product of partial oxidation 306
– two step approach 308
oxidation with N$_2$O, benzene to phenol 223
oxidative dehydrogenation 1
oxidative dehydrogenation (ODH) of alkanes 311
oxidative desulfurization (ODS) of benzothiophene 303
oxidative desulfurization (ODS) process for ULDS production 303
oxidative desulfurization (ODS) processes 301
oxide crystallites 8
oxo-functionalization 1
oxophilicity of metal oxides 55
oxygen chemical potential 17, 19, 21, 23
α-oxygen 225
– formation from N$_2$O 227
– hydroxylation alkanes and aromatics 229
– loading 228
oxygen delivery system 22
oxygen ion diffusion 17
oxygen pumping 27
oxygenates 10, 11
oxygen-based recycle processes 289

p

palladium-catalyzed aerobic oxidation 163
partial bond strength 8
participation of oxygen mobility, explanation 17
Pauling criterion 126
PdHAP (palladium-grafted hydroxyapatite) 163
– catalytic properties 163
peroxo intermediate 15
peroxo species 15
peroxometalates 187
peroxotungstate catalyst (W2) 168
peroxotungstates with P- and As-ligands 189
phase cooperation 3, 26
photoemission, alalyzing activated oxygen 19
photoluminescence spectroscopy 10
photooxidation catalysts 188
[PMo$_{12}$O$_{40}$]$^{3-}$ catalyst, CO oxidation 189
polymer gel immobilized Au NPs
– catalytic performance 109
POM anions, counter cations 200
POM-PEG phase 207

POMs (polyoxometalates) 186, 187, 207, 210, 211
– acid form 207
– assembly properties 201
– green oxidation 194
– immobilized onto anion exchange resins 208
– intercalated in LDHs 210
– structure 186
– supported on apatite 206
– wet impregnation 205
POMs/[AlVW$_{11}$O$_{40}$]$^{6-}$ 189
post-synthesis method 140
[PO$_4${WO(O$_2$)$_2$}$_4$]^3 190
pre-catalyst, transformation process 27
Princeton Advanced Technology 257
propane oxidation 5, 16
propene 13
propene oxide, direct epoxidation of propene 290
propylene oxide (PO) 100
PROX catalysts 51
PROX of CO in excess H$_2$ 86
PS (post syntesized)-Ti-MWW 141
[Pt(Mebipym)Cl$_2$][PV$_2$Mo$_{10}$O$_{40}$]$^{5-}$, methan oxidation 206
Pt/TiO$_2$ catalysts, turnover frequencies 80
pulse reactions 68
[PV$_n$Mo$_{12-n}$O$_{40}$]$^{(3+n)-}$ 189
[PV$_n$Mo$_{12-n}$O$_{40}$]$^{(3+n)-}$/O$_2$/hydrocarbon redox system 189
[PW$_{12}$O$_{40}$]$^{3-}$ 188

q

quinones 61

r

rate-controlling elementary step 14
Re L$_{III}$-edge EXAFS 66
Re L$_I$-edge XANES 66
Re monomers 69
Re/zeolite catalysts, catalytic performances 64
Re$_{10}$ cluster 64, 68, 69
Re$_{10}$ cluster/HZSM-5 (19) catalyst 68
reaction channel 15
reaction pathways 10
reaction-controlled phase-transfer catalyst 201
reactive oxygen 14
Re-CVD/HZSM-5 (SiO$_2$/Al$_2$O$_3$ = 19) catalyst 66
redox-active probe molecules 14
reformer–PEFC system 84

ReO$_x$ supported on Fe$_2$O$_3$, one-step methylal synthesis 64
residence times of intermediates 8
reverse water–gas shift reaction 86
riser concept 16
Ru (hydroxyapatite) catalyst (RuHAP) 158
Ru/Al$_2$O$_3$ 162
RuHAP 162
– spectroscopy 158
– synthesis 158
RuHAP catalyst 158
– catalytic cycle 159
RuHAP-γ-Fe$_2$O$_3$, catalytic properties 160
RuO$_2$, model catalyst 19
ruthenium-substituted sandwich type POM andamantane hydroxylation 192

s

sandwich-type POM [(Fe(OH)$_2$)$_2$)$_3$ (A-α-PW$_9$O$_{34}$)$_2$]$^{9-}$ 206
sandwich-type POM, TBA$_7$Na$_5$[ZnWZn$_2$(H$_2$O)$_2$(ZnW$_9$O$_{34}$)$_2$] 200
selective catalytic oxidation 1
selective oxidation catalyst, microstructure 25
selective oxidation materials, categories 22
selective radical activation 14
self-assembly, hydrothermal syntheses 44
semiconducting oxides, metal sites of oxygen 15
semiconductor metal oxides 81
sheer defects 16
ship-in-a-bottle complexes 125
Si alkoxides 129
^{29}Si(^{19}F) CP MAS NMR 137
silico-aluminates 290
silver, catalyst selective epoxidation 7
^{29}Si NMR spectra 130
α-site formation, mechanism 225
site isolation 3, 11, 22
[γ-SiW$_{10}$O$_{34}$(H$_2$O)$_2$]$^{4-}$ TBA salt oxygen transfer reactions 190
size-controlled metal nanoparticles 207
small-scale integrated productions 270
Sn-Sil-2 150
solidification of POMs 194
[SO$_4${WO(O$_2$)$_2$}$_2$]$^{2-}$, limonene epoxidation 189
spatially controlled reaction environment 44
sticking coefficient 21
STM 83
stoichiometric
– oxidants 158

– oxidation 185
– reduction 185
structural promoters 26
Sumitomo's route to caprolactam 256
supported Au catalysts 79
– gas-phase propylene epoxidation 97
– influence of metaloxides additives 93
– liquid-phase oxidation reactions 107
– mechanism CO oxidation 87
– oxidation of VOCs 93
– performance of CO oxidation 84
– rate of CO oxidation 87
– selctve oxidation of alcohols 98
– selective oxidation of aliphatic alkanes 97
– X-ray photoelectron spectroscopy (XPS) 92
supported Au NPs 79, 107
– aerobic oxidation of glucose 115
– alkene (styrene) oxidation 116
– glycerol oxidation 114
– oxidation of alcohols 108
– oxidation of polyols (diols) 112
– polymers as support 109
supported Cu/CeO$_2$ catalyst 52
supported Keggin-type P/Mo/V/O polyoxometalates 294
supported metal complexes 43, 44
supported mixed-addenda POM catalysts 205
supported noble metal catalysts 83
supported vanadium oxide, catalyst for methane oxidation 220
supported V-dimer catalysts 49
surface lattice oxygen 3
surface science data, predicting catalytic perfomance 30
syngas 289
synthesis of chemicals starting from sugars 317

t

TAP experiments 16
t-butyl hydroperoxide (TBHP) 116
TEM image, AuNPs 79
TEM observation, active surface of catalyst 29
temperature-programmed oxidation (TPO) 92
temperature-programmed reduction (TPR) 92
termination issues, high performance catalysts 28
tetravalent pyrophosphate 32
Ti-beta 136,ff, 142f, 167
– synthesis 137

Ti-beta catalyst, solvent effects   145
Ti-ITQ-7   139, 145
Ti-MCM-41   60, 130
Ti-MCM-48   104, 139
Ti-MWW   137ff, 141, 151
Ti-MWW catalyst, solvent effects   146
Ti-MWW-HM   138
Ti-MWW-PI   138
– UV/Vis spectra   138
Tin-containing silicalite-2   150
TiO_2-supported monolayer catalysts   14
titania   5
titanium catalysts, epoxidation of olefins   167
titanium silicalite-1 (TS-1)   126
titanosilicalites TS-1   223
titanosilicate zeolites   151
titanosilicates   126, 290
– hydrophilicity   130
– Si/Ti ratio   132
Ti-UTD-1   145
Ti-YNU-2   145
Ti-ZSM-12   145
Ti-ZSM-48   145
transformation circus   33
transition metal ion matrix   21
transition metal-based homogeneous catalysts   193
transition-metal-substituted POMs   187, 192
tripodal polyammonium cations   201
TS-1   60, 101, 103f, 117, 126f, 132, 136, 139f, 146f, 167, 264, 290
– catalytic properties   128
– crystallization mechanism   129
– cystallinity   131
– gas-phase epoxidation of propene with HP   305
– increasing Ti content   129
– mechanism of catalyzing epoxidation   135
– propene oxidation   264
– stability   131
– structure   135
– synthesis   126, 134
– sythesis by YNU method   130
– Ti K-edge EXAFS studies   129
– XRD pattern   127
TS-1 catalyst, solvent effects   145
TS-2   150
tungstate catalysts, heterogenization   168
tungsten-based catalyst systems   167
tungsten-catalyzed oxidation systems   187
tungsten-intercalated HT (hydrotalcite)   168
two-center bond   3

u

ultra-deep desulfurization   301
Ultra-Low Sulfur Diesel (ULSD)   301
unsupported POMs   206
UTD-1   145

v

V complex
– ESR spectra   47
– EXAFS- and IR-FT spectra   46
– structure   45
V dimer, DTF calculations   49
V monomer precursor   49
V oxide systems   217
V precursor   46, 49
vanadium catalyst   8, 49
Vanadium K-edge XANES   45
vanadium oxide systems   5
vanadium oxides   5
vanadium oxo species, catalytic operation   30
vanadium phosphates   5
vanadyl group   9
Venturello's catalyst $[PO_4(WO(O_2)_2)_4]^{3-}$ 208
V-MEL   150
V-monomer Schiff-base complexes   45
Vmont(vanadium-exchanged montmorillonite)   172
volatile organic compounds   92
$VOPO_4$, combustion catalyst   32
V-O–substrate   10
VPO-$VOPO_4$ conversion   32
VPO activation   26
VPO catalysts   25
– loss of phosphate   29
– surface analysis   28
– TEM   27
VPO-$VOPO_4$ transformation   33
VS-2   150
VxOy species   33

w

wastes   185, 218, 254, 256, 317
water–gas shift reaction   54, 84, 86
Wells-Dawson type POM   187
$[\{WO(O_2)_2\}_2(\mu\text{-}O_2)]^{2-}$, TBA salt, alkenes epoxidation   190
$[W_6O_{19}]^{2-}$ 188
$[W_{10}O_{32}]^{4-}$ 188

x

XPS spectra of a VPO catalyst   19
xylene oxidation   5

y
YNU method   130
YNU system   131 ff,
– Ti/Si ratio   133

z
zeolite MEL framework   150
zeolites   125ff
– application in oxidation catalysis   222
– coordination chemistry   125
– introducing hetero metals   125
– properties   125
– α-sites   224
Zeolite-supported Re catalysts   64
ZORA-BP86/TZP level theory (DTF calculations)   192
ZSM-5 catalysts, benzene to phenol oxidation   224